The Observant Beekeeper: A Practical Manual

Treatment free, low intervention

Paul Honigmann

The Observant Beekeeper: A Practical Manual
By Paul Honigmann

ISBN: 978-1-914934-85-8

Published 2024 by Northern Bee Books, Scout Bottom Farm, Mytholmroyd, Hebden Bridge HX7 5JS (UK). 01422 882751.

Text and images © 2024 Paul Honigmann unless otherwise noted.
This edition format © 2024 Northern Bee Books.

All rights reserved. No part of this publication may be reproduced, stored or transmitted in any form or by any means electronically or mechanically, by photocopying, recording, scanning or otherwise, without the permission of the copyright owners.

Book design by www.SiPat.co.uk.

Contents

Preface	7
Introduction	8
Part 1: Getting started - understanding bees, hives, and apiary setup	11
Chapter 1: Why Observational Beekeeping - What it is; why beekeeping advice varies; common misperceptions	12
Chapter 2: Bee behaviour - What to expect; aligning your practices with bees' needs	15
Chapter 3: Lifecycle and biology - Castes and variants; how roles change with age and time of year; queens, princesses and supersedure; lifespan; sleeping; bee senses	25
Chapter 4: Hive selection - General principles; choosing one fitting your needs. ~11 specific types discussed	44
Chapter 5: Apiary setup - Site selection; arrangement; obtaining bees	57
Chapter 6: Safety for bees and humans - Avoiding stressors; protective clothing; appropriate behaviour; dangers for bees	69
Chapter 7: Practical points about keeping bees - Common errors; honey basics; propolis; converting from standard to treatment-free style; finding information; costs	83
Part 2: The beekeeper's year - what to expect, check and manage	95
Chapter 8: The annual cycle - Colony rhythms; seasonal beekeeping activities; making your own bee calendar	96
Chapter 9: Swarms - Colony reproduction; types; swarm collection; bait boxes; hiving; care	99
Chapter 10: Harvesting honey - The if, when and how of assessing honey stores; harvesting; processing	132

Chapter 11: Observing a colony - Watching; learning; interpreting; record keeping	138
11.0: Record keeping	138
11.1: External monitoring - Interpreting colony state without opening a hive	140
11.2: Internal inspections - The if, why, when and how of opening hives	176
11.3: Reading comb - What do comb patterns and structure reveal?	183
Chapter 12: Hive management - TBHs; Warrés; Skeps; Logs & Trees; and standard framed hives	200
Chapter 13: Tips and techniques - Key skills: processes and pitfalls	233
Part 3 - Beyond the basics - health, genetics and further reflections	**265**
Chapter 14: Pests, diseases and disorders	266
14.1 - Overview - Patterns across viruses, bacteria, fungi and pests; wild colony health and strategies; problematic beekeeping; historical incidence of disease; research; nutrition; stress and hormesis; heirarchy of needs; commensality; complacency	267
14.2 - Specifics - ID / diagnosis charts; alphabetic listing of over 40 conditions with symptoms, discussion, photos / diagrams and treatment options	282
Chapter 15: Conducting 'post mortems' - Analysing and learning from failed colonies	346
Chapter 16: Bee genetics - Insect DNA; breeds and breeding; family relationships	352
Chapter 17: Reflections on common assumptions - Some standard advice examined in detail	373
Chapter 18: Conclusions - Summing up, moving on	385
Further reading	392
Afterwords and evaluating advice	393
Acknowledgements	396
Citizen Science	397
Index	399

Preface

This book is for those who keep, or intend to keep honeybees, particularly those interested in a less interventionist path. It is a 'how to' book which fills in gaps not covered by most guides. It will be useful to everyone, but especially for:

- those interested in how hives can be managed without frequent opening;
- more environmentally-minded beekeepers;
- explaining why not all advice, which tends to be "one size fits all", is suitable in all circumstances.

When I began beekeeping in 2010, with horizontal Top Bar Hives and Warrés, some long term issues with intensive bee farming were becoming apparent (loss of fertility, new diseases, rising losses etc). Consequently people began to review historical data and techniques, and noted that some modern problems were never even mentioned over a century ago. But specifics of the old methods were hard to find: the nature of beekeeping had changed when varroa arrived two decades earlier, and the vast majority of beekeepers had resorted to intensive management and use of chemicals to survive the varroa mite. This prevented the craft from going extinct and kept our farms pollinated, but by the 2000's conventional beekeepers had largely forgotten the old lore of 'reading a hive' and swarm-based management.

But the internet was taking off, and alternative practitioners gradually found each other online. And it turned out there *were* a few people who had never used miticides, and an overlooked population of wild varroa-resistant bees.

I started a local self-help network, OxNatBees, which grew rapidly. This enabled us to meet and pool knowledge and experience from different mixes of hive type and management. It was very exciting! In time, I was training other beekeepers in arts both lost and new, and I began systematically interviewing the few remaining traditional beekeepers to learn from them.

Other issues have contributed to skill fade. *Link rot* - information on the internet eventually disappears, or is buried under irrelevant posts. And as our lives become busier and more online, people just don't meet physically as much now. Sometimes there's nothing quite like learning that incidental hands-on *tacit* knowledge with a mentor physically present and commenting as you observe a hive. This book condenses as much of that expertise as possible into one volume.

Paul Honigmann, September 2024

Introduction

This book will help you manage your bees in a low stress, low intervention style, appropriate for hobbyists with a few hives. The emphasis is on working *with* the bees' preferred behaviours, enabling colonies to express hygienic habits which are inhibited in hives run for maximum yields.

The Observant Beekeeper approach centers on letting the bees run through their natural lifecycle, watching them and learning as they overcome problems essentially on their own. It can help all beekeepers to gain an understanding of these principles, and reflect on the tradeoffs of standard interventions.

Consider this as a valuable dataset about what happens in colonies *allowed to follow a natural lifecycle*.

I recommend newcomers to read widely, join your local beekeepers' association and take their beginners' course. Even if you come to disagree with their more interventionist approach, you will learn a lot of practical skills and make valuable contacts – if you operate entirely alone, you risk avoidable problems, and are unaware of gaps in your knowledge. As you develop your own style according to your priorities, consider the approaches as *complementary* knowledge sets - each sees things the other misses. You must at all costs keep your mind open to valid criticism and avoid forming an "echo chamber" with others who share your views.

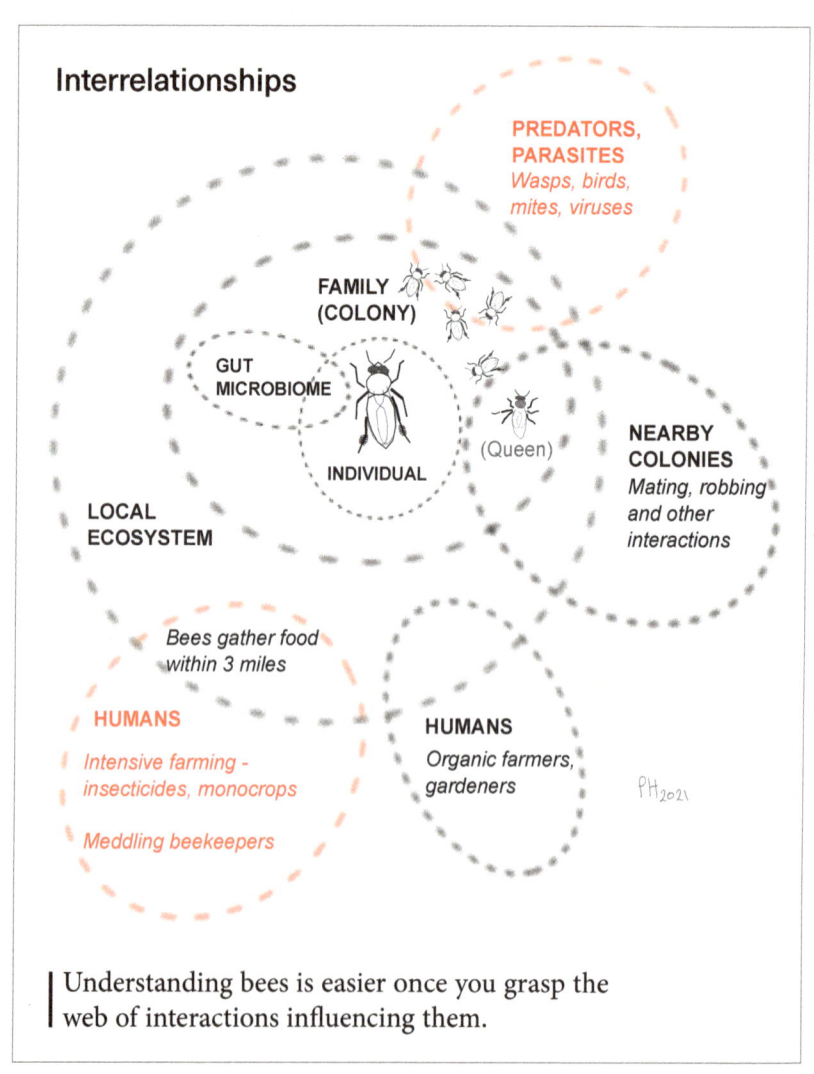

Understanding bees is easier once you grasp the web of interactions influencing them.

Structure of this book

The emphasis is on the reasons behind behaviours of bees, and their keepers, and how some common practices can promote problems.

Part 1 - getting started starts with a review of key bee biology, how to obtain bees, set up an apiary, and safety.

Part 2 - the beekeeper's year is the practical section: How to carry out common operations, manage a variety of hive types and inspect them to determine health and status (which 9 times out of 10 does *not* require them to be opened). This includes many details most training omits: the almost forgotten lore of observing a hive without opening; reading comb; and the advantages of using swarms to populate an apiary.

Part 3 - beyond the basics covers more advanced areas, including disease and post mortems, more radical hive designs, genetics and unintended consequences of directed breeding. Whilst you are unlikely to need this in your first year - it's there when you need it.

Part 1: Getting started

The fundamentals - understanding bees, hives, and apiary setup

Chapter 1
Why 'observational' beekeeping?

What it is; why beekeeping advice varies; common misconceptions

The big secret of beekeeping is that bees survive fine in the wild, unmanaged by humans.

This simple observation has led to a rethink of the dominant intensive-management beekeeping paradigm. A lot of people have begun looking at wild colony behaviour, and trying to find a better balance in hives to enable managed bees to express more survival traits.

A key factor is letting the bees sort their own problems out, rather than 'helping' by opening nests to 'correct' things[1]. But how to monitor without disturbing? How can we enable such survival behaviours? That's what this book focuses on.

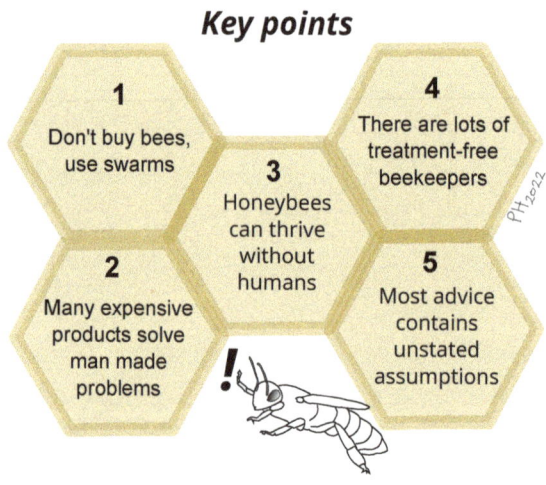

In later chapters, we discuss why wild bees survive whilst hived ones may struggle. But first, some other open secrets are worth mentioning, to give context for many of the contradictory statements you may come across in beekeeping.

1. Locally adapted bees are healthier. Most bees sold by large bee breeders are optimised for migratory beekeepers who keep moving hives to different crops. These are referred to as *commercial bees* in this book. If you put them in a stationary *(static)* hive these fast-breeding strains need periodic propping up with feeding: the forage in a fixed location is intermittent but these need continual food. For static hives local swarms are best, being adapted to an area: adjusting numbers to match forage, and requiring less help than bought bees.

---

1   Doctors have a term for excessive meddling causing harm: *iatrogenics*

2. Standard beekeeping training is for bees in highly artificial conditions. Bees are continually striving to "fix" inappropriate hives to something more suited to their biology, and recovering from the last intrusive inspection. Many common problems are absent in colonies allowed to manage themselves within a suitable hive environment.

3. There are lots of thriving wild bee colonies around. Rumours of their demise from varroa in the 1990s were greatly exaggerated. It was a repeat of an error made when many hives died from Isle of Wight Disease in the 1920s[2]: people assumed all the wild bees died too, without humans to help them. Ironically, because they were assumed to be "extinct", during both outbreaks they were left alone and rapidly developed resistance (if indeed they were ever affected).

4. There are lots of treatment-free (TF) beekeepers, using bees from local survivor colonies, and no miticides. About 20 - 33% of British beekeepers now operate a leave-alone approach to varroa management[3]. In Ireland it may be 30%[4].

5. There are many paths. The rethink of beekeeping led to a spectrum of approaches. The original popular umbrella term, *natural beekeeping,* has become ambiguous - which can be useful - and more specific terms have emerged like *low-intervention, treatment-free, leave-alone* beekeeping - sometimes used interchangeably. *Observational* beekeeping is not specific to one system.

6. No beekeeper knows everything. But some think they do. The common saying "all beekeeping is local" reflects how well-meant advice can be inappropriate.

7. There is no 'Beepocalypse'. This catchy word was coined by the populist press when North America had a series of high losses in the 2000s. It only significantly affected North America, for maybe 4 years, and is basically due to over-stressing the bees to a tipping point[5]. [Refer to footnote for details.] The reality is, whilst there is a worrying decline in other insects, honeybees are doing fine - they have buffer stores of honey during food dearths.

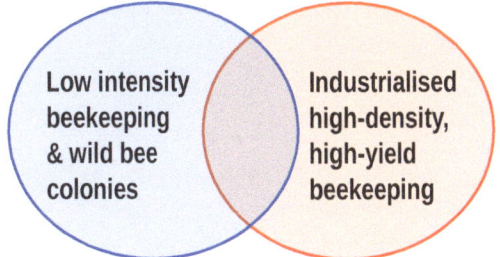

Different approaches - different methodologies

Beekeepers deal with a specialised subset of bee behaviour and issues. There's a lot of overlap between the approaches, and each is true in their own sphere.

---

2   More on IoW disease in chapter 14.2, and on wild nest numbers in chapter 17, *"Finding Wild Nests"*

3   Valentine & Martin, *A survey of UK beekeepers' Varroa treatment habits* PlOS One, 2023 - doi.org/10.1371/journal.pone.0281130. Sample size 3,000 responses, 21% non-treaters, 160 had not treated for 6+ years. The BBKA 2021/22 winter survival survey had 3,802 respondents and 33.1% said they had not treated the previous year.

4   Survey by Prof Grace McCormack, number of respondents unclear.

5   There were particularly high US losses in the 2000s, from various and often unidentified causes. The phrase 'Colony Collapse Disorder' [CCD] was coined but it's just a specific set of symptoms, like saying "he died because he was tall". A label is not an explanation. And not all losses exhibited CCD. Commercial agriculture and apiculture are particularly intensive in North America, where hive losses are typically twice those elsewhere. Their high pressure, migratory beekeeping is intrinsically going to spread virulent pests and diseases, which would otherwise kill all local hosts and burn out; there's an argument that this style of beekeeping actually *selects* for faster breeding, worse pests rather than the natural balance where successful diseases are slow burning and don't kill *all* their hosts immediately.

So why are hives not just left to manage themselves without interference and chemicals? Firstly, if you are running a business managing many hives across multiple apiaries, this takes a *lot* of time. If you rely on hundreds of hives for income, you cannot afford to let each one do its own thing at its own pace, you need uniform, predictable colonies ready to pollinate at certain times of year. You cannot afford to spend 5 minutes watching every colony weekly. Standard commercial practises have evolved to manage hives at scale, whereas amateur beekeepers tend to have a handful of hives they can manage individually. And be aware, going non treatment cold turkey on an existing colony that has become reliant on miticides, can lead to disaster - see chapter 7 for a rough and ready guide to transitioning to no-treatment ("Can I convert from conventional beekeeping to treatment-free?").

Chapter 2
Bee behaviour

What to expect; aligning your practices with bees' needs

Altruism and interdependence

Facts about honeybees don't seem to make sense at first, but once you grasp how they perceive the world and how evolution has shaped them, you begin to see a coherent logic to behaviours like swarming, building weird comb, infanticide, and why they might ignore you on one visit and be edgy the next time.

The key insight which illuminates their behaviour is that they prioritise their colony (family) above themselves. They are so interdependent, that they cannot survive alone.

Grasping their motives allows you to influence their behaviour without resorting to force and intimidation.

Start by assuming bees are aliens

Bee colonies as Superorganisms

Bee colonies are analogous to a ruthless corporation. The organisation has evolved for *it* to survive, as a unit; not for its members' benefit. *Everyone* is expendable once their usefulness is over. Unlike a human group, there is no dissension among the members - they are utterly loyal[6]. Everyone has a stake in the community because they cannot produce children alone - they are too specialised - the young are cherished by the entire tribe. The colony lives longer than its members[7].

Eastern Europeans call colonies families.

A colony's behaviour is an emergent property from many processes like pheromone balances, crowding and forage availability. The queen is a replaceable tool, not "in charge".

Occasionally this lack of a leader locks the bees into a silly behaviour cycle. For example, when they swarm onto a branch and cannot decide on a new home, after a few days they begin building their nest hanging from that branch.

---

[6] I'm simplifying to convey the alien-ness of the bee perspective. They're loyal as long as the queen is healthy and exuding the right pheromones. Colonies can become disorganised, apathetic, or "every girl for herself" once the queen ages and the workers sense her fertility dropping.

[7] Like the *Theseus' Ship* fable - every part is eventually replaced: is it the same ship?

When decisions do need to be made, they are done by a voting system among the most experienced (oldest) foragers, communicated by the *waggle dance* - this specifies a location: scouts go check these out, then repeat the location-dance of the place they prefer.[8] It's routinely used to communicate where exceptional forage can be found.

Reproduction drives behaviour

This diagram shows how bee numbers in a colony vary over the year. Numbers go up when the queen lays, but workers don't live very long, so when she stops laying numbers decline. She adjusts her laying rate to match forage.

The key point here is that swarming is how bees make more *colonies,* and their behaviour orbits round this.

Ultimately, that's because the reproductive unit is a colony, not an individual - so like the free market, evolution has selected for the most resilient, efficient colonies.

Another key reproductive driver is that queens mate with *many* drones from the area. You can't easily control who they mate with. Local bees are optimised for local stressors and if someone imports other bees, it can cause everyone problems as the lines mix.

Beware terminological inexactitudes!

It's worth stressing that a mobile *swarm* is very different to a *colony* living in a *hive* or *wild nest*.

Colonies exhibit a mob mentality. Once enough bees and the queen clustered under this hive's roof, so it smelled strongly like them, this entire swarm repeatedly returned to live in the roof void rather than inside, until the roof was changed. Once committed, bees - like many bureaucracies - lack a mechanism to admit "we've made a mistake".

Beekeepers sometimes use *hive*, *colony* and *nest* interchangeably in casual conversation, but we are very careful not to call these *swarms*.

- *Swarms* are baby *colonies* looking for a new home.

- **A swarm is not an entire colony** migrating[9]: the parent colony has **split into two** and the original nest can persist for years. More on this later, but *newbees* (new beekeepers) sometimes wrongly assume that a *swarm* is a whole colony moving house, and the original hive or nest is now empty. It is not.

---

8 Professor Tom Seeley's book, *Honeybee Democracy* explains decision making very readably. Beekeepers have known decisions are made collectively since at least 1916 (Tickner Edwardes, *Lore of the Honey Bee* 7th Ed, p.47). Yet when Karl von Frisch discovered the mechanism (in *The Dancing Bees,* 1927) he was initially ridiculed by academics for suggesting insects could communicate; he had the last laugh as he won a Nobel prize in 1973 partly for this work. Different bee races have different dialects of waggle dance! Von Frisch discovered many other things about bee perception etc. Interestingly, the Nazis declared him 1/4 Jewish and he would have been purged - but they needed his expertise to address a nosema plague which killed 800,000 German hives in 1941 and seriously impacted agricultural output.

9 Of course, with bees there are always exceptions! African and Asian honeybee colonies often migrate, following forage or if bothered excessively by predators, but in Europe it is very rare.

The mammal analogy

The superorganism has several parallels to a mammal:

* It lives several years[10]
* It has 0-3 children a year
* It keeps its core temperature around 35C
* It has an internal "skeleton" (comb)
* It has a "skin" (hive walls with a sterilising layer of propolis; if a hollow tree, the whole "organism" weighs several tons!)
* It usually builds up resources in its first year and reproduces in later ones

If you think on this scale you see implications of interfering with its lifecycle.

An even better analogy is: a bee colony is like an inside-out animal, with its sensitive parts exposed to everything in its environs – its foragers are exposed to everything in 40 sq km and bring it home.

Don't stretch this analogy too far. In winter, with no babies to nurture the adults keep themselves just above ~18C. Below that they have problems, and below 8C their muscles are paralysed and they appear to be dead - but if they don't completely freeze, *some* paralysed bees may be revived by warmth, even a couple of days later! (Bees often land on you in Spring - you're warm!)

The basics of bees in one easy diagram

Prime swarm heads off with original queen and ~half the flying bees

Sometimes: successively smaller afterswarms (casts) + new queens

Drones booted out after mating season

Population ~40k

Spring　Summer　Autumn　Winter

Notes:
* Bees are part of one superorganism, the colony, which is essentially immortal. It reproduces by splitting, like an amoeba - this is what swarms are.
* Bee numbers increase when forage is available. I tend to see a gap in June, but the pattern will be very different in e.g. Kent or Spain.
* In a good year, the colony sends out swarms to found new colonies. These can be used to populate new hives. It's very weather dependent - some years, there are few swarms.
* Migratory beekeepers try to suppress swarms because they move hives to new forage over summer, and more bees = more honey. Their hives may have 60,000 or more bees. This approach stresses the bees, making them more vulnerable to parasites, and requires more hands-on management.

Embrace the chaos

A strongly conserved trait among some beekeepers is the desire to impose order on a colony - straight combs, neat arrays of identical cells, etc.

But bees are pretty obsessive-compulsive about order and cleanliness too. They just don't have any use for straight lines.

10　Conventional beekeepers sometimes don't grasp the concept of *colony* age because artificial requeening makes colonies appear immortal, whereas low-intervention beekeepers allow them to supersede, die etc and repopulate with swarms. Comparing ages of a natural colony and an artificially maintained one is like comparing apples and ducks. You could argue that wild colonies are millions of years old because they are founded by swarms budded off their parent. Obviously requeening, and running a colony so hard so all founders are dead after 6 weeks, interrupts such continuity.

This is just another aspect of how alien they are to us. *Dunbar's number*[11] for humans is ~150, but for a bee it's arguably ~30,000. And if language affects how you think, imagine how different your worldview would be if you used mainly scents and dance to spread ideas, and big decisions like where to build comb are emergent properties no one person is in control of.

Another insight into colony behaviour

Bees evolved to populate hollow trees (vertical cavities), hollow branches (near-horizontal cavities), and caves so they are pretty adaptable. But just as you arrange your living space for convenience, they follow some general principles too.

This diagram is deliberately not based on any particular kind of hive, to help you understand what they are trying to achieve.

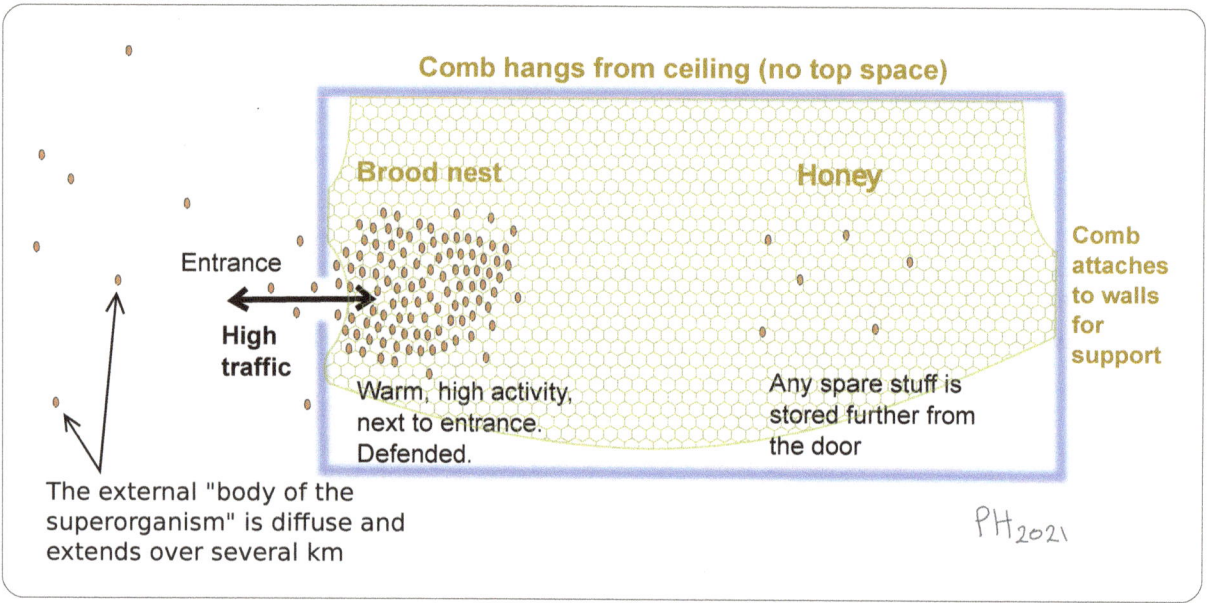

Simplified, schematic diagram of how colonies like to arrange themselves. In Britain, they usually prefer comb 'cold way', edge-on to the entrance.

Most gathered resources go into raising brood. So they place brood comb near the entrance. If your hive type places the nest far away, you may see them try to correct this by moving the nest or building strange comb.

Having the busiest combs near the entrance means any intruders are immediately in a crowd of bees very willing to defend their babies to the death. This deters wasps. But many hive designs have doors far from any comb. (And with populated comb next to the entrance, bees see morning sun and fly earlier.)

Honey is made from *surplus* nectar and unlike brood, does not need to be kept warm. Depending on cavity shape and entrance position it may be at the top, bottom or far end of a

---

11   The number of people you 'know well enough to have a drink with'. Correlates with intelligence, in primates.

hive. If your hive permits accessing the honey without exposing the brood nest to light or cold, you will find these stores pretty much undefended.

In countries which are warm all year round, bees sometimes make free hanging nests without any outer wall. Such nests die in British winters.

In cold climates, as brood laying tails off, bees move honey to the warmest (top) part of the cavity and the colony clusters there over winter. Note how these natural nests have comb right up to the top - any artificially added space above will act as a heat sink.

In heavy nectar flows, there may be lots of surplus nectar and most of the traffic is to the honeycombs. In these circumstances, beekeepers sometimes create a temporary second entrance so the foragers can go directly to the honeycombs without a traffic jam.

Many factors can override the basic layout illustrated above. For example, bees will avoid comb which smells of other colonies. I've seen this a couple of times, a very strong effect. Old brood comb is a swarm attractor, but once inside the swarm prefers to build anew. You may have seen some comb being avoided a few weeks after merging two colonies (when all the living bees are from one queen).

Colonies differ: learning and personalities

With individual workers living just weeks in summer, no one has a 'memory' of things but the colony obviously has other ways of storing information long-term, even though we don't really understand how. Colonies can be trained to be more defensive through continuous rough handling; and they somehow tune their rhythms to local annual forage cycles.

Perhaps this is through epigenitics[12]. The mechanism is irrelevant: what you need to know is that if you alarm a hive by, say, opening it clumsily you may have to stay clear of it for up to 1-2 days; but if you *keep* disturbing them, say once a week, they may treat you as actively hostile whenever you come near.

When keeping bees, it becomes apparent that each colony has a distinct *personality*. This arises from many factors; two sister swarms placed in neighbouring hives will always do things differently, for example you will see them gathering different colours of pollen – they've located distinct forage sources[13]. One may be hived a week later than the other and experience different weather in its first few days, which sets how fast it can gather food and build up. One queen may mate with different drones and her children prioritise building comb over processing honey...

Bee behaviour is different to related insects'

Honey stores permit colonies to last many years. Other social insects tend to persist for only one year[14].

---

12   Or perhaps the previous annual cycle's activities affect the comb's form, which then acts as a template for the next year's behaviour (we've expanded into an area with drone cells, so the queen lays those). A form of writing! Which would imply an advantage to not renewing combs, and natural comb over foundation. Play with offbeat ideas to achieve new insights! But more seriously, learning, and the abstract symbolic communication of the waggle dance, may imply cognition and awareness - which has implications regarding cruelty.

13   Bees (and many other animals) forage with what is known as a *Lévy pattern foraging strategy*, a mixture of long range dashes and short random movements. One benefit is that if an individual forager is lost, the colony can rediscover the food - it is a robust hunting strategy.

14   Some ant colonies also persist many years, the colony outlasts individuals and stores food for winter. Ants are distant relatives of bees.

Therefore honeybee colonies use longer term strategies - these are adapted to circumstance, but typically they spend the first year establishing themselves and building up resources, then in subsequent years they send out swarms to populate the area.

Honey allows honeybee colonies to survive winter *without hibernating*, and last many years – they are basically a tropical species that moved north and refined its honey storing techniques, huddling for warmth in winter[15], so never had to evolve a hibernation metabolism.

In contrast, wasp and bumblebee colonies are one-year wonders, abandoned in autumn when new queens go off to hibernate over winter. They have no winter honey stores and the workers just starve and die when their food runs out.

There are thousands of wasp, bumblebee, and solitary bee species; but only a handful of bee species have a nest structure allowing *easily harvestable* honey which can be stored indefinitely – the honeybees.

Consequences of interdependence

Overwintering: below 8C bees are paralysed, so they cluster together in poorly insulated hives in winter[16]. Just as we shiver when cold, the bees generate warmth by decoupling their wings and vibrating their flight muscles. They use the same technique to warm brood when necessary.

Critical mass: colonies struggle below a certain number of bees. Up to a certain size they are just surviving, living day to day. A single setback like a forage dearth can wipe them out. But if they have enough bees to build up a surplus, they are resilient. This is why we merge small swarms together: experience has shown tiny ones often struggle and die out.

Altruism: most obviously, defending the group at the cost of their own lives, and nursing the queen's progeny. Also expressed in more obscure behaviours: for example, doing unpleasant jobs like removing waste; grooming each other; and bees which are unwell instinctively crawl out of the hive to die outside, reducing the chance of spreading infection.

Meta-altruism: Honeybees will defend their hives against robbers. But in times of plenty, they seem fairly relaxed about which hive (family) a visiting bee is from. Drones in particular seem to have a "free pass" to hives during mating season: they can land at other hives for food and shelter[17], and marked drones have been found in hives many miles from home. This applies to workers to some extent too. This *drifting* seems more common in commercial apiaries, perhaps because the bees are closely related, and tend to be moved to forage on monocrops so the bees all have the same scent. Sometimes bees will defect from a failing hive to another, and they tend to be allowed in if they bribe the guards with a gift of honey or nectar. This behaviour can be utilised to merge colonies.

Worker policing: unacceptable behaviour which puts everyone at risk is policed. For example, workers laying eggs have their eggs removed from cells; noticeably sick bees are prevented from entering the hive.

---

15 These "clusters" contract when cold - huddle closer. Similarly, swarm clusters contract at night. The apparent size of a winter cluster in a hive can vary by a factor of 3 with weather so don't panic if they seem small.

16 Not all bees cluster in winter. In a well insulated hive, they may walk across windows with snow outside the hive. They are active all year in some tree cavities. As Derek Mitchell puts it in *Honeybee cluster - not insulation but stressful heat sink* (2023, doi.org/10.1098/rsif.2023.0488), clustering is a *forced* behaviour: deliberately stressing them is cruel.

17 However, it's not as common as some think - see "Drones- miscellaneous" in next chapter.

Ability to communicate and learn: an individual bee may only live a few weeks, but the interconnected group of bees can pass on geographical knowledge (waggle dance), and retain memories like "stay alert, we have been attacked by something recently". It's suspected they can distinguish between people, perhaps by scent, and drive off known troublemakers. I have read of one beekeeper who wore different scents whenever he opened a hive to avoid trouble.

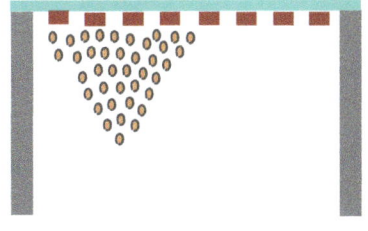

Small colony: living hand to mouth, slow to grow, always on edge of disaster

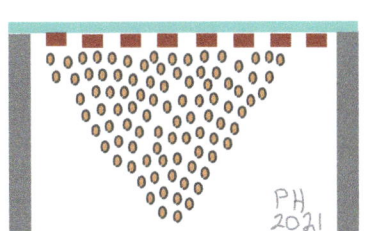

Large colony: excess bees to build nest, fight invaders, forage. A 2x size colony typically gathers 4x as much honey.

A more subtle form of learning revolves around the extent to which a colony expresses anti-parasite behaviours. Colonies increase these activities when parasites are present.

Those are probably epigenetic responses. But they are not intelligent enough to proactively plan ahead[18]. Their responses seem to be reactive and limited, and behaviour changes gradually, for example tuning to new forage rhythms after a move.

Rapid experimentation and adaptation: Mono-brained mammals tend to experience a relatively stable environment and get stuck in a rut of "what works". Contrastingly, every bee is slightly different to others, and relatively short lived - a lack of long term memory is an *advantage* in an environment that can change in hours or minutes, as is the fact that they can try extreme strategies which prove fatal, but the superorganism lives on. They try many things simultaneously and communicate what works[19]. So rapid and powerful is this decentralised system of parallel trial-and-error adaptation, that entire journals are devoted to Swarm Intelligence.

Efficiency: ants and honeybees live in a boom/bust environment with sudden surges of forage, and don't waste energy foraging randomly. Scouts go out, and once they locate a *rich* food source they guide others there. This concentrates the "wins". Compare this with bumblebees, whose forager workforce use a lot of energy flying even on days with little nectar around.

As a result, some people think bees don't forage near their hives. Not true - the trigger is: *is there a lot of food of the same type in one place*. Scouts look for at least one square metre of flowers, which could be a bush, and return with samples of the nectar. Foragers smell the nectar, then go to where the scouts direct them (google "waggle dance") and sniff out the source. A side effect is "flower constancy", foraging on one blossom type at a time. That makes for a very efficient crop pollination

---

18 Or can they? Prime swarms' scouts often pre-choose a destination in the days before the swarm emerges. Margaret Vile observed an extraordinary example of bee decision making in Fringford in 2015, when a newly hived swarm became very agitated, and the entire population decamped to a nearby branch, where they hung in several clumps for an entire day, with much noise, movement and flying. She returned a few hours later and they had gone, but the next door hive's population was greatly increased. Notably, these were sister swarms which had been caught a few days apart in the same bush. We concluded that the first one's queen had not returned from a mating flight and the now-queenless swarm had decided their best option was to join their queen-right sisters a couple of metres away.

19 Examples of extreme behavioural variation I have heard of from reputable sources: flying by moonlight; stealing eggs from another colony to reboot a queenless one; gathering more than one colour of pollen; sucking on fruit during nectar dearths; *European* bees [not *A. cerana*] balling hornets; sleeping in flowers. This helps illuminate why a spread of drone fathers boosts colony health & survival, which we mention elsewhere.

system. Honeybees aren't very interested in wildflower meadows unless they are really packed with one type of blossom. Other pollinators love wildflowers though, so they're worth planting.

Note how if a scout dies because it explores a particularly dangerous location, the colony only loses that scout. Perhaps this helps them avoid the full impact of immediately lethal crop sprays.

Colony level immunity: Honeybees have relatively few genes for their immune system, and no antibodies. Much of their pathogen control comes from cooperation:

- ejecting infected brood, just as our immune systems kill and flush out infected cells
- individuals voluntarily exit hives when dying
- *undertaker* bees remove bodies from hive
- cleaning and propolising
- grooming of nestmates, triggered by a "groom me" request (dance).
- "fever therapy", raising the brood temperature slightly to hamper pests and pathogens
- bees use a form of social distancing if heavily infested[20].

Trophylaxis

Rapid transfer of stomach fluids

Food is communal and bees feed each other (one's long tongue is in the other's mouth). Trophallaxis is also used in honey processing, communicating, moving stores, unloading nectar before rushing off for more forage etc. The downside is it spreads slow acting pesticides throughout a hive in minutes and thousands die.

Bee *colonies* are much more susceptible to disruption from poisons than individual insects, because they rely on subtle communication cues for the whole superorganism to work. A neonicotinoid dose that is sub-lethal to an individual insect can have a serious developmental impact on the colony (affecting behaviour and fertility).

Note that disrupting social activities by opening nests impacts colony health.

Swarming: The queen and workers cannot found a new colony alone. Instead, when a hive or wild nest has accumulated sufficient resources, it splits and about 50% of the flying workforce flies off with a queen to found a new colony as a swarm. A colony may have several swarms in one year (afterswarms are called *casts*).

A very important side effect of swarming is the *brood break* afterwards, where no new eggs are laid until the new queen mates. This acts as a firebreak against parasites like the varroa mite, which almost all target the brood, and parasites tend to have a shorter lifespan than the larger honeybees.

Beware hidden assumptions:

Much published advice assumes you practice swarm suppression, feed etc. Population curves like the blue one are shown in published papers. I think these are based on modelling an ideal spherical hive with an infinite amount of legs, or maybe they really *do* see those curves with Californian bees. You can immediately see how well meant advice about mite numbers and feeding requirements can be wildly wrong! (I think a more realistic curve is the red dotted one.)

---

20   *Honey bees increase social distancing when facing the ectoparasite Varroa destructor,* Pusceddu et al (2021), SCIENCE ADVANCES 2021•Vol 7, Issue 44•DOI: 10.1126/sciadv.abj1398

Bee numbers in my hives (black line) reflect available forage in my area. Without feeding, numbers differ dramatically from "conventional" assumptions.

What I see, in the English Midlands, is the black curve[21] - and an assumption in this book is that you are letting local bees manage themselves, **not feeding** or buying super-prolific queens. Your bee numbers may have a completely different rhythm to mine, but they'll have one, and it won't match the honey farmer assumptions.

Another common assumption is, that you obtain new colonies by buying a nucleus, but this is *effectively already a year old*. So you can harvest it later that year, and it may swarm the same year you install it. Swarm-founded nests, in contrast, are in a race against time to build up enough stores for their first winter. They are unlikely to divert resources to swarming except in exceptionally bountiful years. One doesn't usually take honey from a swarm-founded colony in its first year.

Commercial beekeeping aims to have uniform colonies - it simplifies management. My hives, of varying ages and genetics, all tend to be at different stages of development, honey storage, behaviour etc. Like crop rotation, it avoids parasite buildup.

---

21   I am going by how many combs are black with bees. And yes, I know clusters are denser in cold weather, etc.

Summing up: what have we learned?

Throughout history bees have intrigued us as a bridge to the "Other". Recently we've realised their many-as-one aspect is much more nuanced than we suspected: the lines between the colony superorganism, individual bees and their mutualistic microbiota, and even, as we will see in the Genetics chapter, neighbouring colonies are blurred. Each depends on the other, with stabilising feedback through chemical concentrations and even, remarkably, conscious communication (waggle dance, scent alerts etc).

Review 1

Take a moment to reflect on how interlinked you are with -

- your own microbiota
- your family
- wider society (service providers? The consensus of how we should act?)
- work colleagues
- your home

Where do "you" end? Would **you** thrive as well without one of these?

Chapter 3
Lifecycle and biology

Castes and variants; how roles change with age and time of year; queens, princesses and supersedure; lifespan; dealing with loss; sleeping; senses

Growing up: from egg to adult

General points & context

Everyone starts as an egg. Workers and drones are mass produced in dense arrays of cells: workers in the standard sized cells and drones in larger ones. These are general purpose cells, which can double as food stores if empty. If you look closely you can see these cells are tipped slightly[22].

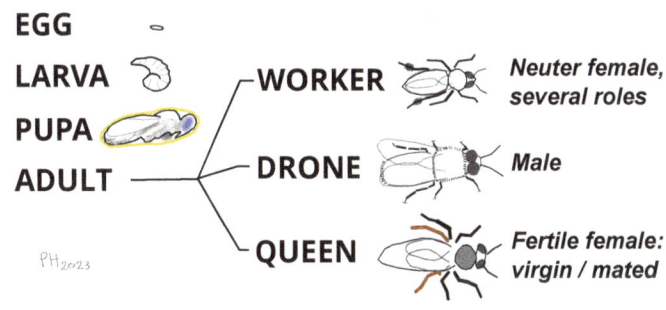

Queens are laid in cells which point downwards. No one knows why. These don't fit in the regular arrays so if the colony plans to raise a queen, large *queen cup* cells are found at the edges of combs[23].

Once the *larvae* are full size they are capped by the workers[24] and spin a silk cocoon to pupate in. These *pupae* gradually metamorphose into the final *imago* stage, adults who can fly.

Bee babies (brood)

Honeybee queens may lay up to 2,000 eggs a day. This is ramped up and down to match forage availability, and as most parasites / diseases live primarily on brood, and breed faster than bees – but die faster - the "brood breaks" provide a natural firebreak on pest numbers outstripping the bees.

---

22 Not so much to stop contents falling out - beekeepers sometimes invert frames and the bees carry on - but to strengthen the comb. *theapiarist.org/comb-fact-fiction/*

23 Although in emergencies, if the queen dies, the bees will promote *any* available female to a princess and alter her cell to a downward-facing queen cell. There's nothing initially special about queen eggs - if swapped into a worker cell they will grow up to be workers. To modify a worker cell into an emergency ueen cell, workers have to rip out cells around her, the result resembles bodged house remodelling. This is the urge which queen rearers use, to trick bees into making excess queens to sell. Sometimes failing colonies only have male larvae, and cannot reboot.

24 The caps are a mix of wax and pollen - they are porous so the pupa can breathe. This gives the worker cap a characteristic flat-domed appearance and light brown colour (drone caps are more domed). **Top tip:** if you take the BBKA Basic Beekeeping Test you **must** parrot that the caps are "*biscuit coloured*". In reality, it varies with pollen colour and is quite different in some countries.

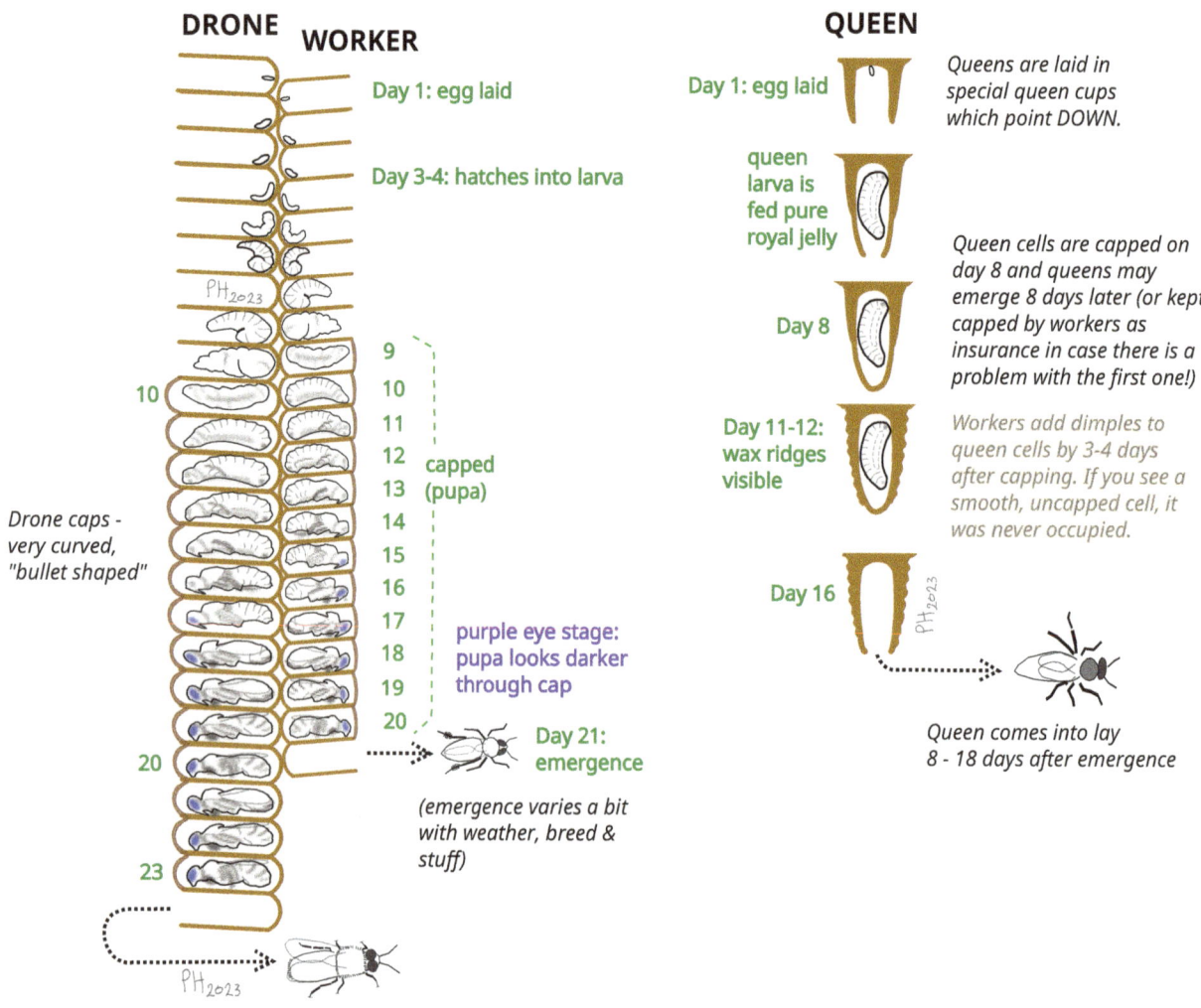

Bee development from egg to flying adult. Hot or cold weather can change timings by a day or so.

The 8 day gap between queen capping and emergence is significant for beekeepers trying to suppress swarms: they open the hive every 7-8 days, to examine every comb and kill queens before they hatch and fly off with a swarm. This extremely disruptive ritual has become confused with "good practise" but disease management, for example, does not need weekly inspections

The larvae, or *brood*, are all fed *Royal Jelly* (a protein rich goo resembling white cream) for their first 3 days. Thereafter, only queens get this and the commoners get *bee bread*. Bee bread is pollen, which has been partially digested (fermented) by adding a little honey and nurse bees' saliva.

Comforting bedtime tales for your own kids: With plenty of spares, bees are ruthlessly intolerant of children they deem somehow 'wrong', like the occasional intersex *diploid drones* who are useless to the colony, and killed and eaten when detected! And, because the children are preferentially targeted by parasites, the adults use spare ones - usually the boys - as disposable traps (Mummy can always lay more eggs), ejecting infested larvae from the nest: throwing out bad children. Oh, and if the colony is starving, they will eat the youngest children so they have protein to make &

regurgitate food for the older ones, who have the best chance of survival. If you think that's tough love, some Japanese wasp larvae regularly cannibalise their siblings[25]. Sweet dreams!

Baby stealing: workers sometimes move eggs around in a nest, for example into supers where there is room for a queen cell. But it's not clear if this is deliberate behaviour or some misfiring of the urge to chuck out bad babies.

Brood nest: the brood rearing area is the core of a colony and is usually an egg-shaped volume spread across several combs.

Given a choice, the bees situate brood combs near the entrance as most of their traffic is to and from these, so you generally find excess honey stored at the "back" of a hive, furthest from the entrance. But brood need warmth, and these requirements sometimes conflict, so where they are laid depends on hive geometry, your climate, whether you use queen excluders etc.

Males (drones) are more expendable: drone brood tend to be laid at the cooler edges of brood combs, and near the entrance, where there is a higher chance of experiencing cold temperatures. If brood aren't kept warm they can die or have a reduced lifespan, which is why when you open a hive, you will see nurse bees thickly covering some combs whilst the honeycombs are sparsely populated.

Larval stages. The youngest (smallest) are blurred because they swim in liquid food. Healthy ones should appear clearly segmented, and "white, bright and curled up tight". The capped ones are metamorphosing into winged adults inside silk cocoons.

A new bee pushes back its cap and emerges

---

25   *Brood reduction caused by sibling cannibilism in Isodontia harmandi,* Tomoji Kendo and Yui Imasaki, Kobe College (2022)

Distinguishing queens, workers and drones

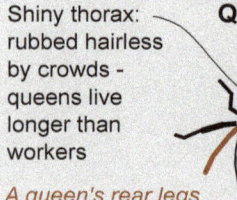
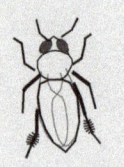

Shiny thorax: rubbed hairless by crowds - queens live longer than workers

A queen's rear legs redden with age

Colouration can vary with age for some races, e.g. orange (virgin), black (mature)

Queens' wings are short relative to body length - flying is hard

Young workers look furry. Over a few weeks their hairs are worn off so old workers have a shiny thorax and can look darker.

Big, boggly eyes which meet in middle

No pollen baskets (corbiculae)

No sting - blunt, stout body, furry bottom. Loud, low buzz in flight, like a bumblebee

Drone on left is larger, has furry round bottom, huge eyes, longer wings than worker, no pollen baskets.

Workers (neuter females)

After hatching as winged adults, workers start as nurse bees, then take on other tasks until eventually they graduate as foragers. They undergo a number of body changes - like our puberty - over a few weeks. Career progression may be summarised as

nurse - builder - guard - forager

Though this is the *usual* progression, if there is a shortage in one occupation bees will switch roles, e.g. to defend the hive.

The adults may live as little as 15 days working hard in summer, or 140 days huddling indoors in winter[26].

Young bees produce wax and royal jelly from specialised glands. As they age these abilities atrophy but their venom glands fully develop, they become guard bees and begin experimenting with flying. The oldest, strongest fliers graduate to foraging. Terminology is loose: *scout* and *undertaker* are sometimes used for specific tasks; I would say a *house bee* is one not yet foraging, others might say guards are not house bees.

Thus it is best to open a hive in the middle of a warm day. The foragers - with the most developed venom glands - are often flying far away; nurse bees are more timid.

Key point: nurse bees can barely fly. If you perform an operation like shaking the bees off a comb into the grass, to harvest honey or inspect combs - many young bees will never find their way back to the hive. They are very weak flyers and have never been outside. Always shake / brush bees back into a hive unless you're deliberately performing a 'shook swarm'.

The nurses are the true mothers in the hive - the queen pays no attention to the young!

Young bees produce wax

Rows of white, lenticular wax scales exuded from wax glands between segments

---

26   *Biology of the Honey Bee*, Winston, first paperback edition (1991) p.55

Winter (worker) bees

The last bees to hatch from the autumn brood, who would normally be nurse bees, have no more brood to nurse and instead gorge themselves with food, storing it in internal fat bodies. Winter bees are basically nurse bees whose development to foragers has been retarded[27]. This lets them live several *months* instead of weeks. Then in the early days of next year they have some nutrients to feed to the first new hatchlings, without having to go out to look for scarce resources.

These bees hunker down for winter, occasionally going out to defecate on good flying days. This is where Italian bees struggle with northern winters - first, they can be tricked out by bright light reflecting off snow into flying when it is too cold (as soon as they land on anything they freeze in place), and also they have not evolved to hold their faeces in for many months so may suffer from dysentery, uncontrolled defecation within the hive.

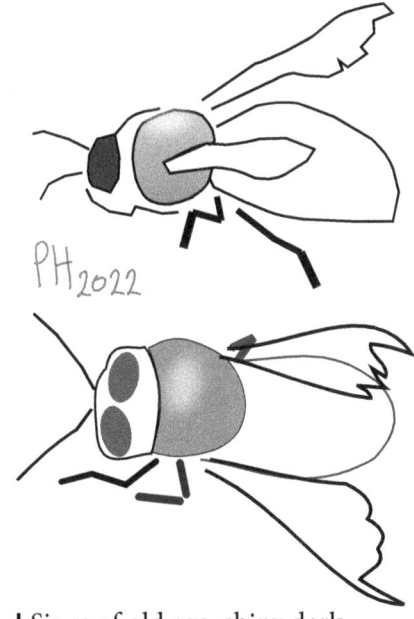

Signs of old age: shiny dark thorax (hairs worn off central section); ragged wings at **ends** (it's **not** DWV). Young bees look furry. Old butterflies have ragged wings, too.

Queen rightness, and drone laying workers

Queen-right means there is a laying queen present, whose pheromones calm the colony. When she begins *failing* - running out of stored sperm from the mating flights of her youth - the workers sense the reduction in brood pheromones and begin raising a new queen if possible.

Although worker egg laying is normally suppressed by brood (not queen!) pheromones, workers aren't really neuter. They *can* lay eggs if a hive is not queen-right. These eggs are always drones, as the workers have not mated: an unfertilised bee egg is male. When a queen dies this offers the colony a last shot at spreading its genes, an elegant evolutionary fail-safe.

Conventional beekeepers hate these *drone-laying hives* as they are unproductive. There are various unreliable tricks to restart a colony like giving them frames of brood[28], but I now let such

---

27   The biochemistry is: winter bees have larger fat bodies, which act as stores of *vitellogenin*, a complex chemical which acts as a hormone in bees and is used as a precursor to egg yolk in many species; and less juvenile hormone. These two substances control how fast bees age (mouthpart, gland development etc) through the usual stages to Forager. In foraging season, the ratio of these substances in the workers swaps, and juvenile hormone accelerates development so workers become expendable foragers, which have just enough vitellogenin to survive. Winter bees' larger reserves allow the production of more brood food from their *hypopharyngeal* glands, so they can feed brood even if there is insufficient pollen when brood rearing restarts at the end of winter. In effect winter bees are reservoirs of nutrients for the brood. Intriguingly, if a colony needs to - say it has to pull back bees from foraging to nursing after some disaster - it can **reverse the ageing** of workers, these hormones swap ratios and foragers regrow the glands etc needed for nursing. Further information in Wikipedia under *Vitellogenin*.

28   Adding brood only works early in the season. It typically takes 3 weeks of transferring a frame of brood in once a week for the colony to accept they need to raise a new queen, because *drone laying workers think they're queen-right*, and by August it is too late for a new queen to mate and get colony numbers and stores up to strength for winter, so all you're doing is weakening another colony. Alternatively, you can *sometimes* reboot such hives by removing the queen and putting in a queen cell - but buying a queen squanders the local survivor genetics your colonies have built up.

colonies die naturally so I can observe what happens (I can afford this luxury as I have other hives) and harvest remaining honey.

Another option is to unite drone laying colonies with queen-right ones. Sometimes you have a small queen-right colony struggling to get by and the extra workers boost it into prospering,

A hive which has lost its queen will last several months because the bees no longer have hard work to do. In effect the workers are winter bees. If you observe the entrance, after a few days, the colony loses its morale – the bees aren't always busy foraging but instead many dawdle on the landing board; they aren't that bothered about wasps and robbers. It's thought some bees in failing colonies defect to nearby queen-right hives.

Sometimes queenless hives reboot inexplicably. Possibly by stealing an egg from another hive[29]; or sometimes an unmated worker lays a female egg - a clone of herself - termed *thelytoky*. This is extremely rare, but there may be 30,000 workers in a hive.

Regarding drone laying colonies, in 2024 natural beekeeper Ferry Schutzelaars in the Netherlands noticed one of his colonies was laying masses of drones, but no workers. Where a commercial beekeeper would have requeened, he decided to wait and see what happened, as flooding the area with these drones would help spread the varroa-resistant traits of his dark bees. To his surprise, several weeks later the combs were solid with worker brood, and no drone brood. Sometimes colonies just decide to do drones for a while.

Drones (males)

Not just flying sperm!

The stingless drones are raised only for half the year, to mate with queens, and are laid in larger "drone comb" cells to allow them to grow to full size. They're as big as a queen. They are really chilled out and do not seem to recognise humans as a threat.

Functions include:

- *Reproduction* - famously they only mate once, and die when their penis explodes with an audible pop. Lots gory videos online if you want the details. The penis is left in the queen and known as "drone sign" when a queen returns from a mating flight with several sticking out of her. More on mating below.

- *Heater bees* - most people don't realise that while workers are out the drones often form a warm blanket layer above the brood nest, until the heat of the afternoon triggers them to fly and mate[30]. They don't fly before 11AM and only in fine weather - because the foragers are

---

29 Gareth John found a queenless hive in his apiary which he inspected very closely – every comb. Definitely no queen in this case and no eggs (he had 40 years' experience at this point so was very sure). A few weeks later he found a single queen cell in it. The next hive was very close and was related. Had the bees borrowed an egg? In time a fully functioning queen developed. Tickner Edwardes relates a similar tale of egg theft to raise a queen in *Lore of the Honey Bee*, 7th Ed, p.77. Other beekeepers report odd reboots of "drone laying hives" over shorter timescales but explain them by saying occasionally queens drift; sometimes a DLQ will have a few sperm left; and sometimes a new, freshly mated queen will lay some drone brood initially.

30 If you search for "drones in supers" you'll find people discussing the "mystery" of why are drones up there - it must be bad, they are eating "my" honey. (A common maxim is "drones can't feed themselves", but beekeepers are adept at believing contradictory things.) The usual response is "drone laying workers in your super" though the posters don't mention other signs like multiple eggs per cell. I even saw one post where the poster remarked on finding drones which had died jammed in the queen excluder which prevented them from getting to the super, all above the brood nest, and apparently trying to get into the super. It's all very puzzling...

Three drones amongst workers. Huge eyes which meet in the middle, blunt furry bottoms. The workers appear to vary in size due to posture and angle of viewing, but are all significantly smaller. Drone colouration can vary within a colony.

out and the babies need blanketing[31]. *Queen excluders prevent them forming this insulating umbrella above the brood area,* another brood-health-optimising behaviour most people don't realise their practises are disrupting.

- *Sacrificial mite traps* - drone larvae are more susceptible to varroa, because they are placed in cooler parts of the nest and give the mites longer to breed, so hygienic colonies end up throwing out more drone larvae[32].

- *Immunological?* - Potentially, drones could spread immunity around because they drift between hives, spreading *low levels* of pathogens which *challenge* each colony's immune system. For example, trophallaxis, where bees share digestive juices, means that drones could regurgitate partially digested foulbrood bacteria to workers. Their digestive system kills or weakens the pathogen but the worker's hive gets a cue that foulbrood is in the area. This spreads valuable news about the wider community. Later, in the health chapters, we will cover Inconvenient Facts like: lateral flow tests *may* be *too* sensitive and show some "false positives", and it seems impossible to infect healthy hives with the dreaded AFB. Appealing though this oral vaccination idea is, I have to point out that nosema seems transmissible orally[33].

- Gareth John has seen drones *washboarding,* i.e. laying down layers of antiseptic propolis.

Drones *can* eat stores from cells, and even gather nectar. I mention this as most beekeepers simply repeat that "drones cannot feed themselves", an assertion based on their "fast charge"

---

31 'The 'Times Bee Master' drew this conclusion as early as 1864 in his book *Beekeeping*. He pointed out drones are only raised in summer when foragers are out, and go on mating flights in the warmest part of the day when heater bees are superfluous.

32 Though varroa expert Ron Hoskins says he's not seen any varroa preference for drones. Some beekeepers deliberately encourage the raising of drones (on drone foundation) and then kill the entire comb. And practise queen wing clipping on drones. It all seems very banal to them as 'it's routine'.

33 *The Honey Bee Parasite Nosema ceranae: Transmissable via food exchange?* Michael Smith, Cornell (2012) - doi.org/10.1371/journal.pone.0043319

mode when they go fly off to mate. For mating flights, it is more efficient for them to be launched in the warmest part of the day[34], whereupon they can fly for about 40 minutes (longer if they catch a thermal). They then return home, are rapidly refuelled by a house bee squirting jet fuel directly into their mouth (trophyllaxis), and sent out again to do their duty. This maximises the chances of the colony to spread its genes.

The curious sex life of the drone

Mating is explosively fatal for the drone. When he grasps a queen his abdomen contracts, causing the pressure of his haemolymph (blood) to rocket and push his sperm out, sometimes with an audible pop. The queen then flies off, with the "drone sign" (part of his endophallus) still in her. The drone dies, much like a worker dies when her sting remains in your skin. The queen typically mates with 13 drones, but some queens with many more[35].

Drone o'clock - rush hour

Mating sites are called *Drone Congregation Areas* - massive genetic mixing pools where drones from hundreds of nests gather on sunny afternoons. They are difficult to spot and few beekeepers know where their local DCAs are[36]. It's suspected drones fly to their nearest DCAs and queens to further ones, thus minimising inbreeding[37].

Drones - miscellaneous

There's an urban legend that drones hop nonchalantly between hives, without being challenged by guards. In reality, it's the exception rather than the rule. Firstly, they're not welcome at **every** hive, only those producing their own drones. Young colonies still establishing themselves don't have resources to spare and their guards vigorously repel drones.

The proportion of drifting drones is easily ascertained by marking them, whence the truth becomes clear. A Russian study[38] found only 1-2% of the drones in a hive are from other hives,

---

34 If we hear the hum in the garden suddenly step up in volume around 3PM, we exclaim "it's drone o'clock!" This synchronises with when queens fly, as they are poor fliers and only go out in the best flying conditions, which happens to be warm afternoons when the drones are no longer needed as brood blankets.

35 Bee sex is mundane by ant standards. Queens of two related species of harvester ants basically rape males of the other species, because they need some sterile hybrid slaves to serve their purebred descendents. Queens of *Mycocepurus smithii* reproduce asexually. And so on.

36 Every source seems to disagree on landscape features where DCAs may be found. All we can really say is, drones like spots around treetop height with thermals or draughts to keep them aloft longer. But if you see a lot of swallows in one place, or hear a sound like a swarm above you - look up. Confirm it's a DCA by tossing a pesbble up through the bee cloud - if a tail of insects follow it down, they're drones.

37 Drones may visit 4 DCAs per flight. Radar tracking of drones: doi.org/10.1016/j.isci.2021.102499

38 This result is from a 3 year research project on 19 hives of 3 races was by I. Livinets in the USSR *Beekeeping* magazine, 1951 #1 pp.25-30, 'About drones entering other hives". I am taking this on trust as I don't read Russian. Source: Пчеловодство_1951_01.pdf (yandex.com) .

and if you move drones to another hive 95-98% return home. Later, Western, researchers concluded drones *often* drift between *hives* but up to 97% stay within their native *apiary*. (Perhaps these later studies were in apiaries full of closely related queens, which promotes drifting because everyone smells the same.)

But hive-hopping does happen, and this allows some to mate with queens more than 40 minutes' flying away.

Karen Giles pointed out an odd behaviour to me. Her drones tend to fly round a nearby tree for 30 minutes and then return to the hive. I've not heard that elsewhere. This is next to the tree her swarms settle on, perhaps there is a lingering smell of queen pheromone. The tree is at the edge of a paddock next to other trees - so probably not a DCA.

Colonies strive for about 15-20% of their number to be drones at the height of summer. If you see far more than this – which generally requires opening the hive to confirm your suspicions – it is probably because the queen has run out of sperm, or because you have drone laying workers. (Though another cause is that if you catch a swarm from a commercial hive, where drone laying is suppressed by use of foundation, in their first year of freedom they overcompensate and produce excessive drones, then settle down in subsequent years. And virgin queens can lay drone eggs, and newly mated queens sometimes lay drones for a while.)

Colonies sometimes produce drones in their first year: colonies divert resources to drone production when workers exceed ~4,000 [39]. Generally speaking, the presence of drones indicates the colony is thriving, and has good nutrition. They don't make drones if pollen is short.

If you move a hive 3 or 4 miles, a considerable number of drones return to the original spot, but no workers. Drones know the map better and fly further. Beekeepers in the Isles of Scilly suspect they even cross short stretches of sea, and are marking their drones with an island-dependant colour code to check this (the *Game of Drones*).

Drones don't have stings because they don't lay eggs: stings are modified oviposters.

It's thought one reason for increased queen failure is a lack of drones these days – fewer colonies, and beekeepers tend to suppress their production.

When packages of bees contain too few drones, the colonies don't settle in or build up as well until drones are present again[10].

Drones are much less hardy than workers. They need keeping warm - but not too warm - and can die if handled roughly.

Alternative uses for drones

I once noticed raiding wasps deterred by a large heap of guards at the entrance. Looking closer, it turned out this comprised some workers on top of drones. Drones can't sting, but they can keep

---

39 Michael Smith, Cornell University, p.24 of 2017 doctoral thesis *Growth, Development and Reproductive Investments in Honeybee Colonies*. This chapter is co-authored by Madeleine Ostwald, J. Carter Loftus and Tom Seeley. He quotes earlier work by Smith et al (2014) who first estimated the threshold of 4,000 workers, and confirms it in 4 colonies he founded with swarms where drone production began at 3671 - 7859 workers, beginning just 39 days after founding. By manipulating worker density he showed this is the trigger to build drone comb. Something that caught my eye about this thesis was that it is in 4 chapters and each one was published in a separate journal. Most theses cover much less ground and are published in only one.

40 *Fly with the Beeman*, page 19 (1989) - memoirs of Robert Couston, a bee farmer and Specialist Adviser in Scotland. Packages with drones "strained out" with a queen excluder performed more poorly.

guards warm, which was useful on this chilly day (cold bees cannot take off instantly); and they fooled me and, presumably, the wasps into thinking there were more guards than there really were, too. Only one hive was using this strategy, it was a large colony and also the only one that had reduced its entrance with a barrier.

Drone larvae are farmed for human consumption in some countries. See teca.apps.fao.org/teca/en/technologies/8778 for preparation tips and recipes.

A mix of drones and guards. Some guards are holding their wings in a V, a threat posture.

Drone exclusion

Around the end of summer you will see them being forced out of the hive by the workers. As the workers control food and have stings it is somewhat one-sided. Some people report the workers initially corral the drones near the entrance, and stop feeding them - this probably weakens them and makes them easier to throw out. The workers don't waste stings, they bite and maim wings. The drones huddle disconsolately and die over a few days.

Drone exclusion is said to happen at the end of August, but it varies between hives. If there is a dearth, they can be evicted early. There is usually one hive that still accepts drones after the others have barred them, so it gets swamped. This can be a sign that the hive is superseding its queen. The bees rarely evict every single drone – a few remain in the hive for months[41] and you may spot drones in winter.

One way of looking at drone exclusion is, it is a good sign - queenless hives don't do this, so the hive is queen-right.

Queens and princesses

The Queen is not in charge. For example, her laying rate can reach 1500-2000 eggs / day, but is controlled by how much the workers feed her. For the winter brood break, the workers put her on a diet; and also for a few days before swarming – to get her down to her flying weight. Mature queens are poor flyers, see how short their wings are!

Think of her as a vital organ in the superorganism. Losing her is awkward, possibly fatal to the colony. That's why she does not get involved in fights: even though her sting is unbarbed

---

41 Personal observation; also *Bevan on the Honey Bee* (1870) 3rd Ed. Ch. 34. Robinson states emphatically (*British Bee-Farming, its Profits and Pleasures* [1880] p.125) that he does see occasional winter drones but **never** in hives with a vigorous and fertile queen.

(multi-use) she does not sting humans, it's only for killing rival queens[42]. When trouble starts she runs towards the dark while the workers drive off the threat. In a swarm or winter cluster, she is at the centre. This behaviour can help find her.

Queens mate once in their lives, but with at least 13 drones (often 20+), giving a range of half-sibling children in the hive. It is thought this gives the colony a greater range of abilities and resistance to pests. This is termed '*extreme polyandry*'.

Day old virgin princess — Furry thorax when young. Wings folded neatly back.

1 year old queen

5 year old queen — Thorax (and eventually abdomen) completely bald and shiny, appear darker.

Wings progressively more tattered with age.

| A princess's wings are proportionately longer, making flying easier.

If you look at a hive entrance and see bees of many different stripe-patterns, that's a sign of a healthily diverse hive.

She stores several million sperm from these encounters in her spermathecra organ and they will be viable for several years.

She produces eggs continuously, but eventually the sperm run out and the colony raises a new queen, a process called *supersedure*.

The commonly repeated wisdom is that queens (and drones) can't, or won't, feed themselves but are dependent on workers feeding them mouth-to-mouth. This is absolutely not true, they have both been observed feeding from cells, and there is even a documented case of a queen foraging[43] (she probably stopped off for a snack on a mating flight). Bee roles are fluid. Queens are however less adaptable than workers - i.e. less intelligent.

| Queens are raised in distinctive large, downward-pointing cells.

Princesses (virgin queens) are raised in small batches when needed. We discuss their behaviour in detail under Swarms, but just to mention -

- They are very nervous and prone to flying away if startled. Being slimmer than queens, they can fly OK (though the wings of queens & princesses are shorter than workers') and are occasionally seen unaccompanied on orientation flights. They've even been seen flying immediately after hatching.

- Princesses scuttle about rapidly and nervously. Mated queens walk sedately.

---

42 The only counter-example I know of is a beekeeper who had been handling other queens, then picked up one queen who stung his hand – presumably because it smelled of rival queens.

43 *Where is the Honey Bee Queen Flying? The Original Case Of A Foraging Queen*, Floris, Pusceddu, Niolu & Satta (2021), doi.org/10.3390/insects12111035. Even in 1870, it was known princesses have to feed themselves (*Bevan on the Honey Bee*, WA Munn, 3rd Ed p.186).

- My princesses change colour dramatically over their first few months of age. Starting out a bright orange-yellow when young, they are easy to spot among the dark workers; they end up almost completely black. Perhaps mating has something to do with it. I have not heard of this elsewhere, it may be a local thing.

Some beekeepers imagine they're in charge.

- Workers largely ignore princesses, probably because they don't smell strongly of queen pheromone until mated.

- When a new queen is needed, the colony raises 6-20 in special queen cells. Why so many? Because princesses sometimes fail to return from mating flights. And anyone who really *looks* at swarms will realise they very often have multiple backup queens (unmarked, not immediately obvious - people tend to stop looking after seeing "the" queen). Surplus princesses are killed or expelled once the colony reaches a new equilibrium after swarming / supersedure.

Supersedure and emergency queens

Supersedure is the term for when bees replace a queen in a planned manner. This may be because the old one's fertility is fading ("queen failure"), because they intend swarming, or simply because she is damaged.

Sometimes the queen is lost through misadventure (usually: not returning from a mating flight, or a beekeeper squashing her or dropping her in the grass during an inspection)[44] and they need to raise an *emergency queen*.

The procedure is the same in both cases. New queens can be raised from female eggs, or larvae up to 3 days old, and are normally raised in special large cells at the edge of combs, hanging vertically. However, *emergency* queens need to be raised from whatever eggs or larvae are available - which tend to be in the middle of combs; and despite the workers rapidly chewing away comb around it and rearanging the furniture, the emergency queen cell, and thus the queen tends to be smaller than a planned one. Bees seem to replace these small queens as soon as they can raise a full size one.

From ~2000 AD, many beekeepers began noticing alarming increases in queen problems - early supersedure; fertility (drone laying); and disappearances on mating flights - which had previously been very rare. This appears to be a worldwide phenomenon. Low-intervention beekeepers (or "natural beekeepers", an alternative term you may see some people use) see this too so it cannot be primarily a management system issue. One wonders if bees are acting as an early warning about pesticide saturation in the landscape.

Supersedure may not cause a 2 week brood break like swarms do, if the fertile queens overlap.

---

44 Opening hives is always a risk. Some beekeepers assume hives need continual opening and checking because their "queens don't last long".

Requeening

Profit motivated beekeepers often raise or buy extra queens to maximise bees, particularly for new crop strains that have been created which flower very early[45]. They might -

- Routinely kill swarms' queens as they think swarms are Bad, then merge the orphaned workers into a honey production colony.
- Stimulate colonies to expand early, split them in 2 or 3 and requeen the new colonies so you have several full strength colonies by summer.
- Replace queens every 1-2 years as they run out of sperm. This is necessary when you stimulate excessive laying. In America this can even be every 6 months (they run huge colonies, and have other issues affecting queen fertility discussed later).
- When you have hundreds of hives, it's useful to have them all develop at the same rate.

I don't buy queens because

- I want the bees to choose and make their own queens to conserve local adaptations and benefit from natural selection.
- Why buy queens to replace ones you routinely killed?!
- I don't stimulate laying, so the queens don't wear out fast. I don't move my hives to forage over the year, so I don't need lots of bees all year.
- If I requeened with a purchased, mated queen there would be no brood break, so varroa numbers would just keep rising exponentially.

I *have* been toying with the idea of "banking" the "spare" queen if I decide to merge two casts in swarm season. But when you merge colonies it seems sensible to let the bees decide which is fitter.

Queen marking and clipping

Conventional beekeepers sometimes mutilate the queen's wings so she cannot fly off in a swarm, which would reduce the number of bees making honey in a hive[46]. The queen is unaware she cannot fly, so sometimes tries to swarm anyway, and falls into the grass below the hive. If she gets lost or cannot climb back, this results in a queenless hive. Queens often raise a back leg when clipped and you can easily chop the end off. I've seen a comment by a Polish beekeeper - "my bees dislike 'disabled' queens and quickly supersede clipped ones".

Conventional beekeepers sometimes put a coloured paint blob on the back of their queens to tell how old she is – there is a 5 year cycle of paints: white, yellow, red, green and blue, so useless if

---

45 The big one in Britain and Canada is Oil Seed Rape, aka Canola or OSR. This crop was improved in the 1970s to create strains which produced an oil edible by humans, the Spring type flowers very early (other varieties flower in winter and are used as a cover / break crop, it's usually too cold for bees to harvest when those flower). This was a **huge** change to our commercial beekeeping, which had declined as clover etc was no longer being farmed at scale. This crop became the major dependable beekeeping moneymaker, it yields lots of honey but it crystallises in the comb very quickly so the OSR season is frantic for commercial beekeepers. Farmers pay for pollination but began depending on neonicotinoid pesticides to protect the crop from flea beetle, then in the 2010s neonics were proven to harm bees, leaving the BBKA conflicted! Neonics were effectively banned in Britain and Europe, our OSR crop collapsed (the pesticides killed natural pest predators like spiders too), and the neonics are proving very persistent in the soil.

46 Just to emphasize, they don't care about losing occasional queens, they view them as easily replaceable. They **do** care about losing 20,000 workers and the trouble involved in recapturing a swarm. If a swarm does emerge, clipped queens are sometimes found in a golf ball sized cluster under the hive.

you're colour blind. If you ever catch a swarm and the queen has a blob on her back, it's come from a commercial colony. Low-intervention beekeepers don't bother with marking queens: the bees will raise new queens as they need them. Again, you can damage a queen doing this.

Occasionally a colony needs to move, say due to flooding or starvation. Queens need their wings.

Quick quiz: A Master Beekeeper found that after clipping the wings of a dozen queens in his apiary, he suffered high queen losses. Why? (*Answer in footnote*[47])

Royal gossip rarely mentioned

There is usually only one queen in a hive. However, perhaps 5-10% of hives have 2 queens: mother and daughter. A handful with 3 queens have been seen; the old queen isn't disposed of while she's still useful. Some strains of bee in marginal forage areas don't swarm every year, but instead routinely supersede with overlapping queens and only swarm every 2 or 3 years[48].

Indeed, entire books have been written on deliberately running hives with multiple queens, to increase honey crop.

This photo by Pat Testerman in New Hampshire shows an unusual dual queen situation. (That's a drone at the top - similar size at first glance, but much larger eyes than queens, his meet in the middle. The queens' thoraxes are worn bald.) These queens are unrelated! Thinking his hive was queenless

Two queens in one hive. Image © 2023 Pat Testerman, The Flying T Farm.

for a few weeks, he bought another queen (marked with yellow dot), then a yellow queen reappeared. They ignore each other: only virgin queens fight[49].

Very rarely, people have reported tiny queens, termed *intercaste queens*. It's thought these arise when the starter egg was more than 3 days old; you work with what you've got. These can be as small as a worker and walk right through a queen excluder. I've seen it written that whilst fully functional, "the colony will replace her with a planned, full size queen after a few months". However, I've since come to realise that source has strong views and is usually wrong, whilst Roger Patterson (who really *does* know about queens[50]) has found queen size is not a good indicator of performance.

---

47 This is a true story from Filipe Salbany, who watched the beekeeper clipping queens' wings. By the time he got to the 3rd or 4th his hands were covered with several queens' pheromones so the later hives balled & killed their own queens. For other tales of clipping problems google "The queen-marking pen killed my queens!" - typically, when someone uses the same queen cage to confine a series of queens being marked.

48 Tickner Edwardes, in *Lore of the Honey Bee* (7th Ed, 1916, p.66) remarks that multiple-queen hives don't seem to swarm, thus have more workers, thus more honey - prompting much interest at the time.

49 Quimby / Quinby (same person, variant spellings), *Mysteries of Beekeeping Explained* (1853). More formally, *Sharing the Throne: Establishment of Multiqueen Colonies* (2023) Withrow. lib.ncsu.edu/resolver/1840.20/40931

50 He summarises some of his *70 years'* experience of queens in *varied circumstances* in *Beekeeping: Challenge what you know*, p.143 onwards.

Pollen starvation can also cause dwarf queens, but then you'd see very small workers, too. Unlikely in Europe's low intensity, pollen rich environment.

Queens tend to be larger when laying at full rate.

Fainting queens: Very very rarely, people encounter a queen who appears to die when she is handled. She may remain motionless and curled up, including legs, for 15 minutes. Really big bee farmers, with thousands of colonies, see this a few times a year. They recover fine. It may be a "playing possum" defense.

An entomologist might say the fertile females, Queens and Princesses are strictly speaking called *gynes*, but I've never heard a beekeeper use that term. It's more useful in species with multiple queens per nest like some ants and wasps.

Queens "don't cross honey". To be more precise - they stick to their job of laying eggs and if they come to the edge of the brood area, they turn back to look for brood cells to lay in. So a 3-5cm band of capped honey (like the arc often above brood) can act like a queen excluder. This behaviour can be useful in discouraging them going somewhere you don't want them, like supers.

However, honey isn't a reliable barrier because she may cross it if -

- she is short of brood cells to lay in
- she only has worker sized foundation to lay in - she will go looking for drone sized cells
- a forage dearth means the workers have depleted the honey barrier
- (I've also found a princess hiding in a honey area, perhaps lost, perhaps avoiding the queen)

Once they have a new queen the old one, if still around, is on borrowed time. Once the old queen is deemed useless she is killed by the workers – by balling and suffocating & cooking her, or by just shoving her out of the entrance. Sometimes they use stings on their own princesses but rarely their own mature queens[51].

Queen pheromones affect colony personality: If you switch the queens of an aggressive and docile colony they sometimes switch personality, within an hour. You can switch them back and see the aggression follow the queen. There is another, genetic effect where a queen's **children** inherit traits like grooming or, to some extent, docility, but that takes some weeks to replace the previous generation.[52]

Doting mothers: Samuel Bagster wrote[53] that sometimes a queen kills her own princesses if no swarm is needed but can only kill 4-6 at a time (runs out of venom?) - which is most frustrating for her!

If you requeen a hive with a non local queen, from more than ~10-20 km away, you destroy any adaptation your colony has accumulated to your area. Plenty of people sell "British" queens but Wales is not Kent, Cornwall is not Lincolnshire. And often these queens are selected to lay eggs prolifically and continuously, on the assumption you are moving your hives from crop to crop to maximise honey.

Ideal queen temperature: they are best kept at our body temperature. So if transporting queens or queen cells without attendants, people sometimes put them in a protective cage in e.g. their pocket.

---

51 Unless the beekeeper inadvertently rubs the smell of another queen on them, for example when marking several.
52 But if the colony has some other stress, replacing queens won't always calm it down.
53 *The Management of Bees, with a description of the Ladies Safety Hive* (1865) Ch. 2

Something to think on: Sara (iyagahoney.com) moved to Rwanda and tried buying queens from local beekeepers. They didn't understand the concept. How can you have a queen without a colony? They're indivisible.

Something else to think on: conventional beekeepers sometimes say naturally run colonies "only last a few years". If you requeen every year, *your colony is never more than one year old.*

Workers seem to specialise in one thing later in life. There are "hit squads" obsessed with hygiene, and others who become particularly good foragers - and it seems the really experienced 20% of foragers bring in 50% of the food. This raises an interesting point: do seemingly minor reductions in worker lifespans due to chemicals, disease etc have a disproportionate impact on colonies?[54]

A related issue is - if wild colonies really do have workers who live well beyond the oft-quoted 6 weeks, such colonies have yet another selection edge over hard-worked ones, because *most food is gathered by old, experienced foragers.*

Beekeepers - the final caste

Bee colony boundaries are vague, within and without. You're part of the family too.

Fig. 35 Worker bees sleeping in cells of their comb.

Sleeping

Bees sleep with their head and antennae down, with the antennae still. Younger bees jam their head in a cell - queens use drone cells[55], whilst foragers tend to sleep *outside* cells, on the edge of combs (less activity). It's sometimes visible through windows. They also take naps in flowers.

| Scanned from *Bee-Keeping New & Old*, by William Herrod-Hempsall (1930)

---

54   I think this was first discussed in *The Apiarist* blog, www.theapiarist.org/workers-not-shirkers
55   *Bevan on the Honey Bee*, WA Munn, 3rd Ed (1870) p.246.

Lifespan of adults

Simple story in many sources		The more nuanced truth
Summer workers	4–6 weeks	Depends how hard they are working. Workers in queenless hives last months, possibly up to a year!
		Genetics is significant. Amm last about 10 weeks on the wing[56].
		Research in 2022 indicates lifespan is half what it was 50 years ago[57].
		Ward & Li-Byarlay found[58] feral colony workers lived 47-57 days, managed bees 28-42 days. This was just 3 pairs of colonies; later we will mention other research seeing similar effects when looking at specific factors like the effect of poor nutrition in managed colonies.
		2019 observations by Torben Schiffer, on the hair lengths of bees in unmanaged colonies in trees, suggest workers in queen right colonies live up to a couple of months in natural conditions.
		Interestingly, Victorians believed their bees lived much longer, for example in Pettigrew's Handy Book Of Bees (1875) he reckons 9 months for a worker, though methodology is not explained. It does seem that bees live much longer in low stress conditions.
		Roger Patterson observes indicators that darker bees appear to live significantly longer than Italians, so produce as much honey with fewer (but more experienced) adults in Challenge What You Are Told p.57-58
Winter workers	4–6 months	Correct. Amm bees, adapted for long northern winters may last 6 months. Winter bees metabolically resemble nurse bees that never grow up. Their extended lifespan appears to simply be due to not donating body nutrients to brood, and being worked to death by commercial beekeepers, over winter.
Drones	A few weeks	Drones are occasionally spotted in hives over winter so perhaps last months if doing nothing, or maybe a few drones are raised over winter. But in summer they wear out in weeks (flying to mate) and are rapidly predated, so 6 weeks is a good estimate.
Queens	1–2 years	Usually 2 or 3 years, sometimes up to 5, the record is about 7 years. Depends on laying rate: once she begins to run out of sperm, the workers replace (supersede) her. Queens in small colonies therefore last longer. Commercial queens laying 1500+ eggs/day continuously over summer run out of sperm in 1-2 years, which is why standard training emphasises replacing the queen every 1-2 years.
		Historically, they lived much longer. Tickner Edwardes reckoned they were at peak laying rate in ther 2nd year, and could live 4 or 5. (Lore of the Honey Bee, 1916). Webster reckoned typical lifespan was 4 years but egg laying usually tailed off in year 3 (The Book of Beekeeping, 4th Ed (c.1910) p.70)

---

56   Beowulf Cooper, pamphlet *Village Bees - the native & near-native bees of Britain and Ireland* (1968). He also mentions *Amm* queens last 36-48 months in full production, longer in small colonies.

57   *Water provisioning increases caged worker bee lifespan and caged worker bees are living half as long as observed 50 years ago* (Nearman & vanEngelsdorp, 2022, DOI 10.1038/s41598-022-21401-2). The discovery was serendiptious as they were initially trying to replicate condirions in **well described** 1970s water feeding experiments on captive bees (ie very reproducible caged conditions buffered from external variables, comparing 46 trials over 50 years), and noticed the discrepancy: worker median lifespan in the USA had dropped from 34.3 to 17.7 days. They noted average US honey yield per hive dropped ~24% as bee lifespan dropped, though that could be due to other factors; and modelling showed shorter lived bees could account for the US's high colony losses. They state *average* queen lifespan was 5 years in the 1960s. The authors mention genetics *may* be responsible but it's more likely increased viral loads like DWV.

58   *The Lifespan and Levels of Oxidative Stress between Feral and Managed Honeybee Colonies*, Ward K, Li-Byarlay H (2021) https://doi.org/10.1101/2021.06.29.450441

Bee senses

Bees spend most of their time among dark combs, and their senses reflect this.

Vision: Despite eventually navigating several miles from home, this is surprisingly poor in some respects, being just good enough for their purposes.

They have two large compound eyes which cannot focus. They're the opposite of 'hawk-eyed': their resolution is about 100 times poorer than ours, so we would consider their vision a low resolution "pixelated" smear. On the plus side, they can detect polarised light[59] (for navigating by the sun on an overcast day) and they are *very* good at spotting movement. They have three extra simple eyes (*ocelli*) on top of their head which are bad at everything except spotting movement, which is useful when a predator is swooping at you.

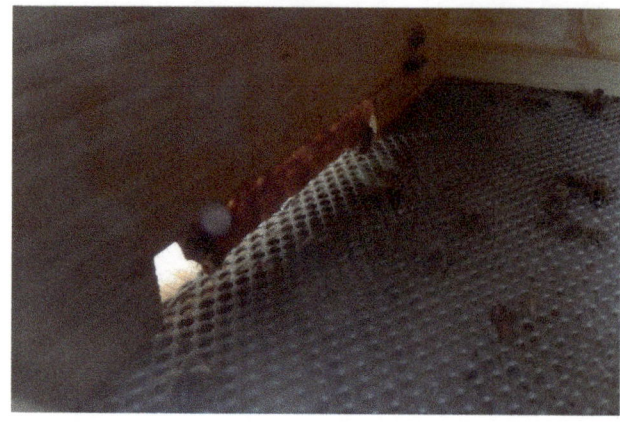

From inside the hive the entrance is very bright. (This one has a defensive propolis curtain above it.) Young (nurse) bees can't fly well, so avoid the bright outside. They have not done orientation flights yet, so do not yet know their way home if they fall to the ground. Don't shake or brush bees onto grass, or assume they will fly home from a few feet away.

They cannot see the colour red; few flowers useful to them are red. It may appear black to them. But they can see UV, and flowers use UV patterns to guide them to nectar.

Their eyes are covered in tiny hairs, *setae*, which help them gauge wind speed and direction.

They don't have big brains to process images so use very simple algorithms to interpret what they see as they fly. If something remains in one part of their visual field, but gets larger, it is on a collision course so they turn away. If you watch them approach a hive they sometimes bounce off the hive wall: one theory is that their vision is too poor to distinguish a uniform coloured wall covering their entire visual field clearly! To measure distance they use *optic flow*, the amount of "change" they see on a route. It's crude, but it works well enough for them.

Smell: their primary sense, perhaps 30x better than ours. They can often tell if an intruder is not from their hive. With two antennae, this is a directional sense.

Touch and vibration: important in communication in the hive by e.g. vibrating comb. The comb resonates at low frequencies and amplifies them, so they are exquisitely sensitive to nearby lawnmowers, and road vibration when you move hives.

(Air carried) sound: only low tones. Unlike vibration bees aren't very sensitive to sound, though if you talk loudly near a hive it can eventually arouse them. Traditionally, beekeepers talk to their hives quietly in low tones which reassures both us and the bees. Try to avoid high pitched excited tones more like the alarm frequency the bees use to warn intruders.

Minor senses: taste, gravity, electric[60] and magnetic fields, temperature air pressure. Sensing static electricity aids weather prediction.

---

59 Nothing to do with bees, but amazing - humans can discriminate polarised light too! Google *Haidinger's Brush*.

60 You can sense electric fields yourself: static charges make the hair on your arms stand up. Bumblebees seem to use their hairs to sense if a flower was recently visited – and its static charge dissipated – in which case it will not have recharged its nectar yet and is not worth stopping at. For more info search for DOI: 10.1126/science.1230883

Review 2

Jot down 5 things you've learned from the previous couple of chapters, to reinforce them in your memory:

1.

2.

3.

4.

5.

Review 3

Jot down 3 things you don't agree with so far, and, crucially, why you and I might observe different things:

1.

2.

3.

Chapter 4
Hive selection

"A house is a machine for living in" - le Corbusier

Why do you want to keep bees?

This influences what hive type to choose - how you want your bees to live. For example, if you are serious about monetising your hobby, or enjoy micromangement, you probably want a framed hive.

Stop and think a minute about what you are looking for.

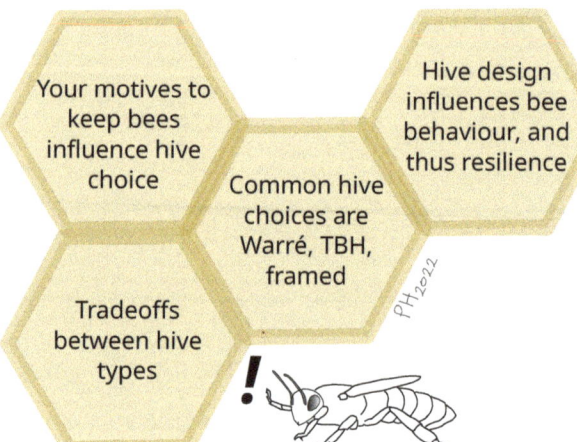

Review 4

Pencil in some thoughts on **why** you are thinking of keeping bees:

Choosing your hive

Some rules of thumb:

- *Avoid second hand bargains* and equipment from glib pressure sellers. They can carry diseases (scorched hives may indicate disease, query history closely); if the wood is warped they can have gaps between modules (entry points for wasps etc); and you may end up with a hive that doesn't suit you. Avoid impulse buys and trust your intuition if you feel uneasy - *once you have a hive you are pretty much locked in to that model.*

- If you have **back problems or trouble lifting heavy weights,** go for a horizontal or long hive where you only need to lift one comb at a time. The basic Top Bar Hive is cheap; if you have a bit more money, there are deluxe varieties with extra features like the Einraumbeute, Lune Valley Long, and Drayton. Horizontal hives require occasional opening, which is quite fun and educational. In the USA, the Lazutin hive is the obvious choice.

- If you prefer a **minimum-management** hive with occasional honey harvests, consider the Warré.

- If you will **never take honey** and want a "conservation hive" which is purely for the bees, consider log, Freedom and tree hives.

- **Get two identical hives.** This acts as insurance if one dies; and you learn a lot more by seeing the variation in what "normal" behaviour is. If they are identical you can swap modules.

- The best wood for man made hives in Britain is considered to be **Western red cedar,** a good insulator and rot resistant. It costs more, but will last 30 years even untreated. In some countries, other materials are preferred or more readily available (like cork in Portugal).

- **Get windows,** if possible[61]. The additional cost is fractional and they are quite useful and fun. They should be glass, you'll find out why when you try to clean the propolis off one, and covered with a good insulator - typically a polystyrene or cork slab is sandwiched between them and a lightproof cover. Thin plastic windows can warp, good luck replacing a window in a populated hive.

- **If you are prioritising honey,** you will want a standard framed hive, unless perhaps you are in the tropics[62].

- If the hive site could be subject to theft or vandalism, don't use an expensive one.

- **Huge hives require consistently good and plentiful local forage.** Bees won't fill a cavern like a Dadant or Langstroth in most areas of Britain. Unused volume just wastes energy (e.g. consumes honey).

- Don't forget the option of **"no hive at all".** You may be at a stage of life where you cannot keep bees (age, changing jobs etc) and some people focus on helping pollinators in other ways, or team up.

---

61 Some hive suppliers claim bees don't like windows. Untrue if the windows are well insulated. Bees will happily build right up to windows. See photos in this book showing comb glued onto them. Get windows.

62 Tropical beekeeping is outside my experience: specialist charities like *Bees for Development, Bees Abroad* and *ntfp.org* can advise here. It can be uneconomic to use precision parts like frames in remote areas, or invest in expensive hives when local bees abscond regularly so average occupancy is low; an unusual type like a woven horizontal top bar hive may be more appropriate. Such bees are usually not European but a local species with different habits, so you really do need specialist advice.

Hive styles

Vertical

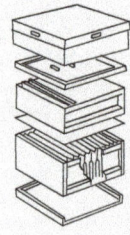

- A stack of boxes.
- Usable in all climates.
- Each box contains combs. To access combs, you need to lift boxes - which can weigh 20kg or more.
- Warre hives are simple vertical hives designed for maximum insulation, minimum intervention. Narrower, so only 12kg / box.
- More complex vertical hives using frames are
- favoured by conventional beekeepers.
- Examples: Langstroth, National (pictured),
- Dadant, Poly(styrene)

Shallow horizontal

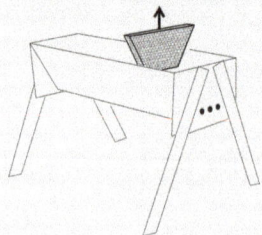

- Combs arranged horizontally.
- To access TBH comb, pull a single one out at a time - max 3kg.
- Cheap.
- Work well in hot and mild climates. Struggle in e.g. northern Britain.
- Examples: TBH; simple "long hives". (Also traditional horizontal log hives and ceramic tube hives - optimised for very hot, remote environments; their combs are sometimes not inspectable, just harvested destructively from rear end.)

Deep horizontal

- Adapted for cold climates - much insulation, deeper combs.
- Not cheap.
- Examples: Drayton, modified Layens, Einraumbeute, deep long hives

...and the rest: a motley crew

- Traditional skeps, vertical log hives; Sun Hives, wild nests in roofs etc.
- Tend not to have movable comb.

Now let's consider some more prominent hive flavours in more detail.

There's no "perfect" hive that suits both bees and beekeeper. They are all tradeoffs and you must choose the balance point that suits you.

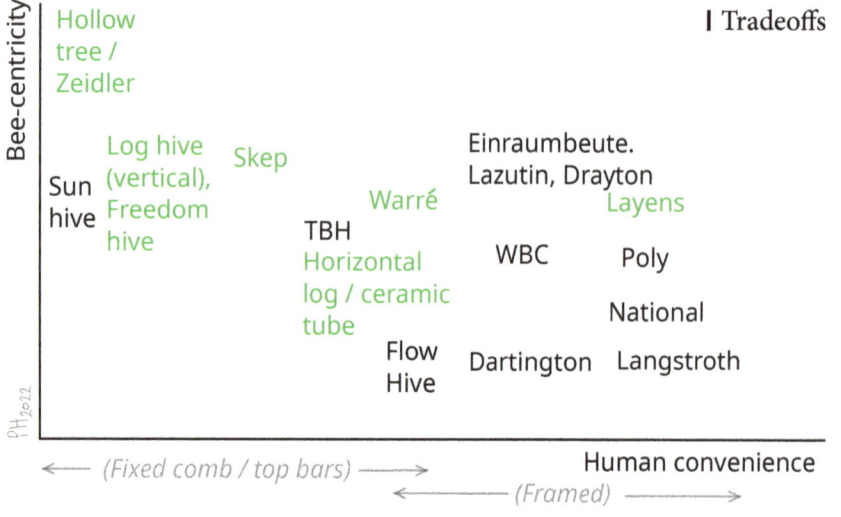

Green hives have exceptionally simple construction

Frames, foundation, top bars - they can't all be wrong?

Beekeeping evolved from fixed-comb hives like skeps, to ones where the comb hung from top bars, to full frames with foundation, sometimes reinforced with wire. Each approach has tradeoffs, and there are blends like semi frames, foundationless frames, or foundationless-but-wired frames.

Frames with foundation

- Reinforce comb allowing rapid handling / inspection; and hive transport over rough roads
- Standardised modular system, suits high volume production
- Combs not destroyed by harvesting, re-usable
- Frame sizes vary eg "14x12", "Dadant" etc - need to use ones that fit your hive
- Some people use black [plastic] foundation for brood areas to see eggs and larvae more clearly
- I advise against Hoffman (35mm) spacing in Britain, this leads to bridge comb between combs; and crowding, which aids disease transmission. Use 38mm.

Top bars

- simple and cheap, easy to make, single piece of wood
- bees build comb to their own requirements
- but require more delicate handling

Foundation's downsides

Foundation is a thin sheet of wax or plastic embossed with hexagonal patterns. Bees build ('pull') cells out from this template.

It promotes straight comb and forces the bees to consistently make cells of the size embossed into it, discouraging drone rearing[63], which is thought to divert resources from making honey. Foundation is also said to reduce the amount of wax needed, which otherwise diverts resources from honey production.

It comes in various types, even versions with extra large cells to *encourage* drone laying, either for a breeding project or because the beekeeper intends removing and killing all those drones – which act as a kind of varroa bait. For a commercial beekeeper, if the bees are going to lay drones, it's useful to have them in one place.

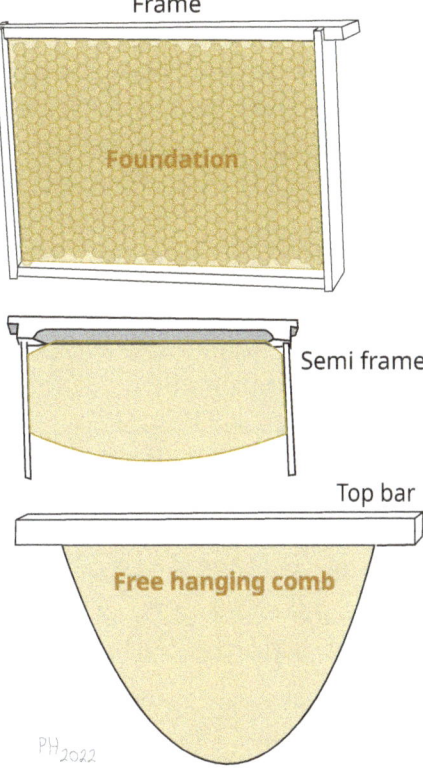

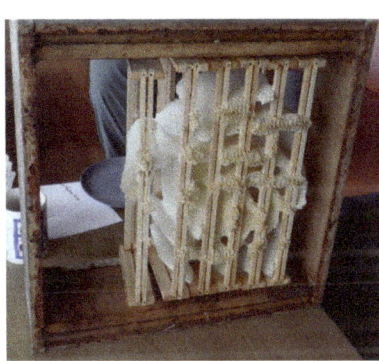

Wiggly comb in a [foundationless] framed hive. This is unusual, though, if frame spacing is correct and there is a sharp edge to build down from.

---

63   False reasoning? Colonies try desperately to make 10-15% drones in summer; if thwarted, they build drone comb in odd gaps in the hive. The beekeeper keeps destroying these and the bees keep rebuilding them. As the bees prioritise drone comb, you surely get less honey.

Plastic foundation is, obviously, non biodegradeable. It needs to be waxed before bees will build on it. Wax foundation sheets are made from **recycled** beeswax. There are 3 grades of foundation wax, from 'pristine organic'[64] to 'don't ask'. This is a bit like re-using syringes, as recycled wax is theoretically able to act as a vector for American Foul Brood, whose spores can survive the manufacturing process.

More commonly, wax accumulates many pesticides. Imagine buying some foundation whose wax came originally from a beekeeper who regularly used miticides, in an area where neonicotinoids and other crop treatments are common. Some beekeepers use powerful antibiotics to treat bee diseases, and these powerful bio-active agents all dissolve into the wax and are passed on to the next hive. Michael Bush notes in his book *The Practical Beekeeper – Beekeeping Naturally* that in his experience, American queens in hives with clean wax seem to last 3 years instead of a few months.

It is strongly suspected that miticide residues in re-used wax help mites evolve resistance to those chemicals.

A dramatic example of the dubious provenance of wax foundation came to light in 2016 when hives across Belgium, Germany, and the Netherlands began experiencing massive brood deaths. The common factor was they had all used foundation from a supplier who had sourced the 'beeswax' from China, where it had been adulterated with stearin to reduce costs. Stearin is used to make soap, and is poisonous to bees.

Despite beeswax and foundation being a product in contact with food (honey), it is generally only subject to food safety regulations when used as food additive E901[65].

Plastic foundation has downsides. Bees like to make one comb free hanging, to act as a resonant *dancing floor*. This allows scouts' waggle dance vibrations to spread further and recruit more foragers to rich forage. But bees can't chew through plastic, so they can't free the edges of combs on plastic foundation to create a dancing floor. Also, plastic foundation cannot retain antimicrobials from propolis. And many people are just very wary of plastic inside a hive.

Embedded wires (to prevent the foundation sagging in larger frames) also prevent creation of this free hanging dancing floor[66].

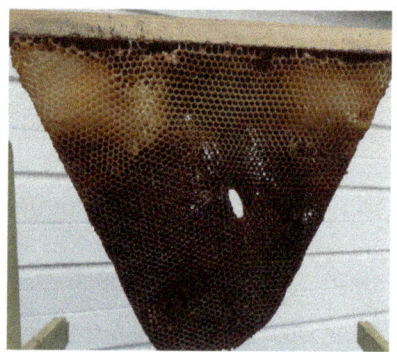

Natural comb (here, on a top bar) permits the creation of access tunnels to avoid cold walls and isolation starvation in winter. Bees can't chew through plastic foundation. Cell size varies across this comb.

Wires sometimes disrupt laying patterns if not fully embedded.

---

64 The purest beeswax is usually from Africa, where a lot of non framed hives are used and the local bees are mite resistant, so miticides are not used. Imports are heat treated to kill pathogens.
65 UN FAO report ISSN 1810-0708, *Good beekeeping practices for sustainable agriculture,* section 9.5.7, p.127
66 John Hewitt, Master Beekeeper, commented his best forager colony was mad keen on dancing floors – BBKA News 228, July 2021, p.231

Finally, foundation *costs money*, which is against the Beekeepers' Creed.

Warm way, cold way, wiggly way

Beekeepers generally fit frames *warm way* or *cold way* (see diagram). This is a topic of endless debating potential and no settled answer.

- Sometimes ventilation / overheating is important and cold way is best.
- In cold climates, warm way may be better.
- It partly depends on hive type. If you have a mesh floor it doesn't really make any difference. I find cold way works better in my solid floored Warrés.
- Some beekeepers rotate the boxes above the floors in winter to switch from warm to cold way. This isn't possible with Langstroths as they're not square - they're always cold way.

Without foundation, comb guides or a straight walled cavity forcing straight lines, bees tend to build curved comb, guiding air as they find best. This can result in some fun shapes, but hinders comb inspections.

Using frames without foundation: natural comb

I don't use foundation. Instead I use bars with a sharp lower edge to act as a comb guide; you can buy foundationless frames like this, or glue something sharp like a thin strip of wood into the slot intended for foundation, or deposit a guide line of beeswax along the underside.

For the latter, the easiest technique is to use an Indonesian wax dripper *Tjanting* batik tool, readily available online.

There's a theory that building natural comb reduces swarminess, an outlet for all that wax they produce[67].

Irregular honeycomb cells on poorly waxed plastic foundation. Image © Mike Castagnetto, 2023

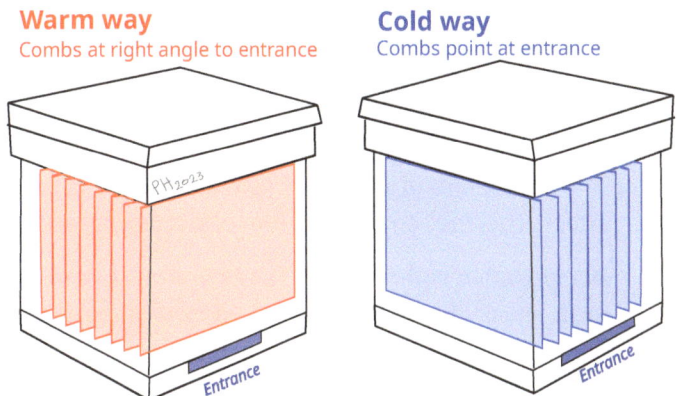

Warm way Combs at right angle to entrance

Cold way Combs point at entrance

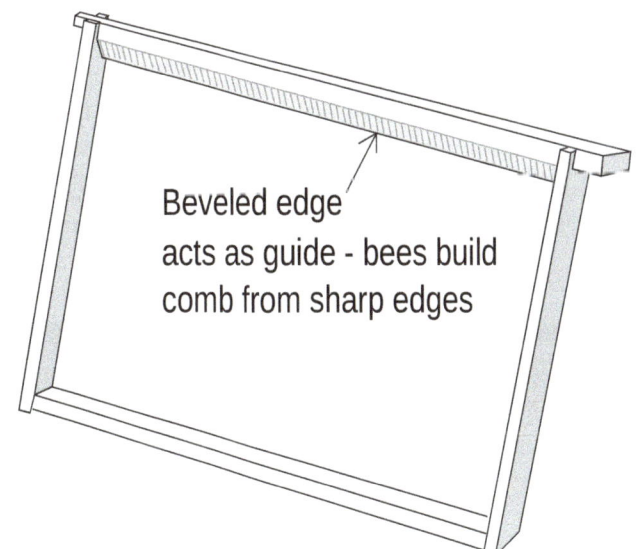

Foundationless frame. Some people use a standard frame with a thin strip of foundation under the top bar.

---

67   Eoghan Mac Giolla Coda, 200 *Amm* hives

Some people use reinforcing wires in foundationless frames, so they can spin them to extract honey. Apparently the bees embed it OK in the centre of the comb, providing it is completely taut.

The bees' view on hive design

We tend to assume hive design is all about the walls, but of course the bees are stakeholders and avid remodellers. Consider this picture of a Warré hive. Warré boxes are often left alone until harvested when the comb is cut out all at once, so the bees have a chance to arrange things as they like.

- On the right, furthest from the entrance, is the thicker, light coloured honeycomb.
- Combs have been attached to walls and each other, preventing collapse, blocking draughts[68] and intruders.
- Combs are twisted slightly, perhaps a ventilation issue, or because they chose to build some larger drone comb[69].

The arrangement of empty space in a hive is significant too. It affects thermal control, defence, swarming and breeding.

An entrance right next to the brood nest is well guarded. A slot in the base of a hive, far from the nest, is easy for wasps to raid on cool mornings, before bees fly.

Forcing uniform flat comb with foundation, and regularly removing frames - thus breaking up the bridge comb they've deliberately placed - denies them the opportunity to optimise conditions.

Bees naturally build down from the ceiling. There's no draughty void at the top to heat.

Ventilation

Adding and removing supers disrupts airflow and thermal conditions. Swarms seem to decide on a "program" of how much comb to build, and where, **when they first enter** a cavity, and modifying the volume disrupts the colony. You can see this in how they sometimes re-organise comb if you add supers above or nadir boxes below. If this isn't an option because you use foundation, they must burn more fuel fanning and heating air to compensate.

Bees dynamically adapt entrances with walls of propolis & wax. In winter, stationary air is a great insulator, whereas draughts strip heat rapidly. In summer a large colony would remove most of this barrier when processing lots of nectar.

---

68   The draught blocking is a significant energy saver. Derek Mitchell calculates top space alone (standard in many hives) increases thermal losses by 70%. *Anthropogenic hive heat loss contrasts with trees'*, doi.org/10.5518/1359

69   Probably because the comb was warm way. I now orient Warré combs cold way and get less twisting.

So, many apicentric hives use small entrances part way up, leading directly to defended comb, and one large volume (*Einraumbeute* literally means "one room hive").

Floor level entrances can be blocked by snow or collapsed comb, leading to suffocation; and mouse guards across them can block with corpses in big colonies.

Many hive designs have a second entrances near the top of the hive. The bees tend to control the airflow through these by building barriers across them, partially blocking them at some times of year.

Simple apicentric hives

By the mid-2000s many Western beekeepers realised beekeeping needed a rethink to cope with increased stressors on bees. One area they could address was hives. Two models have become particularly common:

- The horizontal Top Bar Hive (TBH), originally designed for African beekeepers, was adapted and popularised by Phil Chandler in the UK and Les Crowder in the US;
- Following David Heaf's translation of Beekeeping for All, interest rose in the French Warré hive – a vertical stack of boxes, designed for minimal intervention and to be more like a tree cavity by the visionary Abbe Warré, way back in 1948. Provides a very well insulated vertical cavity which bees love. Keep It Simple!

Both of these are designed to allow harvesting honey with less disruption to the bees. They are discussed in detail in chapter 12.

The TBH (and other horizontal hives) is particularly good if you have back problems as there is **no heavy lifting** involved. It is based around movable combs which allows close inspections, and manipulations – which teaches you a lot about bees. If I had to say anything against TBHs (I have three) it would be that for our cool climate, a deeper comb might be better. For really cold countries, use a deeper and better insulated hive[70]. The Warré is more of a leave-alone hive, which is either a benefit or boring, depending on your viewpoint.

Bees hate open mesh floors (OMFs) and swarms won't spontaneously enter hives with them. They seal such floors with wax and propolis, such as this example from a TBH deadout. The 3 entrance holes have also been propolised down to 1-2 bee spaces for winter. Image © 2022 Gino Sprio.

In some countries, removable combs are mandatory. They prioritise easy inspection for disease. Most hives can be adapted to use frames, but some are easier than others.

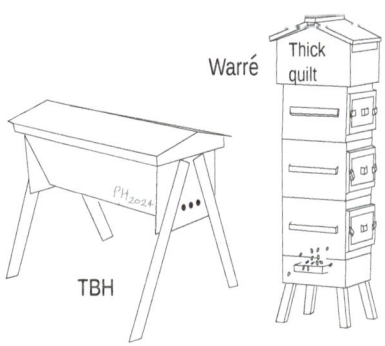

Warré / Thick quilt / TBH

---

70   TBHs were originally for hot countries. Phil Chandler in Cornwall popularised a British version which works fine in my area, the Midlands. In northern England and beyond, I would reccommend a deeper, better insulated hive - but higher spec horizontal hives cost more. Natural nests in cool climates tend to be vertical, and circulating air due to heat rising is integral to their functioning. Marcus Nilsson (south Sweden) made this point to me: *"I found the TBH not working well in our rather long and damp winter climate. It is much* **too shallow to let the bees keep a sufficient amount of honey over their winter cluster.** *The colonies I tried to winter in that hive all failed."*

More recently, more radical thinkers have revived interest in variations on historical designs from Africa and Europe / Russia / Britain which are even more bee-friendly, but some have management issues. People are re-evaluating **skeps, hollow logs** and *Zeidler* cavities carved high up into **living trees,** though the bees tend to affix the comb to the walls which makes management messy. Management is discussed in chapter 12. There are also fit-and-forget **tree hive** systems like the Freedom Hive, which are increasing in popularity.

Apicentric hives tend to be:

- Windowed – adds expense but reduces need to open them – I strongly recommend these, particularly for Warrés
- Unsuitable for migratory beekeeping
- Designed for simple top bars rather than frames
- Operated without queen excluders
- Foundationless - the bees build natural comb to suit them
- Not much kit is needed - identical modules for all jobs
- Honey extraction is trickier and messier than with framed hives

Landing boards help you, not the bees. They slow bees down and enable you to see behaviour.

Deep horizontal apicentric hives

In the 2000s-2010s a number of people independently addressed similar issues and converged on super-insulated deep horizontal hives as a solution. These hybrid designs blend the best of several concepts - easy management, only need to lift one comb at a time, deep enough to permit to allow the bees to optimise temperature / humidity gradients for brood and honey processing. They all have solid floors, optionally removable to allow floor debris inspection / mite counts. They are considerably more expensive than simple TBHs, and most honey spinners can't handle non standard frames, though you use the crush & strain method. They're similar in function, so I advise concentrating on ones available in your country for cost and support reasons.

Question: why do I say elsewhere that a National, Dadant etc is too large a volume for bees to easily heat, yet these are huge cavities?

Answer: Insulation!

Several are developments of the **Layens,** which already matches bees' needs far better than a National or Langstroth. It was developed in Spain around 1870 for its hot, dry climate and its local bees, and is common there and the USA. Its walls are 1.5 inches thick. The Spanish version has a cross-ventilated roof to help cool it and looks like the Lazutin shown here, but with a flat roof.

The **Einraumbeute** (one-room-hive) was developed by Johannes Wirz and Norbert Poeplau of Mellifera eV in Germany. It was introduced to the UK by David Heaf and John Haverson (BeeKindHives.uk), who were looking for a hive with a more tree-like cavity. They found a shorter, 13 frame version is better for the UK where bees will never fill the German version, and sell it as the Golden Hive[71].

---

71 Confusing, as other people use the term *Golden Hive* to mean one with very deep frames in approximately the Golden Ratio to match the natural depth of comb built in a catenary curve.

The **Modified Layens (Lazutin)** was developed for the very cold, **dry** winters[72] of Russia (walls twice as thick, uses frames made from two standard 'deep Langstroth' frames stapled together, and other changes) by Fedor Lazutin in the 1970s (as usual the Russians were decades ahead). Its use was promoted in the well regarded book, "Keeping Bees with a Smile" by Fedor Lazutin, translated into English by Mark Pettus. Lazutin died in 2014 and Leo Sharaskin is its primary champion and expert now; he has a useful website at horizontalhive.com, and a user's experiences can be seen at www.lazutinhives.com.

The **Lune Valley long hive** was developed by a group of treatment-free beekeepers based in NW England, a notoriously damp, cold area. See: www.lunevalleybeekeepers.co.uk/lune-valley-long-hive. The **Long Hive** from hydehives.co.uk is remarkably similar.

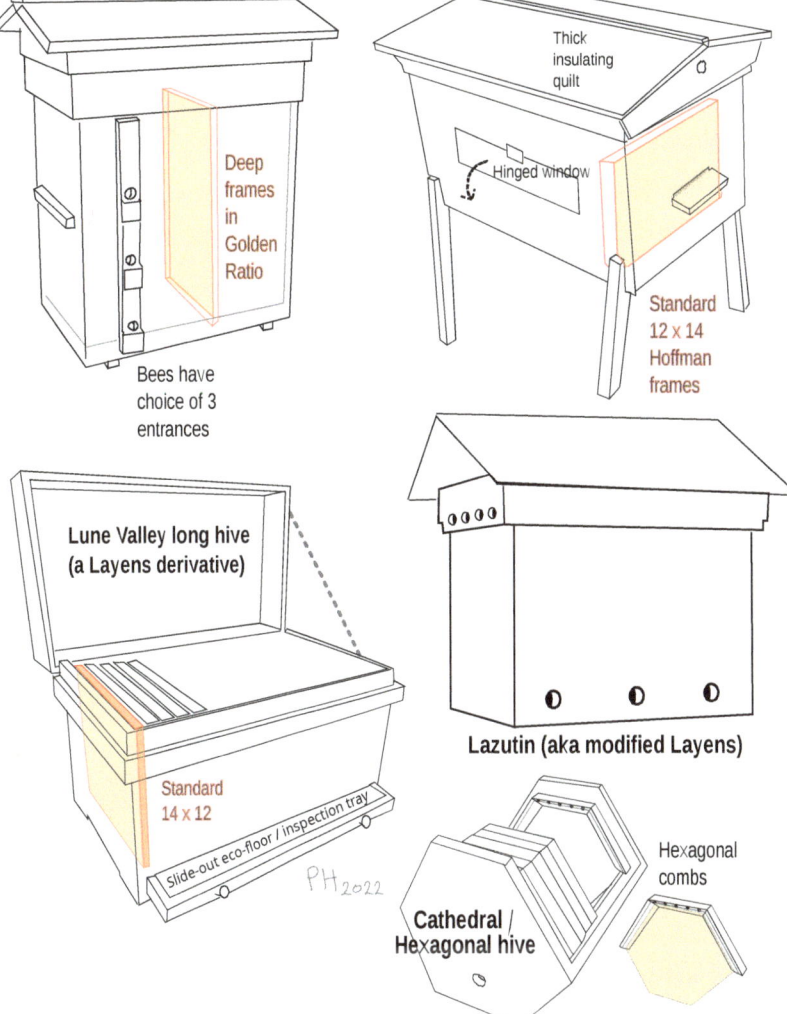

The **Drayton hive** was developed by my friend Andrew Bax in Drayton, Oxfordshire. His aim was to allow conventional and low-intervention beekeepers to experiment with, and blend alternative management systems in one hive. It features double walls, an optional window, accessories such as a queen excluder, clearing board, divider board etc. Its solid floor is held by latches and can be detached to examine floor debris. Sold through Thornes, it has a website with extensive information at draytonbeehive.com. Further reading: *Beekeeping Simplified with the Drayton Hive* by Andrew Bax

The **Hexagonal** or **Cathedral hive** is undoubtedly the most eye-catching - and trickiest to make! Initially developed by Corwin Bell (backyardhive.com) in the USA, a single walled version is available from www.beesathome.co.au in Australia. I've seen a few built as projects by other people - the carpentry is extremely challenging.

The hive's symmetry can be confusing. One inexperienced purchaser had problems, which turned out to be because they had set up the hive upside down. In another case a user fitted the bars upside down (resting them on the floor, not above).

---

72 An Einraumbeute, Drayton or Lune Valley hive is probably better for Britain's cold **damp** winters. Much of America tends to have very cold **dry** winters and the Lazutin works well there.

Concerning large frames

Some people make their own hives... but making *large* frames that don't warp is a nightmare: high precision *and* accuracy is needed. It's easiest to buy such frames - which is why Einraumbeutes use standard Dadant frames rotated 90°, with an edge added at the top as a comb guide.

If foundationless, the larger combs need a reinforcing bar part way down to prevent comb collapse. Some of these hives' frames are too large for British extractors.

The Cathedral's striking hexagonal comb. Photo © 2023 Chris Palgrave.

Framed hives

Covered extensively in other books. Rather than the usual sales pitch for these, let's highlight the flaws others skip over.

If you already have one of these, we discuss using a framed hive in a low intervention style in Chapter 12.

By 1850, designs converged on the basic modern commercial hive concept[73]: a modular stack of boxes using movable frames, foundation, and queen excluders[74]. These took a few decades to completely displace skeps.

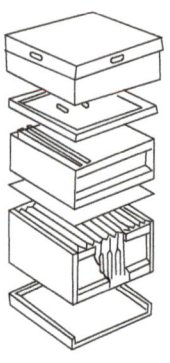 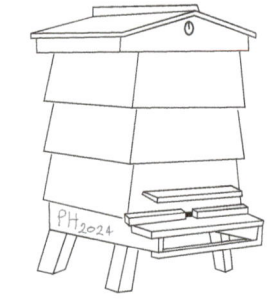

National / Langstroth / Dadant

WBC - a double walled framed hive - very well insulated

Using huge boxes, preventing swarming and an increasingly coercive management style boosted honey yields, but has downsides:

- **Ergonomics:** A full "super" of honey can weigh 50 pounds – about the same as a sack of coal – and you have to bend across to lift them. 'Beekeeper back' is a common problem.
- **Frames invite inspection,** which is normally just needless disruption.
- **Foundation** has multiple issues. Discussed previously.
- **Cold and damp:** the structure requires extra fuel to heat due to:
  ◇ relatively thin walls with a huge surface area (except the WBC)
  ◇ the brood cannot be laid in the warmest position at the top of the hive due to the queen excluder
  ◇ Frames designed for **top space** above the combs waste heat[75]
  ◇ open mesh floor (ventilation thwarts insulation)

---

73 The world's first commercially produced hive was probably the Ukrainian Prokopovich hive, of which about 10,000 were made by 1850. It had three vertical compartments, each with its own door at the back, and looked like a rectangular-shaped vertical box with a gabled roof. Most beekeeping innovations originate in Europe and Russia, a fact ignored in Anglosphere books.

74 People tend not to consider what a massive impact Queen Excluders have on colonies. They totally distort the geometry and thermal properties of the nest. They are more properly thought of as *Queen* **and Drone** *excluders*. Drones can get jammed in them and die. It's generally recognised that the ones with parallel wires are better as the alternative, flat sheets with holes, scrape the workers causing damage. *Fun fact:* very young queens sometimes squeeze through QE's.

75 Discussed earlier in this chapter under *"the Space-Comb Continuum"*

A cold hive requires the bees to work much harder to keep brood warm, weakening them against other stressors. This is one reason conventional beekeepers are continually feeding their hives. The bees are continually flying to gather nectar (because the beekeeper keeps taking their honey and they feel insecure with less than about 10kg[76]). Conventional beeks actually think there is a *problem* when a hive takes time off from foraging!

Should you be the new owner of a framed hive, be aware the frame sizes are different for different models. The most common are:

Britain – National, Deep National, Commercial

Europe – Modified Dadant ("MD"), German Frankenbeute, Austrian Zander etc

America – Langstroth deep, medium and shallow boxes

The iconic WBC is pretty, but rare as it is expensive. It uses National sized frames, but the boxes are shorter so it takes two fewer frames than a National.

Poly(styrene) hives

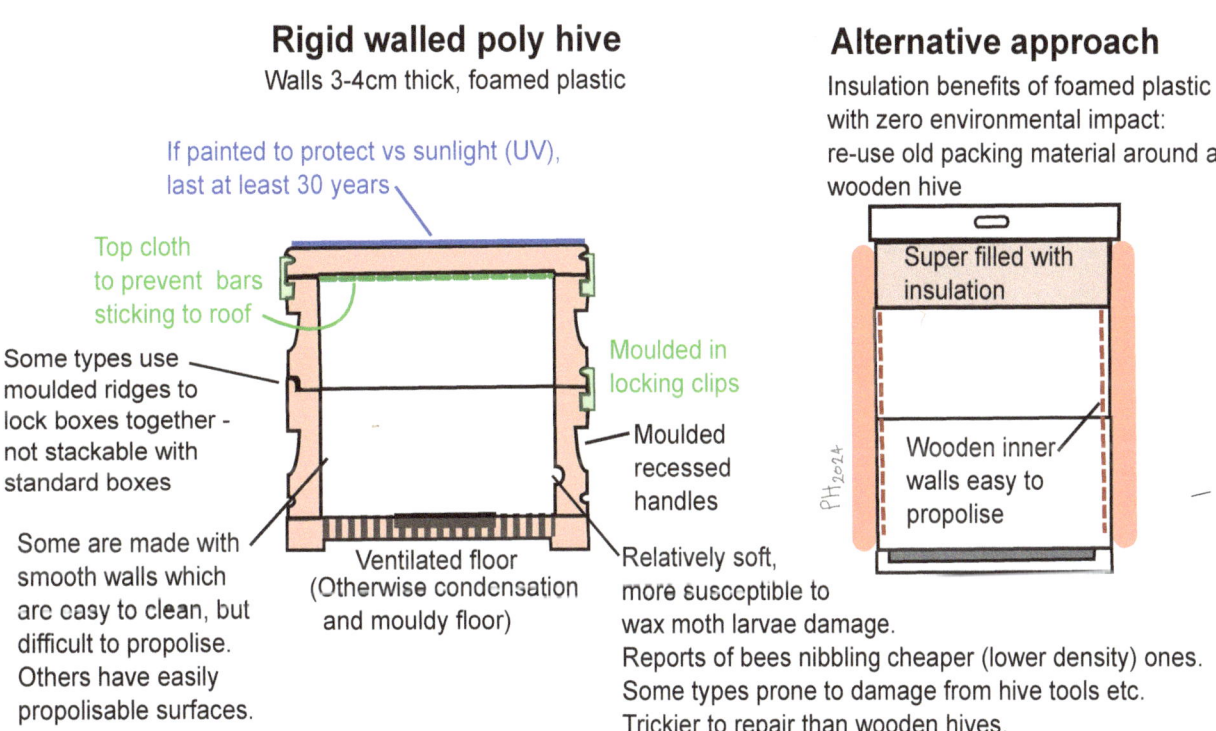

| Poly hives' properties and features vary with make.

---

[76] *The Bee Master of Warrilow*, Tickner Edwardes, 3rd edition (1921) p.122: the bee-master keeps harvesting honey to stop them accumulating 20lbs and switching to other tasks. Again, in *A Modern Bee Farm* by S.Simmins, 1887 p.27 he states a swarm will rarely accumulate more than 20lbs honey in its first year unless you intervene.

Most low-intervention beekeepers are dubious about materials which take centuries to decompose after use[77], shed microplastics and are made with bioactive plasticizers. They are optimised for industrial scale bee farming, being light to lift, and are well insulated - but there are more environmentally friendly ways to insulate. For example, I recycle old foamed plastic material as I recognise its superb insulating qualities, but I make sure it's not in contact with the bees.

Insulation is usually a blessing, reducing the energy required to keep warm or cool, reducing stress and giving bees more options. However the way bee farmers run poly hives is: they populate with queens selected to lay continuously, and use stimulative feeding in Spring. The bees begin rearing bees a month earlier than usual, (so need *more* winter fuel, despite the insulation) and lacking a long brood break, have more varroa problems. They do this because it creates big colonies for early pollination contracts, and the bees make more honey because it takes less effort to evaporate it.

Polys come in many shapes but some types' internal dimensions are not quite compatible with standard frames and accessories, locking you into a single supplier. Some may work in the American climate, but have condensation issues in Britain, where users often mention the internal walls "sweat".

Many other hive variations exist

In chapters 12 and 17 we discuss further modifications, all of which have tradeoffs. Open mesh floors aid damp control but burn fuel; landing boards help our observations... but also help Small Hive Beetle enter a hive; tunnel and periscope entrances are useful versus wasps, but can throttle ventilation; slit entrances help against hornets; multiple entrances have uses; eco-floors may harbour commensal predators, but hide parasites like SHB[78].

There are many, many other hive types, which tend to be variations on the usual themes. That's for another book - this one concentrates on the most practical *apicentric* hives for beginning to intermediate low-intervention "natural" beekeepers.

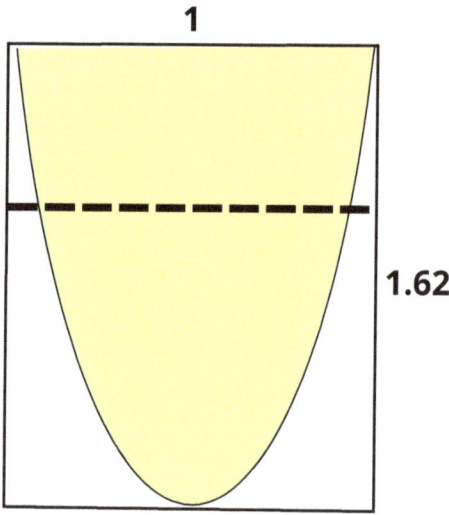

A golden hive is any hive which uses a cavity in the 'golden ratio' beloved by architects. These deeper-than-usual frames permit comb to hang in its natural catenary curve shape. In general, bees' air circulation seems to work better with deeper combs. You can achieve a near-golden ratio by stacking a Deep and Shallow brood box and only putting frames in the top one. The combs are so deep they may collapse without a cross bar (dotted).

---

77   doi.org/10.1021/acs.estlett.9b00532 - experiments on very thin (<1mm) samples showed they break down in 'mere decades or centuries' if exposed to sunlight. Poly hive walls are several cm thick. This work related to plastic rubbish floating at sea. Polystyrene in landfill will take millenia to break down.

78   I'm not a fan of eco-floors: tree nests only have floors covered in debris and bugs until the the comb extends down to within 1-2", whereupon the bees fanatically clean and propolise the floor.

Chapter 5
Apiary setup

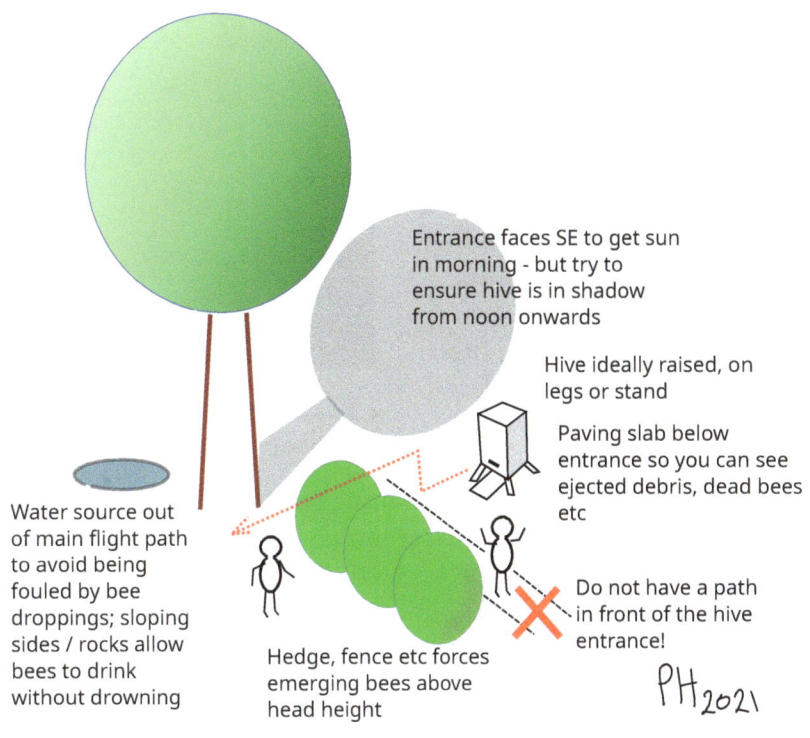

Apiary[79] layout considerations

There may be others for your circumstances - you can never satisfy all of these, aim for best tradeoff. Prioritise safety.

Neighbours - these vary in attitude from paranoid to intrigued. If you think they will spot them, consider explaining swarms will occasionally emerge but are harmless. Use fences, hedges, shed walls etc to ensure the Bee Line of emerging bees is diverted above head height... and hide the hives from neighbours. Prioritise human safety over every other factor. If there's no other way to separate bees from neighbours, point entrances right at a fence or wall to force bees up, even if the entrance is then in shade all day.

Non human neigbours (sustainability). Is forage sparse – will hives soak up too many resources from other pollinators? Are insects already rare in the neighbourhood[80] or are there already free living honeybee nests? A large commercial hive can consume up to 75kg of pollen a year and 3-5 times as much nectar.

Space for operations. Consider where you would place tools and boxes, combs etc during inspections. Ideally you **don't** want to stand in front of the entrance when opening a hive.

Not too near walls. You need room to get round the hive when opening it. Fences flap in storms and can knock a hive over. A minimum of 50cm away if possible. Also, whilst walls can be useful windbreaks, they also reflect heat.

---

79   An *apiary* (American: *bee yard*) is where you keep bees.

80   To give some context here, in my small but pollinator friendly garden I would typically be able to see perhaps 5-10 bumblebees and 1 butterfly at midday in May. The forage in this area is OK but not great. If I drive the car I might see 2-3 bug splats on the windscreen after 60 minutes - when I was a child my father would have to clean his windscreen every week, and that was in a city.

Which side is the window on? Not much use crammed against a wall. South facing gives better views - in other directions, reflections are brighter than the view inside.

Afternoon shade - ideally you want morning sun, so bees see bright light at the entrance and can warm in the sun. But afternoon shade helps reduces the summer heat which can stress a hive full of bees and, in extreme cases, cause combs to melt - this is my experience with TBHs and Warrés (foundationless comb, untreated local bees) and opinions differ - I have

To avoid bees drowning, provide moss, shallow sides, semi-submerged pebbles etc at your watering station.

seen many Nationals with ventilating mesh floors in full sun, using frames with foundation and treated commercial bees: this seems contrary to how bees site natural nests, in the shade of e.g. a tree canopy. Overheating is also a trigger for otherwise unnecessary swarming; and after you've tried opening a hive in full sun in a bee suit you too will appreciate shade!

Avoid trees' drip line - although you want shade, avoid being under tree crowns' edges, where they shed rain.

Beware unstable trees - it's not unheard of for hives to be toppled or crushed when branches fall. Gales topple ancient trees, generally in the direction of the prevailing wind. But woods generally provide shelter and a variety of forage.

How many hives (2 recc). As discussed elsewhere this gives you options. Now consider whether you will end up standing in one's beeline when inspecting another! This will result in stings on your ankles ("ankle biters"), particularly painful.

Raise the hive entrance so pests cannot enter so easily, snow does not block it in winter, floodwater cannot run in, you can pull out floor inspection boards etc. It helps the bees a lot if you can raise it above the damp layer that forms at ground level. Knee high is ideal for most apiaries: too high and you have problems handling high stacks of boxes[81]. Keep the entrance clear to avoid losing bees to spiderwebs.

Water source - bees prefer puddles to clean water: they crave minerals; they sometimes lick your sweat (it tickles). But in droughts they use whatever is available. They like to stand on something solid and suck. They can use gentle slopes down to the water's edge, like a container full of pebbles; but the ideal is e.g. damp moss; they've even been seen drinking from damp carpet. Site away from lakes and deep walled rivers: if blown in bees can't get out. They prefer their source to be in sun.

---

81 *A complete guide to the mystery and management of bees*, 1852 by William White (& Beesley) makes the point that if you raise entrances higher than 18 inches they are exposed to excessive wind. Then again, he also repeats the Roman Pliny's weird recommendation to "place hives away from house free from possibility of an echo".

Consider flooding. Why are you being offered this patch of land? I've known of several apiaries which flooded. The hives floated away and in most cases, the bees drowned. Before placing hives next to tiny streams, be aware they can become raging torrents in winter.

Bee type. African, and Africanised honeybees are considerably more defensive than European honeybees, and need to be well away from humans and animals. This includes crops that farmers need to occasionally weed.

Ground should be levelled. Top bars need to be fairly level along their length, **to ensure comb is built straight along bars.** Warrés, Nationals etc are often tipped forward a few degrees so rainwater won't run in the entrance.

Slab under entrance. Being able to see at a glance that there are an unusual number of dead bees, or some ejected larvae, or crawling bees under the entrance is very useful. Even better, paving under the whole lot, so legs are not resting on earth.

Secure against wind. Winter storms can knock a hive over. Shelter is good, but as I found one year, the sheltering object itself may blow over. Currently, I use huge sections of tree trunk under hives, as shown in the diagram; these protrude forward enough to double as the "debris display pad". Tree sections have an added benefit of raising the hive entrance several extra inches. I don't bother tying down my TBHs, their feet are splayed wide and they have a low centre of mass; but I do secure their roofs, which can blow off, and add the copper tape.

Securing a hive this way also prevents it being knocked over if a deer, cow etc rubs against it.

Consider prevailing wind direction: in most of Britain the prevailing wind is from the SW. If your hives are in an exposed position, consider placing a windbreak (hedge, fence, shed etc) to reduce chances of the hive being blown over in winter storms. A less critical, but desirable factor is minimising winter wind blowing in the entrance: try not to face the entrance SW. If this is impractical, you can rotate or even relocate hives in winter, or just reduce the entrance size in winter.

Don't rest hive on exposed tree roots: these transmit the shaking of a tree in wind. Vibrating a hive upsets bees.

Low frequency vibration: some hives get grumpy near roads with heavy traffic.

Lawnmowers, pruning: some hives get very upset by mowers, especially fume-emitting petrol ones. Can you place the hive away from grass? Consider also any weeding or other gardening you may have to do near the hives. I tend to do this in a full bee suit as sap smells can set off my bees. Some hives are on grass - I cut around them with shears. It's possible, but hot work, so try to minimise maintenance requirements round them.

Forage: Although bees *can* forage 2km+ away, they burn a lot of fuel doing so - they tend to only go that far for really sugar rich nectars. They will make considerably more honey (thrive rather

In larger apiaries, hives are often placed in pairs. Bees can count to three (left, right, centre).

than survive) if there are rich forage sources within 750m - the nearer the better[82]. Ideally this should include a variety of rich pollen sources for nutrition.

Bad smells: Do not place hives near manure piles, smelly chicken coops, cesspits, cow sheds, diesel / paint / chemical fumes. The bees will abscond[83].

Hives too close and / or **hives easily confused by returning bees:**

Landmarks like distinctive bushes near a hive help bees return to the correct hive.

Colour cues like paint probably help bees recognise Home, but remember red paint is probably indistinguishable from black to them.

It has also been known for young queens returning from mating flights to return to the wrong hive and be killed - they need easily recognisable cues.

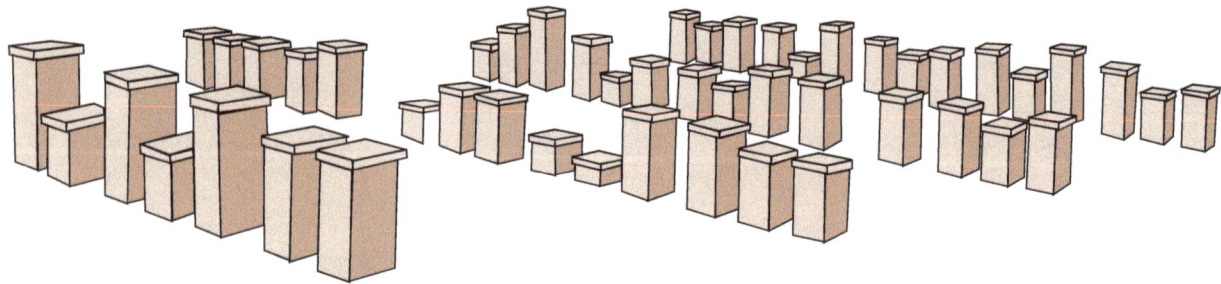

> A commercial apiary in an open field. Rows of indistinguishable, unshaded hives. A lack of landmarks promotes drifting, thus **rapid disease spread**, robbing and high bee losses during dearths. Also, the hives are too near, mixing smells externally - making it difficult for bees to follow unique scents home. A more sophisticated layout would be to place this many hives in clusters, with entrances pointing in different directions.

Domestic animals: most dogs only sniff a hive entrance once, but a few are idiots. Horses can kick over a hive if stung, causing mayhem, so don't put hives right next to them, or a bridle path. Sheep... oddly I have never heard of problems with them, which surprises me as bee legs get stuck in woollen clothing. Some chickens have problems with bees, others don't, best not risk proximity. Cats... sleep on top of hives with flat roofs!

Wild animals: the only real problem in Britain is, potentially, badgers. Sometimes a strong electrifiable fence extending some way below ground is needed. Other creatures can cause issues, but don't impact apiary setup.

---

82 Donald Sims states that they only fly half a mile in poor weather: placing them right next to a crop makes an astonishing difference. *60 Years with Bees* (1997)

83 **Fact check:** I gave examples on a forum of swarms I personally knew of which had absconded after being placed in hives near chicken manure and a creosoted fence, and was roundly mocked by 'experienced beekeepers' who pointed out that African bees (scutella) are very prone to absconding, yet African beekeepers creosote their hives; and it used to be common practise to creosote hives in Britain with the bees still in them. I knew that the examples I quoted were true, and further swarms stayed in those hives after the smells were dealt with, but decided to extend the sample by asking Filipe Salbany about African beekeeping. He replied "Creosote is (was) used and generally the hives were left for up to 3 months to dry. This is heated creosote and hives were dipped in it. It is however harmful to bees and I think the work is by Vorwohl, in German, 1968. It also contaminated bee products as bees of course chew wood. I would not use it and we found that it "leaks" out of wood in very hot temperatures, even after many years."

Apiary siting in valleys: Near the bottom is best so the foragers return with heavy loads *downhill*. Not right at the bottom, to avoid flooding and damp.

Damp. Chalk brood, moulds and nosema are more likely in areas which are damp in winter (typified by frost / damp / mist). Hives should be raised off the ground, and not in excessively shady damp traps.

My garden has 2 levels, one is 15cm below the other and hives in the lower section don't thrive, because damp rolls downhill. Another bad spot is a shady corner where air doesn't circulate and sun never warms. In hilly areas, beware cold air streams rolling down hills.

Entrance orientation: it doesn't really matter if they fact south, east etc. I tried rotating hives 45° to face sunrise and saw the bees emerge 30 minutes earlier each morning... but I couldn't perceive any difference in health or annual honey production[84]. However, if there are no other constraints like human safety, bees do seem to like sunning themselves, especially to warm up in the morning.

Exit strategy: do your executors know whom to contact regarding your hives if you die?

Human predators: Theft and vandalism

Theft – *hide the hive from public view*. Don't assume thieves will ignore your hive simply because it is non-standard and awkward to move - thieves steal bees too, shaking them out of a hive into a box.

Theft is more common (thousands of hives per year) where bees are valuable, like California (almonds) and New Zealand (manuka). Thefts across the whole of Britain are something like 100 a year. It's by beekeepers.

When setting up large apiaries, commercial beekeepers are beginning to use tree cover to hide the hives, and avoid lines of hives - because regular arrays of hive-sized boxes in the open are easily visible on Google satellite images. Thieves are also beginning to use drones to scout out such targets. There are electronic alarms which can warn you a hive has moved, but it only takes 15 minutes for two men to load a dozen hives on a lorry and drive off. These kind of professional thieves just cut chains locking gates. Sometimes they calmly remove hives in daylight and bystanders assume they own the hives as they are wearing bee suits. Do not use expensive hives at an out-apiary.

Vandalism – you hear of a few cases of hives being overturned by drunks; or poisoned or burned by people "taking vengeance for being stung" every year. Sometimes the bees sting them badly.

Urban apiaries

On the plus side, these generally have an urban heat island effect: an earlier swarm season and longer flowering season than the surrounding countryside - some beekeepers in Oxford, 10km from me don't see my rural summer forage gap. They have less pesticide use than near farms, unless you're in the USA (lawn & mosquito spraying). Cities are usually in river valleys, the water table is near the surface so nectar from any flowering trees is usually available. There's a wider variety of plants / pollen from gardens, at least in wealthy areas.

---

84   Scientists in Arizona did a controlled experiment, doi.org/10.1080/00218839.2023.2165769 and likewise saw some changes in colony rhythms but no clear difference between colonies after a year.

On the downside: river valleys are damp traps so hives struggle with condensation in long cool winters, so raise entrances above knee height, further away from damp layer, if possible. (Entrances don't have to be at floor level.) There can be excessive hive density issues (too little forage for lots of hives, though those near parks do OK), and hives may be wasp magnets because alternative prey is sometimes scarce. A high density of humans means more possibilities for problems with neighbours, so emphasize that swarms are harmless and transient, and gift jars of honey before swarm season so bees are seen as Good. Place several bait hives nearby to catch swarms.

Camouflage material on a roof greatly reduces chances of theft / vandalism

Rooftop apiaries

These have extra stressors, you need strong colonies. It's a harsh environment – very windy and exposed - open hive inspections are very stressful for brood. So provide shelter if possible, and anchor hives and roofs *really* well because storm winds are brutal up there.

Whilst bees can easily fly horizontally for miles, vertical flight takes a lot more energy, especially hauling food or water. So it's really important to provide a water source.

There's also a stressor on the beekeeper - hauling heavy kit up there! Elevator access is highly desirable. I strongly advise against ladder access.

Swarms may descend, panic the public and disrupt local businesses. Place bait hives.

Allotment hives

These communal gardens are usually very good sites, but there are some extra issues to take into account. For example, ask yourself why this allotment has not been built on. Often it is occasional flooding. Hives can be drowned or even carried away by floods. Do you need to build a platform, find another site, or move the hives uphill a bit?

Read the agreement with the management committee carefully. Do they require anything unreasonable? It's worth knowing that if you're in the BBKA, you have public liability insurance. Ensure there is a notice with your phone number to report a swarm, vandalism etc.

Educate allotment users about avoiding spraying pesticides on, at least, flowering plants; and point out bonfires should not be near hives. Plotholders will probably see slightly increased crop yields.

Rural out-apiaries

These are typically near crops. Consider hazards from crop sprays[85]; and in exposed areas, prevailing wind direction (are any trees growing sideways?).

Although most people move some hives to an out-apiary once they get above 6, on the assumption they will get more honey, a Northamptonshire beekeeper told me he found he got just as much honey when he placed all 20 back in his garden, and saved a lot of travel time and petrol. And you rarely get the swarms from out-apiaries. Consider also how many extra trips

---

85 Pesticide exposure is higher near perennial crops like vines and fruit trees. *Impact of landscape composition on honey bee pollen contamination by pesticides*, Cappellari et al (2024) https://doi.org/10.1016/j.chemosphere.2023.140829

Example: out apiary. The hives are sheltered by the wood, which also acts as a natural barrier to crops (pesticides) beyond. The hives are not directly under these mature trees (falling branches in storms) and a little way up from the bottom of the hill (avoids floods / damp), but foragers benefit from flying generally downhill when loaded. No humans nearby: just deer, hares, foxes, badgers. A side advantage of this apiary is aesthetics - a beautiful, restful location - the best out apiaries are in locations you want to visit. The major downside is no vehicular access: carrying equipment back and forth uphill is arduous.

may be needed in a bad year when you need to feed them. However, spreading your hives across several apiaries is a form of insurance in case one suffers forage failure.

Forage is the key aspect to consider. You won't be able to monitor these hives as closely as ones in your garden, but you want a range of flowering plants around to provide pollen and nectar for as much of the year as possible. As a general rule, warm well-drained soils yield more nectar than cold heavy ones, e.g. heathery hills are better than heathery swamps[86]; and you want lots of deciduous trees nearby.

Nearby fields of OSR mean large harvests, but the honey will crystallise rapidly so the hives may need to be visited frequently.

A nearby river means plenty of nectar (but consider flooding).

I have occasionally been invited to put hives in peoples' gardens, and decided it wasn't worth the extra trouble for one hive. There are plenty of great out-apiary locations. If you see one, just ask the property owner.

Some rural locations have wildfire issues which can destroy hives and forage over a wide area.

Others lack the forage variety of gardens - look for nearby willow to ensure decent pollen for at least the start of the year.

Out-apiary landowners need your phone number in case there is a swarm, the hive roof blows off etc. Sometimes people find abandoned out-apiaries (and invariably remark "somehow the bees survived without management!"). Perhaps the owner died, or just lost interest. Please put some

---

86   Though bees have a short working life on the moors - it's very exposed.

kind of contact details on hives, perhaps on the inside of the roof, not just a phone number or email address - so beekeepers called in to deal with them can determine ownership.

Make a note of its grid reference or What3Words phrase so emergency services can locate it (if you get anaphylactic shock from a sting - see chapter 6, safety).

Access issues can be significant for out-apiaries because at times you will need to carry extremely **heavy, awkward boxes full of honey** and you don't want to haul them up and down stairs, or a great distance across muddy fields. Is there vehicular access? Is there a locked gate?

A written agreement with the property owner is advisable.

Etiquette: common courtesies and responsibilities

You don't poach a hive spot like, putting your hives on a farm where someone else has theirs. If only because split pastures = split yields.

You don't tell people where hives are. Thieves steal hives.

Consider dangers to the public and livestock, like bee flight lines across paths or bridleways.

Apiary setup: summary

No one factor is vital. Don't fret if you cannot face the hives East, etc. Bees are flexible; about the only thing they won't stand for is smells like manure or solvents. If unsure, prioritise safety for humans and hiding the hives from human sight, followed by working room (access) for the beekeeper. You can compensate for most other factors like needing shade, after a hive is in place.

Which bee race to use?

Commercial queens are selected for docility and prolific laying. However, traits are linked[87] so they inevitably select **out** other, useful traits, like hygiene. This inbred, prolific stock is semi dependent on human help to survive - miticides, feeding etc.

There is a major problem when different races cross. Queens with 50:50 parentage can produce agressive bees. No one knows why, perhaps the queen's pheromone balance is weird. This is called *F2 aggression*.

The original north European Black Bee was largely displaced by imported bees[88], but is bouncing back - luckily everyone *assumed* it was wiped out by varroa and didn't look for nests, it had a chance to adapt without 'help'. It - and its local landrace[89] hybrids - are resistant to varroa and disease in cooler climates, due to a wider genetic pool and lower stress.

---

87 Human eye colour is influenced by a dozen genes

88 Germany imported Carniolans, got lots of ferocious crosses and systematically shifted from Amm to Carniolan in the 1940s: Carniolans' early buildup matched modern crops' early flowering and gave more honey. (Ruttner, *Breeding Techniques and Selection for Breeding of the Honeybee*, p.99-100; Br Adam, *In Search of the Best Strains of Bees*, p.45-48, who refers to the Nazi era *Körsystem* [bee eugenics]). Miticide use has been compulsory for decades. They now lack a population of locally adapted varroa resistant bees, so treatment-free is trickier there.

89 A **landrace** is an animal or plant subspecies adapted to local conditions. The tougher the area is, the faster unfit genes are weeded out and imported colonies revert to this landrace – you will end up with them anyhow! Landraces are sometimes disparaged as mongrels or hybrids, but they are a stable genetic ecotype with a wide range of traits, and inconsistently, breeders often boast about their own variety's 'hybrid vigour' (which only lasts one generation).

Which tool for the job?

Static garden hive - use local swarms, ideally from long established wild colonies or treatment free beekeepers. Already well adapted to local forage, and handling varroa without miticides.

Hives among heather or pine forests - locally adapted swarms, or Amm. These can overwinter on heather / honeydew without getting ill.

Hives near a breeder - if it's their main income, get bees from them. Other drones will impact their liveliehood. The ethical issue is that the person proactively bringing bees into an area is responsible for the consequences to others.

Migratory hives (honey farming, pollination services) - commercial bees. In Britain, the best commercial bees are bred by BIBBA.com members as these are basically native bees which have been selected for commercial traits, *and* are well-adapted to Britain, *and* don't produce irascible crosses with other local bees. In northern Europe, check SICAMM.org for Amm sources.

In all cases, Buckfasts are popular among migratory farmers; Italians give inconsistent results in Britain. Carniolans are rare here. **All** of these will create aggressive F2 crosses with local bees *and each other*. Again, the person bringing bees into an area is responsible for the consequences: your *and nearby beekeepers'* hives will seethe with rage when your colonies cross-mate. I strenuously advise against using these races outside their native area.

Before buying bees, keep in mind...

All races, even *Amm*, will converge to the local strain incredibly rapidly, in about 2 years, if they survive.

Even though the Black bee was the original British bee, and survives in isolated areas, each strain is local. An *Amm* colony from Cornwall may struggle in cold Yorkshire.

All bees from breeders are selected to lay at an unnaturally high rate and will probably need feeding at some point, or starve. Their unconstrained laying leads to excessive swarms unless intensively managed.

Commercial Italian, Carniolan and Buckfast struggle with varroa.

Carniolams are very swarmy.

Lots of bees does not equate to lots of honey, just lots of feeding.

Honey produced depends primarily on the **amount of forage available** in an area.

There is no such thing as a "native Buckfast".

Obtaining bees

Every low-intervention / natural beekeeper I know primarily stocks hives with swarms.

We compare other sources in detail in chapter 9, "swarms", and how to find wild nests in chapter 17, but for now just keep in mind:

- Local swarms are healthier
- Local swarms don't create aggressive crosses

So - where to find them?

The key here is networking. You'll get some swarms most years from your own hives, but more eyeballs = more opportunities.

Suggestions

- Register as a swarm collector with your local BKA.
- Our village magazine has a list of emergency contacts. I asked them to add my details: simply "Beekeeper (swarms, advice etc) - name - phone number".
- We created a page on our group blogsite about collecting swarms in our area and loaded it with keywords.
- Your local network of beekeepers may share spare swarms from their hives.
- Give your details to: your postman; tree surgeons; pest controllers, they come across swarms from time to time; and your parish council.

I've also had calls from random people who heard about me from somewhere - after a few years you become Known.

Once you know the location of some unmanaged colonies, prioritise getting swarms from **them** for your own hives.

Chapter 9 covers how to catch and hive swarms.

Bees and the law; bee inspectors

British regulations

You are free to keep bees in the UK. You don't need a license or any qualification at all.

You own the bees in your hive, and foraging from it. You don't automatically own swarms.

If you catch a swarm with permision of the landowner, it is yours. You don't have an automatic right to enter others' property. You have a duty of care to bystanders when collecting swarms. The BBKA provides insurance covering swarm collection (which is invalidated if you charge money).

There are laws about honey & labelling.

Beekeepers are required to keep records of medical products (like oxalic acid) used for at least 5 years, whether or not the colony has died or been sold.

Neighbours

Everyone has a right to use and enjoy their property. If you behave unreasonably, say by keeping bees near the property line and provoking them, or keeping unreasonable numbers of hives, the law may step in.

In Britain bees are neither defined as wild animals, nor domestic livestock – they're a grey area. This gives judges wiggle room to make appropriate decisions, if required, on a case by case basis.

Judges don't consider 1 or 2 stings a year sufficient grounds for neighbours to prevent you keeping bees.

Bee inspectors and BeeBase

Inspectors have some statutory powers: they have the right to access hives on your property, and inspect them, and order treatment.

In the case of American Foul Brood this could mean killing the bees and burning them, their combs and even the hive, under their supervision. This is why the UK, Europe and NZ have almost no AFB and it is extremely rare they need to do this now. The BBKA's bee insurance will cover at least some of the cost of replacement.

British beekeepers are encouraged to register on BeeBase, a database on nationalbeeunit.com . It's not compulsory and it is thought about one third of beekeepers are not on it [90]. It's a personal choice. My hives are registered. I note it tells me there are 127 apiaries within 10km, which seems unlikely. I suspect many of these locations are no longer active apiaries, because who is going to update the details if, say, a beekeeper dies? The locations are confidential - or I could go steal their hives!

There are certain things you **must** report, by law: American Foul Brood; European Foul Brood; and a couple of extic pests. These are discussed in the Disease and Pest chapters. If an outbreak occurs within ~5km of your apiary they will inspect your hives too. This is why you need your combs to be straight, and inspectable, even if you're not interested in opening the hive yourself.

Don't be scared of bee inspectors. They're a golden opportunity - free consultancy from someone who has opened thousands of hives! Watch how they open the hives, take notes and photos, ask questions, for example about what they are looking for and how to interpret what you see.

Care and feeding of Inspectors: Don't touch the hives before an inspector comes - it can rile bees up. It's thirsty sticky work, drinks and washing facilities are appreciated. They will want to see your hive records (discussed in part 2).

A few bee inspectors are biased against leave-alone beekeeping. One I heard of had a reputation for banging frames around, even conventional beekeepers thought him rough, and when he did this with top bar hives, combs snapped off 'proving' they were poorly managed. The beekeeper began videoing him doing the inspection, "to review later for educational purposes". Suddenly he was very gentle.

Apiaries outside Britain

Every country has their own regulations, these vary a lot, particularly about miticide treatments. Your local Beekeepers' Association will be able to give you definitive advice for your area.

If you're British, beware - Americans speak English and they're all over the Web! Within the USA bee laws vary between states, and even cities. Advice is often... confused.

Examples of regional regulations:

- In Slovenia it is illegal to use any but the native Carniolan bee.
- Southern US states require you to get your bees from a certified source due to fear of Africanised bees (not everyone pays attention to that)
- Bees can affect home insurance - not usually an issue in UK / Europe.
- In NZ and some Australian regions, you need to register hives, and pay for that privilege. Frames are mandatory, so bee inspectors can check hives easily.

---

90 New Zealand tried compulsory registration, thought they had 100%, and then realised that 10-15% of beeks were hiding when varroa spread through the country incredibly rapidly despite controls being put in place. (Source: lecture by Giles Budge of the NBU.) This is a compelling argument against compulsory registration - it doesn't work.

- Americans have sued neighbours over bee poo on cars and laundry.
- National laws about honey sales and labelling vary.

How many hives can an area support?

I have rarely seen more than 4 static hives in urban settings, due to garden size restrictions; and in rural areas 12-20 (not including commercial apiaries where they feed the bees). More trees means more bees: conversely, in intensely farmed exposed areas, 4-6 hives is the most one spot can support. Sheepland is poor forage. Studies show that British cities (gardens and parks) now may provide more, steadier forage for bees than the countryside. Herbicides have eliminated most countryside wildflowers.

If your hives are static (not trucked around to crops) then the upper limit in *exceptionally* good forage areas is said to be about 20-30 per apiary in Britain. Above this density, you will need to supplement their diet with sugar syrup and possibly pollen substitutes. **Hive densities are much higher in cities**[91] and there is an ongoing debate about whether London now has too many hives (estimate: 6600) for available forage due to the number of urban beekeepers there.

Forage has declined dramatically as agriculture changed. In 1887 Simmons wrote in *A Modern Bee Farm* that in a fairly good be area, an apiary could have 150 hives without impacting yield per hive. In a very good area this might be 200 - 300 hives. But changes in agriculture have reduced forage and made it much more sporadic. One memoir, *Sixty Years with Bees* by Donald Sims, states that when he was a lad (1926) near Cambridge, his village supported several apiaries, one of 60 hives and they did well. By 1997 it barely supported 4 hives.

When beekeepers move hives to heather, there is a tremendous nectar flow over a couple of months and 20+ hives in one place is convenient and appropriate.

---

91 2017 OBKA membership data indicated *beekeeper* density was 10x higher in Oxford than Oxfordshire

Chapter 6
Safety for bees and humans

Core principles, warning signs and your behaviour – sound of a hive - troubleshooting guide / list of stressors – protective clothing – stings – other hazards – hazards for bees – safety review

Managing risks, minimising stings

Core principles

- **Bees primarily defend their home** - brood, queen, stores, each other. They are unlikely to attack you if they do not perceive a threat to their home. That's why swarms are so calm.
- Sometimes there are factors outside your control which put guards on alert.
- The alarm trigger threshold varies between bees. Sometimes a few are pre-stressed – perhaps they just fought off a wasp, or had a different drone father. Stresses wax and wane over a year.
- **Respect the hive.** Don't assume it will be calm because it was happy yesterday.

One can argue that aggression has its place. Some beekeepers value 'hot' hives in remote apiaries - they tend to be left alone by wasps, are remarkably vigorous, varroa resistant, and make more honey[92].

Warning you off

Bees sometimes use escalating signals:

1. A fly-past buzzing loudly at a higher pitch than usual
2. Flying at your head, bumping your face[93]
3. Adopting a posture of tail up, with end bent up at the hive entrance. (If the sting is everted and a drop of venom is showing, really go away.)
4. If you smell bananas (alarm pheromone), leave immediately.

---

92 Peter Jenkins (54 years' experience) states that fierce bees **do not** produce more honey. However they are less likely to be robbed - "and discourage meddling, which may be advantageous". *BIBBA Zoom lecture 13 Oct 2020.* Also the opinion of Herrod-Hempsall in *Bee Keeping New & Old*, p.671.

93 Unfortunately this can result in getting tangled in your hair, then stinging you. Thus the importance of veils near hives, especially in flight paths.

The sound of a hive

- Don't be alarmed by sheer volume. Busy workers ignore you.
- A contented worker buzzes around 250Hz, near Middle C. Memorise this hum.
- Their warning tone is an octave higher, 500Hz or B4, shrill and much louder. You will tend to hear this from single bees trying to drive you away.
- But the very loudest, deepest buzzing is harmless (stingless) drones – they're just huge and need to work hard to get airborne. Mid-afternoon, the massed drone flight is very loud. Observe a hive to learn what a drone sounds like so you don't get panicked by one flying past. They sound like a bumblebee – loud but not shrill.

Troubleshooting guide: why are my bees stinging me?

Be realistic. If you have a hive in your garden you *will* get stung eventually, but there are ways of minimising risk.

Cats glide, dogs bounce. Guess which sleeps on top of hives?

How bees spot threats

Let's remind ourselves of bees' long range senses -

- *Smell* – excellent – e.g. petrol fumes from strimmers;
- *Vision* – low resolution; sensitive to **movement, sunlight** and high contrast edges (not shapes);
- *Vibration* – comb forms a large "ear" which resonates to low rumbles like motors.

So unlike mammals, bees might ignore you standing motionless next to a hive, or even touching them gently. You're just too large for them to see clearly and they treat you like furniture. But they are put on guard by jerky movement ("wasp!!!"), unfamiliar scents like stale sweat, or rumbling vibration.

Smells

I have concluded the key here is to avoid a **sudden <u>change</u> in <u>any</u> scent.** Masking alarming scents with smoke, or less alarming smells *may* help.

Thus crushing plants near hives signals: "a large creature is near our nest and is a threat". (Always wear protection when weeding near hives.)

I sometimes shower and brush my teeth before a hive manipulation.

The bees know the state of the hive through pheromones, and in particular Alarm Pheromone (actually a mix of chemicals) is released when they **sting** something, or when a bee is **crushed.** It smells like bananas or pear drops (aldehyde).

Bees have evolved over millions of years to sense their ancient foes, lawnmowers and strimmers.

This means once one bee uses its sting, others are in a heightened state of alertness and are more likely to use theirs – a cascade effect, a rapid rise in aggression by a mob of bees[94]. Sometimes you can mask the smell with another. I find **dilute** peppermint oil on the site of a sting works. Other scents I've tried have just enraged them further. So:

- Wear clean clothes, and wash your bee suit occasionally. (*Hand wash* **veils,** they will be ripped in a washing machine.)
- Don't open an edgy hive first, it will set off the others!
- If you ever smell bananas / pear drops near a hive, leave immediately.
- When opening a hive, don't breathe into it. They recognise mammals' breath in the nest as a threat[95].
- Many beekeepers rub their gloves, at least, in plants to mask traces of alarming scents. I would recommend herbs they use as food like rosemary. Don't overdo it.

Apart from their own alarm pheromone other smells said to instantly enrage them are:

- Isopropyl alcohol-based **hand sanitiser** (on gloves)
- **Paints and thinners** – many contain Methyl Amyl Ketone (2-n-Heptanone), the major component of honey bee mandible venom.
- **Bad breath** - from poor oral hygiene / not drinking on a hot day.
- **Garlic** on your breath
- Eating **blue cheese** provokes persistent stinging round your mouth
- **Some plant saps** - if I'm gardening and pull out a common weed, Herb Robert (*Geranium Robertianum*), within a few metres of a hive, a guard bee will buzz me. So I wear a veil when weeding. Catmint is another plant with alarming sap, often planted near hives for the bees – beware walking on it! Even to our noses, the sap from their crushed stems is pungent. There will be other trigger plants where you live.

Smells which *sometimes* aggravate bees include:

- **Perfume**
- **Bananas and pear drops**[96]. "Never eat a banana before opening a hive" is a common saying. But I have yet to come across an example of this in real life.
- **"Bad smells"** – very old books warn against siting hives near these, probably meaning cesspits and dung heaps.

---

94 Roger Patterson, with 70 years' experience opening thousands of hives, disputes that alarm pheromone attracts more stings. *Beekeeping: Challenge what you are told*, p.71-71

95 Because bears and badgers close their eyes, stick their snouts in and begin chewing. Ever seen the remnants of a wasp nest eaten by a badger? Tough critters. I still remember the time I reflexively blew at a bee on my glove, straight into an open hive, and hundreds of bees rose as one and plastered over my veil. Good learning experience.

96 Again, Roger Patterson (UK) asserts the banana smell thing is an unfounded myth, having tested it with many bananas, and I myself have seen someone eat a banana over an open hive without effect. Their sweet smell is isoamyl acetate, aka isopentyl acetate, a constituent of alarm pheromone - and in very low levels, some perfumes. The trigger sensitivity to this is genetically controlled, and varies between bees so usually only a few guards are very upset by it. However America's infamous Africanised honeybees (AHB) all have a very low trigger threshold so once a single bee is annoyed, the entire colony attacks. Anecdotally, this trait seems to be gradually dying away in AHB, presumably natural selection finds it wasteful to lose hundreds of bees per attack.

- **Stale sweat** – fresh sweat is not a problem, and bees sometimes land on you to drink this nice salty stuff. Problems occur when bacteria have had time to digest it and create the characteristic body odour of Unwashed Sweaty Human, or if your bee suit smells of stale sweat. Old books warn to clean your hands scrupulously after handling horses.

Other stressors, and how to mitigate them

Move slowly & calmly.

Bee docility varies with temperature. **Any bee will be docile at 30C.**

Talk quietly. Many people talk to their bees to calm themselves while opening hives. This is best done in a low, quiet tone. If you get excited, your voice gets louder and higher pitched... towards the bee alarm tone...

Starvation, and robbing – a forage dearth, or beekeepers stealing too much honey, triggers desperate robbing of other hives and increased defensiveness of their own. It's remarkable how hives switch mood between glut and famine. Even drones may be denied entry in a dearth.

Feeding sugar syrup just stimulates laying, and spreads the smell of food coming from weak hives, amplifying the problem.

Opening hives too often, and treating with miticides – this trains bees to associate humans with threats. After my first 2 years' beekeeping (regular inspections, occasional miticides) I switched to a low intervention regime and was amazed how calm my bees were thereafter: 4 hives now calmly ignore us in our modest garden. And I use random swarms, not specially bred placid bees.

Genetics – some bees, especially crosses between two strains are more defensive than others. Such hives are sometimes relocated to remote apiaries; it is also possible to requeen colonies. If I encountered an extreme example now, my strategy would be to try to kill the queen but allow the bees to supersede, rather than give them a new queen from a breeder. This would retain the local mite resistant genetics and as we shall see, aggression tends to just last for one generation.

Disease – I've heard this can make hives more defensive, but have no experience of this.

Guards can see you – sometimes when hives are having a bad day, the guards are looking for trouble and will launch themselves at nearby movement. Placing an object in front of the hive entrance to block their line of sight can stop this.

Poor entrance design - by making the entrance more defensible (a short tunnel is ideal) the guard bees feel less anxious.

Do not block Bee Lines in and out of the hive. The bees may not be looking for trouble but interupting a stream of foragers will lead to occasional incidents like bees in hair. Their wings are V-shaped so they cannot reverse out of hair. I wear a veil when I am in a hive's flight path. One solution is to face a hive entrance at a fence, forcing bees to fly up above it – thus they set off foraging above head height and do not interact with us.

Low frequency vibration (1) – hives near roads with heavy traffic (especially buses and lorries) sometimes turn grumpy, but are fine if moved a few hundred meters away. Comb resonates and amplifies such rumbling.

Low frequency vibration (2) – the buzzing of power tools like chainsaws and drills sometimes triggers immediate defensive behaviour from hives. It doesn't help that many vibrate around the alarm frequency of bees and are coloured black and yellow like a wasp!

Low frequency vibration (3) – **lawnmowers** can distress bees. Petrol ones' fumes are an additional irritation. I wear a veil when mowing near my hives, and if they begin attacking I switch to a full suit. Interestingly the attackers, if any, only tend to come from *one* of the four garden hives, and not every time – showing how hives' moods vary over the year. I used to have one otherwise calm hive which hated the lawnmower, *until it swarmed*. Within a few weeks, the new queen's children ignored the machine. If they do attack, watch and learn from how they more or less ignore you and try to kill the lawnmower!

Low frequency vibration (4) - tree roots - if your hives are on these, and the tree is liable to rock in the wind... (Excessive brace comb is a sign of sustained vibration.)

Our comb is resonating, and a large black and yellow insect is emitting an alarming noise. Attack!

Monofloral diets - Hives in the middle of large areas of monocrop can get mildly grumpy. Probably hungry for some nutrient. Old varieties of OSR were notorious for this.

Unfamiliarity with humans – hives in remote out apiaries tend to be much more defensive towards humans than ones in gardens. (Regular, *non disruptive* inspections can be OK).

Who knows? - Janet's hives at an organic farm suddenly got aggressive for no obvious reason. She moved them 2 miles and they were fine. Sometimes you never find the reason!

Colony size - larger colonies can afford to sting more – they have spare workers – and people often comment their hives seem more defensive after their first year[97]. Yet another reason not to buy an over-prolific queen.

Swarms can change after a week, once they raise brood and have something fixed to defend.

Height and wind - Apiaries on tall roofs are windy. And bees need to work **hard** to gain height. These factors tire them. Similarly, if the bees have to haul their loads home by flying over trees they can get grumpy.

Weather – bees **really** don't like you opening a hive when it is cold, about to rain, or thunder. So don't. If you chill or otherwise threaten the brood, you will trigger defensive behaviour. Ideally you want the weather to be hot, dry and windless – so the bees barely notice when you open a hive.

Leaving the hive open more than a few minutes will worry the bees. You get faster with practise.

Excessive, hot smoke.

Pests:

Slugs sometimes enter hives. Copper tape round the hive legs repels them.

Ants – discussed in chapter 14.2

---

97   The flip side of this is that colonies which suffer a trauma in their early days, like a swarm which almost starves in its first few weeks, seem very subdued for at least their first year as they build up slowly and cautiously. This may be over-anthropomorphising, but I wonder if "mild strains" of bees are just continually intimidated and disoriented by e.g. regular requeening, regularly losing their most defensive members who die stinging bee suits etc.

Parasites - Wax moth, varroa etc – check hive floor for excessive detritus. If a weak colony seems in danger of being overrun, dispose of (burn) old comb not in use, which tends to harbour these.

Predators

A **general rule with predators** is that once they discover hive contents are delicious, they come back for more, so ideally you want to make it difficult for them to get that first taste, and don't want hives near other attractors. I clear up windfall fruit near my hives as it attracts wasps. You can generally block large animals' access, and they will find food elsewhere.

In general **wasps** are the most common problem, because you can't block their access, and **Asian hornets** are expected to infest Britain soon.

Chapter 14.2 discusses badgers, bears, woodpeckers, wasps etc in detail.

Beekeepers mimicking predators

Bees don't have high resolution vision, but they see high contrast well. We tend to wear white bee suits (to stay cool) with dark mesh over the face (to minimise glare) so sometimes grumpy hives mob your face veil.

The **bear nose effect**: a dark object like a mobile phone, camera or walking stick, on a light background mimics a predator's eyes / nose and can trigger an attack.

Black humans don't get attacked any more than white ones – it's the contrast which triggers the response.

Black and yellow stripes are another trigger. Don't use black and yellow gardening gloves near a hive.

Roger Patterson has inspected thousands of colonies (with only a veil and smoker) and concluded[98]:

- Many triggers are not universal rules
- Defensiveness is due to a queen whose father is a different race to her mother. Colonies whose workers have many colours (fathers) are either all gentle, or all grumpy. Change the queen and the colony temper changes.
- Always keep your smoker lit, it gives you options (most colonies just need a few light puffs, but when you need heavy smoke you need it immediately).
- He's been called out many times to see "bad tempered" bees which were fine with him…

Your shadow across open hives - bees are acutely aware of the sky for navigation, and if your shadow falls across a hive as you inspect it, the bees will view you as a bear looming over the hive and attack. Hands moving slowly across them do not seem to be an issue.

Calm, slow, smooth movement calms both you and the bees – you become a landscape feature. Moving in quick bursts resembles a predator.

---

98 *Beekeeping: Challenge what you are told*, R. Patterson (2021) p.68-69

Cycles of violence

Like a bad marriage, beekeepers can get locked into behavioural patterns like excessive smoking, bashing the hive or "rolling" the bees[99], which bees learn to react pre-emptively to. These unhealthy feedback loops can be broken by one party deliberately giving the other more room. Bees aren't intelligent and flexible like us - we're the ones with options, so it's up to us.

Example: my bees no longer treat me as a threat because I no longer rummage through their brood nest every fortnight, but they *are* now familiar with me peering closely at the hive entrances without ensuing problems. And my colonies are founded by the toughest, most feral swarms I can find.

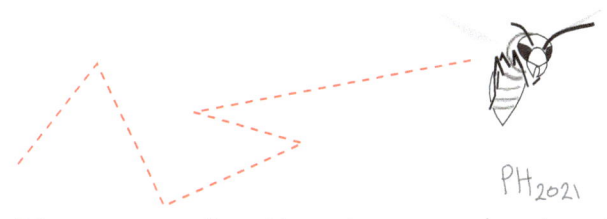

When near prey, like a hive entrance or an insect, predatory wasps move in sudden darts, sweeping out a volume of space to triangulate on visual and scent cues.

When not targeting a specific prey, wasps scan back and forth above the ground, searching for food:

This flight pattern is punctuated by sudden jumps further away - a "Lévy pattern" foraging strategy.
Bees use this pattern too, seeking pollen and nectar.

When someone tells you all wild bees are aggressive, what they're really telling you is something about themselves.

Escaping irate bees

Bees' eyes are excellent at detecting movement, but poor at distinguishing shapes like animals. If a bee buzzes your head, your instinct is to slap it away. That's exactly the wrong thing to do, it will consider you a bigger threat. Instead, freeze if possible. After a few seconds the bee will usually fly off.

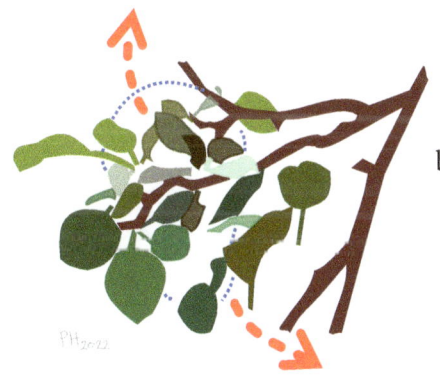

If buzzed by guard bees and you have no veil, an old trick is to put your face behind a bush. If the bees are persistent, shake the bush. Their low resolution vision cannot see you. In extreme situations, break off the branch to cover your face while escaping.

If you need to retreat, people say "get into shade" or "get inside a house", they won't follow. This is not true in my experience - I think you need to get out of line of sight of the hive, and if there are still one or two persistent "followers" you have little choice but to squash them. They are probably the most irritable members of the colony and will set off others if you return[100].

---

99 "Rolling" the bees occurs when you pull out frames which are too close together, and the bees are scraped against each other. Sometimes for beginners it's better to have one less frame per box. Bee brushes tend to roll bees and wind them up, but the traditional goose feathers seem less annoying (softer?). Pushing bees with fingers can roll them. Tapping a hive tool between boxes to break the propolis seal can set them on edge, etc. Blowing [your breath] on bees will aggravate them (CO_2? Smell of a mammal?), try not to breathe into a hive.

100 Confusingly, some academic introduced the obscure term *soldier bees* to distinguish the more trigger-happy guards from calmer foragers, as if they are a separate caste, like soldier ants.

As a last resort, defensive hives have been known to be calmed down by putting a scarecrow, or a washing line with the family's clothes flapping on it, in front of the hive for 3 days. This presumably causes some distress to the bees.

Protective clothing

What kind to use?

Veils - good for casual observations when not opening hives. Keep mesh away from your face, with baseball cap if needed. Some think round veils too big, catch on things, catch the wind.

Half-suits - not recommended. A half suit is worse than useless if the bees get agitated - they'll crawl up inside.

Full suits[101] - Hot but pretty effective.

Ventilated suits - expensive but very desirable for working in full summer sun. Also, bees cannot sting through them and don't get their stings caught in them.

Wear as much protection as makes you feel confident and calm.

I *always* wear full protective gear if opening a hive if I am working on my own, as I have had occasional anaphylactic reactions. I carry an epipen.

Fear of bees – at some point you will come across a hive that unnerves you. There are two main ways people get over this. One is knowing you are invulnerable in a good bee suit; the other, is by mentoring beginners who are reluctant to open their own hives. Helping others overrides your own fear.

"Fitting like a glove"

Dexterity: have you ever noticed how gloves never fit? There are many shapes of hand, and glove manufacturers always seem to assume your fingers and thumb conform to some kind of mutant anatomy. You end up with floppy bits at the end of fingers and this is a real problem as reaching into a hive requires dexterity – if you crush a bee alarm pheromone is released. Look around for a type that fits at least your thumb and index finger *properly*.

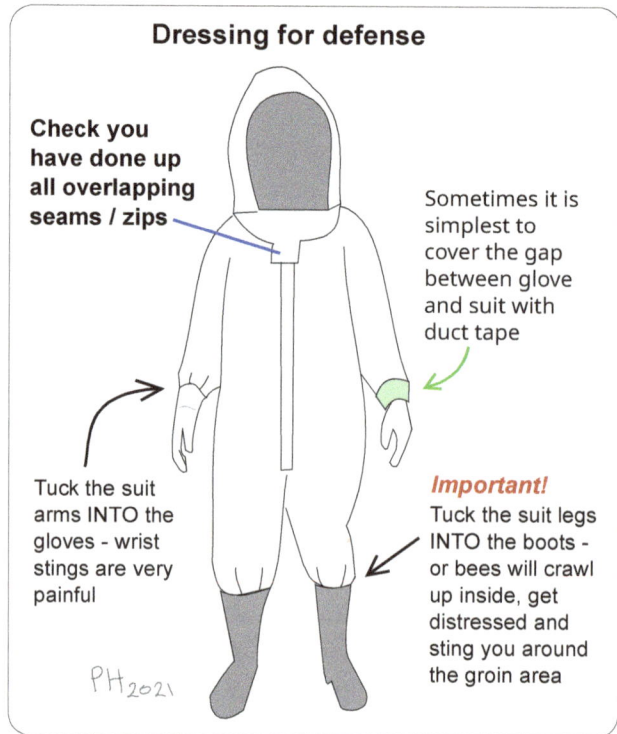

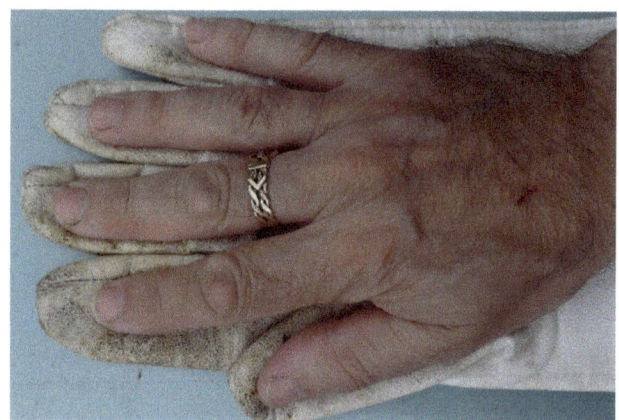

Problems: (1) wearing ring. (2) floppy bits at end of fingers make you clumsy (3) leather is difficult to clean / sterilise / remove alarm pheromone from (4) poor protection because bees seek the seams and sting through those, because they fight wasps etc by stinging between their segments.

---

101   People just used veils until Sherriff introduced the full-coverage bee suit in 1966. It was a welcome breakthrough, but enabled rough (lazy) handling.

Long white **leather gauntlets** are Bad because they are difficult to sterilise, and clumsy so you end up hurting the bees. They get sticky and are difficult to clean quickly while a hive is open. Also I found they eventually cracked and provided no protection at all.

Rubber kitchen gloves are the normal choice. These generally need either a rubber band or duct tape at the wrists to ensure they don't expose your wrist. If a bee tries to sting you through these, you may feel a tickle from the sharp sting – but the rubber squeezes the sting tight and they cannot pump venom through! I have small hands for a bloke and find Long Medium Marigolds work well for me.

Thin disposable latex gloves are used by bee inspectors and some bee farmers.

Bare hands are the preferred choice of really experienced beekeepers, for dexterity and tactile feedback. I don't do this when working on my own, but a side benefit is that hands are easy to wash frequently and get much less sticky than gloves. But if you develop contact dermatitis when you touch propolis, you need gloves.

Problematic clothing

Wool: this can get tangled with the pollen-gathering hairs on the back legs of worker bees which stick to wool like velcro. Trapped bees cannot pull free.

Static electricity? - Roger Patterson has seen bees attack clothing of various colours even after being pulled off, colour varied but typically fluffy man-made fibres, which can build up a huge charge. This should be testable...

An invitation to be stung. The bicycle clip does stop bees climbing up inside trousers, but exposes a high contrast black band of thin material over the ankle near the hive entrance. This WILL be attacked by guards and ankle stings are very painful.

Colours: inconsistent responses imply that these effects only kick in if the bees are already riled up by rough handling -

- **Black** – several writers warn bees associate black clothing with predators, and attack[102]. But I frequently wear black near hives without problems.

- **Blue (no longer true):** In the 1960s it was commonly known never to wear blue near bees, to avoid stings. The dyes changed, but it's still true that if you wear bright blue in Spring, you attract (non aggressive) bees and wasps[103].

---

102   E.g. *Beehive Design for the Tropics*, G.F. Townsend (University of Guelph, 1984) page 4

103   Bees are most attracted to a mix of yellow and ultraviolet, dubbed "bee-purple" (Daumer, 1956 - the papers are in German so few English speaking beekeepers are aware of them, though it's also mentioned in Winston's *Biology of the Honey Bee*, p.166-7). Although blue jeans have used the same main dye, indigo, since the 1870s there was an additive, now known to be carcinogenic and no longer used, in 1960s jeans. I'm guessing this was a benzedrine azo dye - azo dyes are sometimes added to jeans and this subtype was discontinued in the 1980s due to cancer concerns. Azo dyes reflect UV. *Sources:* conversations with Gareth John, Filipe Salbany and much digging on the internet; Robert Couston reminisces about dark, and dark blue clothing on pages 9 & 10 of his 1989 memoir *Flying with the Beeman*; but no one seems to have researched this topic specifically.

Stings

We all get stung, generally when we get overconfident.

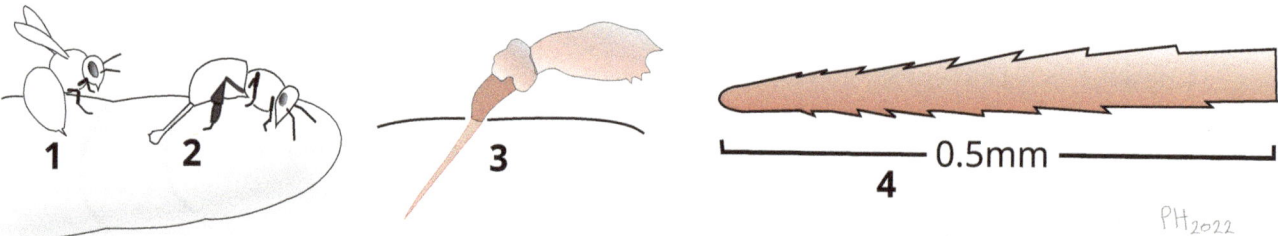

1: bee prangs finger. **2:** bee crawls away leaving sting in place, pulling own guts out. Bee will die in an hour. **3:** close up of sting in skin: external venom sac will continue pumping more poison into you unless removed quickly. **4:** close up of sting showing barbs.

"You aren't a real beekeeper until you've had your first sting" grin the old ones, and proceed to relate horrific tales like how they can't feel more than 20 stings because after that their arms go numb. I confess I am a complete wimp and find 6 a year adequate. But enough banter, let's look at some oft quoted facts.

Myth: honeybee stings are one-use.

99% true, they generally die after stinging us, though -

- Queen bees have unbarbed needle-like stings and could theoretically sting you repeatedly, like a wasp; however they are programmed to leave fighting to the workers while they seek cover.
- Also, given time, workers can sometimes *unscrew* their sting and walk away (the barbs are arranged in a spiral). Search YouTube for "honeybee unscrewing sting" for examples. However, one's reflex when stung is to brush them off or rapidly shake your hand, which tends to detach and disembowel the bee.

If stung:

Remove stings as soon as possible before the attached poison sac pumps in more venom. The faster you remove stings (it doesn't matter how) the less it will hurt.

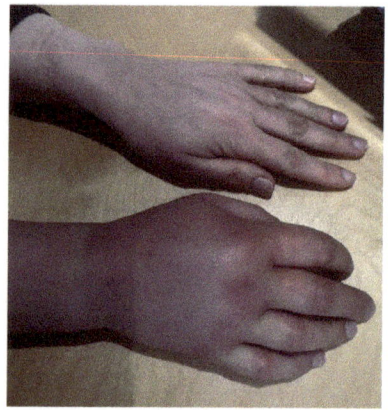

Swollen hand after being stung. REMOVE RINGS BEFORE HANDLING BEES.

Reactions vary: I would barely notice the same sting on my own hand, but, rarely, I might get an allergic (anaphylactic) reaction. A bad local reaction to venom is different to an allergy.

Mask the pheromones from a sting with something pungent, like smoke, or crush a plant for its sap though obviously some are irritants. This helps prevent the cascade reaction where one sting provokes more.

Sting reactions vary with...

- age of bee: older bees have more venom
- where they sting you[104]

---

104   I find thin skinned areas hurt most: scalp, wrists, ankles - which is why I stress protecting these. Michael Smith won an Ig Nobel prize for researching which areas hurt most. He figured nostril, upper lip and penis shaft. (Honey bee sting pain index by body location, 2014, doi:10.7717/peerj.338). See also the whimsical *Schmidt Sting Pain Index* in e.g. Wikipedia. Inexplicably, all such experiments seem to be by men.

- person being stung - some humans' bodies react more extremely
- whether you experience an *anaphylactic* reaction (see below)
- (probably) how deep the sting penetrates
- how quickly you get the sting out
- How often you are stung (most beeks build up a tolerance)

You'll either get better or worse as you get more stings – after a year you'll know if you need to withdraw from beekeeping (like my wife had to) or if stings are now trivial for your body. Accumulated tolerance fades a bit if you aren't stung for months, so a common saying on being stung is "well it helps keep me immune".

This is another reason to do the BBKA introductory course. You're not guaranteed to get stung, but at least if you do you'll have some idea of what to expect, and there will be experts around you for support.

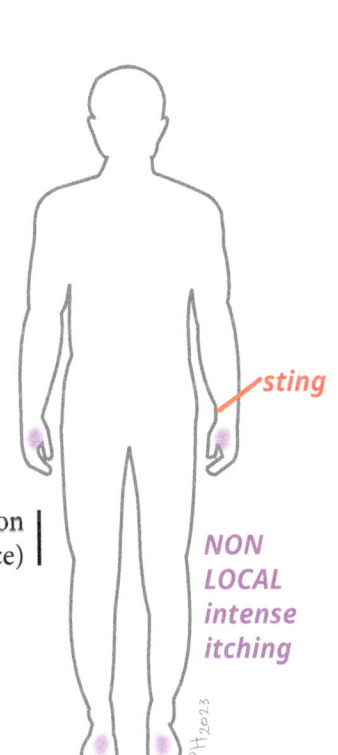

Thin skinned areas hurt more; but nostril, lip and penis shaft are particularly painful (Smith).

Anaphylactic reaction (personal experience)

sting

NON LOCAL intense itching

I have heard of two instances where someone got 40+ stings, and felt manic or euphoric for a few hours. There could conceivably be other non-standard effects.

The generally accepted amount of bee venom to kill someone is 10 stings per pound of body weight, so most adults would survive 1000 stings! Death from bee venom poisoning is unheard of in the UK, though occasionally Africanised bees mob and kill people in the Americas this way.

Anaphylaxis

Enough venom can kill by poisoning. But there is another mechanism by which even a single sting can kill you: if you are allergic to the venom you can get anaphylactic shock. This is where your immune system begins reacting to the sting venom, then gets confused and begins reacting to the chemicals *it itself* is pumping out. This feedback reaction spirals out of control, flooding your system with histamines, and it is estimated that every year 2 – 9 people die from bee or wasp stings in the UK this way.

The characteristic sign of an anaphylactic reaction is a **non local reaction** to a sting. For example, intense itching in the palms of your hands and soles of your feet. If you ever experience this, **back off and get someone to watch you for half an hour** in case you collapse, and **talk to your doctor** as you may be developing an allergic reaction to bee venom.

The really dangerous symptom is if your throat begins swelling so you cannot breathe. I've never had this but I have experienced some mild symptoms which were nasty enough, like a feeling of panic.

Even if you are allergic, you may not get such a reaction to every sting. But you may. It is like rolling a dice. That's why it is important to take such reactions seriously, and assess them and discuss options with your doctor.

Your doctor can prescribe an Epipen (adrenaline injector – gives you an extra 15 minutes for the ambulance to arrive) and refer you to get a desensitisation course. After initial testing to confirm you are allergic, an immunologist will give you regular injections of bee venom[105] and then monitor you closely. I have been on such a course – I had to show up every 6 weeks for 2 years, a serious commitment. About half the patients at the clinic did not complete the course! I've heard they can be done much faster now.

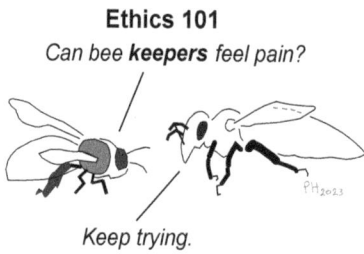

Ethics 101
*Can bee **keepers** feel pain?*

Keep trying.

Some areas of the body, like fingers, have few *mast cells* to pump out histamines, thus are less irritated by stings.

Everyone seems to have a neighbour who is convinced their children are 'allergic' to bee stings. They rarely are – it's just that some stings hurt more than others. And good luck explaining that to a parent in protective mode.

Effects of stings on animals

Bee stings are most effective against their primary enemies - other insects. But:

Dogs and **horses** are particularly sensitive to stings. Horses are protected somewhat by their large body mass but are vulnerable if a colony takes exception to their smell (not unknown), especially as a horse's instinct is not to run away, but to kick the hive over. Both these animals tend to move jerkily which attracts guards.

Birds vary wildly in sensitivity. Sparrows and, surprisingly, chickens and geese have been killed by a single sting whilst the tiny great tit predates them! Swifts eat bees on the wing, and the bee-eater specialises in predating bee and wasp nests.

Toads seem virtually immune and have been seen lurking by hive entrances eating bees. Frogs are more vulnerable.

Infections from stings

Apart from venom, bee stings can give you an infection. It's rare, as they are generally clean creatures, but not unknown.

Dead bees can sting

If you step on a dead bee in bare feet, it may still prang you.

There are often dead bees in the grass near hives, don't let kids play in bare feet near them.

If bees get in your house, and die there, they don't always end up on windowsills. Sometimes they are hidden in carpet.

Sting remedies

Antihistamines. That's the important type. These will reduce the itchy aftereffects. Can make you feel sleepy. Antihistamines can interact with other medicines so ask a pharmacist for advice.

The pain of many bee stings fades so quickly, that by the time you've applied *anything*, they are already fading away. So there are all kinds of cures touted, like honey, baking soda and vinegar, running hot water over the site - but you could probably rub the sting site with a pencil eraser

---

105 Bee venom is valuable for use in some medical therapies, so you will be completely unsurprised to learn it is sometimes counterfeited or adulterated with other water soluble white powders like salt or sugar.

and get the same "effect". Toothpaste probably works the same way it makes your mouth feel cool – by numbing nerves; and meat tenderiser's papain enzyme is said to break down the venom but surely it will do the same to your cells' proteins too? Aloe vera and witch hazel seem credible. There is also a homeopathic remedy called Apis Mell 30 cH. For people who normally have a big itchy rash, a vinegar compress taped over a sting is said to help a lot. Suction devices work well for wasp stings, but wasps don't leave barbs in your flesh. I don't usually react badly to stings, so can't judge these cures' effectiveness.

Biting and hair pulling

Dave Cushman[106] recorded occasional colonies of *Amm* and Russian bees with individuals who would bite his skin, occasionally drawing blood; or brace against his skin and pulling hairs hard enough to be almost painful.

Other hazards in beekeeping

Smokers – sometimes you see hives in fields of tinder dry grass. I do worry.

Working alone in remote areas – always tell someone where you're going.

Lifting heavy things – the occupational disease of beekeeping is back problems, from lifting heavy supers.

Overheating / heatstroke – most hive inspections are at the height of summer in full suits – take fluids with you. You can drink through a veil.

An American told me they found a black widow spider in one hive.

Wax has a dangerously low flashpoint

Collecting swarms - just walk away from ones next to traffic, up high trees etc.

Don't feed honey to babies under 12 months old. (Google "infant botulism".)

Review 5

Write a short poem or haiku on this chapter, to help reinforce it in your memory

---

106 www.dave-cushman.net/bee/biting.html (page created in 2001)

Safety review

The bees' viewpoint:

- Defend the nest
- They are alarmed by changes in scent
- They then home in on moving threats with vision
- Opening a hive and exposing the brood to cold, rain or wind also viewed as a threat
- Like yourself, colonies can get tetchy when stressed.

Stings

- Remove stings from your skin as quickly as possible to minimise venom injected
- If you get a non-local reaction (itching in palms of hands etc) back off, get someone to observe you for half an hour, discuss with your doctor

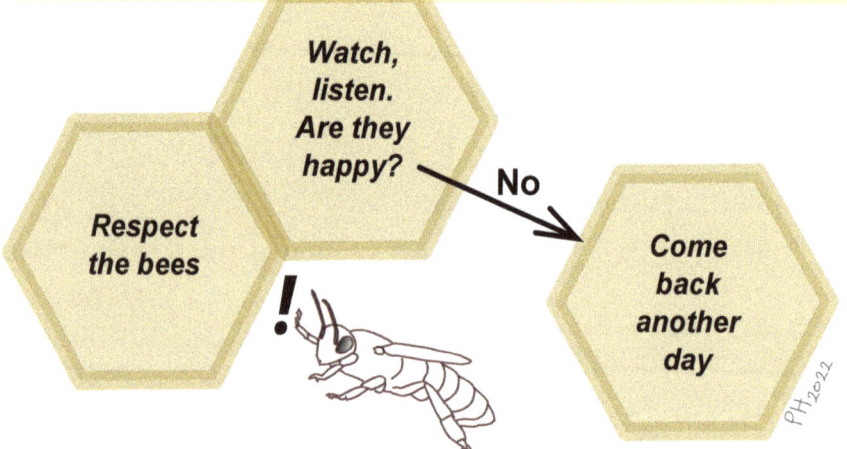

Chapter 7
Practical points about keeping bees

The beekeeping year begins in Spring. If you start preparing and training too late to get a swarm, don't worry, it just means you have an extra half year to research, plan, and choose a hive. Beekeeping requires patience.

Common beginner errors

- **Not uniting small swarms,** less than say 3-4 fistfuls of bees. They soak up immense time and only about 1 in 3 survive their first year. Beginners are loathe to reduce colony numbers, but these lack the numbers for resilience and if they hit a problem like wasps or a forage gap, they really struggle. Very small swarms will have virgin queens and many fail to mate leading to a dwindling colony. Small colonies going into winter often lack the stores to survive. If a queen is failing, brood care is neglected and damaged larvae are not ejected, which is an open door for disease.
- **Opening hives too often.** Very disruptive, weakens colonies.
- **Not questioning well meant advice.**
- **Not using local bees,** purchased bees require more time and money (initial purchase, ongoing feeding and treatment) to manage.
- **Tendency to rush in** to sort stuff out, and get stung.

General advice

Like any business, success isn't guaranteed, not every colony will flourish. There are factors you can't control like weather, and beekeepers importing disease nearby, but you can tilt the odds in your bees' favour by providing a low stress enviroment, and spread your own risks by running several colonies and having a local "bee buddy" or network for mutual support.

Every year will be different. Luckily "normal" bee behaviour covers a very wide spectrum - they're very adaptable creatures, and you will often hear the phrase "they've never done that before!"

Varroa resistance depends as much on management as bee strain. There's no magic bee that does everything.

Sometimes a struggling or ferocious hive will thrive or calm down just by being moved a few meters. The reason may be obvious in retrospect, like too much shade, but often it is not. This is related to the saying *all beekeeping is local* - management advice may be invalid 30 miles away.

Honey 101

There will be more on honey later, once some other background has been filled in. But before that level of detail, we need to discuss how it is made and why.

What is it?

Honey is essentially a very strong sugar solution, only about 20% water, with a few trace components like minerals.

Different flowers, different honeys. Honey can vary between combs in a hive.

Because the water content is so low, yeasts and bacteria are inactivated by osmotic pressure, aided by some antibacterial properties of honey (it's mildly acidic and contains traces of hydrogen peroxide). So it can keep for years.

How is it made?

By concentrating nectar in a process involving both the bees' digestive system and evaporation.

Nectar contains 20% to 40% of various sugars in water; bees use digestive enzymes to convert these to a standard glucose-fructose mix whose properties suit their purposes.

The bees spread nectar across cells and fan air across with their wings. This dries and concentrates it. They also suck it up, mix it with digestive enzymes and regurgitate it - it's bee vomit.

Why do bees make it?

It's a food store for winter and gaps in forage.

Why do honeys differ?

Primarily, different flowers give different tastes and aromas. There are many other factors, like: "mouth feel" can vary with the sugars' different microscopic crystalline structure because the bees made them at different temperatures; how the human intermediary processes the product; pollen content.

If there is a big flow of one nectar type, perhaps because you moved your hive next to a field of lavender, you can gather combs filled purely with that flavour of honey. Generally speaking, spring honeys are liquid and light coloured, and autumn honeys are thicker and darker.

Sometimes neighbouring hives have completely different honey. Maybe one swarmed so had fewer foragers when a particular tree came into bloom. Were the hives harvested at the same time? And neighbouring hives can simply find different nectar sources – you sometimes see hives a metre apart bringing in different coloured pollen. Other possible factors are, hive type (Warré honey is dark, from re-used brood comb); and did you feed with sugar syrup.

There is a growing problem with fake honey, where suppliers use cheap industrially processed sugar blends. Various tests have been created, and each time the counterfeiters find a way around. Real honey has pollen in, but this makes the honey set faster so many beekeepers go to great lengths to filter it out. Western consumers prefer clear, runny honey.

What is organic honey?

You're not permitted to describe UK honey as "Organic". It's a *reserved term* and bees are assumed to forage over several miles and pick up pesticides in nectar from crops or gardens[107]. Many gardeners don't realise that spraying their garden plants affects non-target species.

Also, many beekeepers deliberately add pesticides to their hives - miticides. These end up in the wax, though by careful timing they try to avoid doing this while bees are processing honey. Organic standards exempt miticides if they consider the chemical Organic (!)

What is raw honey?

It is straight from the hive with no processing other than coarse filtering. It contains pollen. It has not been given the mild heat treatment (Pasteurising) which commercial beekeepers sometimes use to kill yeasts and break up crystals, to extend shelf life. Heat treatment tends to drive off volatiles, and thus taste. So most people prefer the taste of unpasteurised, 'raw' honey direct from a hive. This is why comb honey tastes so good - all the volatiles are still there.

What's the significance of water content?

The bees work hard to evaporate the nectar, but have difficulty in cool or humid weather.

If you put honey with 21% water content in a sealed jar, it will begin to ferment in a few months. The lid will bulge! The taste is... off, and there is a definite yeasty smell.

| A hand held refractometer.

20% is considered the maximum acceptable moisture content (USA: 18.6%). I find 20% honey lasts about 1 year, but it's best to aim for 18%, which gives several years' shelf life. At 17% it will last indefinitely. People sometimes use dehumidifiers to reduce water content, but my bees usually manage 18% and with extended hot, dry weather 17%.

Moisture content is measured with a refractometer, you put a drop of honey at one end and look through the other[108].

Honey is hygroscopic

It absorbs moisture from the air. So don't leave the lid off.

Crystallisation

Most honeys start out liquid, then gradually solidify - it takes a few months. The process can sometimes be reversed by gentle warming. Some types of honey are notorious for crystallising

---

107   Pesticides are **everywhere.** Some persist in soil and are re-expressed by later plants in *their* nectar. Most honey samples have *at least* one pesticide residue [at levels below *human* harm] and some have over 15 [source: the British National Honey Monitoring Scheme, honey-monitoring.ac.uk]. For scientifically sound but easy-to-read summaries of pesticide issues see The Pesticide Action Network, pan-uk.org and pan-international.org

108   You can calibrate a refractometer with extra virgin olive oil (27% water) or liquid paraffin (24.5%). They are also sold for brewing, jam making etc which use different scales: be sure to get one suitable for beekeeping.

ridiculously quickly, because certain sugars are more prone to this: the Oil Seed Rape (canola) season is extremely busy for commercial beekeepers because they get huge harvests of OSR honey, but they need to get it out of combs ASAP before it solidifies and cannot be extracted!

Personally, I like set honey, I can get more on my toast.

Example - my own honey often sets solid in 3 months. But I have a batch from September last year which is still gloopy months after other batches set solid, and it's just crush & strained from a Warré hive. Water content is 18%. It must just be a different blend of nectars from usual. I could find out what by pollen analysis, but it wouldn't change anything.

Do NOT feed honey to infants!

Honey can contain tiny amounts of botulism bacteria. Adults' stomach acids will kill these but the digestive system of infants under 12 months cannot handle them.

This is a very serious point. A retired pediatrician tells me that every year, he saw a couple of cases of children who suffered paralysis because they had eaten honey. This can be permanent. Google "infant botulism" for more information.

We'll look at the practical aspects of extracting honey, and some more technical stuff, in Chapter 10.

Propolis 101

Propolis is plant resins gathered by bees - like frankincense and myrrh. It smells divine.

It is a multipurpose constructional material used to block holes, sterilise areas, reinforce comb, and entomb parasites.

It suppresses mould growth in the hive. It is an antiseptic versus bacteria. And it reduces the number of varroa mites that mature[109] - it's a natural acaricide. But because it is sticky and gets in the way of handling frames, conventional breeders have been deselecting for it since the 1800s[110].

Key point: one batch of propolis is different to another. Bees can gather it from many different plants and trees. Its melting point is usually 60C - 70C... but might be 100C.

Bees in my area gather brown propolis, but bees also gather green, red, pink and black propolis.

---

[109] Honeybees "polish" brood cells with a propolis layer before egg-laying. Researchers used artificial cells, coating some in propolis. **19% of newly hatched mites from propolis coated cells died, versus 6% from untreated cells.** And of the surviving mites, **only about half as many went on to reproduce** from the propolis-treated cells. *Honeybees use propolis as a natural pesticide against their major ectoparasite* (2021), Pusceddu, et al, doi.org/10.1098/rspb.2021.2101. Another finding in this paper is that **propolis coating in brood cells also reduced the symptoms of DWV,** though it's not a complete cure, and the paper makes it clear it's not because propolis kills DWV directly, but a side effect of the reduced mite load. The paper also references other work on how it directly affects **AFB, Chalk brood and Nosema.**

Marla Spivak (major propolis researcher), however, contests this as a small effect and found no significant anti-varroa, antiviral or anti-nosema effect from propolis in her own researches. What she *does* agree on is a **strong** antifungal and antimicrobial action, reducing e.g. chalkbrood; bees exposed to lots of propolis show weaker expression of immune system genes (because their immune systems don't need to work as hard). Spivak, quoting another researcher says even low levels of propolis helps reduce AFB significantly.

[110] I've seen one bee farmer extolling the virtues of soda because it dissolves and cleans off all the awful sticky propolis. Marla Spivak, a prominent propolis researcher, says wild bee nests are completely enveloped by ¼" - ½", and American bees have generally had the trait bred out.

Medical implications

Medical properties will vary from batch to batch. It can cause allergic reactions[111]. It is an anticoagulant; don't take it if you're on Warfarin.

There don't seem to be any conclusive studies that show it has significant medical benefits *for humans* - such as large scale double blind trials; and what research has been done has shown conflicting results. **iprg.info** are collating research here.

Propolis for medical applications is gathered using food grade plastic meshes placed in hives. The bees propolise this, the mesh is removed, placed in a plastic bag in a freezer and then flexed. The frozen propolis falls out into the bag.

Propolis properties

Its properties vary with temperature. **When warm and fresh**, above 20°C (inside the hive) it's sticky. Foragers collect it in the warmest part of the day when it's soft. It's also fire retardant in this state, which is exploited by some African bee species which live in wildfire areas, to make a fire barrier which protects the core of their nests. (Honey **does** burn.) Above 10°C, propolis vapour is present, and acts as a disinfectant and mould suppressant. But at the bottom of cold hives (mesh floors!!!) there is no such vapour, thus more mould / mildew issues.

When cold and old, it's hard, brittle and essentially not sticky. So when you separate boxes there is an audible "crack!" as a hard propolis seal is broken. And although *Amm* bees gather a lot of propolis, it doesn't make their hive / frame manipulations a problem, because *Amm* lives in cool climates. Solid propolis does burn.

It's not waterproof, as originally thought: it's **semipermeable** to water vapour. People used to wonder why bees coated the inside of hives with it. Wouldn't it lead to condensation? Torben Schiffer showed it's like gore-tex: it lets water *vapour* through but repels water droplets.

It dissolves - eventually! - in pure alcohol or strong ammonia solution.

Bees add it to combs to strengthen them, shown in pictures showing in Chapter 11.3.

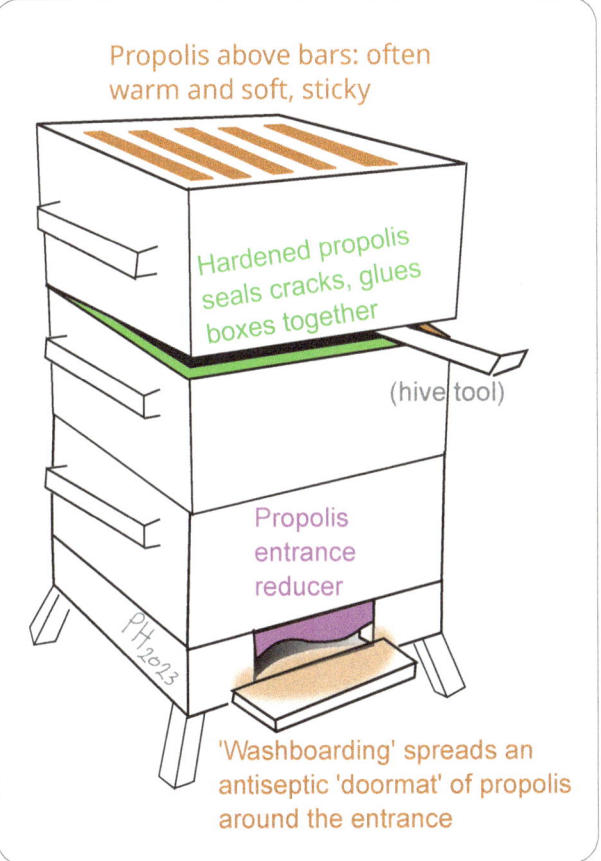

Propolis is a multipurpose antiseptic cement.

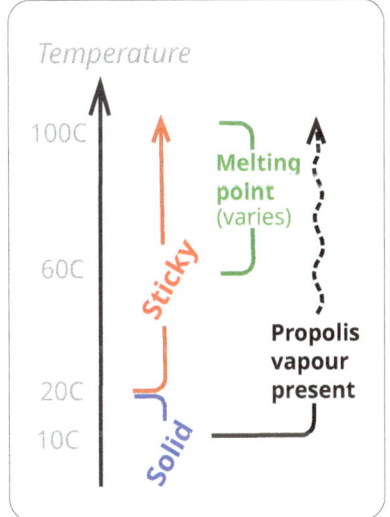

---

111   Some beekeepers develop *contact dermatitis* and have to wear gloves to avoid touching it.

Insulation 101

Insulation reduces stress both in winter (less fuel needed to keep warm) and summer (less effort diverted to fanning etc). It slows transport of heat in either direction. Heavy walls also add *thermal mass*, which stabilises temperature, but make hives trickier to move.

Framed hives throw away a lot of hard-gathered energy (nectar) through poor design. This diagram shows some loss paths, based on work by Derek and Elaine Mitchell[112]. Simply laying a thin sheet of something above frames, to shut down convection in "top space" saves the bees a remarkable amount of work. You can imagine how chilling empty supers above a nest are.

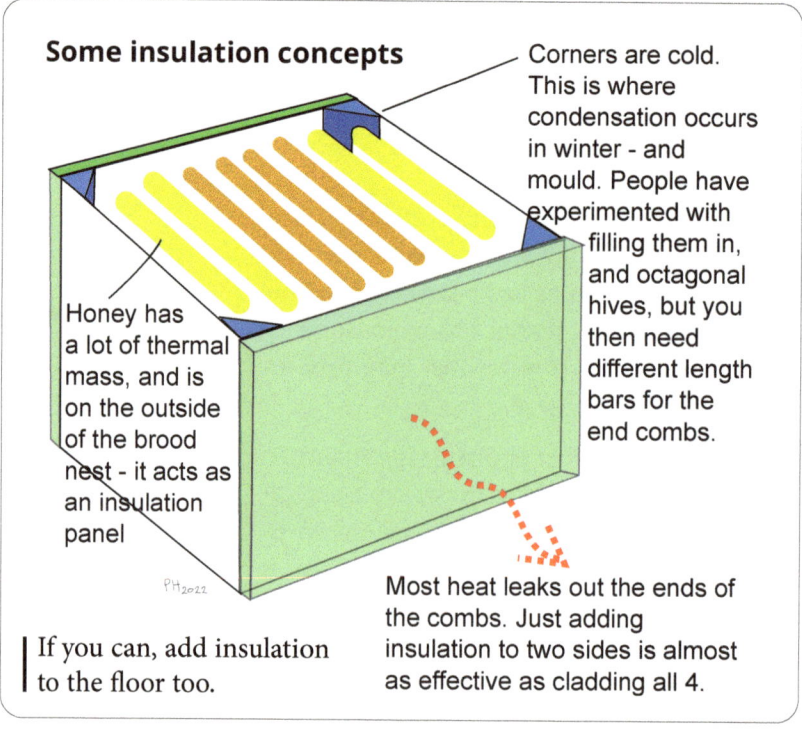

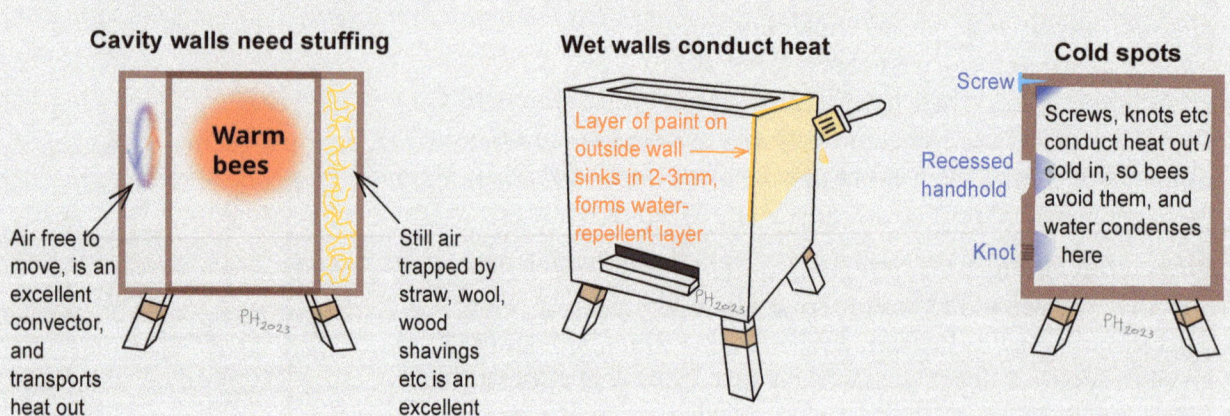

Double walled hives need their air gaps filled - if air can convect, it is a surprisingly poor insulator. Beware as a cavity wall is a potential parasite resevoir for e.g. Small Hive Beetle. Paint is useful for certain woods, damp walls promote mould; but some paints poison bees. Avoid antifungal "wood peserving" ones and "boiled" linseed oil, and ask other local beeks what they use.

---

[112] *Are Man-made Hives Valid Thermal Surrogates for Natural Honey Bee Nests*, Derek Mitchell (2024). His computer analyses were backed up by experimental verification for his PhD. Very thorough work. This is just one aspect of his PhD, etheses.whiterose.ac.uk/34266/ *Thermofluid Design of the Honey Bee Nest,* which seems likely to thoroughly disrupt and reform hive design.

Heat losses in National hive

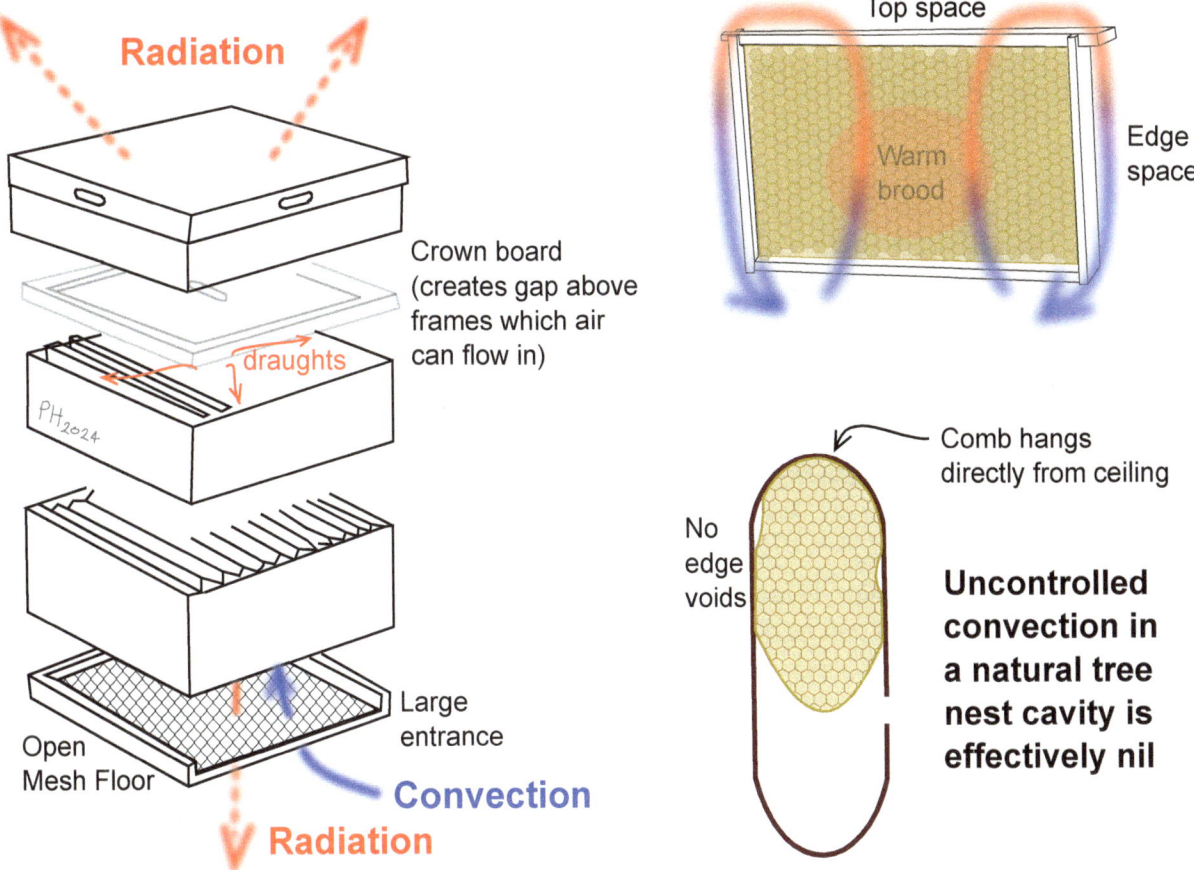

Counterintuitively, radiation losses are enormous. About 25% of a hive's heat is lost by radiation through an open mesh floor - you can feel it if you put your hand there. This can be blocked by putting a thin sheet of silvered material, like the foil placed behind radiators, on top of open mesh floors. The top of a hive benefits from being polished metal as this reflects excessive heat during the day and reduces losses at night[113].

Wood has an unusually high emissivity (losses by radiation) and a hive can lose 50% of its heat this way on cold nights. Aluminium has an exceptionally low emissivity, so simply wrapping a hive in aluminium foil or silvered bubble-wrap helps a lot.

Mitchell, an expert in computational fluid dynamics, created a lot of controversy pointing out bees don't naturally cluster in well insulated tree cavities[114] unless external temperatures drop towards -20C. This means lots of previous research has been done on bees in unnaturally stressed conditions and is suspect.

---

113   A clear night sky is incredibly cold, which is why you get night ground frost even when the air temperature is above freezing. Polished metal has far lower radiation losses than wood. Obviously you want an insulator between the metal and the bees. But there are other considerations: metal roofs are very noticeable, making a hive more likely to be stolen.

114   *Honeybee cluster – not insulation but stressful heat sink*, **Derek Mitchell (2023)** doi.org/10.1098/rsif.2023.0488

Experimental silvered follower board (both sides) for a TBH, reflecting heat back into cluster over winter and conserving resources, but the empty part of the hive is colder, promoting mould formation. Image © David Overbury, 2024

He found there's an optimum separation when the gap between bees is "0.3 bees" where convection is basically nil and heat conduction through the bees' bodies is minimised. Crucially, this is about the density of bees on a busy comb - if bees are clustering more tightly it's *not* so the outer "mantle" layer of bees can "insulate the core" but because the outer mantle bees are freezing to death and desperate to get warm by snuggling up to the cluster's warm core. A **dense** cluster is actually a **poor** insulator. This was not received well by beekeepers who considered clustering in their cold hives normal.

See chapter 12, *"Running a standard framed hive on low intervention principles"* for specifics of how to improve a National hive.

Weather vs climate: 'honey days'

British weather is very unpredictable, being next to the Atlantic. Annual honey harvests are erratic. Typically 2/10 years are amazing, 2 dire, 6 so-so. This is partly why bee farmers are relentless in their optimisation.

Many plants only produce nectar above a critical temperature. Many need rain to express nectar. Wet days are bad foraging days too. Low temperatures and high humidities hinder nectar evaporation --> honey.

Consequently there are sometimes just a few days a year here when the bees have ideal honey-making conditions (unless you move hives around). You'll know these boom *honey days* by the insane level of activity at the hives, and thrumming through the night.

A fascinating corollary of this is that the common saying "you get more honey from a large colony" isn't true in Britain *most years, unless* you are a migratory beekeeper moving them to continuous forage. The brood of prolific bee races like Italian and Buckfast eat their honey almost as fast as they produce it here - they're adapted for southern climes.

Can I convert from conventional beekeeping to treatment-free?

A common scenario is that someone starts beekeeping with conventional framed hives, then becomes disenchanted with casually brutal husbandry techniques.

It is quite possible to use the techniques in this book in a standard framed hive, but there is a risk your bees will die if they have to suddenly start looking after themselves - they may be overly dependent. It's like converting a farm to organic practices, it takes a while for pest predators to re-establish. In this case you want to convert in stages over a year, or maybe two, or **trial things with a fraction of your colonies,** replacing losses with swarms from local wild or TF colonies. Let's assign scores to some factors:

Factors influencing conversion to treatment-free beekeeping	"Fitness score"
The bees are pure Buckfast, Italian, or Carniolan, bought direct from a breeder	-4
The bees are from a local swarm, or Russian (Primorski) bees; OR… …The bees already keep their floor clean (indicator of hygienic behaviour re: disease - the Mewis Method)[115]	+4
The bees are bought from a supplier of varroa resistant stock who is >80km away (tends to be a single, unstable trait)	+2
Intermediate race (they were bought from a breeder a year or more ago and have since swarmed and open mated with locals)	0
Foundationless, natural comb	+1
Highly insulated hive	+1
You open the hive fewer than 4 times a year	+1
The bees have survived 1 year without miticides	+2
You have not fed the bees for a year (and they experience significant forage gaps like winters / dry seasons, i.e. they have long brood breaks)	+2
You live in the home area where this race of bees evolved (e.g. Italian bees in Italy, or Amm in northern Europe, Britain, Russia)	+2
You live in a country where beekeeping is very regulated and miticides are practically compulsory	-4
The bees are near a lot of other hives managed primarily for honey	-3
They gather massive amounts of propolis	+1
Total	

---

[115] The root cause may be: spare capacity → clean floor and general hygiene. **Kevin Mewis** spotted the clean floor = [hygiene and no varroa or DWV] correlation in 2015. An additional, **and key observation** from hive experimenter Gareth John is that bees only seem to keep **warm** floors clean.

Scoring

- 10 or more: you can probably switch directly to treatment-free.
- 4-9: you should introduce low-impact beekeeping practices gradually over a year or so, winding down feeding etc to make them less dependent. You may have to address starvation / skyrocketing varroa mites with a little feeding / shook swarming, so close monitoring is advised.
- < 4: best to plan a 2 year transition.

To put this in context, early results indicate about 18% of wild bee colonies in British trees die each winter. In Germany the figure is 75-80%, perhaps because Germany has replaced most of its native stock with Carniolans and Buckfasts, which are then routinely doused in miticides[116], and may have less varied forage.

If you already have sunk costs in a bunch of hives, *don't be afraid to fail fast, inexpensively* - what Silicon Valley calls "failing smart". Try running a handful of hives with local swarms, treatment-free, foundationless, minimal management while your other hives act as controls. Most failures are temporary setbacks and don't forget, they weed out unfit colonies - equivalent to the active directed breeding some beekeepers practise, but here, you're letting the bees do the work of discovering which genes to select.

Finding information

- **Networking:** the **most** helpful thing is to find another enthusiast near you. Search for "natural beekeeping <area>" and "Treatment Free Beekeeping <area>".
- **Reading and web sources.**
- **Training courses:** Google is your Friend. Your local group may offer training or mentoring. For beekeeping, in-person courses are best.
- **Online forums, mailing lists:** Some good, some bad, plenty of choice.
- **Social media:** Beware. Generally, poorly informed echo (ego) chambers reinforcing unbalanced views. The signal-to-noise ratio is very low. Designed to absorb as much of your time (watching ads) as possible, so their algorithms feed you stuff similar to what you've already viewed - but what you actually need is to *broaden* your knowledge.
- **Your local beekeeping organisation:** conventional beekeepers have a lot to teach you, not least common sense and confidence in handling bees.

A network is a huge distributed apiary with many observers pooling thoughts and helping each other. And if you have a disaster and your bees die, people tend to share excess swarms next year.

---

116 Comparing results of similar investigations by Filipe Salbany in Britain and Torben Schiffer in Germany. (From a conversation with Filipe.)

Equipment suppliers

Hive makers seem to come and go over the years. Three tips:

- **Suits** – Sherriff and Mann Lake have good reputations. If you have a big budget, ventilated suits are vastly superior *and* thick enough to be impervious to bee stings.
- **Comb knives** – tricky to find - thorne.co.uk → search for product M5131.
- Second hand equipment carries a risk of carrying disease, so ask the vendor about its history.

Part 2:
The Beekeeper's year

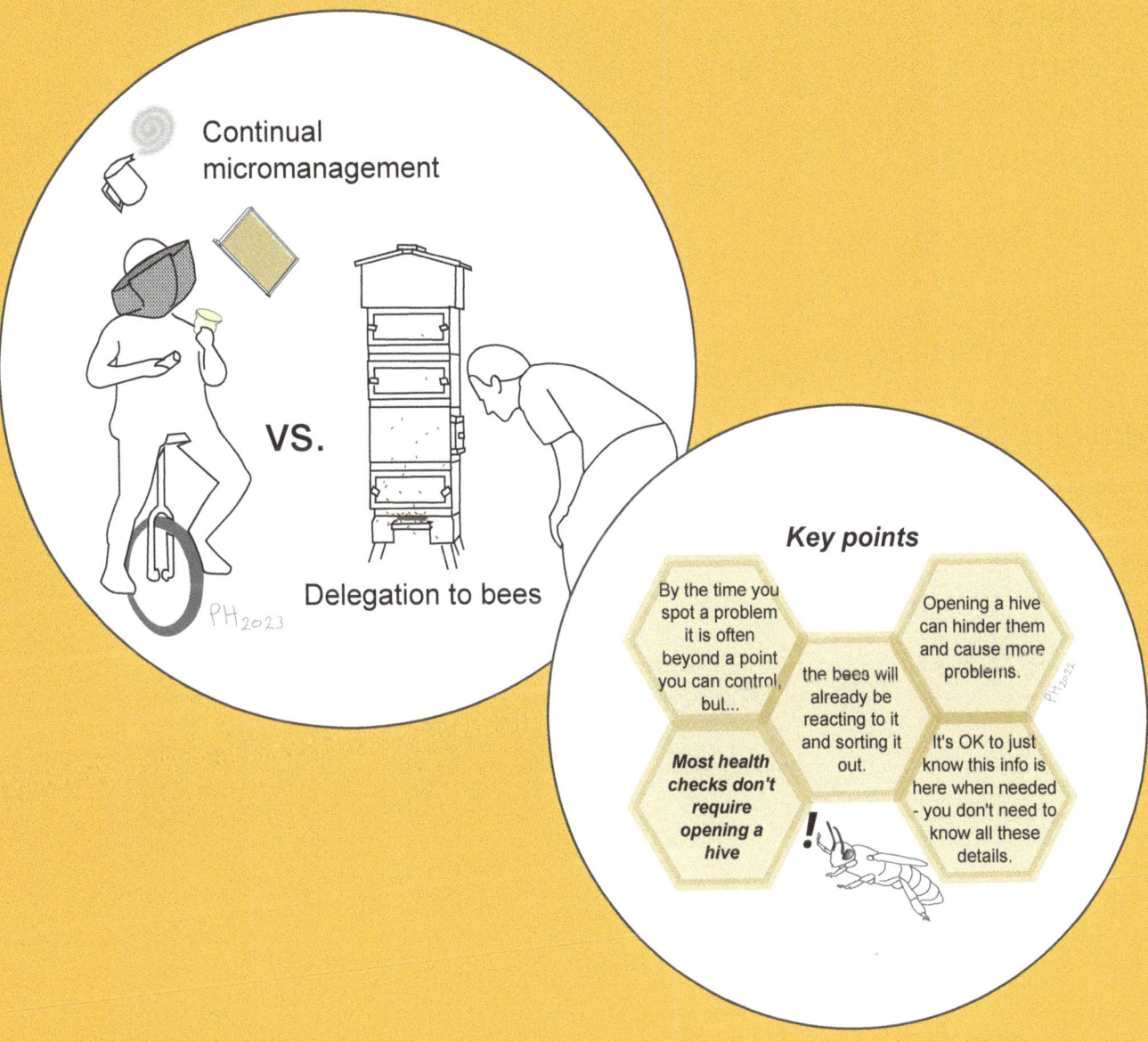

Part 1 covered how to set up an apiary. This is about managing it over the annual cycle: the physical skills (*manipulations*) you may need, and interpreting what you see during inspections.

Chapter 8
The annual cycle

Bee calendars

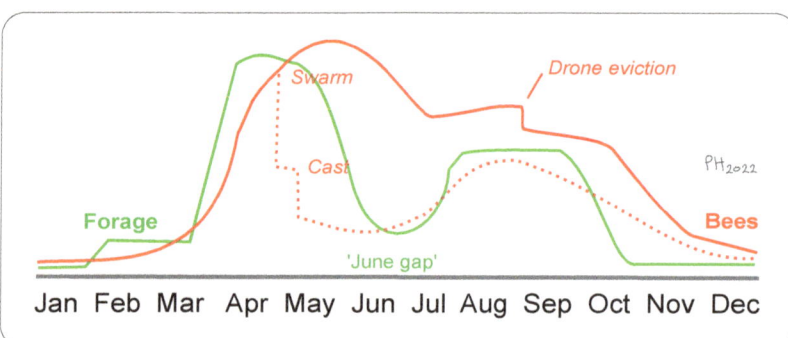

Bee numbers in my (static) hives. Sometimes they swarm (dotted line). The key point is the forage timings will be different everywhere: in my area spring trees, and autumn brambles / ivy are significant but not crops. This population curve is very different to the usual one you see because I don't feed my bees.

Calendars are of dubious value outside the immediate region they were compiled for.

It is also clear, looking at old ones, that climate change has shifted everything forward by as much as a month[117].

Here's what seems to happen in my apiaries - take the dates with a pinch of salt!

---

[117] William White & James Beesley wrote *A Complete Guide to the Mystery and Management Of Bees, To Which Is Added A Practical Monthly Bee Calendar* in 1844-1852. They lived in Banbury - about 10km from me. This states (p.72) that if weather is favourable, swarming begins in May. Now (2022) I see swarms start in April if weather permits.

[A mild winter means a long pollination period, March to May. A long harsh winter concertinas everything together.]

Deliberately omitted:

- NO winter checks - why? Highly disruptive.
- NO feeding (except exceptional circumstances). Feeding accelerates development and permits earlier, and more, honey harvesting; tradeoffs include higher likeliehood of early starvation if weather turns bad.
- NO mite treatments
- NO regular internal inspections to kill queen cells etc (swarm suppression)
- NO messing around with queen excluders.

Winter preparation

Bees prepare for winter. They move stores around so they will be accessible by the winter cluster. They begin raising long lived "winter workers", and late syrup feeding can trick them into delaying this so they end up going into winter with short lived workers, and the colony dies before Spring.

Important for winter survival are mouse guards, insulation, stores (heft early Sept; Warrés and small nucs in poly hives do NOT need 20kgs), and local bees.

Mould is the enemy, not cold.

Fungal infections can't survive at our body temperature. But insects, being cold blooded, are vulnerable to fungal diseases. *Damp* cold is ideal for fungi.

So the colony must stay above ~12C but also dry. A colony will transpire several litres of water over winter. Thus:

- One approach is to accept that a cheap, thin walled (standard framed) hive will have condensation on its cold walls, and increase ventilation, e.g. with an open mesh floor. This works - but the colony has to burn a lot of fuel to keep above 12C! The walls are little more than a windbreak. That's why some hives need 20kg of stores to get through winter.

- Alternatively, a well insulated hive has warm walls - little condensation - and with an interior above 10C it is full of propolis vapour, which suppresses mould and sterilises any condensation - which is thus a *healthy* source of water for the bees.

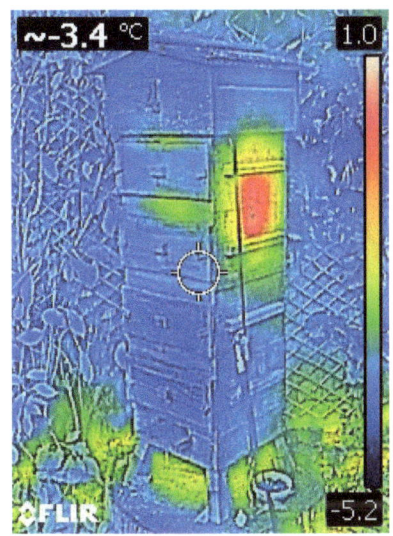

Infrared image showing the warm winter cluster.

(The 'warm' glow below the hive is due to the hive sheltering this patch from the cold sky.)

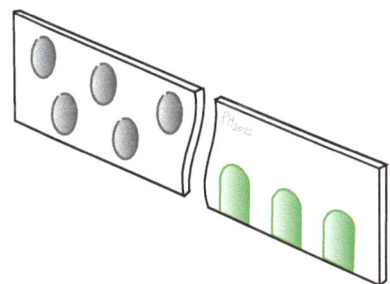

Mouse guards, often just holes in a metal sheet, are pinned over entrances for winter. The holes block mice - but also drones and queens, so must be removed in Spring. The simple hole type (left) also prevents bodies being dragged out, so are best only used when bees are clustering. The green design is better as it does not scrape wings or brush pollen off legs. You can use a queen excluder as a mouse guard.

Both approaches work.

If you do decide to feed, it's important to finish by the end of September, so that the colony shuts down laying before winter and the last generation of workers switch to the long lived "winter bee" metabolism.

Over winter you may notice:

- Cleansing flights on sunny days.
- Clusters will draw tight in extreme cold but can expand and be quite loose other days.
- Apicentric hives are toastier than Nationals and Langstroths so the bees don't need to cluster until temperatures are lower – behaviour is different.
- Laying restarts at a low level at the equinox, as the days lengthen again.

To confirm your bees are OK in winter, without opening the hive - stick a stethocope (or any flexible hollow tube held to your ear) in the entrance. You should hear them rustling. If you don't hear this, tap the hive.

Special considerations for large colonies, and damp winters

These are more vulnerable to starvation and generate more condensation (mould).

- Keep an eye on their floors and quilts for dampness - they may benefit from open mesh floors, especially in damp climates.
- Remove unused boxes on **big** colonies: their walls will go black with mildew over winter because they're not next to warm bees.

Two common causes of colony winter death:

- **Snow blocking entrances.** Bees can breathe through a little snow, but not once it freezes into ice.
- **Queen excluder not being removed.** Prevents cluster accessing stores.

Something to ponder: apart from the queen, every colony is a new thing each year.

Chapter 9
Swarms

What are they?
What do they do?
How do I catch one?
What do I do with it?

What are swarms?

Swarms are how honeybee *colonies* reproduce - queens are too specialised to start one on their own. Workers go off with a queen to found a new colony; these swarms can be used to populate empty hives.

Bees emerge from a hive in a swirling cloud, then settle for a while as a **cluster** as they scout for a new home, which can take anything from 40 minutes to a few days. Once they have decided where to live next, they fly there. They are following a "program" and it is often very easy to place the cluster into a box before it gets to its new home. You can then put them into a hive of your choice - they think they chose it.

But if you relocate them **after** they complete their program, i.e. pull them out of their chosen cavity and put them in a hive, they'll likely abscond ("this isn't where we live!"). So if you want a swarm, **it is essential to act quickly** before they set off on the second and last leg of their journey. Don't tarry to finish your lunch - go!

| Classic swarm cluster.

Once you see a swarm it is unmistakeable, on a larger scale than other bee flying phenomena. There's a loud buzzing, but I wouldn't describe the noise as a "roar" as many writers do. If you stand in the cloud, some will land on you to rest, but they ignore you; the only danger is if they get caught in your hair.

Unlike humans, where the adult boots the young out of their house when they grow up, with bees the first "prime" swarm is headed by the original, mated queen of the hive[118]. She leaves

---

118   Except for at least one African subspecies, where the original queen stays home and every swarm is headed by princesses. There always seems to be an exception to Bee Rules.

behind larvae in queen cells who hatch about 8 days after she leaves[119]. [*Important:* but if you clip queen wings so they can't fly, all swarms will have virgin queens - even prime swarms - I'm assuming you don't clip.] So after the prime swarm leaves you may see one or more **casts (afterswarms) start leaving about 8 days after the first, prime swarm.** The casts are headed by new, virgin queens (princesses).

That's the plan, but if weather or beekeepers bottle the bees up in the hive after the process begins, the delay causes the prime swarm / afterswarm processes to overlap, like all the number 10 buses showing up at once. And chaos reigns. Possible outcomes of the palace intrigue include:

If weather prevents spare princesses leaving, they may fight it out[120] for control of the entire hive, so there are no casts. Or...

The workers may keep spare princesses in their cells, as insurance, until the new one has successfully mated - then

Important: this book assumes you **don't** clip queens! This has a huge impact on prime swarm behaviour. If you clip queen wings so they can't fly, **all** swarms will have (possibly multiple) virgin queens and be less cohesive - **even the first, prime swarms.**

---

119 Occasionally a few days longer.

120 Conventional beeks talk about queen fights a lot but low-intervention beeks see few, and generally if queens ARE killed we either see one submitting to another's sting, or the workers ball her. Actual fights where both are armed with a lethal weapon would be very risky. Workers do not, as a rule, sting queens, but it has been seen during **usurpation swarms** (swarms moving into already-occupied cavities and fighting the residents) and when running virgins in to hives. In these cases of course they are killing a *foreign* queen; I am not aware of a case where workers stung their own queen. Perhaps conventional beekeeping stresses the bees in some way (swarm suppression?), or promotes the raising of too many queens, so queens are more *likely* to turn on each other. But queen fights *have* been recorded for centuries, e.g. Pettigrew reports queens rolling out onto the landing board as they fight in his *Handy Book of Bees* (1875).

kill the spares[121]. They actively herd the hatched princess away from the reserves. The incarcerated princesses will eventually escape, and some fighting between rivals is inevitable if the first one doesn't mate within a few days. Or…

Competing princesses try to attract workers to fly with them, and several casts emerge, typically one a day, sometimes stripping the original colony of almost all its workers - which leaves it undermanned and vulnerable - until one of the princesses kills all her rivals in the hive. Thus you can get casts with 2 or more queens as several flee at once. Because none are mated yet, they don't smell strongly attractive to the workers and there is no clear leader to these swarms. Gathering multi-queen casts can be tricky because these queens are reluctant to fight, preferring to avoid each other. The bees aren't in nice coherent clusters.

Early swarms are desirable because they have all year to build up whereas, a late swarm risks a dearth of food just as it is trying to set up home. The bees need to build up to a "critical mass" of numbers and stores to survive the next winter. This is why you don't take honey in their first year.

Prime swarms hit the ground running and build comb and numbers rapidly; afterswarms (casts) have unmated queens, have fewer workers, are later – and are more likely to die in their first winter.

Swarms act as a firebreak against disease – most bee pests attack the larvae, but only flying adults are in a swarm[122]. The mother colony also experiences a brood break so pests are reduced at both ends. (Natural beekeepers consider this a Good Thing, whereas honey-oriented beekeepers prioritise Maximum Bees.)

Unlike purchased bees, the genetics will be adapted to the local area. Even if the swarm is from a commercial colony, the queen will soon mate with local drones[123].

Low-intervention beekeepers emphasise propagating survivor genetics and we particularly prize queens and swarms from unmanaged colonies. Some commercial beekeepers catch swarms, kill the queens and use the workers to reinforce hives with queens selected for honey yield, but not survival traits. This is like throwing away a gift and keeping the wrapping paper[124].

The first cast doesn't take over many of the remaining bees. But subsequent casts really weaken the colony because the remaining queen can't lay immediately and they may not be able to rebuild sufficient stores for winter. Or get overrun by wasps or wax moth. So later casts are best reunited with the hive (see *Uniting*, chapter 13).

121　An immediate suspicion is, maybe the factions of half-sister workers in the hive try to promote their full sister princesses to queen, and bump off rivals? That would be the human thing to do! But so far no one has found any evidence for this, there seems to be no rivalry between sister groups.

122　The flyers in a prime swarm are of all ages. Casts tend to have more young bees as most of the older flyers have already left - so are less likely to sting you.

123　Even if it's a prime swarm with the original queen, because **swarms rapidly supersede old queens.** As mentioned, crosses with commercial bees can turn aggressive, another reason to prefer swarms from local colonies. If the queen has a coloured mark on her back, she is from a conventionally run hive; she may still be local genetics but it is perhaps more likely not. View coloured marks as a warning of possible F2 problems. Low-intervention beeks don't mark queens.

124　Whereas Ron Hoskins, a rational conventional breeder famous for his varroa resistant bees, collects swarms and evaluates them for a while. Sometimes he concludes "this colony has useful traits, I shall use it to widen the gene pool of my stock."

Why swarms are the best way to populate hives

Pros	Cons
Swarm	
Locally adapted Likely to have survivor traits. Best choice if you prioritise health over honey Free Very low varroa & disease load **Brood break** in mother hive and new colony - natural firebreak against disease and pests Gut microbiome suits area (eg can digest honeydew) Fits in any shape of hive **You learn far more** - they vary more than uniform colonies of poory adapted near-clones	Should not take honey for first year - let it establish comb, reserves etc Very variable outcomes - I actually enjoy this and learn more than by buying bland, identical queens Approx 15% queen failure due to unsuccessful mating in my region Some swarms are small, but they can be united with others to improve survival rates Some people only value things they pay for Less predictable than buying a nuc or splitting - so less suited for most large scale commercial operations
Nucleus (mini hive)	
Comes with stores on frames Lots of honey even in first year Lots of bees (reduces risks) Balanced age profile Reassuringly expensive	Pre mated queen with non local genetics, no survivor traits, F2 crosses can be aggressive in year 2 F2 crosses can also make neighbouring beekeepers' bees aggressive Frames don't fit in some hive types Unrestrained breeding, high management required Standard advice is 'check health before buying to avoid buying in pests or disease' (Americans mention fake nucs, 2 frames with good laying pattern from another hive added to make nuc with unproven queen look good)
Package (an American abomination)	
Likely to get lots of honey in first year ...if it survives Fits in any shape of hive Lots of bees Reassuringly expensive	If pre mated queen: non local genetics, no survivor traits, F2 crosses likely to be aggressive in year 2 F2 crosses can make neighbouring beekeepers' bees aggressive Usually supplied with mated queen, but if unmated princess she will not lay for up to 3 weeks by which time the workers are getting old (no brood hatching) so nurse bees are now forager age, not a balanced colony. Many such virgins are superseded leading to further buildup delays and high risk of colony failure. NOT a colony. Stressed bees from typically 4 colonies are shaken into a box with a strange new queen, she's in a cage to protect her from them! Temper may change once her progeny dominate. Bees treated as "fungible assets."

Pros	Cons
Package [continued]	
	High failure rate in first year ("pump and dump" sales strategy).
	In Britain, usually imported, which risks importing disease
	Highly stressed during shipment
	Unrestrained breeding, high management required
	Seller's treatment of bees reflects their attitude to business. Difficult to inspect packages before buying. Variable quality. Caveat emptor. If the package is created (shaken into box) too early or late in the day, it contains a lot of old forager bees who die off quickly.
	Common advice is to assume they arrive with mites and treat them with miticide immediately!
Queen (i.e. requeening)	
Cheap - the first year	Pre mated queen with non local genetics, no survivor traits
High honey yield	You are locked in to this race or risk aggressive crosses
Fits in any shape of hive	You make neighbouring beekeepers' bees aggressive (crosses)
	Unrestrained breeding, lots of varroa etc, high management required
	Sometimes tricky to introduce to your colony
	Can carry disease e.g. queen imports is thought to be how varroa spread to Europe and Australia
	Often superseded rapidly[125]
Split (from own hive)	
Free	No brood break (disease firewall) in the group left with the prime queen
Like a swarm, but a rather more predictable outcome	The new colony still has comb, which can carry parasites etc, so be sure to only do this with a healthy colony
Locally adapted	
Known colony history	Different mix of bee ages to a swarm - more young bees who cannot yet fly
Lots of bees	

It's almost as if there is a correlation between how aggressively marketed the product is and how aggressive its progeny will be in a year or two. Remember, you have a duty of care to your neighbours not to create aggressive crosses.

Common misconceptions

Swarms are not dangerous. Temperament during collection, and of hived swarm is gentle, if handled gently. Our group has probably collected around 200 swarms now and we only had one vicious swarm – from Kidlington, known for its mad bees, and where at least one beekeeper mixes

---

[125] *Beekeeping: Challenge what you are told*, Roger Patterson, p.175

Swarms 101: textbooks versus real world

Swarming is driven by simple impulses which adapt to weather etc.
Behaviour varies, but the drives are consistent: optimising reproduction, survival.

1. Ideal natural sequence

Colony judges it is safe to reproduce, based on nectar flows, stores, crowding etc.
Lays new queens, current queen starved to flying weight, scouts check nearby cavities.

New queens capped. Flying bees (half the colony) gorge on honey and leave in the first, 'prime' swarm with the OLD queen - queens need workers to found a new colony, typically in April / May. Young nurse bees remain to raise brood & new queens.

Swarm clusters to decide on new home. At this point it is easy to catch and rehome in a hive.
Once it is in a new cavity, they consider it home and the swarming 'program' is complete, rehoming them is very difficult.
Prime swarms build up incredibly rapidly - occasionally managing to swarm again the same year.

8 days later, back in the mother colony the virgin queens hatch. Characteristic "tooting and quacking" noises indicate queens challenging each other, but workers keep them apart and fights are rare. One becomes queen of this colony. Spares fly off with afterswarms ('casts'), which tend to be smaller than primes, and the queens need to mate before laying, so casts build up slowly. Each cast bleeds off more of the limited remaining workers so too many is bad.

2. But a common scenario in British weather...

Same...

Prime swarm delayed. Queens can't fly in wind. Rain and cold can kill entire swarms.

Some time in May, the weather finally clears and all queens emerge - multiple swarms from many hives all over a wide area - a "swarm day". Your phone does not stop ringing! Some swarms have multiple queens and behave oddly (indecisive, no clear centre, trickier to catch).

3. Commercial beekeeping disrupts the cycle further...

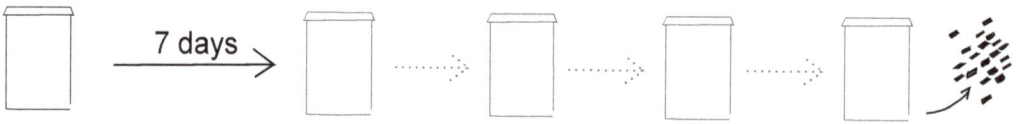

Same...

Swarms reduce bee numbers, thus honey crop. Beekeeper repeatedly disrupts swarm preparations by killing queen cells weekly, clipping queen wings, use of queen includers. (Bees get crowded, grumpy, varroa levels shoot up.) Sometimes the BK misses a queen cell and the bees swarm anyway... this may be a late swarm (eg June) but it is really huge... and likely to be varroa laden, agitated, and abscond. Conventional beekeepers report a lot of queen fights.

You will hear inconsistent and simplistic accounts of swarming and queen behaviour from people who repeat accounts they read 'somewhere'. Most textbooks focus on how to suppress swarms and ignore variations.

PH 2022

Buckfast and Carniolan, and lately switched to Italian. The biggest detractors of swarms are users of non local bees - they're talking about swarms which hybridise with their own bees! *Exception:* a *queenless* swarm *can* be aggressive. If a swarm acts agitated, be very cautious, and look around for a queen under a hive etc.

The swarm does not follow the queen. Watch them. The queen can barely fly[126]. The workers are herding her, she does not know where to go; and she has to stop and rest fairly frequently. At these points some bees cluster round her protectively until she is ready to go again, while others form a protective cloud in case some bird or other predator gets ideas. Kevin Mewis, a careful observer, reckons the workers signal the queen "over here!" with a characteristic loud buzz.

Swarms are not naturally "swarmy". Although there is a genetic component to this (some lineages are definitely more prone to swarming, especially if they have Carniolan genes) the main factor here is probably crowding. If you have an over-prolific queen and keep feeding sugar to stimulate laying...

If the queen is "old", this is not a bad thing, the bees will just supersede her. Prime swarms are actually highly prized for their vigour, and those queens are a year or more old.

The swarm doesn't just contain old bees. Just look. New furry ones and old shiny ones. 70% are <10 days old.

The swarm is not just females, it contains a few drones. Maybe they are there as a meat shield for the queen, maybe they are just caught up in the fun. There are often spare queens.

Swarms don't need treating for mites. Very low mite loads.

Swarms rarely carry disease. Roger Patterson reports in 70 years he's only ever heard of (never seen) one instance of foul brood in a swarm. He's seen a lot more in bees which have been bought[127].

Don't feed swarms. Good way to propagate disease. Discussed below.

Swarm psychology: what are swarms collectively trying to do?

Once you understand their aims, you will be able to catch and hive them.

The queens are not in charge. The workers are not in charge. Each influences the other.

Key difference between Primes and Casts: the workers encourage the old queen out when THE WORKERS want to swarm; but casts (afterswarms) are led by new princesses fleeing combat, and tend to be less prepared – more chaotic, no clear leader, leave very suddenly with less food.

Prime swarms occur because the **workers** have decided the hive has the resources to found a new colony. There are various triggers, typically crowding at a time of plenty of stores and forage. The workers <u>begin scouting for new cavities</u> (you will see why this is significant shortly) and ensure the queen lays eggs in the queen cells they've prepared. 8 days later, these queen cells are capped and the workers are ready to go. During this period they reduce the queen's diet, from her usual "own weight in food each day" to make eggs, because she needs to slim down to fly.

---

126   Queens' wings are short. European honeybee swarms tend to fly no more than ~2km. However *Africanised* honeybee swarms will fly 20km a day(!) which is how they spread from Brazil to America so fast. Older queens - who need slimming to fly and have worn wings - are weak flyers, virgins seem ready to fly almost immediately on hatching and go on strenuous mating flights.

127   BIBBA webinar, 24 April 2021. In the Disease section we discuss how swarms *can* spread FB, if they come from diseased apiaries, but FB is now very rare outside America.

They wait for a warm, sunny, dry, windless day - partly because queens are rubbish flyers (tiny worn wings and never go out!). Then the hive goes weirdly quiet for a few minutes as flying ceases and everyone stuffs themselves stupid on honey, until both stomachs are full[128]. Prime swarms are very heavy[129]. Then suddenly the flying bees begin streaming out, a great rush and the queen is more or less herded out with them.

This prime queen may not be too keen on leaving her home, but the workers begin flying in a huge mob. Everyone is carried away by excitement, it's basically sex (the colony reproducing).

Only once I have seen a queen refuse. We had a colony that kept swarming then returning to the hive within a couple of minutes. It did this several times over a few days. This happens when the queen is damaged somehow and unable to fly. A week later they swarmed successfully, which would, unusually, have been a first swarm with a virgin queen.

At this point the swarm typically clusters on a nearby branch, confers for a while then flies to a new cavity they have chosen. Prime swarms, being planned, often fly **straight from a hive to their new home without an intermediate cluster.** Scout bees have pre-checked nearby cavities in the previous weeks. Bill Anderson has even seen them *walk* to a new hive like a moving puddle of treacle across his rooftop apiary[130]. Other times they only cluster for 40 minutes before flying on[131]! So if you want these highly desirable prime swarms, place bait hives near your apiary and have your collection kit ready to go at a moment's notice.

An important behavioural factor for prime swarms is that the queen is already fertile and **smells strongly of queen pheromone.** They act as a "classic" swarm - typically one big cluster hanging conveniently from a branch and acting as one cohesive unit. Once you have boxed the queen the others follow her in as a unit. Catching a prime swarm may take as little as half an hour.

The queen is programmed to **take no risks** and if you put her in a box her instinct is to go into the **darkest** area, towards the top, where workers will cover her for protection. They will then begin fanning Nasonov pheromone to summon the others.

Survival chances: small casts are best merged

The first cast is usually large and should be fine when hived. Even in the wild, at least 50% survive a year or more[132]. But -

---

128 A bee can just about fly while carrying its own weight. So gauging the number of bees in a swarm by weight can be out by a factor of 2. Another consequence of this extreme gorging, is that prime swarms can feel sticky when you touch them! Something to ponder next time my mother in law makes another record attempt for the amount of honey that can go on toast.

129 I once collected a swarm while my wife stood below holding the ladder. I cut the branch and suddenly was holding the full weight of the swarm & branch, a few kg, and overbalanced. Instinctively I dropped the swarm, right onto my wife's bare head. She got just a single sting. This illustrates several things: position a ladder so you don't need to reach far from the centre of gravity; and swarms are *extraordinarily gentle*, because they don't have brood / stores to defend and anyway, they're often so stuffed, they can't physically flex to sting you.

130 *The Idle Beekeeper*, Bill Anderson, p.200.

131 I have my suspicions that one reason swarms abscond, it is because they already decided on another home nearby. The implication is that if you catch a prime swarm, it is best moved to an apiary a few miles away so they don't find their pre-prepared Escape Pod - the opposite of the usual advice to keep bees local. You could probably move them back once they'd committed to their new hive, begun building comb etc. This is just a theory so far, I've only caught ~60 swarms so far, too few to confirm this pattern or gauge how common it is.

132 There's a lot of debate around this. Helen Tworkowski claims only 20% survival rate for primes (BBKA News, April 2024, p.26) and perhaps this is true for her bees in Devon. In Oxfordshire, I see wild colonies swarm into walls, roofs and trees and most survive 1-3 winters, sometimes much longer. I find large casts have over 50% chance of survival in a hive, the usual killer is queen failure. I systematically monitor local colonies.

- Very small casts lack the numbers to perform all necessary tasks, and usually die if hived alone.
- Mother hives sometimes "swarm themselves to death" by excessive swarming, leaving too few bees for the original colony to be viable[133].

I strongly recommend *merging* (chapter 13) any casts after the first one, reinforcing either the mother hive or combining it with other swarms you've just hived.

Quimby (a.k.a. Quinby)[134] recommended merging **all** casts unless your experience indicates there will be plenty of nectar for a while. I.e. **late** casts should **always** be merged with another colony. He also pointed out that if they are still piping they will probably swarm again.

Afterswarms (casts)

Casts behave very differently to Primes!

This impulse to swarm arises in the queens, not in the workers. Their motivation is that there are several rival princesses in the hive so they need to get out, or fight to the death. They go to the entrance and try to attract flyers to come with them. Rather than waiting for ideal flying conditions, they want to escape the duelling ASAP.

Unfortunately for new virgin queens (princesses), they don't smell strongly, so attracting bees is tricky. And several rivals may have the same idea at the same time. "She's getting away with all the workers!" So you can end up with:

Multi-queen swarms, which behave much less cohesively than a prime swarm. **This is common for a cast.** Filipe Salbany found a swarm with 9 queens – in Britain! Sometimes they settle in two clumps. You might collect one, and wonder why bees keep flying back to that other branch. Some people say that if you get one queen in a box the other won't go in so you only collect half the swarm... I am not sure if that is true (with little queen pheromone present, the bees and princesses are attracted to the mass of other bees) but it is another reason to have multiple boxes to hand. You can merge the casts later to form one decent sized swarm with a better survival chance.

Princesses with **so little smell,** the workers are unsure where she is. Though the bees can smell a mass of each other, so they will cluster when cold. This is basically a disorganised mob and loose cloud where no one knows what's going on or where they're meant to be going. You may box the princess and the bees still prefer to hang on the branch with no princess, because that cluster smells more strongly of Bees!

Several casts emerge simultaneously on the first good day after a long period of bad weather. You literally see them melting and reforming on nearby branches as workers defect from one to another, then back again. And even give up and everyone goes back in the hive!

Mixed prime / cast with multiple queens - if the weather has prevented the strong smelling prime queen and prime swarm emerging for over a week, princesses join the prime swarm. The swarm acts cohesively around Mother and other bees more or less ignore the princesses. You can view these princesses as insurance in case Mother fails. A lot of beekeepers are unaware that prime swarms' queens are often rapidly superseded.

---

133 Probably due to producing excessive queens, trying to compensate for the beekeeper killing them as part of their "swarm control"

134 *Mysteries of Beekeeping Explained* (1853) p.129

Note that most of the flying bees left with the prime swarm (except whatever foragers were out when the swarm occurred), so casts have a much less experienced set of bees, and are *usually* significantly smaller - the first might be 50% the size of the prime; subsequent casts are progressively smaller. (Though odd weather conditions can result in the first cast being bigger than the prime.) It is said that the mother hive can lose so many bees it will "swarm to death", though I think it more likely that they just suffer occasional queen failure and this is misattributed.

Casts have not pre-scouted for possible homes, so may stick around for up to 3 days before flying on.

Casts may not be carrying as much food as prime swarms - it depends how well the hive was stocked originally[135]. If swarms seem exhausted and out of energy, they may need feeding. But another reason can be heat exhaustion, as swarms happen on hot days. So on really hot days, first try perking them up with just a spray of water from a mister, which is why it's in your swarm kit.

Quimby (p.127) and Edwardes (*Lore of the Honey Bee*, p.132) state that afterswarms are not as particular about the weather, which I interpret as the virgin being desperate to get away with a workforce before being murdered by a homicidal sister. They are sometimes well outside the usual 10AM to 3PM window too.

Afterswarms, both workers and their princesses, *are all sisters,* (their mum went off with the prime swarm). They should form a tight family unit, but tend to act dispersed and disorderly. That's partly because unmated queens don't have a strong "queen pheromone" smell, and often because there is more than one such queen in the cast. So there is no clear centre for everyone to clump around.

It is also possible to pick up swarms, even big ones, which have **no** queen or princess: she got lost along the way somewhere, perhaps because her wings were clipped. You can sweep them into a box and some stay, others wander out... they take ages to gather... some fan but I think it is more because the mass of bees is in the box rather than a queen. I've picked up a few swarms which must have been queenless, usually small casts but once a very large one, and they all either dwindled over a few months, or absconded within a day. Such swarms are best merged with a queen-right colony to reinforce it.

Swarm clusters: scout bees and decision making

If swarms misjudge the weather and are exposed to rain and cold, they may die. And they only carry 3 days' fuel, max. So they are under pressure to move on.

If you look at the surface of a swarm, you will see some bees moving. Look closer - they are moving in the figure-8 patterns used to communicate location, more commonly seen when reporting forage. These are the most experienced forager-scouts, a few hundred bees who are checking nearby chimneys, roof cavities, tree hollows, bait hives, compost bins etc to find potential new homes[136]. The details are fascinating and covered in Prof Tom Seeley's excellent book, *Honeybee Democracy*, but the practical application you need to know when collecting swarms is this:

If there is just a bit of movement on the surface, you can take your time collecting them calmly.

---

135   Quimby (p.104) states the number of queen cells a colony creates for swarming (2 to 20) depends on stores. Implying you can expect a heavy hive to produce loads of casts.

136   Historically, people compared bees' famously orderly and selfless society to monarchy, communism, etc to prove their human rulers governed in the best possible manner. If only our MPs made decisions with competitive dance-offs.

If there is a lot of movement, they have just about made up their minds and **you have just minutes,** or less before the cluster suddenly dissolves and swarms off to its new home. So knock it in a box immediately.

Sometimes they cannot decide where to go, feel more or less safe in a dense tree and end up building free hanging comb. Apparently this can work in dense fir trees, but in deciduous trees the leaves drop in winter, the comb is exposed to the elements and the colony dies.

Meanwhile at the far end of the dances - cavities they are checking out - you will see bees going in and out and hovering outside. If you ever see a dozen bees doing this, they are seriously interested. If you see 100, the scouts are decided and just memorising its location before going and getting everyone. Sometimes, scouts from rival swarms fight at the entrances to bait hives.

| Swarm settling into tree

Preparation 1 - advertising for and hearing of swarms

Some techniques I use:

- As a member of the BBKA I signed up to their swarm collector service.
- Our beekeeping group has a swarm page with contact numbers. Careful keyword use like "swarm" and "Oxfordshire" ensure it is a top hit when people search.
- Our group alert each other when we have spare swarms
- I am listed in our village magazine along with emergency services
- My postman covers a large area and has my card

- Our own hives give us swarms each year
- If I spot a wild nest I introduce myself to the property owner
- Local pest controllers and tree surgeons know I am looking for swarms
- A couple of times I've been rung by people who say they heard of me through other routes - word gets around!

About the only route I do not use is, I'm not networked to my local BKA as I don't use social media. But I'm not sure how well their bees, managed for honey, would thrive in my no-treatment regime. The swarms I prize most highly are ones from a colony that has looked after itself for years!

I typically catch 10 swarms a year and distribute most among other treatment-free beekeepers. I am often gifted a swarm by someone before our own hives swarm, which helps avoid inbreeding.

Preparation 2 - equipment

First, ensure you have a hive ready to accept a swarm! Every year I offer a swarm to someone who says "oh I haven't got the hive assembled yet". And check it does not already have a nest of wasps, hornets, mice or bluetits in it - I am not making this up. Make sure all the bits are assembled in its final location, and it is scented to smell attractive to bees - see *Bait Hives* below.

Basic swarm catching kit needed on permanent standby in car from April to June:

A   Ventilated **swarm catching box** (actually two is better, you sometimes find there's more than one cluster). See diagram on next page. A woven laundry basket or skep will do, you seal those by wrapping them in a sheet.

B   old brood comb

C   Duct tape

D   knife or scissors[137]

E   large feather[138] or bee brush

F, G   secateurs / pruning saw

H   Piece of cardboard or flexible cutting mat to sweep bees onto

I   smoker (don't forget lighter **J**) or...

K   scent spray[139]

L   water bottle

M, N   Bee suit, veil, gloves etc

O   white sheet

The only really crucial equipment here is a box and something to seal it with.

---

137   Swarm collection boils down to driving round the countryside with duct tape and knives, abducting girls.

138   Goose feathers can be picked up near some rivers while they raise goslings in summer.

139   E.g. very dilute peppermint oil (encourages stubborn bees off branches). Safer than a smoker. Lacking this at the crucial moment, Ann Welch dissolved an Extra Strong Mint in water for this purpose; old bee books talk of wiping nettles on branches to move bees off them. Most items can usually be improvised or borrowed at the scene.

Other useful kit you eventually accumulate:

Collapsible ladder - but you would be amazed how often you can borrow a ladder from the householder, or even a neighbour. Just explain and ask.

Spray bottle filled with fresh water to revive dehydrated bees

Swarm collection boxes

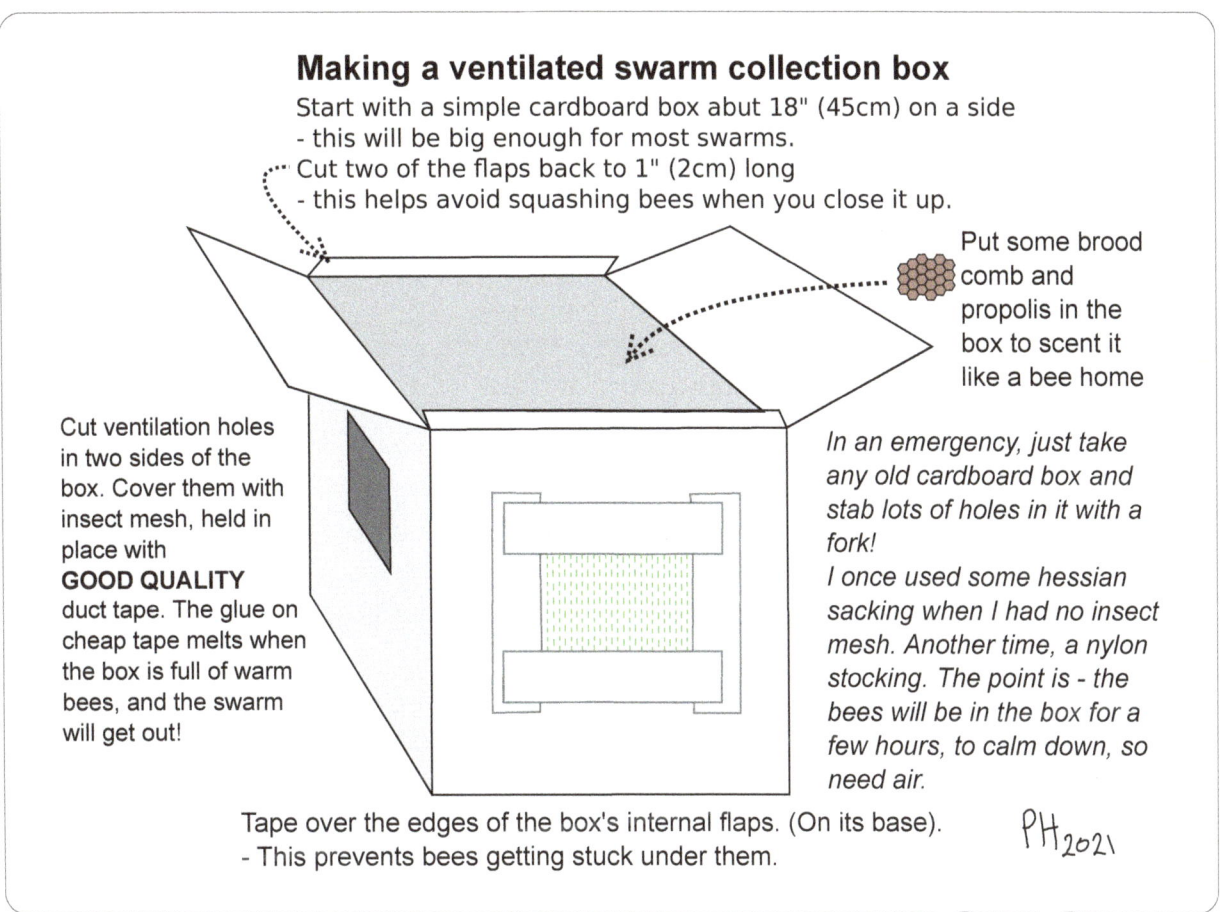

If you don't have any comb or propolis to scent it, rub it with resinous branches (fir trees, etc). Propolis is basically plant resins. An alternative, and very effective lure is the crumbled wax left after wax moth larvae have chewed up comb. Their digestive systems sterilise it.

The reason for 2 grills on right angled walls is because once you begin driving, the swarms get agitated and generate heat, and they're likely in partial sun on the seat of your car. If you're careless you might place one grill against the back of the car seat if they were

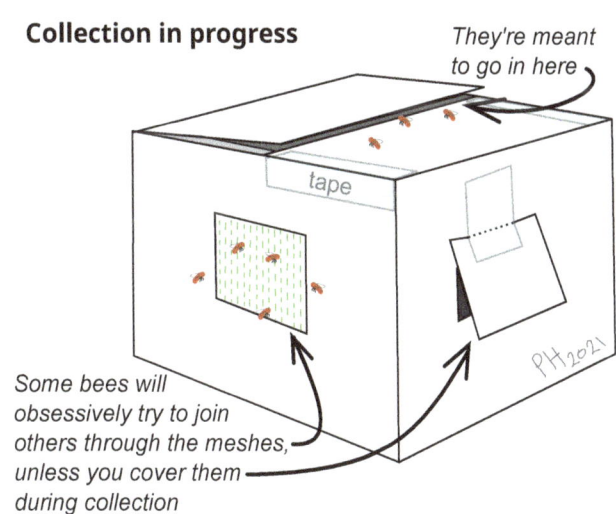

on opposite sides. Large swarms need very good ventilation and can set up a through draft if needed by fanning.

Keep a couple of pieces of cardboard in the box. These will act as **temporary** flaps over the ventilation holes during the actual collection. This ensures -

- The inside of the box is dark. The queen likes dark.
- Avoids the problem of some bees trying to join the others *through the grill.* They can smell their queen through it.

Checklist for a swarm cell

Is it really a swarm? Is it feasible to get it?

Remember the **caller is probably stressed:** stay calm and polite even if they demand you come.

When a member of the public rings, sometimes they have a problem and just want someone to sort it out; so asking them "is it honeybees?" will always evoke the reply "yes!" So never ask a question which can be answered with a yes / no. Always ask them to describe what they can see. Confirm species **colour, shape, furriness, behaviour, how many, size of cluster.** *(You can sometimes advise them usefully if it is not honeybees, but don't commit to doing anything yourself. Even the Bumblebee Trust doesn't move bumblebees. Wasp and bumblebee nests only persist a year and they are great for gardens.)*

Don't take swarms from Foulbrood areas

The Bee Inspectorate publishes data on where foulbrood is

In Britain it's at nationalbeeunit.. com/diseases- and-pests

Confirm it really is a swarm and not an extant nest. See "cut-outs" in chapter 13. Decline these requests.

Get a **clear description of height above ground** eg not "6 feet" (it will be 12 feet) but "head height"

Ask **how long it has been there** and it is not "going in and out of a hole" (ie it has moved in and is no longer capturable)

Confirm it is still there. I've had calls from people unaware they move on.

Has another beekeeper been called. Some callers leave messages on answerphones resulting in 2 showing up at once.

Consider accessibility. Sometimes people say "it is in my back garden but I am going to work" - get their mobile phone number and a clear description of where it is in their garden and how to get in, and whether you need to take a ladder.

If not a swarm, explain why, and potential hazards – probably none – and whether they should leave them be.

If it's a swarm get their **name, address, phone number**

Is it settling in a chimney? This is bad news – advise them to light a smoky fire to drive the bees out immediately. A nest full of wax and honey up a chimney could be a fire hazard.

Don't go for swarms high up – it's not worth the risk. There will always be another swarm.

How to actually catch swarms

The following points are so important they bear repeating:

If a swarm is in a dangerous position to collect, like very high up or hanging above a pond, **leave it.** There will always be another swarm.

The process of boxing a swarm and moving it, particularly by car, physically shakes the swarm up and agitates it. Once you have the swarm boxed and at your apiary, **let it settle for** *at least* **an hour,** ideally in a dark cool place like a garage, then hive it an hour before sunset. If you try to hive a swarm immediately, it will probably abscond. If possible, leave it in a cool dark garage for up to 2 days to lose the swarming impulse (only do this if it is heavy with stores).

90+% of swarms are very calm. If someone tells you they are vicious, you are talking to a "dominator" who handles bees roughly.

Because they may move on very quickly, **it is important to go get a swarm as soon as you hear of it.**

A water cooler bottle on a repurposed telescopic pole for a window cleaner. Collapsed, it is short enough to fit in a car boot. A pool net works, too.

Bees instinctively crawl upwards, and into dark spaces. You can often use this impulse by placing a box above them. But beware - bees on ground will crawl up your legs so at least **tuck trousers into socks,** ideally tuck them inside wellington boots. Otherwise they crawl inside your trousers and conduct some impromptu apitherapy.

On arrival:

Examine the swarm for a minute or two. If it is very active (lots of figure-8 dancing on the surface) and you hear lots of high buzz tones, it is getting ready to fly off, you don't have much time, though you may delay it a few minutes if you spray it with cold water. You may see other things of interest - any drones? Pollen? Ages of bees? You may see different dances on different sides. Every few minutes, a queen will pop up to the surface. Are her wings frayed (old prime)? A blob of paint on her back indicates the swarm is from a conventional hive. Are there multiple queens? Signs of disease (e.g. lots of very dark shiny bees) or varroa hitchikers? How uniform is the bee colour?

Look below the swarm. A lot of dead bees on the ground imply it is exhausted, may have been there days, may need feeding on hiving.

A lack of coherence - several clusters - indicates multiple queens and probably some trouble gathering everyone into one box.

Swarming is highly stimulating for he bees and excitement can tip over into aggression. Some fighting among the bees is not unusual! It's more likely in multi-queen swarms, or if the bees were pre-stressed by an interfering beekeeper before swarming, or because the swarm is formed by swarms from several hives merging (this happens in larger apiaries when swarms launch simultaneously and intermingle). Fighting amongst themselves *should* die down with minimal casualties.

Look round to see if you can find the mother colony. It's usually within line of sight. Inform property owner you are interested in further swarms.

Smoke doesn't calm swarms. Even light smoking agitates them. ("We can't smell our queen!").

Ideal case: "low hanging fruit"

Sometimes it is as simple as this:

Place the collection box as near the original position as feasible, so bees on the branch can smell where it is.

A collection like this might just take a few minutes. Or an hour, depending on how rapidly the stragglers enter the box. The more bees you catch, the better the swarm's survival odds; but at some point there will be as many scouts leaving as stragglers entering, and it is time to seal it.

Often some remain on the branch because if the swarm, and queen have clustered there it smells of them. And there will be a couple of hundred bees out scouting for suitable cavities, which will return to the branch. If they won't move, you can -

- mask that smell with e.g. peppermint spray (this will annoy them, wear protection)
- cut the branch and lower it into the box

If the swarm is near the ground, place the box on the white sheet. Bees' vision is poor, this helps them spot the dark box. If the swarm is high up, sometimes it's better the box is wedged in branches within a metre of the swarm.

When a swarm is fanning strongly to signal stragglers, you can sometimes smell the lemony Nasonov pheromone.

Large prime swarm on a tree.
Image © 2024 Gareth John.

Case 1: ideal situation: swarm on easily accessible branch

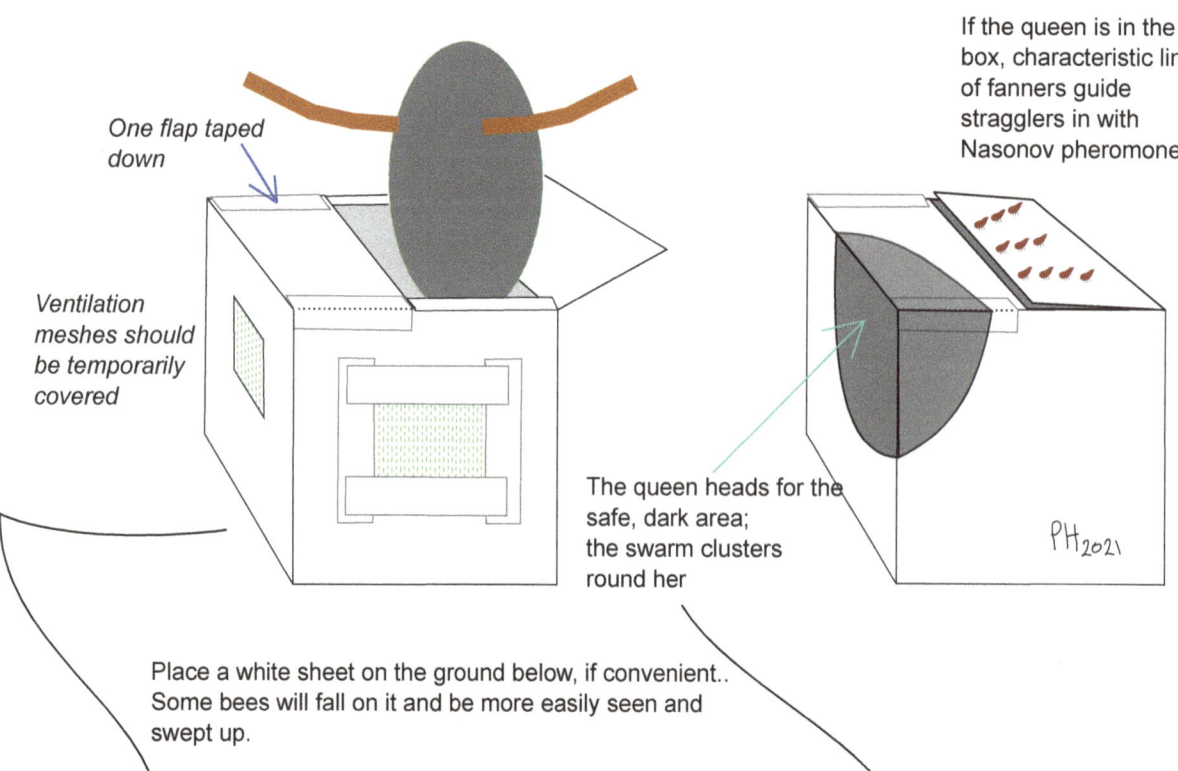

Case 2: swarms on walls

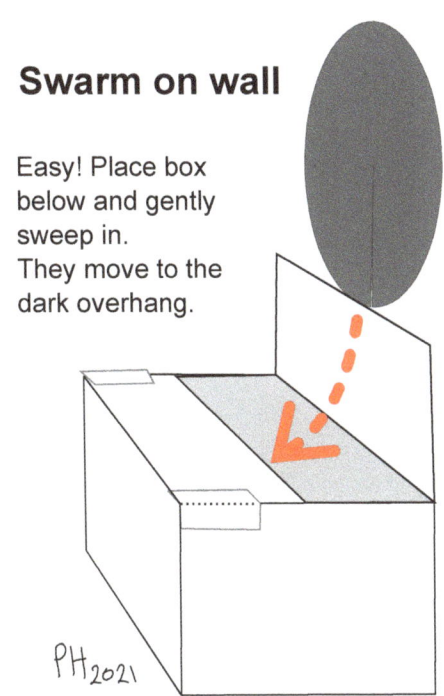

Case 3: swarms on fence posts

Here's how I handled one, placing the catch box above the swarm, whose natural tendency is to move up and into shade. About a third moved up in 5 minutes, then they stalled, and I had to use a goose feather and scoop handfuls of bees in. Luckily this was a cohesive swarm. It took about an hour.

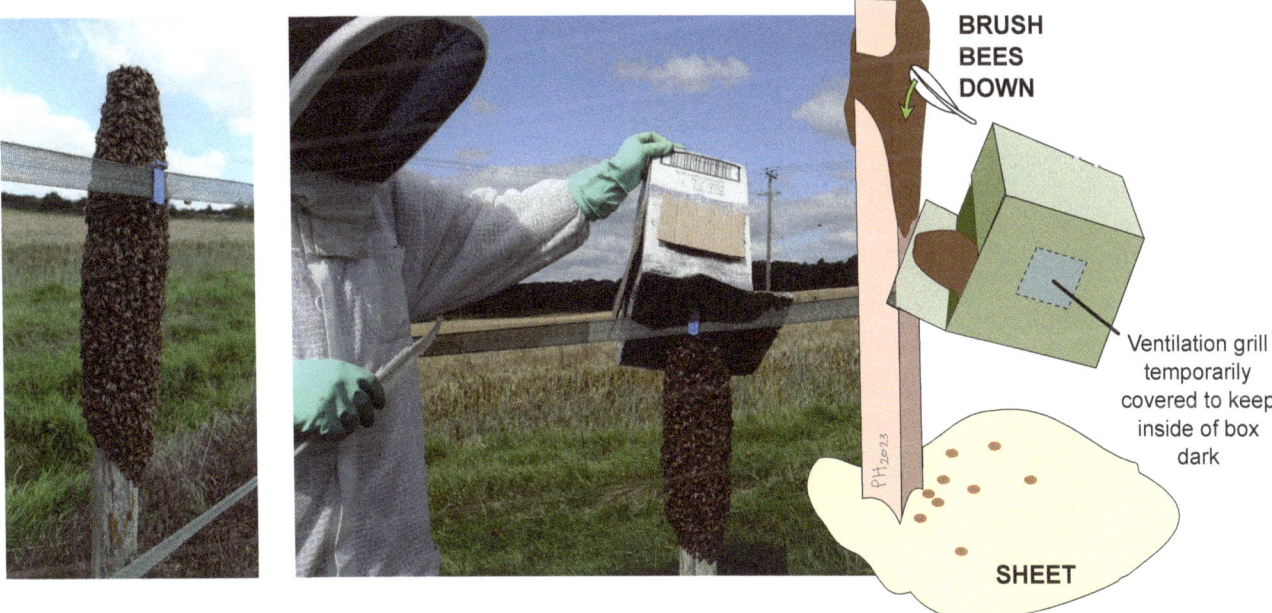

Note the temporary cardboard shield over the ventilation grill, so the box looks dark.

I've since found brushing them down into a dark box is much easier and faster.

Case 4: swarm in grass

Some queens are poor flyers and the swarm clusters round her on the ground. You have two main collection options, see diagrams. I recommend option 1.

Cold, wet and apparently dead swarms on the ground may revive if warmed[140].

They will walk up onto a piece of brood comb.

Option 1 in action. Guessing the queen was in the densest clump of bees I transferred it into the box, and they started walking in. These days I would know to use brood comb and look for more queens!

Swarm in grass - option 1

This is tricky because the bees can't see through the grass and it slows them down. You can't brush them through grass.

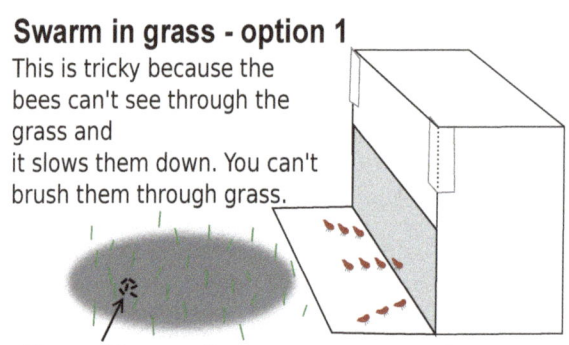

There will be a small pile of bees somewhere in the mass - a cluster over the queen.

Pick the queen & cluster up and place them at the back of the box. She will run up into the dark, the bees will begin fanning "she's in here!" and the puddle of bees will gradually flow in.

Swarm in grass - option 2

Put a box or skep above the bees.
Tip one side up so it is light at the bottom, dark above.
(Otherwise the bees stay on the ground!)
Return in the evening, the bees *may* have moved up and clustered to stay warm.

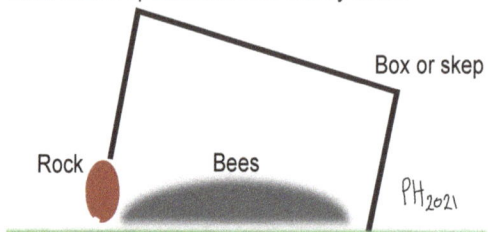

Case 5: swarm in dense bush

I have no clever suggestions for gathering these other than, grab handfuls and put them in your box. If you get the queen the others will eventually follow, but you will probably need to overwrite their scent on the branches with something else, like crushed leaves. Sometimes the property owner will let you cut branches.

This picture features a bare hand actually inside the swarm. This trick is useful in reassuring onlookers that swarms are harmless.

Swarms in dense branches can't just be swept off, and present problems

---

140 I know of one case where a hairdryer worked.

If you try this -

- Don't wear rings
- Ensure hand is clean, not smelly (maybe smoke it)
- Insert hand very slowly and gently
- Don't do this if they seem agitated.

A piece of old brood comb is a good lure for awkward swarm remnants deep in a bush.

Case 6: not actually a swarm

A bee or not a bee, that is the question.

You will sometimes be asked to deal with:

Bumblebee nests:

It is illegal to move bumblebees in the UK except the usual culprit, the Tree Bumblebee which is not native and can be identified by googling images of *bombus hypnorum*. It is the only UK bumblebee that can get aggressive - it does not like vibration near its nest.

Bumblebee nests are too fragile to move unless already inside a container like a bird box – even specialist bumblebee experts do not move them.

If you move a nest during the day, foragers will return and be lost.

Wasp / hornet nests:

No one relocates these.

If you suspect Asian hornets (aggressive invasive species), they need immediate dealing with - see chapter 14.2 - but normal hornets are **beneficial** to an area, controlling wasp numbers.

Hornets can sting through normal bee suits.

Solutions:

Advise the property owner to just leave them to die out in the Autumn and point out they pollinate, and in the course of a year a wasp nest will eat tens of thousands of garden pests like caterpillars and aphids.

In my experience wasps are remarkably chilled out, despite their reputation. Once they begin to starve at the end of Autumn they *can* become desperate but before resorting to poison, consider that other animals like mice will then eat the bodies and spread that poison up the food chain.

If coming in windows, suggest fitting stick-on insect mesh, easily available online.

Established bee nests in trees / walls / roofs / chimneys:

See "cut-outs" in chapter 13. Relocating a nest can take many hours and the bees have about a 50% chance of survival. There could be significant property damage. Working at height is dangerous. Walk away.

Advise the property owner that it is illegal to simply poison them because of the risk of contaminating the local ecology's food chain. You *can* poison them if you can somehow block all access to the nest, but that is tricky which is why pest controllers try to pass the problem on to beekeepers.

Bees don't chew through wood like wasps. They generally keep to themselves. Why not leave them there? If they've only been noticed because they produced a swarm - they have probably been there 1+ years already without causing problems!

Free living bees are tough local survivors. **You want this stock.** Give the property owner your contact details and explain you want any swarms emerging.

Left-behinds: Some bees are always off scouting when you catch a swarm, then return to where the cluster was after the swarm moves on or you box it. Although some people claim these bees will eventually return to the mother colony, this is obviously untrue. I have monitored some left-behind clusters, and had calls from people after I picked swarms up on their property. They stay at the last known spot for several days, in a sad mini-cluster, gradually dropping to the ground below and expiring. Once they enter swarm mode, it is like deleting a computer program – ther map memory of where they came from is reset.

When I find such clusters, I try to sweep them into a small box with a feather. With 100 bees there's no need for a fancy ventilated box. I take them to a queen-right hive and brush them onto the landing board that evening, and within seconds they turn and walk in. There is no fighting.

Is it a Prime or a Cast?

Primes tend to be **large**, **cohesive** swarms which are **easy to catch and hive** (a mated queen smells strongly). Some casts are cohesive too though, so this is not a definite sign.

If the branch where a swarm was resting is sticky, or if the bees themselves are sticky, it is probably a prime swarm. They are so stuffed they are oozing honey. Primes have access to lots of honey stores before flying off.

Swarm remnant on a fence after main cluster has moved on. These bees will stay and die here if not collected.

Examine any comb they built in the collection box for eggs. A fertile queen - indicating a prime swarm - will lay by morning.

Older prime queens may exhibit some fraying at the edges of her wings.

Older Queens walk in with the swarm. Virgins are more nervous, flap about, walk over others, huddle in weird places.

Primes are more likely to sting because they have older bees[141].

Primes are about 70% young furry bees. Casts are all young bees.

Less reliably:

You may be able to tell by how they build comb. The Times Bee Master claimed[142] that "A prime swarm begins building comb in the middle top of a skep, whereas a cast begins comb next to a wall". Brother Adam noted comb in horizontal cylindrical Jordanian hives was built lengthways along the cylinder (cold way) by large primes, and in circular sheets across the end entrance by smaller casts. I don't know how this plays out in other hives.

---

141   Rudolph Steiner in a 1923 lecture
142   *Bee Keeping,* written in 1864, with British Black Bees, in skeps. No idea if it is still applicable. Probably a function of swarm size.

Although young queens are stronger fliers[143], a swarm's distance from the parent hive isn't a good indicator of parentage. Weak queens land frequently en route to distant cavities.

Gareth John has observed there seems to be a tendency (not a rule) for prime swarms to cluster low, casts high.

If the queen's wings are clipped or she is marked with a coloured dot, she's a queen from a commercial hive and presumably pre mated (prime). These tend to be very large swarms and you may want to kill her and use the workers to reinforce another colony, perhaps after treating them with miticide[144].

If it is very late (August onwards), it may be -

- an absconding colony;
- a starvation swarm;
- a swarm from a swarm (ie from a colony established very early which grew unusually rapidly - this swarm will have the original prime queen from two hives ago!);
- a really huge one may be a swarm from a commercial colony where the beekeeper tried to suppress swarms, but eventually they escaped;
- a swarm from a commercial colony where massive syrup feeding during a warm spell of weather has tricked the bees into thinking it's spring / swarm season.

Swarm left in wrapped box in car overnight escaped (top right of photo). Didn't want to leave car. Chaos ensued.

This swarm came down a chimney! It clustered next to the window (light source). The householders, non beekeepers, were taken aback. We lit a smoky fire in the fireplace to drive remnants out of the chimney.

Loose scatter of bees which did not enter box until night fell. Probably completely queenless. This swarm died out after a few weeks.

What can go wrong

Eventually we all come across something unexpected and need to ring our mentor for emergency advice. Here's some situations I've come across.

---

143   Virgins are not yet at full weight and have to make strenuous mating flights.

144   Assume zero hygienic behaviour and consequent high mite load. And I refuse to hide behind euphemisms like "requeen": if she is commercial stock her drones will dilute desirable tough survivor bloodlines, so kill her and check for princesses hidden in the swarm too, or you doom entire colonies to a lingering death. If this does not appeal, give the swarm to someone who wants it.

I once left a large swarm in a box in my garage overnight. The heat of their bodies melted the glue on the cheap duct tape holding the box together. In the morning I found them all over the garage, in cobwebs etc. The bulk of them escaped and absconded. Lesson: use good quality duct tape.

The swarm down the chimney was trapped in a sunny room for hours before being discovered. By the time I arrived, perhaps 1,000 dead(?) bees were scattered across the carpet. In retrospect, I suspect these could have been largely revived and recovered with a simple water spray - they were probably dehydrated. That swarm failed about 2 months after hiving for lack of bees. Live, learn, move on.

Liz had a swarm cluster **under** a hive – she'd overscented the hive cavity with lemongrass oil – I had a similar experience. Solution: swap the boxes for unscented ones.

When to walk away

Some swarms are too risky to collect.

There will always be another swarm.

Aggressive swarms:

Most swarms don't sting, though some can get excited - these can often be calmed by distracting them with a few wafts of lavender smoke (**not** heavy smoking).

If they begin seriously attacking you, leave them alone. There's some kind of stressor - disease, someone's crossed races and produced an aggressive hybrid, etc. You don't want these in your apiary.

For a swarm high up a tree:

Usually more trouble than it's worth, but -

Throw a weighted line over a nearby branch.

Lower the weight, replace with brood comb, raise the comb to near the bees, which sometimes move onto it.

Ropes can *sometimes* pull very long, springy branches within reach. Usually it snaps loose releasing an agitated cloud of bees who resettle even higher up.

Hiving swarms

Preparation

The hive should smell like a home: **wax and propolis** are key here. If you lack these, rub a resinous branch inside - propolis is basically plant resins.

Swarm by road. Extremely dangerous collection. This swarm was hived but did not survive, having had many bees squashed by passing traffic, and displaying symptoms indicative of poisoning from the road surface. Photo © 2021 Marc Sheikh

A very unwise collection in 2016. The cluster kept reforming higher up the tree with each attempt, well away from the central trunk. The property owners left, leaving me alone should I fall. I got the swarm eventually with the aid of an extensible pool net - but it was so agitated it absconded the next day.

You **absolutely must** give the bees time to calm down after catching a swarm. If they are still agitated, they **will** abscond the next day, or just refuse to stay in the hive. The swarm collection box needs to rest at least an hour in the shade, and if you took it home in a hot, vibrating car, 2 hours. If necessary leave them overnight in your garage.

If you can leave them for 24 hours, great- they are very unlikely to abscond then.

If hiving within 24 hours:

Aim to hive **one hour before sunset.** This is enough time for the procedure, and when the temperature drops they have a big incentive to stay hived.

Even more reliable - hiving after 48 hours in a cool dark garage:

If they have just been caught, they have plenty of food so can be left 2 days, long enough to forget the swarming impulse. In this case you can walk them into a hive even in the morning and they will just walk in, happy to have a home and very unlikely to abscond.

Remember to fit all bars along the top. Otherwise the bees build comb straight onto the roof. Nightmare to fix.

Remove most or all the bars from lower boxes. A large, uninterrupted cavity is more attractive to large swarms - they may abscond if they feell constrained. Also - you want them to start building comb at the top! After they've committed (1 week) replace the bars before the first combs get too long.

Smells like a home

Old brood comb

Propolis

Disintegrated wax (wax moth remnants)

BAIT HIVES ONLY - no more than 2 drops lemongrass oil OUTSIDE entrance (attracts scouts). Too much makes hives smell overpowering, unattractive.

PH 2021

Preparing a hive or bait hive for occupation. Traditionally, the old brood comb needs to be the "right way up" or bees sense something is wrong and avoid! The old comb is just a lure and should be removed once a swarm is acquired.

The hive **must not have am open mesh floor.** This is a big incentive to abscond, perhaps because the scent in the hive is not intense enough. (Packages often abscond from hives with OMFs too.)

They also don't seem keen on **shiny metal roofs**[145].

There are two techniques for transferring bees from the box to the hive.

Quick method – tip them in:

You can simply turn the catching box upside down over the top of an open hive, give the box a sharp knock, and the bees fall down into the hive – then close the hive, leaving the open box with any remaining bees exposed to light and next to the hive entrance.

Providing you get the queen into the hive this way, which should happen as she will be in the centre of the bees, this usually works fine if done in the early evening.

---

145   Example: Eric Asher made a beautiful copper roof for a Golden Hive, put a swarm in, they absconded. Changed to a wooden roof, next swarm stayed! Possibly some electrical effect, but more likely: due to the roof being polished. A couple of years later, he tells me the copper has a patina now and bees are very happy with this non-reflective roof. Bees seem to get confused by the reflection of the sky, and tend to crash upside down on them. This *dorsal light response* is common to many insects, which assume "up" is the brightest part of the sky. Hartmut Münch has observed that bees emerging from a hive when snow lies on the ground will turn upside down and accelerate downwards until they hit it.

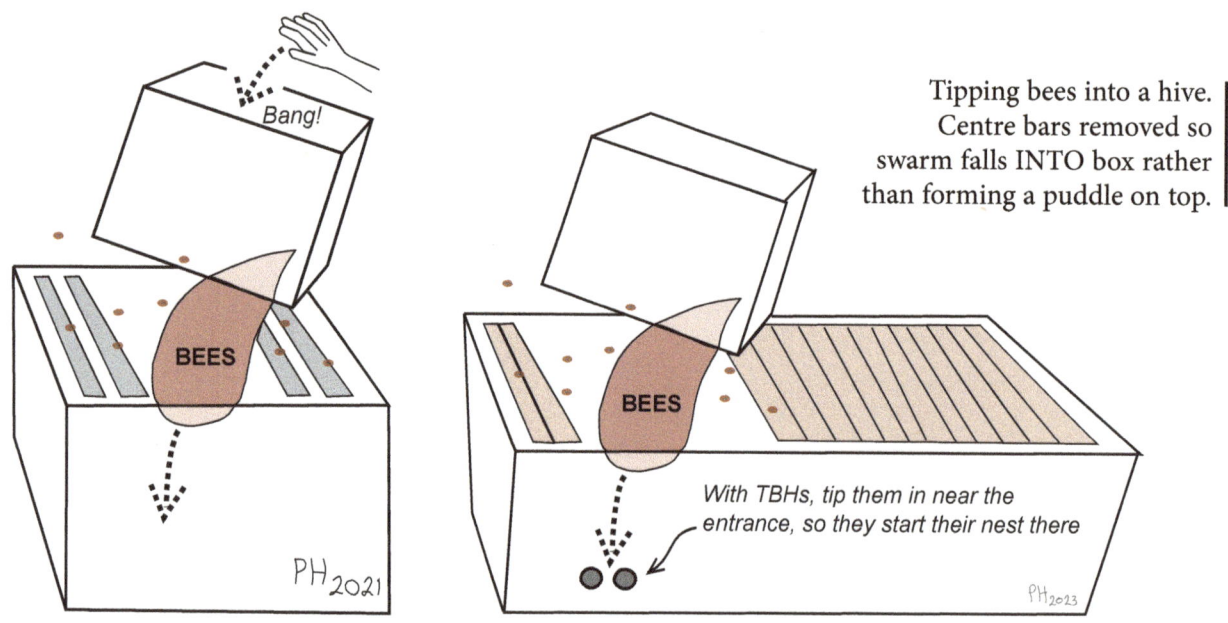

Tipping bees into a hive. Centre bars removed so swarm falls INTO box rather than forming a puddle on top.

With TBHs, tip them in near the entrance, so they start their nest there

The workers then appear at the hive entrance and fan "come hither" pheromone into the air – you can see them stick their bums in the air and fan their wings in rows at the entrance. This calls any stragglers into the hive.

What can go wrong:

The bees have several conflicting drives when suddenly exposed:

If unsure, or cold, go UP

Many workers go towards light to deal with threats

Queen heads for darkness (safety)

Many workers cluster round queen to protect her

So the trick is to get everything prepared and then act FAST. Open the box (slice tape) FAST, dump them all in, and while they're confused, replace those top bars, top cloth (if any) and hive roof FAST. If you take too long - say 1 minute - a lot of bees will be in the air trying to defend the colony, or return to the swarm collection box (last known site of, and now smells strongly of, the queen).

You will not get every last bee out of the box. Give it 2-3 sharp raps to dislodge most of them, then remove the box and close it while you close up the hive. Once bees begin fanning at the entrance of their new home, open it again and bash / brush the last few out next to the entrance.

Slower more fun method – walking them in:

Not recommended for Top Bar Hives, their entrances are too small and high up. Problems with traffic jamming and ramp construction. End up with bees outside overnight, trauma all round.

If the weather is cold or wet don't do this, it takes too long.

Alternatively, you can walk the bees into the hive up a ramp – emptying the bees from the catching box onto a white sheet which is laid partly on the ground and partly on a ramp leading up to the hive entrance – the bees are inclined to move upwards towards the dark entrance.

Pros:

Walking them in is quite a spectacle to watch when it works well. It resembles a puddle of treacle flowing uphill.

Some people say this is better because the bees actively choose to go in, and there is no chance of the queen falling outside of the hive.

Thousands of bee feet walking through the entrance leave a strong scent of "our home".

Informative. Can you distinguish young and old bees? Is there a big spread of colouration among them? Are any varroa visible clinging to bees' backs? Are any exuding wax scales? You will see a handful of drones and may see multiple queens[146]. The bees will sort out who's boss – they may **want** multiple princesses as backups lest the first doesn't mate successfully. A queen with a coloured mark on her back indicates the swarm is from a conventional hive.

Cons:

Can take at least 30 minutes.

I have not observed any difference in how many swarms abscond whether you tip or walk them in - that's invariably attributable to other factors like the hive being too small for a huge swarm, the hive smelled wrong, swarm not given time to settle before hiving.

Potentially more problems, like lots of bees ending up under the hive.

Be careful with the sheet – the sheet may fall off the hive ramp, it has to be secured properly.

| Walking bees in. Remove box ASAP or bees will return to it.

Very experienced beekeepers are quite certain the bees are not damaged by tipping in.

Traffic jam at entrance: [See diagram below] if the bees crowd here they can block the smell of the queen inside, and the march in stops in confusion. A gentle sweep with a goose feather may help, or you can temporarily pull the second box forward to create a much larger entrance gap.

Bees clustering, refusing to move as night falls: if air temperature drops rapidly you may need to sweep bees into a small container and physically tip them in the hive. But opening the hive roof in low temperatures aggravates bees so more rise up to defend it. So check weather forecast to ensure a mild evening before walking in!

Bees will move towards the darkest spot. If your shadow falls on the sheet, sometimes a few head for it.

Bees are reluctant to walk into a white box, but happy to walk into something tree coloured.

---

146   Filipe Salbany saw 9 queens in one UK swarm in 2021. Roger Patterson mentions one with >19 queens in *Beekeeping: Challenge what you are told*, p.193

Once hived: don't feed: force new comb building

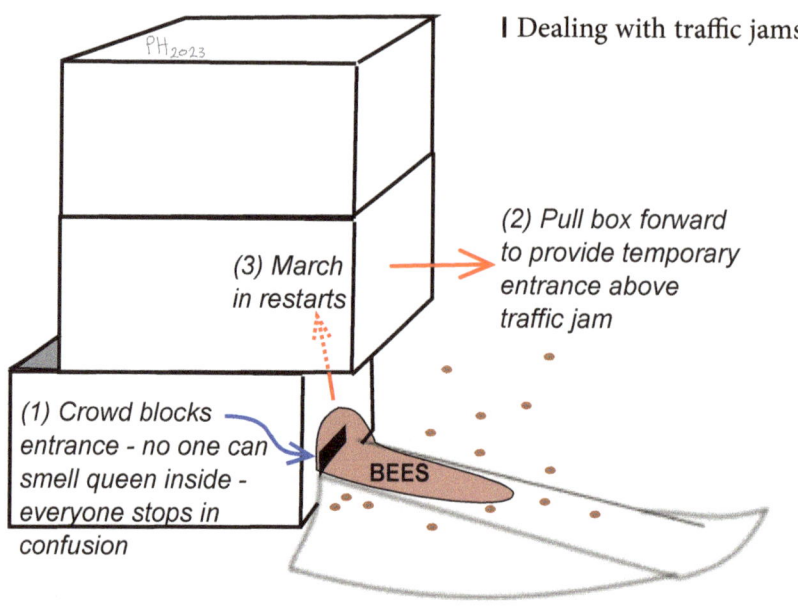

Before swarming, they gorge themselves on honey and carry three days' supplies with them. A swarm bee weighs about twice an empty bee!

There is a possibility that they came from a colony with a disease which is carried in this honey. *You want them to digest all the honey they came with. This will destroy spores etc.* If you feed them in their first three days, they may end up mixing their imported honey with your syrup and storing it for a rainy day.

Advice from the National Bee Unit is **don't give them drawn comb**; I would add, it is usually a good idea to **remove the bait comb** to minimise the chance of disease transfer[147]. They'll probably avoid the old comb anyway. Force them to use up their stores making new wax.

Swarms basically *should not be fed* in Britain unless

They have been hanging in a cluster for more than 3 days, or

There is a weird forage-blocking weather event like a week of continuous rain or cold immediately after you've hived them.

Some swarms occur just before the June Gap – many hard core low-intervention beekeepers do not feed these as they are poorly adapted to the local forage rhythms and best not propagated. Such poorly timed swarms are probably best used to reinforce other colonies by merging them. This is equivalent to the conventional technique of breeding from your best queens.

David Heaf points out in countries with endemic AFB it might be prudent to not just *not feed them* for 3 days, but then *destroy the comb they made* in that period and *then* feed them.

Exceptions to the do-not-feed rule

In **changeable weather,** it is quite possible that, while not actually starving, the parent colony may be so low on food that the swarm does not get properly 'topped up'. Such swarms will need immediate feeding.

Hunger swarms absconding en masse

Swarms that have been hanging on a bush etc for 3+ days

---

147   I have broken this rule in exceptional circumstances, like hiving colonies which absconded too late in the year to build their own comb. These are guidelines, not absolutes.

Behaviour once hived

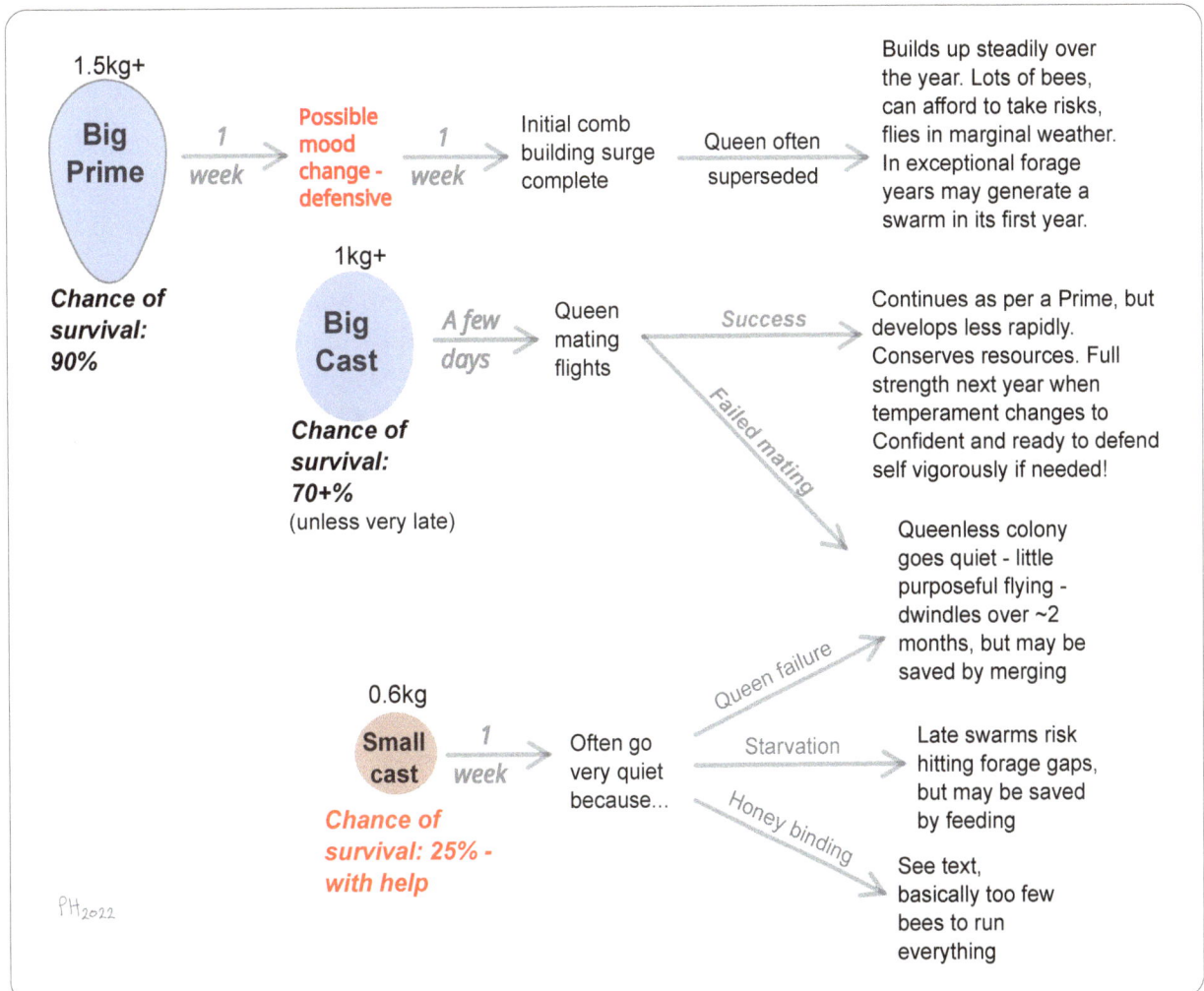

> Typical swarm development in my region - a race to build up a buffer of resources for dearths and their first winter. Small casts are best merged or will likely perish. Our local bees form small winter clusters needing few stores. Any colony surviving its first winter will probably live several years thereafter.

Princesses will probably be mated and laying after 1 week, and beware - at this point a previously mild colony's temper can change, once they have brood to defend. If bad weather prevents mating beyond ~4 weeks she has missed her fertile window and the colony will dwindle and die.

Swarms often shrink or appear to go quiet for a few weeks after hiving as old bees die off and remaining personnel cluster on brood. Then the new brood hatch and the colony booms.

First combs: Casts make less comb. First combs are general purpose. Gilliane Sills noticed that swarms hang in a corner, supported by walls, until there are combs to take its weight - working out from the corner.

Swarms moving into deadouts spend considerable effort removing dead bees, rotten comb and debris. You sometimes see scouts preparing a chosen nest site before the swarm moves in: Gilliane Horrocks saw one swarm hang on a branch nearby for a week preparing the cavity; in this case she had reason to believe the previous colony died by poisoning. If comb is intact, it is often left, *but not used,* probably as it smells of another colony.

Honey binding

This term refers to how a colony can sometimes run out of comb. In particular, if there is a big nectar flow the bees may fill up every cell available with nectar, honey or pollen (*pollen binding*) - leaving nowhere for the queen to lay. (New cells are built on the cool edges of combs, unsuitable for brood.)

Although this is normally discussed as a problem for mature colonies backfilling a brood nest with honey, promoting swarming, it is also a common problem for **small,** and sometimes medium sized swarms who are "lucky" enough to hit a strong nectar flow and over-commit it all to nectar processing before they have much comb built. Nectar comes in faster than they can build comb.

The colony sort of stalls, as everyone frantically converts nectar into honey while the flow is on. Food collection is every species' priority if supply is intermittent! By the time things slack off, there is no new generation coming on-stream to replace losses from predation, old age etc and the colony may have slipped below the critical mass where there are enough bees to nurse, build comb, defend, process nectar... The colony struggles for a few months then usually succumbs to a random stressor (wasps, starvation etc).

The phenomenon can lead to a significant and sometimes fatal delay in build up, especially if the weather turns poor during the rest of the summer.

The different age profiles of casts vs prime swarms may play a part.

Note that you can inadvertently cause this problem by overfeeding a colony!

Absconding

In addition to reproductive swarming, colonies sometimes abandon hives altogether, usually due to stress. This is termed *absconding*. It may happen immediately after hiving a swarm, but it also happens with established colonies. When entire colonies abscond, they act essentially like swarms but can appear at odd times of year and tend to be quite large.

Absconding after hiving: this seems to be something mainly large / prime swarms do - though it's been noted it is more likely if your floor is open mesh instead of solid.

If you put a swarm in a hive and it acts very agitated, with an enormous number of flyers for several hours, it's probably not going to stay.

One instance I'm aware of is when a nearby fence was being creosoted.

I've seen it when I tried hiving a swarm too soon after driving it home, it was still agitated from the bumpy drive.

Another time I concluded they thought the hive too small.

Sometimes prime swarms appear to have pre-scouted where they are going and if you hive them locally, they decamp to their preselected home.

Entire colonies absconding: They may even leave brood behind. I suspect this is much more common than people admit. Causes include:

- Frequent clumsy inspections
- Excessive use of miticides[148]

---

148 Bees will abscond from hives where comb is contaminated by excessive miticide use. This "mysterious disappearance of the bees" coupled with the usual pests not touching contaminated stores, would tend to be diagnosed as Colony Collapse Disorder – as if the beekeeper was not at fault.

- When a colony runs out of food (sometimes due to harvesting too much honey). Such abscondings are known as *hunger swarms* or *starvation swarms*. Ones at the end of winter are very small, move restlessly from one location to another and may enter unusual places.
- Overwhelming wasp attacks
- Mouse in hive
- Overheating (poor hive placement) is a major trigger.
- Leaky hive roof (mould inside hive)
- Leaving a queenless hive for a queen-right one.
- During California's 2019 wildfires, people reported colonies desperately abandoning hives despite having nowhere better to go - after days of intolerable smoke, and light so weak that cars needed headlights to drive.

African and Asian honeybees are much more prone to absconding than European ones.

How much does a swarm cost?

People often just gift them to others, prioritising recipients with no bees at all.

However there are occasional freeloaders who take advantage of this generosity ("I'll have 5 swarms and work them hard for honey thanks") and a small charge prevents this becoming excessive.

I sometimes charge a nominal amount to cover petrol and time used.

I have heard of scammers charging half the cost of a nucleus for swarms. This is a red flag. Avoid.

If you are *collecting* a swarm, beware:

If a property owner offers to pay you for removing a swarm - decline. It invalidates your insurance. Suggest they donate to a charity, like your local BKA.

If a property owner asks *you* to pay *them* - walk away.

Bait hives (catch boxes)

By setting up small hives around an apiary, you can get the bees to collect themselves.

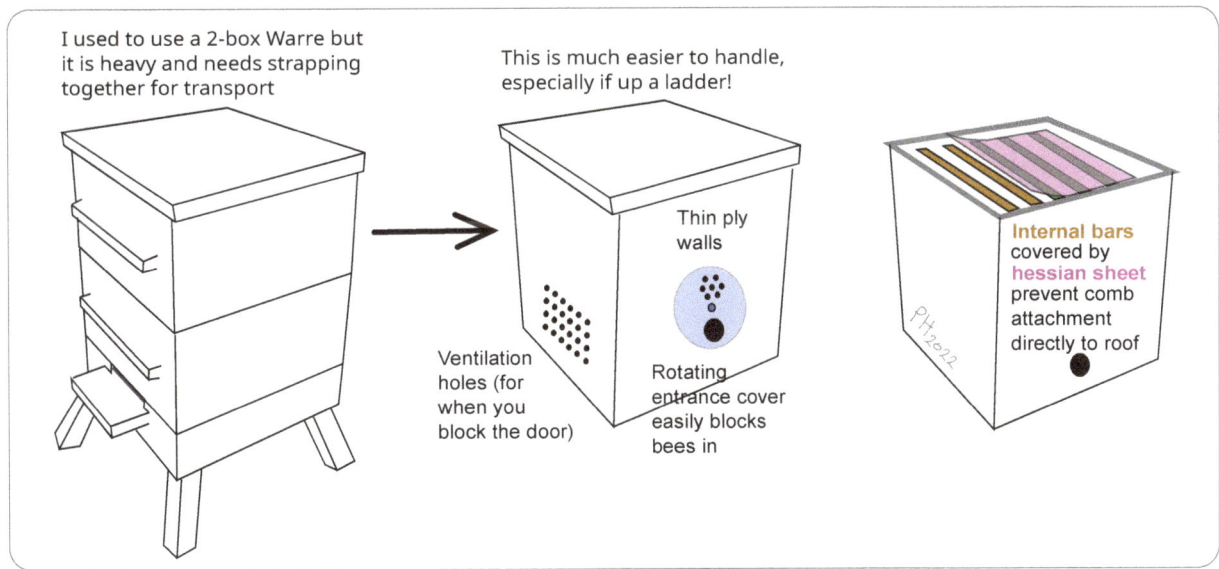

If baited with the scents of brood comb and propolis, and placed correctly, this attracts swarms. I put 1-2 drops of lemongrass oil at the entrance to attract scout bees[149] so they're aware of the box.

If the size and smell is OK, the other key factor is height. Bees hate ground level entrances but are strongly attracted if the bait box is on a table. Higher is even more attractive.

Swarms sometimes abscond from bait hives due to **overheating once the sun falls on the bait hive.** They are thin walled and people sometimes place them in sunny spots (especially when high up). The bees go in then a few hours later, leave. Imagine the elation followed by disappointment!

Tom Seeley has determined other factors but those only seem significant if the bees have a choice of cavities. A dry, snug bait box like this near your hives ticks most of the boxes.

Watch your bait hives: in swarm season, you will occasionally see a bee go in or out, this is just scouts keeping an eye on potential new homes. One day you will see 10+ at one time! Look around for clusters in trees, bushes etc. If it is an easy catch, box it - you can't guarantee the swarm will choose your bait box over a neighbour's chimney.

Maybe you are out at work or away for a weekend when a swarm moves in. In this case you will see entrance activity, including pollen going in.

Tips: Check the bait hive is not full of spiders. Or a wasp nest. And bear in mind, bait hives don't contain *swarms*, the colony is no longer looking for a new place to live, having completed its "house moving program". It will behave subtly differently to a swarm - if you just transfer the bars / frames of bees to a new hive it is unbothered, but they will be agitated if you knock them out of the bait box or try to walk them in to a hive.

Russian scions and bee bobs

If there is a wide open space in front of hives with no obvious clustering point, you can sometimes entice them to settle on a convenient *Russian scion* placed directly in front of the hives.

This is a pole, usually with a detachable object around head height like a horizontal disc, or cross-bar which the bees perceive as convenient, and easy to cling to. They seem to prefer oak.

This scion in a hedge facing hives has attracted a swarm.

To maximise chances of attracting swarms:

Hang from a tree branch - swarms like to cluster shielded by leaves;

Bait it with beeswax and propolis (e.g. staple on an old top cloth from a hive);

Allegedly, squashing a dead queen on them works as a powerful lure.

You can of course use an old frame or bar with a comb attached.

After the cluster has settled, you simply pull the whole thing up or, if you've got a removable bit, detach that and shake / brush the swarm into a hive.

Scions aren't guaranteed to work but save a lot of time when they do. Note that unlike bait boxes, you still have to box the swarm before it flies off.

---

149 Said to mimic the "come here" pheromone. It certainly attracts them. You can overdo this and drive them off so just use 1-2 drops. Don't bother buying queen pheromone. Wax crumbs after moths eat comb are a strong lure.

Bee bobs are similar: in this tradition a board of wood, or a sock filled with material to resemble a swarm cluster, is covered in melted *slumgum* (the grungy leftover coccons, propolis, dark wax etc filtered out when purifying wax) as a lure and hung in a tree. *Further information*: search for "*Russian scion*" bees images.

Swarms - miscellaneous

"End of life swarms" - It's suspected that in totally unmanaged hives, with no human requeening or comb rotation, healthy colonies eventually abscond en masse early in swarm season, possibly because the internal diameter of brood cells has reduced to some critical point so they leave, letting old black comb be broken down by wax moth. The hive can then be repopulated by another swarm. I have heard of this happening to a couple of 6 year old colonies in Oxfordshire hives[150], but I'm aware of much older, continuously populated cavities in e.g. trees and church walls, so I don't think it's a general rule.

"Mating swarms" are more frequent than is often realised. Their function is to signal the returning queen. They act like a small swarm, with bees milling around in the air for about 20 minutes before all returning whence they came. Bees may settle at random, over anything in the vicinity (including observers), sometimes they don't and sometimes a cluster may form on a bush or post, only to evaporate later.

"False swarms" refers to when a swarm forms and hangs round in a clump for a few hours, then goes back into the hive; sometimes due to unsuitable weather, dodgy queen etc. Obviously this term overlaps with *mating swarm*.

Starvation swarms are sometimes young swarms that only managed to make a few scraps of comb in their first home. See also, *absconding*.

In addition to forcibly entering an occupied cavity (*usurpation*), swarms have been seen peacefully entering hives with failing queens, and peacefully entering multi-entrance cavities (walls, TBHs etc) next to other swarms who have just moved in (possibly from the same mother colony).

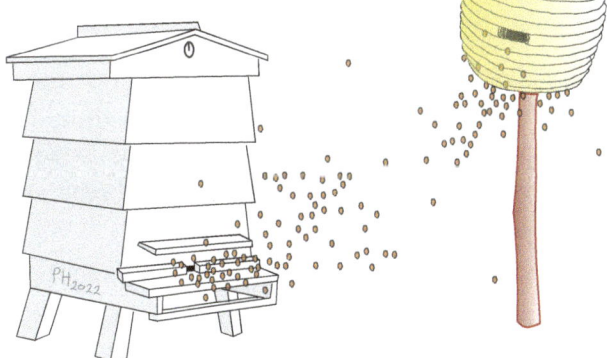

It's generally considered good sensible practice **not** to take honey from a swarm within the first year it was hived.

And **don't** poke around in a newly populated hive. They're delicate, they may abscond and you won't change anything.

One argument against suppressing swarms is, the bees will prepare for one anyway; then when the swarm finally emerges it can be immense, because bees kept hatching - and are now able to fly off. Such swarms can strip a hive of rather too many bees. (This can also happen if swarms are held back by weather.)

Help! They're swarming right now! - When a hive begins to swarm stick a skep with 2 drops of lemongrass just in front, on a stick. They sometimes go straight in. You could even use a veil in an emergency, but getting the bees out of those is tricky. Some people place insect habitat nets over hive entrances they think are about to swarm in the next day or so, but this blocks foragers.

---

150 Observations from experienced beekeepers - the hives were completely abandoned in one go, the colony did not die over a matter of weeks due to left-behind virgins failing to mate. It's currently quite rare for a colony to survive 6 years in a hive.

The number of swarms is falling dramatically. We see it in Oxfordshire (which may simply be *Amm* traits reasserting) and it's very noticeable in South Africa. Pesticides are suspected.

Observations

Gareth John shared the following:

He's seen hives throw a cast and NO prime – decided to supersede, then used spare queens for casts.

Swarming bees can be literally bursting with wax. It ends up everywhere.

You can tell which hive swarmed by dusting the swarm with powdered sugar: white bees will appear **at the hive that has swarmed,** because one function of the cluster is a way of sorting out who's staying, who's going.

Don't panic if you see lots of drones after swarming. After swarming hives can produce huge numbers of drones and, since worker numbers are low after swarming, one can think the hive has only drones. If multiple casts have been produced it can take a long while before a queen becomes mated and worker brood is produced. Hives have the ability to "hold" unmated queens in reserve for far longer after the prime swarm than counting calendar dates would suggest.

If a swarm forms an elliptical cloud it's about to fly off very fast and high, herded by what Prof Tom Seeley calls *streaker bees,* flying fast through the cluster from the rear to the front to direct it to a new home. Such swarms may travel a mile, faster than you can run.

Removing a swarm far from an area denudes the area of those locally adapted genes - especially if a beekeeper kills the Queen.

Bees will sometimes ball spare virgins when a cast is being hived. They are not necessarily killing them. It could be protective, at least initially.

Swarm suppression

There are many conventional techniques like wing clipping to discourage swarming, without considering whether one **should** and the **impact.** They're all highly disruptive and prevent the beneficial brood break and genetic re-tuning of swarming.

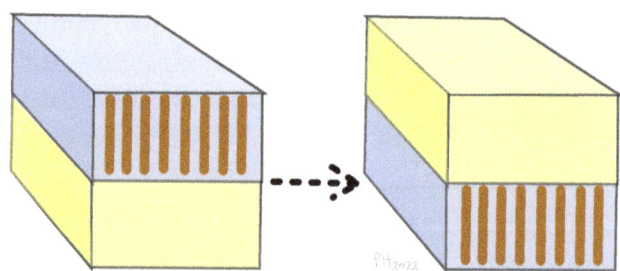

A typical ritual for swarm suppression: 'rotating' or 'reversing' (swapping) brood boxes in Spring. Easy and quick, but low success rate - bees expand down.

I myself add space (boxes **below** the brood area) to avoid the bees feeling crowded, but if they get up to 4-5 boxes and still want to swarm, I let them. There is a school of thought that one should not change a hive's volume more than ~20% after the bees move in, to avoid disrupting their heating and ventilation, so a single **shallow** super above the bees is all you should add at once.

See also: checkerboarding, under "chilled brood" in chapter 14.2.

Luring swarms in flight

In the 2019 docu-film *Honeyland* a traditional Macedonian beekeeper, Hatidže attempts to coax a flying swarm into a skep-like hive whilst calling *booooop booooop booooop boop-boop-boop* in a high tone. She's mimicking a queen.

You can force a swarm to land by throwing dust into it in flight. You can *sometimes* make them land with a mirror, reflecting the sun at them and confusing them into which way is up[151]. Tanging - beating a metal object to make a high pitched clanging noise - is an old wives' tale (a misinterpretation of a custom to claim ownership of a swarm) and is ignored by the bees, though sometimes the swarm settles anyway.

Swarms will join other swarms.

On swarmy days, several swarms may emerge almost simultaneously. Members can get confused and sometimes go off with the other one. I've seen this a few times, for example as one bee cloud settled into a swarm box another flew past above it, leaving a significantly smaller cluster in the collection box than I'd seen on the branch a few moments before. This probably only happens with virgin queens who lack strong pheromone smells. It's a spontaneous uniting of colonies.

Swarms: summary

Many beekeepers regard swarms as a problem (!!!), are focused on suppressing them, and have a very simplistic understanding of this highly adaptable phase.

Review 6

How would you explain what is going on to a property owner who has asked you to remove a swarm?

List four differences between a swarm and purchased bees:

Why might a swarm be difficult to hive - what can you do about it?

---

151 Mirrors and dust were mentioned by Virgil in *The Georgics*, Book 4 *Apiculture* around 38 - 32 BC.

Chapter 10
Harvesting honey

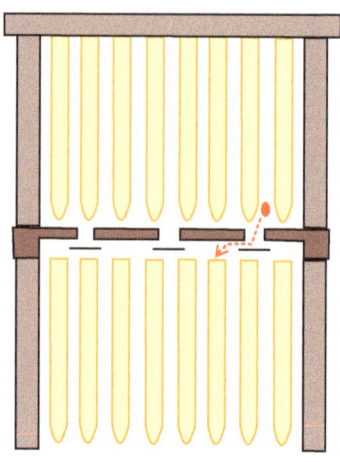

Your local Beekeepers' Association is the best place to find out about local **labelling** and **legal** requirements. They often lend equipment; their core purpose is to promote and teach honey production.

1. Check for surplus honey and brood

Heft the hive or box to confirm there is enough honey that they will still have sufficient winter stores if you take some (usuallly 10kg-20kg, depends on climate and hive type). Inside the hive, combs with a thick covering of bees have brood - don't harvest these. You are looking for sparsely populated excess combs of capped honey.

Clearer boards incorporate one-way gates such as *Porter Escapes* so bees can exit one box, but not return. Over a few hours most bees leave.

2. Remove bees from honeycombs

The bees may object.

For vertical hives you can avoid 95% of the disruption with a clearer board, which uses one way bee escapes. Come back next day, and the bees in the isolated top box will have wandered down to the box with the queen. Take the entire box.

For Top Bar Hives, remove combs with a comb knife (see p.241) then brush bees off individual combs - be sure to do this over the hive as queens and very young bees may get lost in grass.

(You can also smoke them off, but excessive smoke flavours the honey.)

You can shake them off frames. Don't try this with top bars - the comb will snap off.

Warré hives need a special technique because you often can't inspect or remove individual combs to brush bees off. See chapter 12 for explicit diagrams.

If bees remain obstinately on a comb in a thick crowd, or crowd round and try to get in through a bee escape, it is because there is a queen or brood there. It is important that stray queens, in particular, are returned to the hive.

3. Temporarily storing the combs

The smell of honey will draw every nearby bee and wasp to the harvested combs, so they need to go in a container with a lid. This could be e.g. a nuc box. If you just lie them flat you can end up with squashed bees and honey everywhere.

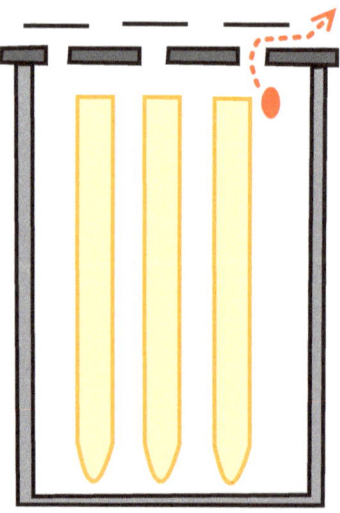

Principle of a clearer box.

I made a special clearer box for my TBH combs. This may be overcomplicating things - after removing bees **most people just cut out combs next to the hives (easy when foundationless!) and put the frames / top bars back in the hive,** and put the combs in a container there and then.

Anyhow, here's my method. After removing most of the bees there are always some left on the comb. My clearer box is based on a black plastic box, with Porter bee escapes at the top. Remnant bees head to the light from the bee escapes. It was trickier to make than I expected - the combs had to hang with bee spaces between them to allow passage, and needed brackets to prevent them sliding around and falling off the support rails when I picked it up. But it was worth it, because by leaving it outside for a couple of hours, I only get a couple of bees in the kitchen.

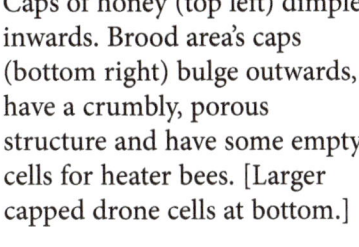
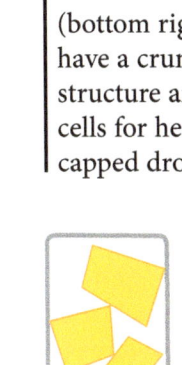

Caps of honey (top left) dimple inwards. Brood area's caps (bottom right) bulge outwards, have a crumbly, porous structure and have some empty cells for heater bees. [Larger capped drone cells at bottom.]

4. Remove unsuitable comb

Process only CAPPED honey. Or moisture content will exceed 20% and it will go off[152].

You **must** cut away comb which has **glistening liquid** in cells - **even if the other side of the comb is capped honey.** This is unripened honey, or nectar. (See Chapter 11.3, *Reading Comb*, for reference images.)

Pollen will jam your filter, so cut that away.

You don't want brood in your honey.

Cut away empty comb. It will sponge up honey when you cut cells open.

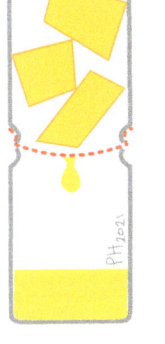

Small amounts can be processed with two jam jars and a muslin sieve.

Any cut-away pollen, nectar and comb which you have processed but still has traces of honey, can be fed back to bees. Put it in a bowl a good distance from your hives. The bees will smell it and mob it. If you do this next to the hives, though, it will trigger robbing. The comb will however be reduced to flakes of wax in a couple of days: if you end up with pretty intact combs in frames, you may prefer to re-use these in a hive rather than expose them to this robbing frenzy.

<u>Scale of harvest issues</u>

Whenever you filter honey, some honey gets stuck to the equipment and is lost. Crush-and-straining 4 combs will give you more than twice as much honey as 2 combs! It pays to process large amounts in one go.

Cleaning equipment afterwards takes a while, another reason to do few, large harvests.

Don't bother harvesting boxes which feel light. Lots of effort for little honey.

5. The actual extraction

In all cases, you need to break open the honeycomb cells. This can be as simple as uncapping

---

152   If you use **frames** with a mix of capped / uncapped honey, shaking a horizontal comb will only make nectar drip out if it is above 20% water. If it stays in it is safe to use even if uncapped. Top bar combs will not survive this!

them with a fork. Extraction works best in warm rooms so the honey flows faster - this speeds up processing enormously - but don't warm honey above 40C, the peak temperature it sees in a hive, or you will lose flavour. Some people strain their honey several times using smaller mesh to remove pollen and 'bits' – but I find one crude strain sufficient and removing all the pollen removes taste.

<u>Cheap and simple techniques:</u>

Uncap and drain - after uncapping you simply let the honey drain out. However, honey takes a long time to drip out of uncapped comb – it's almost as if the cells were designed to stop honey inadvertently escaping.

Crush and strain - you mash up the comb. Faster than cut and drain. However, you end up with more particles of wax and pollen in the honey so it sets in weeks rather than months (they act as seed crystals). The extra agitation can contribute to losing the volatiles that give honeys their distinctive aromas. I find the mesh immediately blocks with wax so I use the *Rapid Crush and Drain* method below.

Both these methods require some hours for the honey to drain.

Crush & strain honey contains more pollen, is better for allergies than regular honey.

<u>Fast / high volume techniques:</u>

Rapid crush and drain - Using disposable gloves and a kitchen colander above the fine filter. Crush the comb in the colander in your gloved hands. Takes about a minute per frame. No time wasted uncapping. Most wax ends up in colander rather than blocking the fine mesh. Less messy than uncapping.

Pressing - for very thick honeys like heather. Combs are crushed in an expensive custom press, or some people convert cider presses etc. Usually a huge and expensive machine. Can be combined with e.g. cut and drain to get the last bit of honey out. I've considered investing in a honey press myself, because I'm failing to extract up to 40% of the thicker honeys crush-and-draining by hand.

Spinning - for high volume commercial production. Only comb in frames is strong enough. Much faster than other techniques, and removes all the honey - at least 15% more yield than if same comb was cut and drained. Spinners can be hand cranked or electric (even the hand cranked ones are hundreds of pounds). Physically large (storage issues). Comb can be re-used by bees so they waste less resources on building comb, make more honey. Spinners can be *radial* (for large operations - big machines that spin many combs at once) or *tangential* (smaller, suit hobbyists). Spinning cannot extract crystallised honey. Spinning is only suitable for complete combs of pure honey because you have to balance each side of the spinner with equal weights of comb: if you cut a large patch of nectar out of one, you will have problems.

Spinning foundationless comb needs care or it will break. You need to spin gently on both sides first, then repeat faster for fuller extraction. It must be a tangential extractor. What breaks the

Crush and drain

Medium amounts can be processed faster with specialised plastic tanks with a built in metal seive.

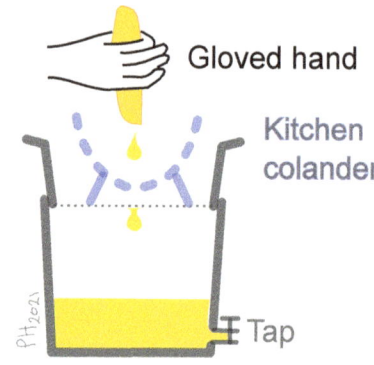

Rapid crush and drain

comb is the weight of honey on the inside cells. The honey can't escape so it presses on the joining wall, bursting it. An initial gentle spin extracts enough honey that breakage is unlikely.

The old fashioned way:

Comb honey: You can simply cut honeycomb out in blocks to give portions of edible comb, complete with wax. Sometimes the colour – honey flavour – varies across the comb but in general new white wax combs with white waxy cappings is considered most appetising / saleable. Top Bar Hives are good sources of fresh white honeycomb, whereas all Warré comb was brown brood comb once.

Don't use brood comb for this, because if you use old, dark comb the 'wax' will be unpalatably crunchy rather than soft; the wax will be strongly flavoured with propolis; and will contain old cocoons, which are chewy. It's also more likely to contain wax moth eggs (see unappetising picture).

You can still extract honey from brood comb by crushing and straining – it will contain much more pollen than honey from white honeycomb and have more taste.

6. Moisture content check

Use a refractometer to confirm the moisture content is 20% or less. If it's 21%, you only have a few months to eat it.

Important: **honey is not safe for children under 12 months** (see Honey 101)

Factors influencing honey harvests

The most obvious factor affecting honey yields is proximity to forage. Whilst bees *can* fly 3 miles for food, but they burn a lot of energy and time doing so. Most foraging is within 1km or less. And significantly, it's thought the main factor limiting worker lifespan is that their wings seem to wear out after about 800 km of flying[153] - they literally can't fly home. So it makes a big difference if your apiary is 250m or 1km from the main forage source.

"Cut comb" section.

When cut comb goes wrong: wax moth. Freezing cut comb sections kills WM eggs. You can never unsee this.

---

153 *Dependence of the life span of the honeybee (Apis mellifera) upon flight performance and energy consumption*, Neukirch. Journal of Comparative Physiology B 1982; 146: 35-40

If your hives don't thrive, there may be too many in one place. Move half out and see if you get a harvestable excess.

It's long been known that hives need to be distinctive, so returning foragers - and queens - don't get confused and enter the wrong one, and get killed by guards. This effect can reduce bee numbers dramatically. Solutions include ensuring hives are more than a metre apart; facing entrances in different directions; ensuring the hives have unique objects like a tree or wall nearby[154].

Derek Mitchell's work on hives showed very clearly that more insulation = less honey wasted in thermoregulation efforts. In Britain you want afternoon shade to avoid heat stress.

Many beekeepers don't appreciate how much honey they lose by opening hives! Every time you open one you lose several hours' production. (Unless you can inspect a hive as fast as a bee farmer, in 60-120 seconds.) It's often said the real reason aggressive hives make more honey is, the beekeeper disturbs them less.

More dubiously, there's a theory that facing entrances towards the rising sun means the bees get up earlier and work longer. I've tried it. They began flying 30 minutes earlier, but I didn't see any increase in honey harvest. It didn't change how fast the plants warmed up and expressed nectar!

Horizontal TBHs are designed with a relatively small cavity that gets filled quickly: they are intended to be harvested "little and often", but in Britain this might only mean 2-4 times a year as our forage and weather are not optimal for making honey. Remove honeycombs to make space, always leaving some.

Warrés are intended to always have a box at the bottom of the stack for the bees to expand into. This may be completely empty, or have empty comb in.

Dave Cushman mentioned[155] that long landing boards sloping down to ground boosted honey yield up to 25%, because so many returning foragers fall short of the entrance and waste time climbing up and in. Beekeepers also find it best to avoid damp hollows, they promote chalk brood and bees need dry air to cure honey. There are many, many other clever tricks, many of which (feeding, requeening, swarm suppression, re-using comb, foundation etc) distort bee behaviour so much their health suffers, but this book's focus is on what you *can't* find in standard texts.

Crystallisation

Honey starts out very liquid, then becomes thicker over a period of weeks or months as crystals grow in it. This process is accelerated by low temperatures[156]. A jar of set (hard) honey can be softened by placing it on a radiator; a gentle heat is needed to avoid affecting its taste (*don't* use a microwave).

Some nectars, like ivy, crystallise quickly due to high levels of relatively insoluble glucose, whereas most nectars are mainly fructose. In mixed-forage areas, bees tend to mix all incoming nectars together when creating honey, so all the honey can be contaminated with seed crystals and set hard: only honey

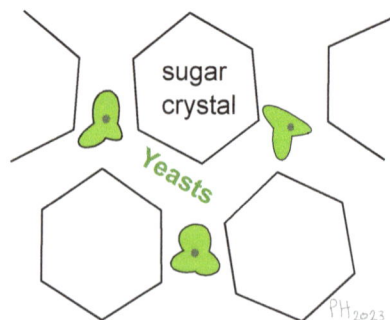

When honey crystallises, yeasts can grow in the liquid between crystals.

---

154   The common technique of giving hives different coloured paints is questionable. Usually the painter forgets bees can't distinguish red from black. Unique, simple **patterns** are better.
155   www.dave-cushman.net/bee/beeflow.html
156   Some sources also mention exposure to light

which is already processed and capped (*ripe*) will be immune to this effect. OSR honey can crystallise in days[157]. Once solid, the honey cannot be crushed or spun out so honey farmers have a frantic time harvesting this as soon as it is ripe.

Getting rock solid honey out of comb by crush-and-strain is not feasible; the trick is gentle heating. One technique is to place the comb in a honey straining bucket in a warming cabinet, the wax melts and *floats to the top of the honey*. You can then open the honey gate at the bottom of the bucket and pour out the honey. You could also try a melting tray.

Crystallising is arguably a sign of quality - the honey hasn't had all the pollen filtered out or been adulterated with syrup!

When honey crystallises, excess water is expelled from the crystals into the fluid between them. As this fluid rises above 21% water, yeasts activate and ferment it, causing bubbles and pushing plugs of honey out of the cells.

| Ivy honey crystallising and fermenting

The fermentation is retarded by low temperatures and if the crystals are very hard, so bees can allegedly eat rock-like ivy honey in Spring without apparent ill effects - though tellingly, if fresh nectar is available in quantity, they'll discard the solid honey, as little white crystals outside the front of the hive! However, if you ever notice a jar of honey which has a domed lid and seems to have separated into two layers, smell it. It's not awful, but it's offputting! Fermenting honey can be used to make mead.

Further information about honey processing can be obtained from your local BKA, and YouTube is a good learning resource for this physical skill.

---

157   I've heard of it remaining liquid in Warré hives for months, but it solidifies quickly in cooler Nationals

Chapter 11
Observing a colony

11.0 Record keeping

11.1 Inspections: external monitoring

11.2 Inspections: internal observations

11.3 Reading comb

Chapter 11.0
Record keeping

Don't be overwhelmed by the list of things you *could* check. My own hive records are often simply:

> 20th June 2020 – 2PM – 20C, sunny, dry, no wind - pollen going in – bee behaviour purposeful, calm – traffic 1 bee / 5 seconds - all OK.

The point is to log **variances from normal.** If you see something weird, take a photo if you can.

Almost all my records refer to external, non-invasive observations - they're more detailed when I open a hive, but that's very rare.

To avoid confusion, use separate records for each colony. I use a ring binder with a section for each colony.

In addition to dated records, I have a one sheet summary of each colony stating key facts such as origin and health history. This helps a bee inspector assess it rapidly.

Why keep records?

- *Predicting.* To be able to spot / anticipate / address significant problems, and to reassure yourself as to their ongoing health. Many bee lifecycle events have known delays which can be used to infer queen health, or predict (e.g. watch for afterswarms emerging 7-8 days after the prime swarm).
- *Getting to know your colonies' baseline behaviour.* Each has its own "normal".
- *Learning (self training).* Even if you do not understand the significance of what you observe at the time, record *everything* that strikes you, *especially* things that make no sense; this will help you as you reflect back, and help you - and others you ask - diagnose problems and spot them in future.
- *Bee inspectors* will also be reassured that you are aware of the state of your bees and they are cared for responsibly. They will then be less inclined to rip the entire nest apart looking for problems. They are particularly interested in the summary sheet as it saves them lots of time - when disease breaks out in an area they have to race around as many colonies as possible in a few days. Include why the previous colony in the hive died: this can help them identify ongoing problems.

- **[Legal requirement:** *UK beekeepers must keep a record of any veterinary medicinal products administered to colonies for at least five years, whether or not the colony has died or been sold. Irrelevant for non treaters.]*

Ignore phone apps and conventional record systems. They are not flexible enough – they're focused on queen manipulations, medication and honey yields, and have no guidance on how to actually find and recognise disease when you open a hive. Such systems don't record much detail, you don't learn much from them, and how many apps do you know which are still supported 5 years later?

Large scale bee farmers typically use a concise chart like system which can be analysed rapidly.

Note: I don't cover **electronic monitoring** of colonies. It's currently very expensive, limited, over-claims, and goes rapidly obsolete. After a career in electronic instrumentation, I'm acutely aware that you tend to spend more time fiddling with the kit than actually gathering information. This situation may evolve as technology improves.

Review 7

How will you organise your written records?*

What is most important to you to record?

How will you store & organise photos?

---

* Consider what's important to you to record, and how you will look back and spot patterns. For example I record details of unusual behaviour and have just a handful of hives; other people are more interested in breeding or honey yields and have dozens. Anything software based is likely to be lost, corrupted or an unreadable format in a few years, and very inflexible; you can't make sketches of ideas or things observed. Use paper.

Chapter 11.1
Inspections: external monitoring

It is a good idea to monitor your hives in case of problems. This should be done with minimum disruption of the colony, and it can be fun – not an ordeal for either party. It's not complicated, you're mainly looking for signs that something is amiss that requires human intervention, but normally bees are the best beekeepers.

Passive observational techniques often don't require more protective clothing than a veil and are ideal for hobbyists with just a handful of hives.

These techniques are not unique to low-intervention beekeepers but have dropped out of the standard lore of conventional beekeeping.

In chapter 11.2 we will review the very few occasions when it may be really necessary to open a hive for inspection. Both approaches are useful, but if you overdo the open-hive inspections you can cause problems and train the bees to recognise you as a threat.

- 95% of inspections don't require opening a hive
- Small, young colonies don't behave like books say
- Hive records are really useful
- **Learn what's normal**
- Internal inspections are overrated. Less is more!

Observations - overview

The key thing is to watch your hives every few days to learn what their baseline behaviour is. It is useful to have two or more colonies to give you an idea of what is within the normal range of behaviours.

Key areas to observe are the entrance, through the windows and debris on the baseboard / below the entrance.

Following an inspection, make a note of what you saw. Even if everything is OK, it can be useful to look back at a log to confirm "they were not doing this 2 weeks ago". And make a note of what you expect to see **next time** - this will help you think and anticipate instead of just reacting.

Performing and interpreting external observations

Before starting, consider if you need a veil. If the bees seem unusually agitated, or you have lots of hair they can get caught in, or have to stand in a bee line, pop one on. Don't stand in the bee line.

Record - date, time of day, weather; nearby unusual human activity.

Inspecting a hive. 95% of inspections are non invasive. These colonies all differ in age, origin (genetics) and size so differ in behaviour – the small new colony's traffic is lower, another is superseding etc. Each has its own baseline personality.

Index of external signs and behaviours

General

A1 Surroundings
A2 Temperament
A3 flying patterns

Traffic related

A4 traffic levels
A5 sudden traffic surge
A6 large colony quiet during flow
A7 colony stops working
A8 frenzy (robbing)
A9 heavy traffic between hives
A10 sudden cessation of activity
A11 only one colony flying in Spring
A12 foragers flying strongly but no apparent harvest

Other

A13 pollen going in
A14 excessive pollen
A15 propolis
A16 swollen abdomens
A17 kissing
A18 bees expelling bees
A19 grooming
A20 bees returning very heavily laden early morning
A21 oddly coloured and "ghost" bees
A22 bees entering & exiting are different colours
A23 unusually dark bees
A24 bees of unusual size
A25 fanning
A26 bearding
A27 >4 guards
A28 shimmering
A29 Traffic jam
A30 excessive licking of arrivals
A31 unusual visitors
A32 propolis barriers
A33 wasps
A34 hornets
A35 purposeful behaviour
A36 very loud buzzing
A37 agitated, groaning
A38 indolent loitering
A39 swarm returns to hive
A40 bees enter empty hive
A41 washboarding
A42 appropriate numbers of drones
A43 ragged wings
A44 deformed wings & bees
A45 K-wings
A46 balling
A47 Q enters with Drone Sign
A48 bubble-butt
A49 moths
A50 drone expulsion
A51 response to rain
A52 hitch-hikers
A53 bees stuck upside down to landing board
A54 night flying
A55 bees under hive
A56 flies entering

Entrance behaviour

A1 Outside the hive - before approaching it, look around for external influences. Weather? Trees in bloom? A swarm in a nearby bush? Collecting **water** (monitor level of their source) indicates a nectar dearth, it's needed for cooling & diluting stores.

A2 Temperament – if unusually defensive, check for predators or robbing by e.g. wasps, birds; any odd smells (from hive, or nearby grass mowing etc). Bees are notoriously edgy when a thunderstorm approaches, perhaps due to static, or simply because all foragers are present. Queenless hives can also be more irritable (or apathetic, basically a variance from ther norm).

HIVE ENTRANCE BODY LANGUAGE

- Tail horizontal - *Chilled*
- Nasonov gland. Fanning tail up, end bent down Nasonov gland showing - *Come here*
- Wings blurred, hunched down - *fanning, to ventilate*
- Front end down in submission (as seen in dogs) - *Please may I come in*
- Tongue out, licking floor, often in sync with others - *"Washboarding"* (very content)
- Guard facing e.g. wasp at entrance - wings up, legs splayed, mandibles apart - *I am big, I see you, go away*
- Tail up, end bent up Antennae up Wings out - *Go away*
- + stinger everted, drop of venom at end (smells of bananas) - *REALLY go away*
- Reared up on 4 back legs - *Go away* (don't come in)

A3 Flying patterns - in calm weather bees go fairly straight, but in windy weather they take sheltered routes, like hedges, rising fast when that runs out then diving into the next ditch etc.

A4 Traffic levels – very weather dependent; don't expect a steady high traffic rate except when there is a nectar flow on. Foraging in bad weather is a warning of low stores! A large established colony might have several bees a second coming and going, but this varies enormously. Small, new colonies may have one bee every 5-10 seconds, only flying when they have spare workers they can divert from raising brood. Traffic levels plummet for a couple of weeks after swarming. I've rarely seen the same traffic level from 2 hives at the same time.
Each colony has its own timetable. It will vary with colony development, entrance orientation, when shade falls on the hive etc. With time you will learn the rhythm of each colony's daily flying habits – and how it changes with the weather and season.

In forage dearths, the bees conserve energy until scouts locate a food source. Blossom may be visible, but droughts and low temperatures inhibit nectar creation.

A5 Sudden traffic surge

- **Orientation flights:** New bees come out in hundreds for side-to-side practise / orientation flights once or twice a day – easily confused with traffic jamming (**A29**), a constant cloud of bees when the entrance needs widening.

- A **nectar bonanza:** bees clambering half way up front of hive before taking off (due to congestion) - nectar flow discovered by scouts, often immediately after rain recharges nectaries., then you see a flurry of frantic activity as the bees scramble to get it.

- **Toilet flights:** bees trapped inside by bad weather for days will emerge for mass toilet flights during breaks in the rain / wind / sunny calm period in winter.

- **Swarm:** tide of bees pours out of entrance and forms a huge cloud...

- **Earthquake** imminent! Comb resonates at frequencies similar to initial shocks, and just before earthquakes, beekeepers sometimes see a great commotion at hive entrances as bees leave en masse.

Part 2   Chapter 11.1 Inspections: external monitoring

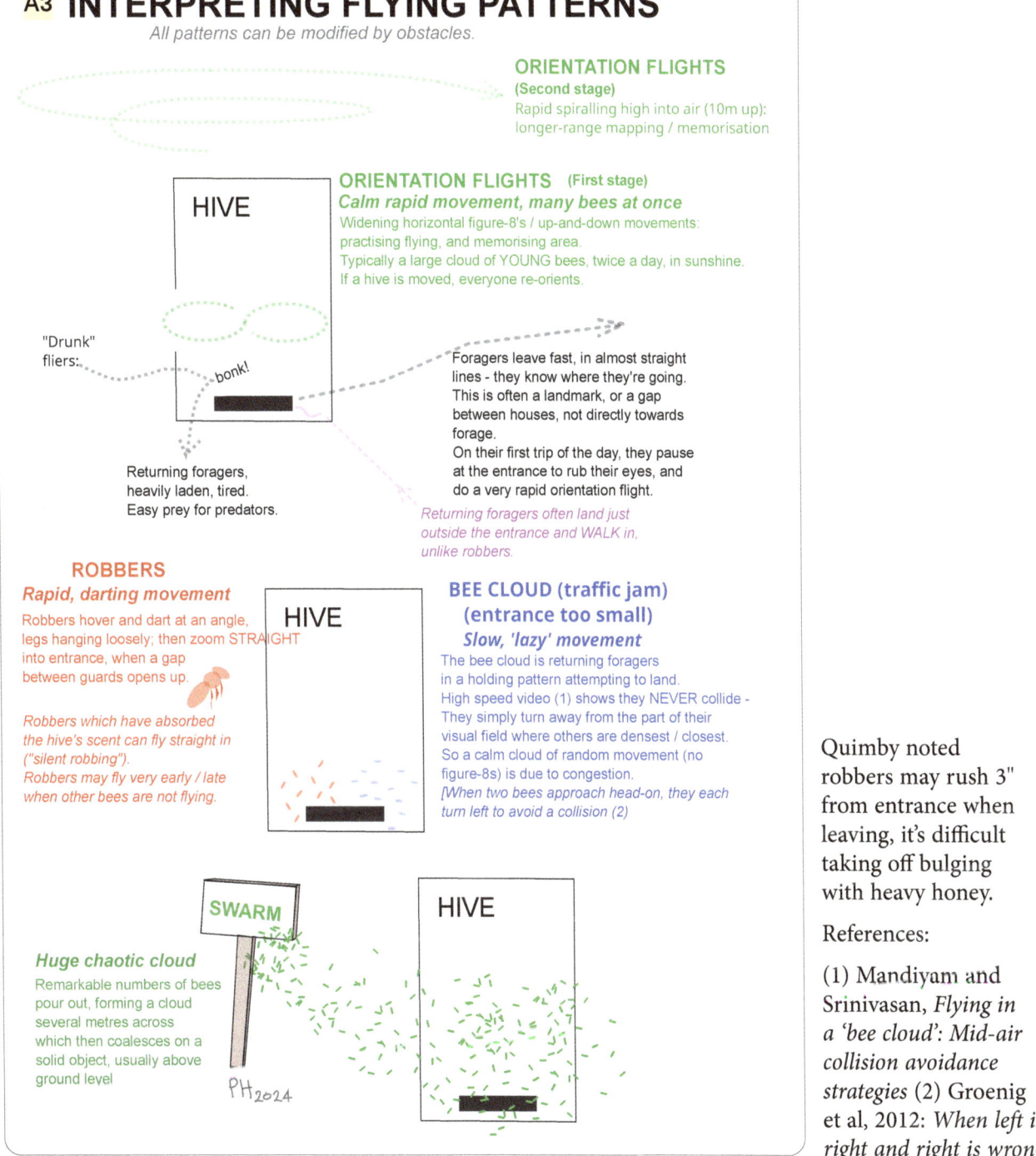

Quimby noted robbers may rush 3" from entrance when leaving, it's difficult taking off bulging with heavy honey.

References:

(1) Mandiyam and Srinivasan, *Flying in a 'bee cloud': Mid-air collision avoidance strategies* (2) Groenig et al, 2012: *When left is right and right is wrong*

A6 Large colony is quiet despite large nectar flow -

- **Age distribution of bees.** If the majority of bees in a large colony are nurse bees, there aren't enough bees to go foraging.

- **Starvation.** Typically, bodies of bees around front of hive; no fanners or condensation at entrance; feeble flying. Colonies can get locked in to harmful patterns, sometimes described as losing morale. Big colonies need big reserves to raise brood. If you take too much honey, they get locked into a state where they consume all reserves, even the broods' jelly and lack energy to reboot. Emergency feeding needed or the colony will die.

Other signs of starvation: A rap test will not evince a brisk, lively buzzing if the hive is out of energy. You will see some bees lying on the landing board or below combs, these can be revived but still try to leave (forage).

In late summer, if you place your fingers in an entrance you can feel humidity if they are evaporating nectar. (This risks stings with some hives.)

A7 Colony "stops working" but seems content and hive is heavy - Good! Although this may indicate queenlessness, it often just means the hive is brim full of honey or simply feels it has enough stores for winter. They turn to maintenance and hygiene behaviours, or begin backfilling brood comb with honey in preparation for swarming. Historically, beekeepers[158] viewed this as awful because "bees should be working" and termed this a *degenerate* hive. They would "solve the problem" by harvesting honey, spurring (stressing) the bees to forage until they had enough stores again - and interrupting the vital hygiene behaviours. A modern commercial beekeeper might spot a huge colony is no longer foraging in a nectar flow, check for backfilling and if so, split the hive to head off swarming and restart foraging.

Humans aren't as sensitive to weather as bees. Sometimes they just don't like the temperature, cloudy sky etc.

A8 Huge number of bees and noise, bees wrestling at entrance, bees flying with body at an angle and legs hanging, dart straight in entrance rather than walk in - robbing by another hive, or wasps. Often accompanied by wax crumbs at the entrance. The legs probably hang down to absorb the target hive's scent, thus fooling the guards. In extreme cases, body hairs are rubbed off in the frenzied crowd - or contact spilled honey which dries and renders their hairs brittle - and bees look black. Through windows you will see robbers running extraordinarily rapidly to the honey stores, to avoid being challenged.

If you move the hive - far enough that the robbers cannot locate it via scent - the robbers inside can leave and you can try to reset the situation. The weak hive being robbed needs help to survive - block the entrance immediately for 2-3 days, but make sure they can breathe. Then construct a restricted tunnel entrance they can defend more easily. **Further advice in chapter 14.2 under** *Robbing*.

Robbing tends to happen at the end of summer after beekeepers have removed honey stores(!), or late Spring when some beekeepers split hives and feed them; it can be triggered by the smell of feeding syrup - which is why you should **only feed at night,** when no one's flying.

Guards bite wings and legs to cripple, before escalating to stings.

A9 Heavy traffic between hives - sneaky robbing! Possible during main nectar flow when guards are not on high alert. The only case I've heard of in detail were between sister hives.

---

158 E.g. Thomas Wildman (1770). Simmins explains in *A Modern Bee Farm* (1887): Brood rearing and honey production tails off by September *unless* the queen is new that year, *and* there is forage, in which case it continues into late October. *The whole of the stores accumulated by a swarm thus left to itself seldom exceeds 20 lbs, but let the reader compare this with the product of a swarm worked as explained under "General Management"*, and he will find that there is but poor economy in the "let-'em-alone" policy.

Tickner Edwardes implies something similar in *The Bee Master of Warrilow*, working with largely Amm bees in southern England circa 1920. He mentions in chapter XVI, "*Bees and their masters*" - "a balance of stores of about 20 pounds" (not kilos!) "weight at the end of a season will safely carry the most populous colony through any ordinary winter. But from the bee-master's point of view it means practically a lost harvest."

A10 All activity of a strong colony ceases abruptly - swarming imminent - "the calm before the swarm". All scouts return to the hive and foragers don't leave, so as not to miss it, and inside the hive the bees are gorging on honey.

This might last 5 minutes. The swarm emerges, and it's all over in 10 to 15 minutes, after which the hive behaves completely normally. Many beekeepers never see a swarm - they're at work.

A11 (Italian) bees flying in large numbers while other colonies are quiet (March) - they have bred too early for a cold climate and are desperately seeking food. Probably Italian, or selected by a commercial breeder to breed prolificly very early in the season, for pollination contracts. Many foragers will die on long, fruitless trips. You could feed them, or take the long view - let them die and restock with local bees better matched to the local climate and seasonal variations.

A12 Bees flying as if foraging strongly, but do not seem to be carrying anything - lime (linden) trees are flowering, but not yet yielding nectar - they need a high water table and heavy, humid weather for that so some years are good, others not. The linden was European bees' dominant food source and its blossom excites them.

A13 Pollen going in - brood are being fed. Pollen is the protein source for growing larvae. The colony likes a buffer of 7-10 days' supplies, so if there is a gap in pollen collection longer than that it indicates a brood break - possibly a queen problem, but more likely just reduced laying due to a forage dearth.

Examine the pollen baskets on the rear legs. Some pollen is very difficult to spot, for example dark grey types. If it's a really big lump, the queen is **probably** OK (but due to the lag in curtailing collection, this is not a definitive indicator). If just small amounts are going in, it could be due to cool weather (they get chilled and return early) but according to Patterson may also indicate a failing queen or drone laying workers. Small "smears" on the legs is probably just pollen picked up incidentally by a nectar forager.

Pollen collection naturally varies a lot, so don't panic if it seems to stop. Foragers prioritise it in early Spring. I have seen pollen gathering pause for 3 weeks in late Spring, due to a drought – there was no nectar in flowers, so no spare food to raise extra mouths; and dry pollen doesn't stick easily to workers' pollen baskets. And in my area, we see little pollen in June. It's important to be aware of the environment before jumping to conclusions. And if a colony is having a brood break, typically after swarming, very little pollen goes in – because they have enough already. Another common cause of brood breaks is thymol mite treatments.[159]

Structures on the rear legs form "pollen baskets" (corbiculae) which can grip huge wads of pollen

There are colour charts online to identify the plant from the pollen colour and month – not very useful, but fun.

It's gathered primarily by the oldest foragers, and sometimes gathering seems to tail off after 10AM - this is simply because flowers are now warm enough to express nectar, so some workers divert to that. Pollen is usually in shorter supply than nectar, so they start gathering it early before competitors get it. (In extreme shortages, bees resort to stealing it off the bodies of other pollinators - a rare behaviour termed *cleptolecty*.)

---

159   Roger Patterson, *BIBBA Monthly*, Dec 2020 - bibba.com/bm-8349

"Normal" pollen use varies between colonies. Large commercial ones with continuously-laying queens, moved regularly to new forage, need lots, continuously. My static colonies have occasional extraordinary pollen binges, especially in Spring, but often only one in 10 or 20 foragers is carrying some in.

Different colours of pollen being gathered. The large drones lack pollen baskets and don't gather pollen.

Underside dusted with a very fine grained, sticky pollen.

Bees build up about 10 days' supplies of pollen, then slack off.

Pollen being fetched in November implies they are still raising brood - an Italian trait, undesirable in northern latitudes - this will disrupt the winter cluster.

A14 Excessive pollen on landing board / below hive entrance - enlarge the entrance / remove mouse guard.

A colony needs 20 - 80kg pollen a year (estimates vary wildly, depending on colony size, pollen type etc). As not every day is a flying day, a ball park figure is ~0.5kg a day for a strong colony.

Pollen loads under hive entrance in spring. Time to remove mouse guard.

A15 Propolis going in - small **shiny** lumps of brown, green or red on rear legs instead of pollen. This is a good thing. Easier to collect when warm, so tends to be seen in the afternoon. François Huber described it best in 1798 as small, glistening globules "the colour and lustre of a garnet".

A16 Bees with swollen abdomens going in - water carriers. Ensure there is a water supply near the hive.

Water is used for:

- evaporative cooling in hot weather and may be an indication of a nectar dearth (nectar is mainly water): but in very hot weather they may need water from both sources;
- to dilute honey stores during dearths, and winter;
- in brood rearing.

Bees at a bird bath is another indicator of a nectar dearth and associated need for water.

Some diseases cause distended abdomens, but we're talking about returning foragers here.

Wildman claimed[160] you can squeeze them and see if they regurgitate fluid(!)

Bees are translucent if held up to the light, so conceivably one could use this feature to determine what they are carrying.

Confession: I can't differentiate between a bee which is larger because it's tanked up on fluid, or one that's simply larger (older) than another. I often see small, young bees on the landing board but my bees' colouration varies so much I can't say "that one is darker because it is full".

Another sign of a heavily loaded bee is said to be it **walks low to the ground,** but I've not seen that either.

A17 Bees kissing on landing board - during periods of heavy foraging, returning nectar foragers often offload their payload to a house bee via trophallaxis. The house bee then takes the nectar in and stores it. The house bee is not yet a strong flyer, so she frees up the older, stronger forager to go get more nectar immediately.

However, this is usually done *inside* the hive. If outside, it implies a temporary excess of storage bees, perhaps because a nectar surge is winding down, who will soon be reassigned to other duties.

Pollen and propolis gatherers do not use intermediaries.

If a hive is not queen-right, workers have been seen to gradually defect to other hives, often bribing the guards with honey this way. Others remain as drone laying workers.

A18 Bees expelling bees - dead bees are dragged out by one or two *undertakers*. They attempt to fly them away to dump them remotely - but carrying their own weight they tend to fly down at 45°. Higher entrances help them to dump bodies further away.

Related behaviours: guards fighting robbers (**A8**) or the expulsion of drones / old workers (**A50**). Bees with DWV damaged wings have occasionally been reported as being forcibly ejected, and hives suffering from CBPV often repel returning infected foragers.

A19 Grooming and rubbish collection - general purpose workers at the entrance of a strong colony pick up dropped pollen loads, and groom returning foragers covered in pollen dust[161].

Propolis collector. Photo © 2024 Christopher Wren.

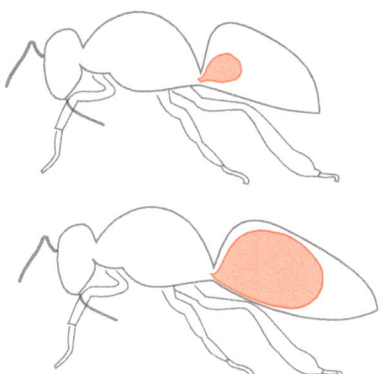

Bees transport water and nectar in one of their stomachs, visibly swelling.

Guards check new arrival, probably due to non standard scent.

---

160 Thomas Wildman, *Management of Bees* (1770) mentioned if you squeeze a forager at night, they regurgitate honey.

161 *Allogrooming* of others is mentioned by Tickner Edwardes in *Lore of the Honey Bee* (1908). More common on sunny days. Bees also groom themselves. *Allogrooming* also means grooming parasites off each other in the hive.

A20 Bees returning very heavily laden early morning; summer; little pollen; many rest on landing board before entering; lots of fanning and condensation; many shiny black bees (but not crawlers, i.e. not a paralysis virus) - honeydew harvesting from conifers and / or deciduous trees. Hairs get covered with honeydew and rubbed on pine needles during collection; it dries, making the hairs brittle; the hairs snap off --> black bees.

A21 Coloured patches on returning bees - pollen. Harmless. Even if bees are just taking nectar, pollen can rub off on them. Yellow foreheads implies nectar foraging on a brassica like mustard or OSR. Himalayan balsam[162] rubs very fine, white dust on bees' backs, creating white "**ghost bees**". Sometimes it's just on their foreheads. One species of orchid with sticky pollen creates **two small club-shaped "horns" which** remain on the bees for days. **Yellow spots on waist** is pollen from clover, cereals, yellow toadflax, daisies etc.

"Ghost bee" with stripe of Himalayan Balsam pollen on back. Photo © 2024 Christopher Wren.

A22 Bees entering & exiting are different colours - bees have empty honey crops when leaving a hive, thus can appear darker.

This is probably because the tergites (segments) over the abdomen concertina together when the honey stomach is empty, covering the lighter colours.

This effect is only discernable on bees with very consistent colouring, probably from a breeder. I've never seen it because I value outbreeding.

A23 Unusually dark colouration -

Unusually "dark" bees can be an illusion due to:

- Lighting (your colour perception varies with ambient lighting, background colour etc)
- Viewed from behind, you see partially under the tergites (segments) of the body so the same bee looks darker from this angle.
- Young bees' thoraxes (the 'shoulder' area behind the head) are covered in hairs: they look a bit furry. These get worn off by the end of their lives so older bees look darker and shiny. (Queens live years, not weeks and become quite bald.)
- Dead bees tend to look darker.

Dark, greasy bees - a lot of dark, shiny bees clustered outside the entrance may indicate a paralysis virus, especially if many have wings at odd angles or are trembling. Refer to chapter 14.2.

A24 Workers / drones of unusual size or shape -

Often an illusion because if a worker is hunched up while washboarding / cleaning the landing board, it seems much shorter. But:

---

162   The nectar of OSR and Himalayan Balsam are exceptionally concentrated sugar solutions, so are especially attractive to pollinators. HB is an invasive foreign species and is so attractive that native wildflowers can suffer from a lack of pollination. HB is the most common cause of ghost bees in Britain, but a few other plants create this effect, like a type of cultivated poppy, though wild poppies yield blue-black pollen.

- Summer foragers leave empty and return to a hive with distended abdomens filled with nectar. They also look about 10% longer. Winter bees are the opposite, they leave for evacuation (toilet) flights.

- Curled up dead bees look smaller than living ones, heart no longer pressurising their body.

- According to the literature, and conversations with bee researchers, bees **don't** grow after hatching and workers are of remarkably constant size compared to other hymenoptera. Nevertheless, I see (rare) half size workers, always young & furry, perfectly formed. And see photo - new hatchling seems ~10% smaller. Cannot explain.

Bee on right hatched a few minutes previously. Its wings and hairs are still damp and stuck to its body so it does not yet look full size.

- Small bees can result from starvation from long term pollen dearth. Indicators: colony is very defensive. Low pollen stores / few brood. Larvae are dry - should be swimming in milky liquid. No drones (expelled). Small larvae with hole in body, internal juices sucked out (by nurse bees prioritising feeding older larvae).

- Small bees can result from too few nurse bees for the number of larvae.

- Small drones: eggs laid in worker cells (drone laying workers, or no drone comb for queen). If it's laying workers you will see **many small drones** - far more than a healthy 15-20%. Such colonies will die in 2-3 months. See chapter 3, *Queen rightness, and drone laying workers*

- Old brood comb's cells gradually fill, the resulting larvae cannot grow to full size. Discussed in the *Reading Comb* chapter, under *Old Comb*.

- See also *Mutants and Mixes* in chapter 16.1

A25 Fanning – usually seen in summer. Often entire lines of bees shifting air. The draught is easily visible with smoke. Fanners are regularly relieved.

Fanning

Lemonbalm smell of Nasonov "come here" pheromone

Tail up Signalling **Tail down** Ventilation

- Tails down, accompanied by a sweet smell: nectar is being force-evaporated. Loud humming, even at night. You may also notice returning foragers' abdomens are rounded, bees aren't collecting water and the hive puts on weight rapidly.

- Tails down, no unusual smell: cooling, i.e. thermoregulation (you may see bees drinking water at nearby source); or just lack of oxygen. Can you enlarge the entrance?

- Tails up, lemony smell: signalling to a swarm or a queen on a mating flight.

Generally a good sign, unless *hundreds* are doing it and many bees are bearding below, indicating overheating, which may require temporarily shading the hive, and possibly enlarging the entrance, and ensuring a nearby water supply. If due to overcrowding, add a box if possible.

If you can see *inside* the hive, e.g through a window, you will see more bees fanning in there.

If ambient temperature drops below ~13C, it's too cold to process nectar or syrup into honey, and fanning stops.

Not queues to get in! Two lines fanning, wings blurred

A26 Bearding

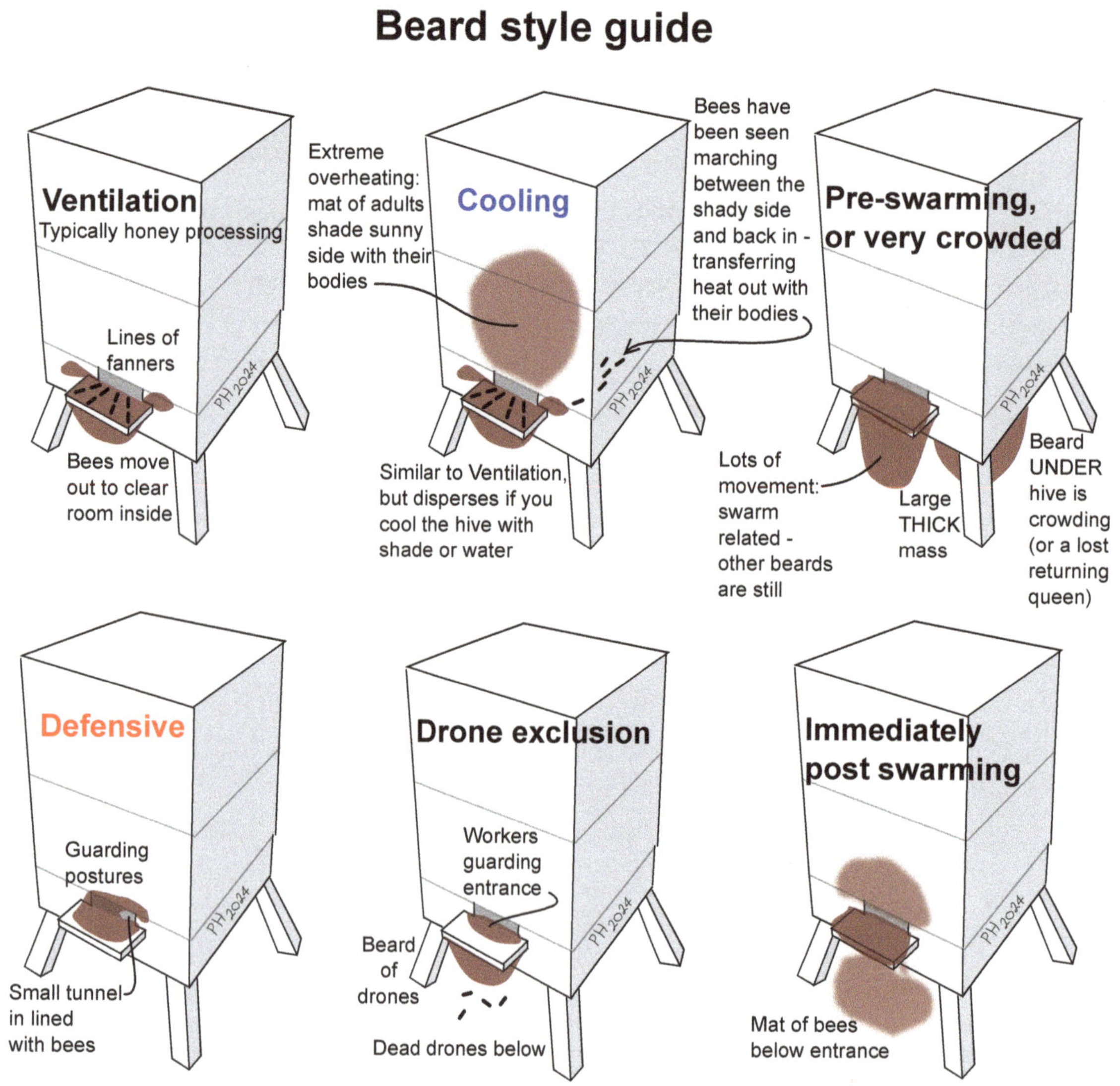

Beard style guide

Ventilation — Typically honey processing. Lines of fanners. Bees move out to clear room inside.

Cooling — Extreme overheating: mat of adults shade sunny side with their bodies. Similar to Ventilation, but disperses if you cool the hive with shade or water.

Pre-swarming, or very crowded — Bees have been seen marching between the shady side and back in - transferring heat out with their bodies. Lots of movement: swarm related - other beards are still. Large THICK mass. Beard UNDER hive is crowding (or a lost returning queen).

Defensive — Guarding postures. Small tunnel in lined with bees.

Drone exclusion — Workers guarding entrance. Beard of drones. Dead drones below.

Immediately post swarming — Mat of bees below entrance.

| The major beard types

Be aware:

- There are so few drones in managed hives using foundation that they can't be used to distinguish types of beard.

- Bees **do not hang around on honeycomb,** perhaps to avoid warming and collapsing it.

Bearding (1) - a mass of bees outside, usually hanging under the entrance, sometimes because the hive is **overheating** (in which case the beard covers the sunny side, acting as a heat shield - consider shading the hive) but often, particularly in humid weather, **simply to clear room on combs to process honey.**

Beard of bees below entrance and extensive lines of fanners pointing at the entrance. Cooling beards like this don't contain drones.

You can determine the cause by:

- Shading the hive - if it is overheating, the beard will shrink within minutes
- If the hive hums at night, smells sweetly and is bearding in the cool morning, the bearding is to provide ventilation and make room for traffic, not due to overheating[163].

The photo shows a healthy hive processing a lot of nectar during a heavy flow; the shadow is due to swing over the hive so no extra cooling was needed.

If due to overheating, don't 'help' the bees by assume they need **uncontrolled** extra ventilation in the form of opening mesh floors, raising the roof etc. Large extra holes will disrupt the air circulation and evaporative cooling keeping the brood from cooking above 35C. What you *can* do is:

- provide shade
- widen / enlarge the entrance they have
- adapt the hive to have 2 modest entrances (see chapter 17, *alternative entrances*). The bees will draw cool air in the top entrance and fan hot humid air out the bottom. (The opposite to uncontrolled convection!)

Bearding was quaintly referred to as *the loafing habit* in some early literature. Beards can persist overnight in hot weather.

Bearding (2) - Some sources claim that bearding is a precursor to swarming in the next few days. This is a **weak** correlation which is true for some colonies, in some locations, but is not a reliable indicator. But if you rule out the other causes, it can be an indicator of a swarm in the next few days, particularly if it grows huge over a few days.

Gareth John has noted (in windowed Warré hives) that beards often grow *inside* an empty bottom box or several days before a *prime* swarm emerges. It seems likely such beards are a way of preselecting who goes in such a swarm[164]. He's not spotted internal beards before a cast, perhaps because those are more spontaneous - so perhaps external beards are only a reliable sign of an imminent prime swarm, for a hive with no internal bottom space.

Bevan stated that if drones, and particularly pollen carriers, join the beard rather than go in, they are about to swarm imminently.

Early morning beard on a warm, humid day. Making room in a crammed hive for honey processing - a steady hum is audible from fanners. Beard disappears once foragers emerge. Note surprisingly small air channel at lower right and fanners in front. Normal for large colonies in warm, humid weather.

Bearding (3) - **crowd on landing board or small beard, fanning** *outwards* **with raised abdomens** (typically in May) - new queens need to go on a mating flight. To guide her home, the colony may deliberately fan come-hither pheromone out from their Nasonov glands. You may note a lemony scent. See diagram above for characteristic posture.

---

163   You might get more honey if you add a box or super.
164   Gareth John summarising a discussion with Tom Seeley. Gareth has noticed swarms from such beards often don't form an intermediary nearby cluster but shoot off a long way, possibly directly to a pre-scouted new home.

Bearding (4) - bees on front of hive despite moderate temperature; some fanning, but not as many fanners as when doing "emergency cooling"; strong smell of honey, typically July / August – *either* the hive is full of honey, it is even being packed into the brood nest. Make room immediately. *Or* they are indeed making room for honey production but simply because the weather is very humid – they may stay out even in light rain. Once the weather becomes drier they go back in the hive. *Bees don't generally hang around on honeycomb.*

Overheating. Entrance still has gap so hive can breathe. Photo © 2022 Helen Nunn.

Beard of drones (5) - beards don't normally contain drones (maybe a few in swarm beards). A mass of drones in late Summer or Autumn, indicates drone exclusion. In Autumn, they may huddle for warmth below hive. But you see a drone beard in, say, June i.e. there are still princesses around to mate with, it implies there is something wrong - it can be a sign of **starvation** - heft it or otherwise check stores. Recently in my area we have seen some hives initiate drone exclusion in July, a month early (climate change; plants are a month early too).

In Japan, a beard of *Apis cerana* drones is said to indicate imminent swarming.

Bearding (6) - Bee numbers have grown until they've run out of room, you need to expand the hive, e.g. add another box. During summer they also need extra comb area, not covered by bees, to process nectar into honey.

Bearding (7) - If it's young bees at 7AM, they are simply making room for traffic and ventilation.

Bearding (8) - entrance blocked by plug of bees - in Spring, this is simply conserving heat in the hive. An initially similar phenomenon is a traffic jam (**A29**) but that is much more frantic than a beard of parked bees.

Bearding (9) - defensive mat over entrance - see **A27**, **A28**. Sometimes bees cover the entire front of a hive after a clumsy inspection (cold weather, excessive smoke etc) - a threat display. If hornets are nearby, you may see a co-ordinated 'Mexican wave' warning ripple across the beard: "keep away!"

Not a beard: the mass of bees below shows it is a failed swarm. Queen is unable to fly and has not left the hive (in this case); also seen when queen wings are clipped. The entrance is not blocked by protective bees and they are fanning at the entrance, pointing in to entrance ("get back in here"). The bees will return inside within minutes.

Bearding (10) - after a swarm occupies a cavity - occasionally many bees remain outside as a beard for a few days. This is usually observed when there is old comb requiring cleaning already present.

Bearding (11) under hive or on unusual side of hive with open mesh floor - a newly mated queen has returned and is confused about where the entrance is, she's following the scent of the colony. This is also seen if a colony attempts to swarm but the queen's wings are clipped - she can't fly and crawls back to hive and is stuck underneath, trying to return home, perhaps thinking the scent coming out of a mesh floor indicates the entrance.

A27 Lots of guards (>4) at entrance, indicating **raiding and robbing** by other creatures – typically wasps or a larger bee colony – small colonies may require help defending, like reducing their entrances. Reduce the entrance (but if bees begin fanning, you have overdone this) or place branches in front. Chapter 14.2, *robbing* discusses more anti-robbing strategies, like moving hives.

Guards identify outsiders by scent. The home hive scent is a bit different for each colony - it depends on which nectars they have gathered (and genetics, people say, but I am not sure there is evidence for that). Guards are less particular about visitors if there is plenty of food, and particularly if they are all gathering the *same* nectar. Commercial beekeepers, who place many closely related hives next to the same crop, report much higher levels of *drifting bees* between hives than small scale amateurs see.

A28 Shimmering / sudden surge among bee beard when disturbed - guards reflect light off wings and move *en masse* when a threat like a flying hornet approaches. Until 2022 this was thought to be purely an Asian bee behaviour - *dorsata* and *cerana* ripple reflected sunlight at threats in a coordinated "Mexican wave". It's now been seen in Amm when European hornets from a nearby nest get too near[165].

A29 Traffic jam at entrance - enlarge the entrance.

A30 Excessive licking / inspection at entrance, guard clusters round some bees - **(1)** a forager has returned smelling oddly, but not enough to trigger the bouncers into full-on aggression. Perhaps she found an unusual nectar source, or has been robbing another hive and now smells faintly of strangers. **(2)** There are general purpose workers at an entrance who pick up dropped pollen loads, and groom arrivals covered with pollen dust.[166] **(3)** Young bees on their first flight extend their tongues fully, are surrounded by groomers then after a good licking go on their first flight.[167]

A31 Bumblebee entering hive - Weird huh?[168]
More common early in the season when bumble queens are looking for nest sites. Extraordinarily, bees often ignore them entering. However, I have seen a guard - enraged by a lawnmower - attack a large yellow and black bumblebee sunning itself nearby, possibly thinking it was a wasp.

A32 Entrance reduced by barrier – smaller colonies tend to build ramparts or curtains of propolis and wax[169]. It reduces heat loss, and the area their guards need to defend. It's not necessarily a sign of attack, it can be precautionary. François Huber[170] in Switzerland noted propolis walls being made versus Deaths Head Moths, forming zigzag entrances which were too narrow for them, but only in colonies in regions with DHM and at the time of year the moths actively raided. The propolis also helps sterilise incoming air, in hives where there is not yet any comb built behind the entrance (which is normally heavily propolised).

---

165 Filipe Salbany, Blenheim forest, 2022. The bee nest is near a hornet nest and has a permanent mat of entrance guards. Rather than a ripple, the videos show an instant reaction to approaching hornets across about 5-10cm of the beard, as if a large animal's hackles have risen.

166 *The Lore of the Honey Bee*, Tickner Edwardes, (1908) p.90 and p.154

167 Tickner Edwardes

168 Personal observation - bees were heavily foraging and content; Gareth John has seen it and Ron Brown writes of it.

169 The word "propolis" comes from Greek, meaning "before the city". The Greeks 2000 years ago realised the bees used it to build a barrier in front of the 'city', to reduce entrance size making it more easily defensible. Barriers can easily be 1cm thick and rock hard. Although everyone calls these *propolis barriers*, a more accurate term would be *cerumen* barriers - cerumen is a mix of wax and propolis, used by stingless bees to construct honey pots, nest walls etc, though if you google "cerumen" most hits will point at earwax.

170 Huber (1750-1831) was the first major methodical, scientific bee researcher and his contemporaries were in awe of his powers of observation and deduction. He didn't let a little thing like being blind stop him – his wife Marie, his servant François Burnens, and his son Pierre acted as his eyes and hands. Very interesting fellow.

Not all colonies exhibit this behaviour. It is evidently a trait that has been inadvertently bred out, but reappears in stocks allowed to revert to "wild" bees, along with a tendency to propolise entrances and walls near the entrances.

Propolis curtain across hive entrance; granular composition visible. This one was added by a strong hive in Autumn - previously they kept the entrance wide open to prevent traffic jamming. Sometimes the barrier has more propolis giving a melted look, sometimes it covers the floor too.

These walls are dynamic. Colonies may remove them as their numbers peak and they need ventilation, and during honey flows, then build them back when their numbers fall after swarming. They may act as mouse guards in winter.

A33 Wasps - they will probe hives and go for the least defended - not always the small ones. Some small colonies have strong morale (lots of queen pheromone?) and guards are much more tenacious than larger colonies. Wasps can fly in lower temperatures than bees so can raid with impunity early morning and early winter. See chapter 14.2, *the joy of wasps*. Use a veil as guards will be defensive.

Guard posture, legs splayed, wings extended, in front of propolis barrier (calm, ignored me).

A34 Hornet attack - see chapter 14.2, *Asian Hornets*

A35 Purposeful behaviour - hive is likely to be queen-right. For example: foragers flying straight in and out, not dawdling at the entrance.

A36 Very loud buzzing accompanied by crowd on landing board, massed fanning and a lemon smell: queen is on mating flight and they are signalling "home is here".

A37 Agitated, worried bees running in front of hive; a special buzz ('groaning') - colony has lost its queen. Mentioned in many old books.

A38 Indolent behaviour, bees loitering on landing board, few flying - do the rap test to determine if hive is queen-right (healthy, laying queen). If queen-right, the hive is simply happy, has enough honey for winter. (A conventional beekeeper would remove honey to stimulate them to gather more.) I simply leave the hive and watch and learn. The bees will be doing internal maintenance - propolising, raising brood.

However if queen wrong, the colony will die (drone laying queen / laying workers). At this point you may wish to simply observe the process. I personally might merge with a queen-right colony, which is simple, but if you favour complex interventions see *requeening / forced supersedure* in chapter 13.

A39 Bees swarm, then immediately return to hive - queen not flying. She may be relectant to fly, or not light enough yet (needs slimming), wing damaged and unable to fly, lost on the ground (though then they would cluster over her).

I once saw repeated aborted swarms from the same hive over a few days, followed by a successful one about 8 days after the first attempt. Clearly the old queen was unable to fly so in this case the first swarm was a *cast,* headed by a virgin queen.

Warning: the swarm cluster can get very tetchy before returning to the hive, if the queen cannot leave with them.

A40 Bees entering empty hive - scouts checking it out as a possible home for a swarm. If you see 20+ bees doing this they are *very* interested. 50 means they've probably chosen it as their new home[171]. Don't panic when all 50 disappear - they've gone to get the swarm (though sometimes this can take a few days). Watch closely and you occasionally see scouts fighting rival colonies' scouts to defend particularly desirable homes.

A41 Washboarding – continuous licking of the landing board area is a sign of contentment. Sometimes you see dozens of bees doing it in synchronous waves. They are thought to be cleaning and spreading a thin layer of sterilising propolis[172].

These are always young bees, there is a joke that the older workers tell them to go do this just to get them out of the hive and stop bothering everyone.

Torben Schiffer has filmed wild colonies doing this on inner walls too, and theorises that if beekeepers are not stealing honey all the time, the bees have time for desirable secondary tasks like keeping their home disinfected. *Every* internal surface of a healthy hive has a microscopically thin propolis layer.

Gareth John has seen drones washboarding.

A42 How many drones visible – Established colonies (usually older than one year) raise drones, often a precursor to swarming, and these tend to all fly off at the same time on a sunny afternoon. At that time of day the entrance traffic seems dominated by them and some people worry about drone laying queens / workers, but unless that happens all day, don't worry about it. New colonies focus on establishing a nest and do not generally have many drones – they can even be seen actively ejecting drones who, presumably, have drifted in from other colonies. Conversely, you occasionally see a lot of drones for a short period when a newly mated queen starts laying, before she begins laying workers.

Swarms from commercial hives (where drones are routinely culled, or limited by use of worker-sized foundation), raise a huge number of drones in their first year of freedom to build natural comb[173]. After this first year they realise you can have Too Many Boys and settle down to a normal 15-20% drone population.

The last hive accepting drones in Autumn, after others excluded theirs, gets swamped for a few days until its workers too finally lose patience.

If you see **lots** of **small** drones, see chapter 3, *Queen rightness, and drone laying workers*

A43 Ragged wing ends – old bees.

A44 Stunted bees, short abdomens, wings just stubs – varroa damage while they were larvae & pupae in the brood cells.

A45 K-wings - and asymmetric wings. Tracheal mites and some viruses can dislocate the wing muscles, affected bees can't fold away their wings properly. **Poisons** (including **stings**) can also

---

[171] I have only once *not* seen 50 be succeeded by a swarm: I discovered later, another beekeeper had caught (boxed) a nearby swarm, hanging from a low branch, at that time.

[172] Guy Thompson observed that after a torrential shower, his hives were thoroughly wet - but the areas which had been washboarded shrugged off the water, and his bees could land without getting their feet wet.

[173] Confirmed by Gareth John & Guy Thompson

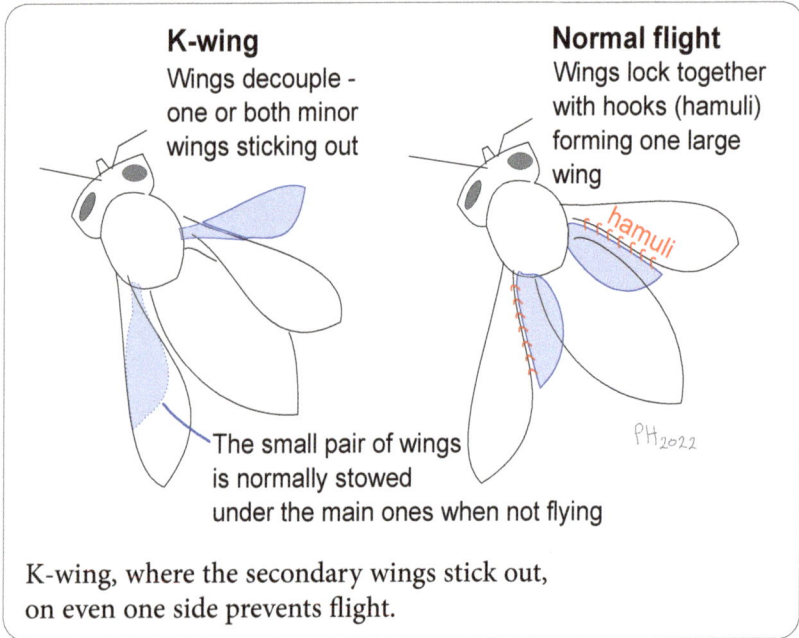

K-wing, where the secondary wings stick out, on even one side prevents flight.

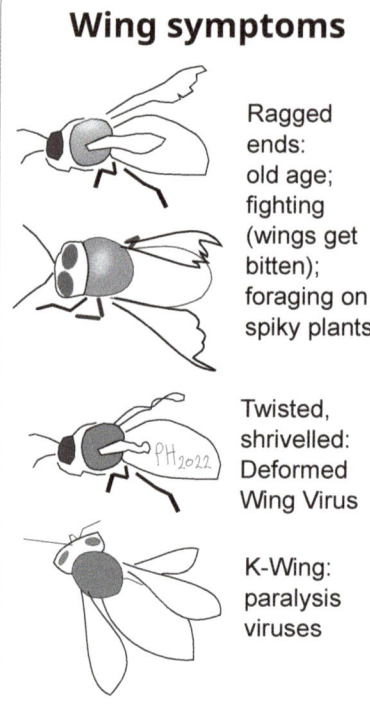

Don't panic about the odd K-wing, if the bees can still fly they just folded their wings awkwardly. Old queens and old foragers' wings show signs of wear at the wing tips, but DWV damage is much more severe.

cause this, and so can **old age.** You sometimes see one or two like this, which otherwise act normally, and then there are none the next day.

Occasional bees hold their wings out wide, but symmetrically and only one pair of wings visible - ignore them, they're fine, just weird. But if you begin seeing K-wing regularly, or a LOT of V-wings, watch their behaviour closely. Are they distressed? Trembling? Unable to fly? Do guards refuse entry? If so, take photos and refer to chapter 14.2 (diseases).

A46 Balling - a golf ball sized mob of bees smother a hornet, wasp or queen and kill them by overheating and stinging. Sometimes princesses returning from mating flights land at the wrong entrance, or smell of *other* queens after beekeepers touch them, or they are killing spare princesses after one successfully mates. Sometimes the ball is to *protect* a queen. You'll find the cause when it disperses. European bees balling wasps is rare and hornets, very rare - but it's been seen[174].

A47 Queen entering or approaching hive with thin white thread or point at end of her abdomen - this is "mating sign", the queen is returning from a successful mating flight with the white remains of a drone's endophallus visible after a successful mating flight. Do not block the queen's path.

Storch wrote (p.37-8) they - very very rarely - ball this queen, possibly because there are drone laying workers.

Not a disease: note leg and body fur. No further individuals like this seen at this hive. Simply a dark bee damaged on one side. No danger to colony.

---

174 See e.g. *Honey Bees*, Arndt and Tautz, p.78-85 published by the Natural History Museum

A48 Gelatinous blobs hanging off bees - "Bubble butt" - bees that have been squished during an open hive inspection. Their internal organs are hanging out and they will soon die.

A49 Grey butterflies around entrance - wax moths. A symbiont.

A50 Drone expulsion (August onwards) - no longer needed after mating season. Also sometimes seen during severe dearths. Hundreds huddle nearby for a day or two, gradually dying from cold and predators or trying their chances at other hives. Workers bite wings to drive them out - wings may look tattered. Any remaining drone larvae are also chucked out. Queenless hives, or ones undergoing a late supersedure, don't chase the drones out. If drones had stings you would lose many valuable workers in the fights!

Balling a wasp on a landing board, England, 2022. Image © 2024 Tanvir Mukhtar.

A51 Storm about to break: bees rush home - reason obvious, but an interesting sight amid the ominous calm. Insects caught in rain shelter under leaves. Bees driven to earth by heavy rain sometimes survive and revive in sun the next day, but there is a risk of losing a significant number of foragers to sudden showers[175].

Mating sign

A52 Passengers on bees - you may conceivably see **varroa mites** on bees on the landing board, but as returning bees move fast it's unlikely. **Thrips**, millimetre-long green insects like thin aphids, are a crop pest and sometimes hitch-hike on bees[176] to get to new plants. **Pollen mites,** considered harmless to bees, are sometimes seen in hives, having presumably been carried in.

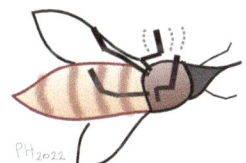

A53 Upside-down, exhausted bees on landing board in morning - wings stuck to condensing water running out of hive during heavy nectar flows.

A54 Night flying rare in Britain, but bees can fly by moonlight. An African species does this to avoid predatory wasps. *A. cerana* in Borneo fly very early in the morning to avoid *A. dorsata*[177].

A55 Bees under hive - a typical reason is exhausted fliers. The photo shows a mix of very young bees and returning foragers laden with pollen, having a rest before trying to take off and get in again. If they have a path to crawl in, they'll walk up that - it's easier - you can see one on the stick.

Tired bees are easy prey for wasps etc.

A56 Flies entering hive - dead hive. Flies are laying their eggs in bee bodies on hive floor. Expect maggots.

---

175   According to Pettigrew, in cold Scotland

176   *Thrips on bees: more than hitch-hikers?* Infante, Ortiz, Goldarazena & Sanchez, Ecology (2021) doi:10.1002/ecy.3551

177   Conversations with Filipe Salbany and I Made Setiawan

External debris

Early morning ejecta are clearest - before wind, ant and bird clean-up crews have scavenged evidence!

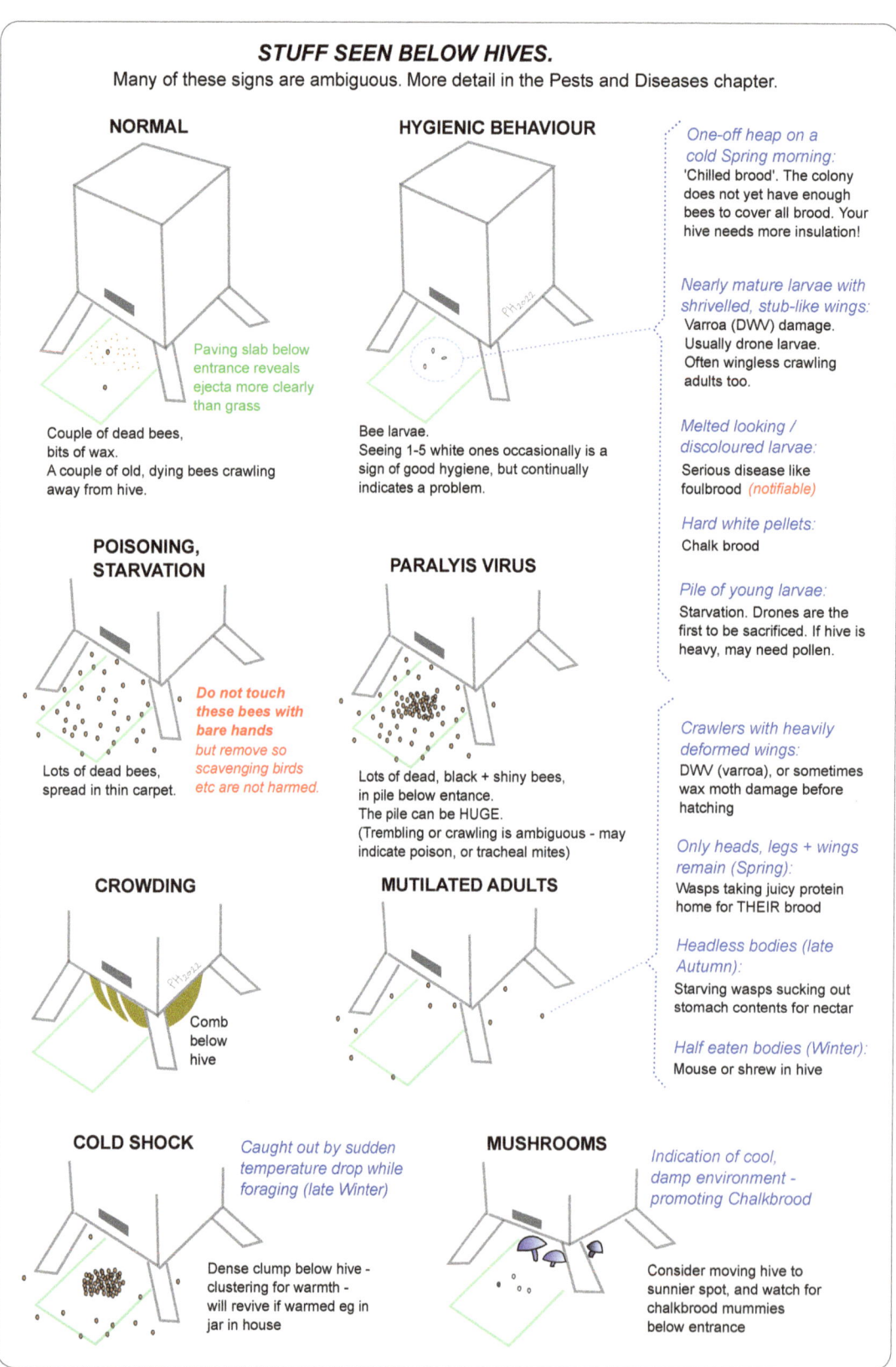

Other **non specific** clues sometimes seen:

- **Tongues protruding** - ambiguous symptom, could mean e.g. starvation, discussed in *Pests & Diseases* under Poisoning
- **K-wings** - can be viruses, tracheal mites, or just mechanical damage from a fight, spiderweb etc.
- **Rotting pile of corpses** - sometimes seen in spring as a colony clears out hundreds of bees who died over winter in one go. Cause ambiguous, possibly just old age.
- **Balling** - wasps, hornets, queens (sometimes they smell wrong, sometimes protectively). Once the ball breaks up you will see.
- **Mutilated wings**[178] - decades before varroa, ascribed to wax moth damage to larvae, along with **part of leg eaten off**, or **legs tied together** (with webs). Storch p.30 says[179] stunted wings and malformations can be due to chilled brood or lack of food.
- Watch for **animals below hives**: mice, slugs, ants, wasps, spiderwebs. See *Pests and Diseases* chapter. Toads are also mentioned as a problem in old texts (rare now). Many tired foragers end up below the entrance, so animals lurking here have a disproportionate effect.
- **Lethargic crawlers** - starvation, viruses, or cold; or just healthy young bees who have fallen out of the hive and, never having flown, don't know their way back.

Dead bees

- There are always a few dead and dying bees below an entrance – undertakers remove bodies and dying bees instinctively leave the hive. **The question is, do you see an unusual number?**
- Some "dead" bees in February or March are just cold and can be revived by breathing on them in the hand. They may have been chilled gathering cold water to dilute honey stores.

Ejected larvae

- **Ejected white larvae** – bees, often drones, with stubs where the wings should be.
- **Ejected white or brown larvae** - bald brood, development disrupted by wax moth.
- **Ejected thin white grubs** - wax moth larvae. Bees dealing with it. If the hive contains a lot of unpatrolled empty comb, remove it or the moths will multiply rapidly.

On a gruesome note, often the pupae aren't dead when they're thrown out. When collecting them for analysis, advanced ones often wiggle their legs feebly for a day or so.

The one thing notable by its abscence is cocoon fragments. Bees have difficulty biting through silk, so leave it in cells, though perhaps the cell caps contain some.

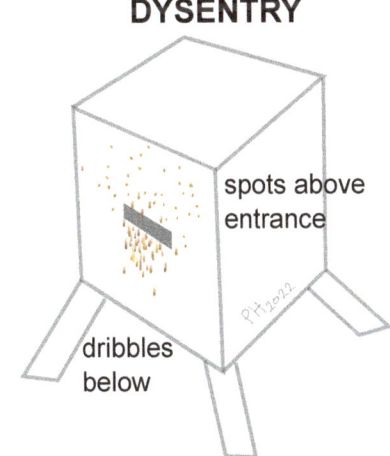

Yellow / brown faeces on the landing board - if there are streaks mainly below the entrance, and spots all over the front of the hive: dysentery, see chapter 14.2. A uniform brown stain is propolis, which is a good thing.

Many small, white crystals on landing board (Spring) - housecleaning: bees are discarding crystallised honey (or sugar?) from cold edges of combs.

---

178 Quimby, *Mysteries of Beekeeping Explained* (1853), p.161
179 *At the Hive Entrance* (1985), p.30

These crystals dissolve in water and are seen even with hives that are never fed, it's probably ivy honey. As 'syrup never crystallises' [180] this is one reason beekeepers in heavy ivy flow areas feed syrup in Autumn even when hives are heavy. It's not clear how bees survived before beekeepers.

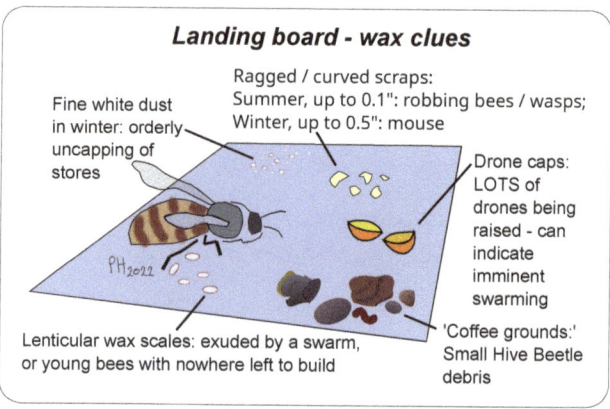

Condensation / ice at entrance in morning - this is the breath of the bees condensing as it leaves the warm hive. In winter, you may see ice. In Spring, the amount of water indicates how much brood is being raised. In summer, condensation is mainly due to forced evaporation of water from nectar, to make honey, and is related to how much honey is being made. Only seen with solid floored hives.

Ice at an entrance suggests that the hive is in a frost hollow. Not good. Frost hollows can be very local, the downslope side of a garden against a barrier may be one where the upslope side is not.

Honey on landing board - comb collapse, probably due to hive overheating. If so there will be bearding. It will taste sweet and could attract robbers.

Random foreign objects, like this grass that end up in a hive may be ejected from the entrance.

Dead queen below hive / on landing board, but colony not acting oddly - the colony has raised a new queen(s) and disposed of surplus princesses or the failing mother-queen. An old queen's body still attracts attention from bees, due to pheromones which a virgin lacks. This is an opportunity to familiarise yourself - compare queen & worker bodies with a magnifying glass and consider how you would spot her on a comb. Take photos. A virgin implies the old queen died or swarmed away 2 weeks ago.

Hard, rounded wads of pollen on landing board, often white on one side (Spring) - bees ejecting mouldy pollen - it gets cold and damp and mouldy over winter, if too far from the warm cluster.

Batch of chilled brood on landing board, after sudden cold weather.

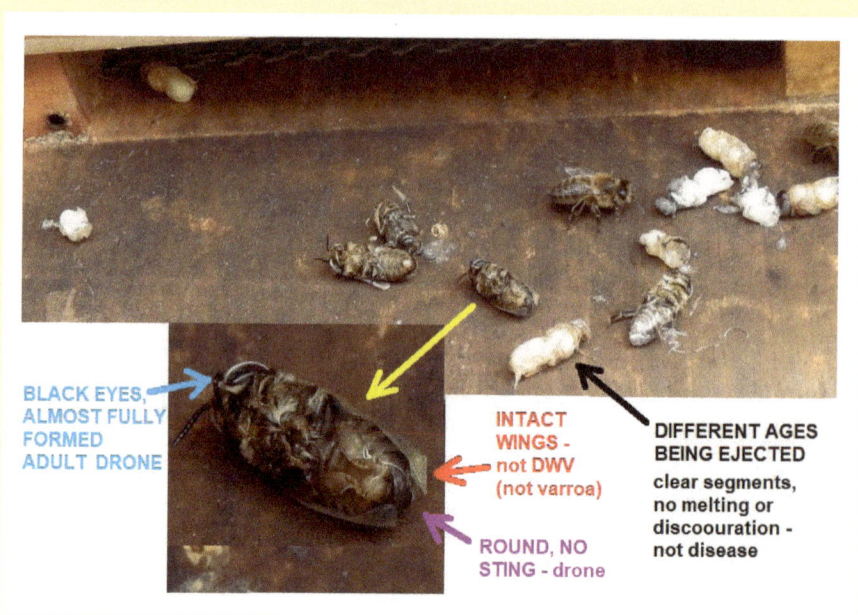

---

180   I can't verify this common assertion, but have yet to hear of an example

Examples (continued)

Hollowed out dismembered corpses are generally due to ants scavenging on dead bodies, or fly maggots - a natural clean-up crew.

Lack of wings on a pure white ejected larvae is normal, not DWV. Wings only form in the last few days of pupation, as the eyes darken. This is chilled brood.

Bee bitten in half by a wasp in Autumn to suck sweetness from its stomach. This half-bee has not died yet.

Alternatively, wasps sometimes chew the heads off to access stomach contents.

Photo © Elizabeth Robinson, who witnessed the wasp attack.

Early morning condensation (huge nectar flow being processed, May 3rd). Bees not yet active (cold morning). A wax moth larva, a few dead bees and some other debris (quilt stuffing fell into hive!) have not yet been cleared away.

Wad of chewed up and spat out bees: a skunk or possum has raided the hive. They lure out bees by scratching at the entrance, leaving scratch marks.

Photo © Joseph R. Rickle, Michigan

These late stage (dark eyed) ejected worker larvae show stubs where wings should be - DWV damage. All antennae are missing - pulled off as nurse bees pull pupae out. The colony is exhibiting hygienic behaviour.

Photo © Elizabeth Robinson.

Internal debris (hive floor / inspection tray)

Some hives have slide-out floors designed to permit examination of what is falling from the combs above.

If covered by a varroa / ventilation mesh, the bees can't clear away this detritus[181].

Wax cappings: among the numerous small white crumbs, are some larger wax caps which indicate the activity above:

You may see debris like this under a mesh floor. However - hygienic bees keep accessible solid floors clean!

- white caps - uncapping honey **stores**;
- light brown caps - hatching **worker brood**;
- large domed light brown caps - hatching **drone brood**

Thus you can tell if a colony is having to eat its stores prematurely, and if they are hatching young adults, without opening the hive.

You may see distinct bands of wax matching the position of the combs above, with most of the debris being under the brood.

Faeces informs you about pests. Insect poo is hard and cylindrical, but healthy bees don't defecate in their hive; the two types in this picture are probably **wax moth** and **ant**. Wax moth ones are about twice as long as their diameter. Wax moth faeces varies in colour - white to brown and black, depending on the colour of the comb they are consuming - and the size depends on the age of the moth larvae. Mammals like **mice** leave soft droppings which are pointy at each end.

White crystals (sugar) - honey at the cold edges of combs crystallising; the bees lack sufficient water to eat them and need a liquid feed. Mentioned by Storch.

Honeybee eggs on the inspection tray may indicate drone laying workers. (DLWs lay multiple eggs per cell, other workers haul out excess ones.) It may also indicate starvation (heft hive to check stores).

You will also see:

- **Pollen**
- **Propolis**

---

181   ...and wax moths lay eggs in this debris the bees can't clear away; pests are attracted to the smell of the hive; small hive beetles can get in through wider meshes; the hive loses heat and humidity; ventilation is disrupted. Natural nests have solid floors.

You may see:

- **Varroa**

- baby varroa, larval antennae - see "signs of hygienic behaviour" below

- **Wax moth larvae** (thin white grubs - lesser WM; fat ones - greater WM)

- **Condensation**

- **White pellicle fragments** (tiny torn bits of carapace shed by growing larvae)[182]

- **Tiny moving pollen mites** - harmless - probably absent in hives treated with miticides.

- **Fine, dark powder** in early Spring - old propolis from old brood cells before re-lining those cells. Once there is brood there will also be chewed wax from brood comb, which looks similar but is slightly lighter brown and is less fine[183]

- **Black mould** or other fungi on the baseboard indicate that zone is too cold and damp and, of course, if such a floor is below a mesh, the bees can't clean and propolise it. Consider adding insulation below the baseboard.

- Coffee grounds, sawdust, fungal spores and other pollen sized grains[184] - bees are known to gather these **if pollen is unavailable.** The nurse bees discard these as they can't predigest them into bee bread[185].

- Flour is sometimes gathered from mills in pollen dearths, but they *can* digest that, it's even used as pollen substitute in some "pollen patties".

Some beekeepers are experimenting with trays of litter (*sumps* or *eco-floors*) populated with book scorpions or predatory mites, to control varroa. As bees scrupulously clean and propolise the floors of well insulated cavities, this seems unneeded - and unwise. If you examine such sumps, the bees desperately try to propolise the litter.

Eggs (white lines) are smaller than the yellow pollen balls. Image © 2020 Will Hanrott.

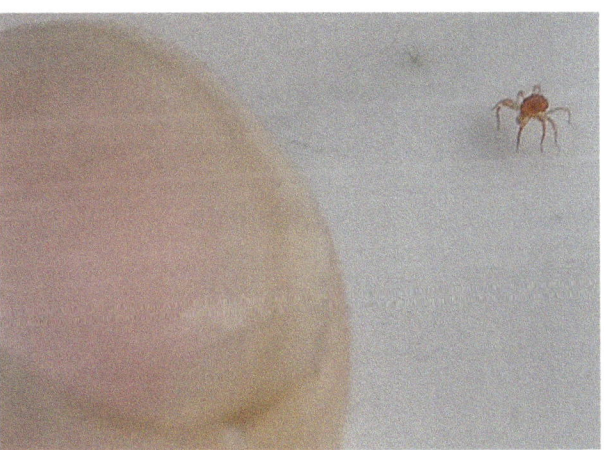

Seen near pollen.

---

182 *Bee Keeping New & Old* (1930) by Herrod-Hempsall p.68
183 Observations from Will Hanrott
184 See e.g. honeybeesuite.com/the-honey-bee-diet-requires-many-amino-acids - an article discussing intensive American beekeeping. This sounds like too many hives for available forage. I have never seen these gathered and my bees ignore them, and mushroom spores, if offered. Bees only need 10 amino acids, plentiful among the European plants they co-evolved with. The coffee grounds may simply be gathered because bees like caffeine! *Caffeine in Floral Nectar Enhances a Pollinator's Memory of Reward,* Wright et al (2013), doi/10.1126/science.1228806
185 One theory is this indicates nutritional stress and the bees need pollen substitute, but fungal spore gathering despite plentiful pollen is reported not infrequently, as is drinking exudates on the surface of fungi, so a more likely reason is for self-medicating.

Record curious findings

Identifying ambiguous detritus without a magnifying glass

- Wax is soft, floats on water, and melts easily
- Sugar crystals are hard, and dissolve in water
- Pollen balls smear into a powder when crushed. Pollen disperses in water.
- Propolis is sticky
- Sticky liquid below brood cells may be condensate off honey cells or, in very hot weather, honey / nectar leaking due to comb collapse. Ants would be attracted to this, but not to condensing water.

There is talk about tasting stuff to identify it. This seems unwise as there could be all kinds of bacteria remaining from faeces, rotting bodies etc which have since been removed[186].

Wax scales shed by comb building bees (large translucent flakes) amid more normal crumbs, and a couple of varroa. Photo © 2020 Will Hanrott.

Signs of hygienic behaviour

These signs of nurse bees ejecting infested larvae were first spotted by Ron Hoskins in 2007, using a magnifying glass to inspect his varroa-resistant hives' inspection trays.

The first picture shows miniature varroa, about 1/6th adult size - immature mites which fall from a pupa as it is pulled from its cell.

The second picture, of tiny transparent rods, shows pupal antennae, which are pulled off when hygienic adults remove infested pupae.

Another sign of hygienic behaviour is seeing mites which have dented carapaces, or missing legs. These wounds are due to bees biting them[187].

These are good things to find.

Odd one-off things turn up in hives. This is not a beetle, but a dead bumblebee (4 wings not 2 - but 2 are just stubs now) among a crowd of workers on a floor. Its colourful hairs have been worn away, leaving its black carapace. It was probably too large to haul up to the entrance 40cm above.

---

186 Though on beehomeblog.wordpress.com, two blind beekeepers mention using taste to distinguish wax / pollen

187 There was great debate over the significance of dented carapaces and I may be out of date in my assertion. Ron told me he's read even an apparently minor wound, like a missing leg, can be fatal as the mite then bleeds to death. You can read more about Ron's reseach and findings at www.swindonhoneybeeconservation.org.uk/research , but Ron died in 2023 and the website won't be there forever.

Larval antennae. Image courtesy of Ron Hoskins

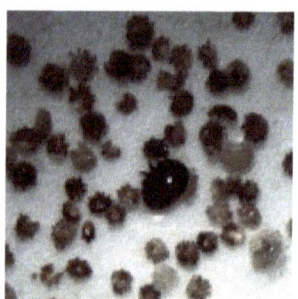
baby varroa. Image courtesy of Ron Hoskins

The futility of mite counts

Although mesh floors allow you to count the varroa mites falling from the colony above, it isn't very useful data.

Their absolute numbers are irrelevant – as long as they stay stable, the bees are dealing with them. To put this another way: as a no-treatment beekeeper, either the bees survive, or they don't and you restock with survivor stock. Either way, counting mites takes a lot of time and doesn't change the outcome.

If the numbers spiral out of control there are two likely causes -

- You're using commercial bees that don't have mite resistance traits, or
- A nearby hive with commercial bees has died of varroosis (a "mite bomb") and your bees are robbing its honey and inadvertently carrying home parasite passengers.[188]

At this point you need to consider if you want to resort to chemical treatment. Or a *shook swarm*, described in chapter 13.

Window observations

Unless you have a LOT of hives, windows are worth investing in. Viewing the flat side of the comb tells you more about the state of the colony's stores; but most people use them to view comb edge-on so they can monitor how a colony is building up and whether they need to give the bees room to expand.

Note how rapidly bees move within your hives. Runny or calm? A change may indicate eg queenlessness. Runniness varies between colonies, some do this if sunlight falls directly onto combs.

Tradeoffs affecting window orientation. Excessive reflections can be an issue. Other constraints include comb orientation (cold way) and how near walls the hives are. To get a good view, I have to stand in the flight path which is non-ideal.

Window clues

Stores / brood - Here is lots of capped honey, in a hive in late Autumn. They only put capped honey (or brood) against the cooler window once the comb is absolutely full, so this box must be crammed with honey.

---

188 Some folk are sceptical mite bombs even exist as they don't take out every colony in an area. See e.g. parkerbees.blogspot.com/2019/08/a-much-simpler-explanation-of-why-mite.html ("Some hives are just bad at dealing with mites"). Tom Seeley takes the view that they exist, but they occur when **conventional** colonies collapse and are robbed out. His simulated wild colonies, with small nests and low hive density, never get high mite levels. He has really good evidence of this, eg a 2 year study looking specifically at the impact of nest size on mite levels. [arboreal-apiculture-salon.libsyn.com/salon-no-22-with-thomas-seeley]

The Observant Beekeeper

Capped honey against window. Bees use gaps as runways.

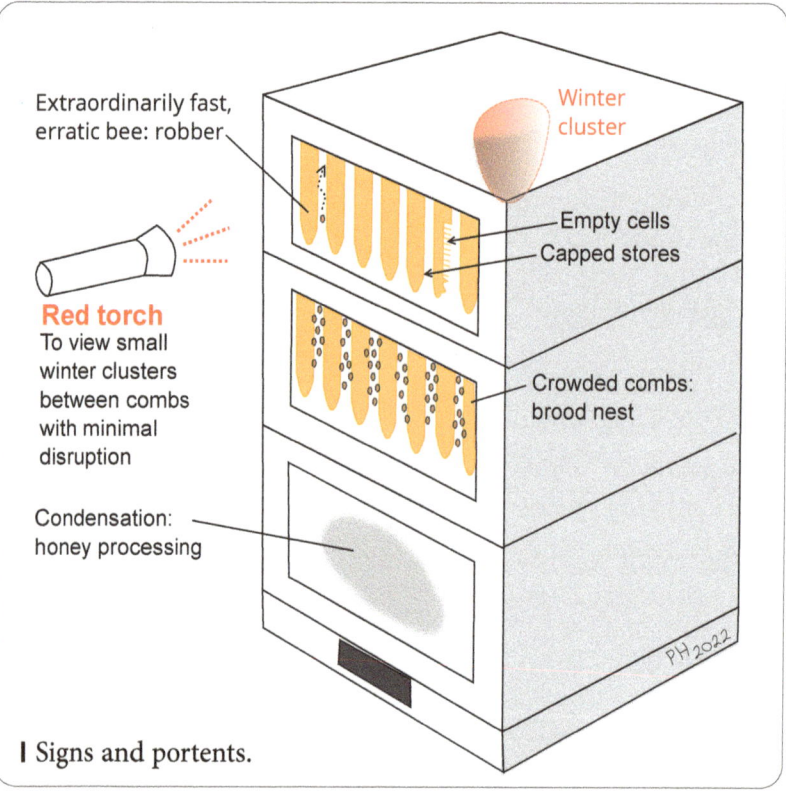

Signs and portents.

Chaining

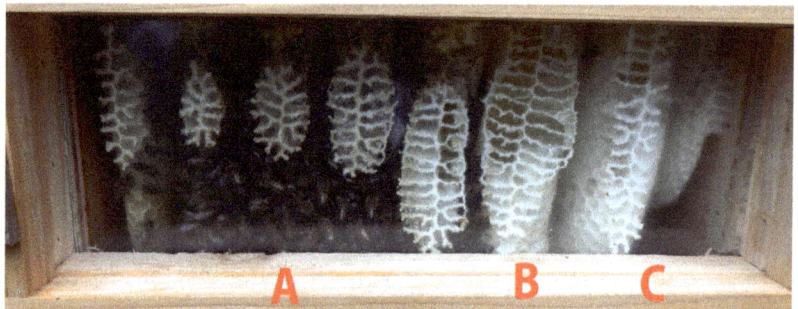

December. Cluster **A** is big enough, but lots of empty comb **C**, colony will starve over winter. Above **B** you can see how much deeper honey cells can be. Access runways are also visible.

Queen cells - Queens (and drones) are often raised at the edges of combs (possibly requiring a slightly lower temperature), you occasionally see queen cells through windows.

Winter clusters can be ridiculously small – a couple of fistfuls of bees – yet the colony will be fine in Spring, which illustrates dramatically how BBKA advice that you 'need' 3-5 frames of bees to survive winter is quite divergent from how bees optimised their survival over millions of years. Clusters contract when cold, and then suddenly seem 3x larger on a warm day.

Chaining ('festooning') below combs - Sometimes described as 'bees on a string' - no one really knows, various ideas have been put forward:

- *Comb building guide / scaffold*: The bees often hang in a catenary curve, the same shape a chain makes when it hangs - and the same shape as comb is initially made in. This implies a connection and it does seem to occur most when building comb. Sometimes the chains are a thick rope of bees, which is thought to help them to heat the wax they exude, making it easier to mould, or stimulate wax production.

- *Blocking draughts from the entrance*: chaining can be a way of conserving heat for brood
- *Guiding airflow* like a dynamic rudder. Derek Mitchell humourously refers to this as "solving differential equations", making the point that they may **plan** where best to hang comb to e.g. block cold draughts and optimise airflow.
- *3D web of guards*: Torben Schiffer, using an endoscope, observed chains below comb in a wild nest with an entrance below them. A wasp pushed into the entrance was immediately mobbed by bees from the chains.
- *Artefact of frame use*: After studying behaviour in tree hollows, Tautz & Arndt suggest bees mainly use chaining to form 3D webs to hold heat in to small volumes where they are comb building or raising brood.

Thinking outside the box - external factors

Sometimes bees' behaviour is due to things well beyond the hive. Imagine you are a bee. You will be *acutely* aware of smells, air pollution, temperature, humidity, wind, sunlight.

Bees acting oddly may mean nothing if a rainstorm is coming, but could be significant if there are no other apparent external stress factors.

- What's the weather like?
- What is the forage situation in your area? Drought or cold can lead to dearths. Sustained wet weather prevents new queens getting out to mate.
- Bees working honeydew implies no good nectar sources.
- What can you smell nearby? Petrol fumes are toxic to bees.
- Is a neighbour mowing grass? Having a bonfire? Is anyone painting nearby? Are there roadworks or builders creating dust, noise and vibration?
- Has anyone sprayed crops or used lawn treatments?
- Are there lots of wasps this year?
- Watch for wasps entering via unguarded cracks.
- Look under the roof - warm dry place for other creatures - occasional ant and mouse nests. Check it is dry.
- Any signs of other animals nearby, such as tracks in snow
- Evidence of vandalism

And has the colony recently swarmed? How long ago? This affects baseline behaviour.

Smell

- If you've smoked the hive, that's all you'll smell.
- Humans' sense of smell varies widely but is poor, we're said to have just 10 basic smells.
- Hives never smell of fish, mint, gasoline or almonds.
- The default smell of a healthy hive is a sweet and flowery, with a heavy propolis note. Every hive smells a bit different, for example harvesting OSR gives a more acidic smell while it is processed into honey. Some people can identify the smell of different nectars.

That being said:

- **Sweet smell**, particularly noticeable on humid summer evenings, accompanied by loud humming (fanning) – honey processing

- **Pungent smell** - The smell of goldenrod being ripened, if you have a large bloom in your area, has a musky sour sell which people sometimes think is foul brood. A few people describe goldenrod nectar's smell as having a vinegar or ammonia tone. Asters are also notorious for pungent nectar.

- **Unpleasant smell like glue, or a sour odour** - rotting brood, possibly foul brood; rotting chalk brood mummies; or deaths from overheating, or rotting adult bee bodies on the floor which have not been carried out. See diseases section.

- **Bananas / pear drops** - alarm pheromone. Leave immediately unless wearing protection.

- **Yeasty** - "fresh baked bread" is pollen fermenting into bee bread, i.e. there is brood to feed: the queen is laying, all is well.

- **Strong alcohol smell** - Conventional beekeepers warn of a strong alcoholic brewery smell, in Spring, accompanied by yellow-brown stains on the front of the hive (dysentery). This indicates fermenting unripe honey. This may be because sugar syrup was fed too late in Autumn, or may be due to a late flow of ivy the bees never got round to consuming befoe clustering. (Nectar can't be processed into honey if air temperature is below about 12C.) If noticed, you need to open the hive and investigate further: consuming fermented stores will weaken them and give other problems a foothold. Remove and discard spoiled stores.

- **Lemons / lemongrass** - the "come here" pheromone calling to e.g. a queen on a mating flight

- **Ammonia, urine, mouse smell** - mice in hive (only happens in winter). Check floor for mouse droppings to confirm. Stick stethoscope in entrance, rap the hive & listen for scuttling. Mouse will kill colony - must be dealt with.

Sounds

You will come to unconsciously diagnose hives by their distinctive tones.

There's no one frequency bees buzz at. The hum is generated by wings moving air[189], and they have to work harder when carrying a heavy load back to the hive than when zooming off empty! It also varies a bit with temperature.

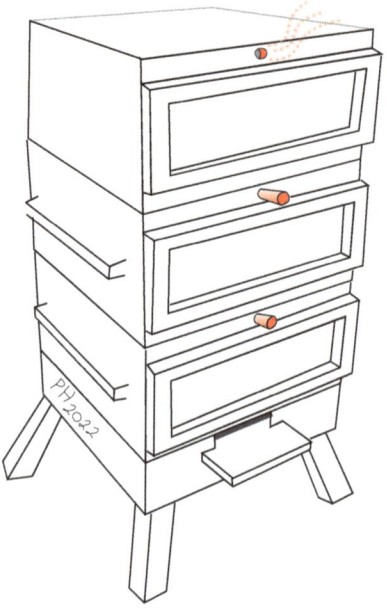

Geert Steelant (delachendebijenkast.be) adds a 14mm hole above each window, allowing users to remove a plug, insert a piece of garden hose and smell. Push mosquito net into one end and use a funnel at the other.

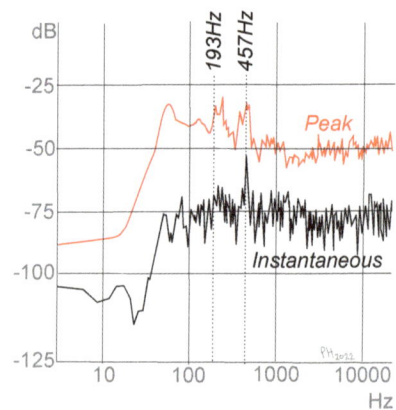

A spectrum analyser app shows how many different notes make up the sound of a (content) hive. But you've evolved to interpret sound with your ears, not eyes - your ears will instantly warn you of shifts in dominant tones, volume etc - a much more intuitive tool!

---

189 Tickner Edwardes claimed bees also make sounds with their spiracles (breathing tubes in their skins), *Lore of the Honey Bee*, 7th Ed (1916) p.112-113. This has been disproven by experiments like using helium rich atmospheres and noting no change in sounds, i.e. they depend on wing oscillations, not flute-like resonances.

So the sound at a busy hive entrance is actually a medley of tones, and your marvellous ear-brain signal processor will just *learn* what a happy hive sounds like and alert you instantly when something sounds "odd".

External sounds

Baseline buzz outside the hive, sound of content, busy flyers - a mix around 200Hz, a "drowsy, dreamy song", like a chorus humming G3.

This choir-like mix of tones, around the same frequencies as deeper human voices, creates the characteristic relaxing buzz of a calm hive[190].

Loud, high pitched buzz about one octave above normal tone - "Go away" - upset guards sometimes try to scare you away by flying at your face with this characteristic warning buzz, before escalating.

Loud, low pitched buzz or "clatter"- Drones also buzz very loudly, but a much deeper tone (a... drone!). Get to know the difference, to avoid unnecessary panic. They have no stings. A spectrum analyser shows their noise has the same mix as worker, but with the addition of a broad buzz below 100Hz.

Deeper, louder song - swarm issuing. Sometimes preceded by a peculiar low throbbing note like a locomotive brought to a standstill, then a silence just before issuing (as the bees gorge on honey?)[191].

Very loud low buzzing - as if lots of drones are flying - can mean a swarm is about to emerge from the hive, especially if actually accompanied by lots of drones flying[192] and, especially, bees still carrying pollen emerge from hive.

However, in the moments before a swarm departure, many hives in an apiary will show increased entrance activity, making it tricky to know which one to stand in front of with a baited skep ready to attract the queens! Sometimes the swarming one will have a beard, or show a drop in activity immediately before the swarm issues.

Piping - particularly noticeable in the evening. Two noises, often called *tooting* and *quacking*. Queens in the hive are calling to each other; the quacks come from queens still in cells, and are more muffled.

Description: the noises are a sequence of pulses, like a mobile phone's vibrator buzzing shrilly. Bevan described quacking rather well as an *ark ark ark* noise.

Tooting has a long first syllable followed by subsequent short ones, whereas quacking usually has a short first syllable followed by longer ones[193].

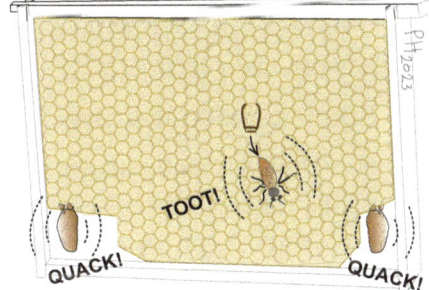

Queen piping
Loose queen toots imprisoned queens quack back

---

190   A spectrum analyser app shows broad peaks around this fundamental frequency and the first harmonic - this overtone is characteristic of a resonant system. That's because wing movement comes about by flexing their thorax – this wiggles the wings - the wing / elastic thorax system resonates like a spring. There are no muscles connected to the wings! The resonant frequency will be different for a forager stuffed with nectar.

191   The silence is mentioned by both Wildman (*A Treatise on the Management of Bees,* 1770) and Edwardes (*Bee Master of Warrilow*, p. 49, ~1920). Edwardes also mentions an uncannily still beard of bees and, at the end of the distinctive low vibrating murmur, piping as the queen works herself up.

192   Storch, p.34

193   *Acoustic and vibration monitoring of honeybee colonies,* Uthoff, Homsi & von Bergen (2023), *https://doi.org/10.1016/j.compag.2022.107589*

Significance: an afterswarm (cast) will emerge soon[194].

However, Tickner Edwardes mentions (*Lore of the Honey Bee*, p. 134) that if a mobile queen's challenge (piping) is answered by a hiss, it means the workers don't want afterswarms and once she has mated, she will be allowed to kill her imprisoned rivals.

Bevan states that if you don't hear piping between 9 – 13 days after a prime swarm, there will be no afterswarm.

Workers near a queen freeze in position whilst she pipes - if one pipes on a comb you're holding, the entire frame goes still for a moment. It is usually heard from virgins, but occasionally defintiely-mated queens shipped with others have been heard piping.

Loud continuous high tone - fanning, typically simply for ventilation. Around 150Hz (B2).

Loud humming at night and many fanners at entrance, late summer - honey processing. Don't disturb them, guard bees can see better than you at night; but the powerful thrum is impressive. Around 150Hz (B2).

Continuous angry buzz - also described as a "queenless roar" or "shrill chaotic clamour" or "keening" or "warbling followed after a few hours by a roar"[195] - hive is not queen right. She's missing or infertile. Old books consistently describe hives who lose a queen as roaring, and setting up a great commotion, running around in a frenzy; but many modern beekeepers (who requeen regularly) say they have never heard such a roar.

My own hives get apathetic if the queen *fails* - loses fertility - but I've yet to see what happens if she disappears suddenly. Re-reading the old texts, and after discussing with Gareth John, it seems likely the roar and running around only occurs *when a queen fails to return from a mating flight*, and the bees are trying to create a homing signal of scent and sound to guide her home. This frantic fanning, which Gareth has seen involve an entire cast on the outside of a hive for days when a queen fails to return from a mating flight, eventually subsides. He's also heard the sound from established hives that have become queenless when opened.

'Screaming' (1) - The Death's Head Hawkmoth enters hives to eat honey, and uses a scream to deter attacking guards. Actually it's more like a loud chirp or mouse-like squeak and sounds like a noise queens make which makes bees freeze.

'Screaming' (2) - Asian honeybees, *A. cerana* make a warning noise some call a scream[196] (a series of high pitched buzzes modulated with vibrato) when giant hornets attack their nest, but it has not been recorded for Western honeybees, yet.

Sounds inside the hive

Heard with a **stethoscope** placed in the hive entrance[197]. A stethoscope is particularly useful in winter when there is no entrance activity. Even a small cluster will make detectable noise this way. You can get an idea of cluster size from its volume.

---

194   Eva Crane, in *The World History of Beekeeping and Honey Hunting* (1999, p.169) says that present-day Egyptian beekeepers "call" the queens during swarm season. Paraphrasing: "The beekeeper... imitates the sound of piping by young virgin queens... the sound is like *kak, kak* or *kak, kak, kaak*, or *ee,ee* ... if any virgin queens reply he knows the colony will swarm unless he takes action". It is suspected that a 4,500 year old Egyptian depiction of beekeeping shows the same practise.

195   https://doi.org/10.1016/j.compag.2022.107589

196   *Giant hornet (Vespa soror) attacks trigger frenetic antipredator signalling in honeybee (Apis cerana) colonies,* Mattila et al, Royal Society (2021). This has links to at least one audio recording of the 'scream'.

197   You can make a simple Laennec stethoscope (as used by midwives) by pressing one end of a simple wide tube, like guttering, against the hive and your ear at the other end.

Baseline hiss - a colony normally makes a quiet rustling hiss, easily audible[198]. This is probably wings brushing by each other, because if you rap the hive the volume rises for a few seconds. For large colonies, you can hear this by pressing your ear against the wall or roof.

Rap test – put your ear against the hive and bang the hive sharply. The hiss of the bees will rise to a "roar". If this roar goes on for just 1-3 seconds, the hive is queen-right, i.e. there is a healthy queen and the bees are calm and purposeful. If it goes on for 30 seconds, they are queenless.

To be more precise, the hive *thinks* it's queen-right. A queenless hive of drone-laying workers think they're QR. What it's actually checking is *"do we have a fertile queen OR brood"*. So don't trust this test 100%.

This test is easy with TBHs but for Warrés you may need a stethoscope or listening tube inside the entrance to hear what's going on in there. Some people use a stiff rod pressed between the hive wall and just below their ear, to hear by bone conduction.

This applies to open hives during inspections too - queen-right ones respond to knocks with a momentary roar. A lack of response may indicate starvation. But it may simply mean they are *temporarily* queenless after swarming.

The rap test has been investigated by scientists[199] using thousands of randomised knocks. Their findings broadly agreed with the above, and they noted the bees' response (roar / loud hiss) was most noticeable in winter, when they were clustering.

Occasional loud purr / sporadic thrumming - bursts of a low frequency drumming sound: scouts resonating the *Dancing Floor* comb with the *Waggle Dance*, informing foragers where to go for food. This may be the sound some people have described as *growling*.

One bee making a continuous loud buzz tone for maybe a minute - no one knows. Rare; quite striking. Not fanning, because that's silent. Seems to come from deep within the combs. Heater bee?

25Hz (G0, just below the lowest piano note) - 'Keeping Warm' buzz, louder on cold nights and over winter.

High pitched *beep!* - a multifunction noise[200] like English's "oi!"

- apparently a noise of surprise when two bees bump head-on in the darkness of the hive;
- also used to signal "stop doing that" to e.g. waggle dancers;
- and a danger signal
- It's much shorter than other worker piping noises, typ 0.14 secs
- Workers tend to freeze when they hear this, or piping

Clicking - chewing and moulding wax with their jaws. When a lot of comb is being built it can sound like the "snap crackle pop" of Rice Crispies, or crunching up a crisp packet[201].

---

198 Asian honeybees (*A. florea, A. cerana*) hiss loudly at predators. One source states "Hisses are produced when many workers move their bodies and vibrate their wings synchronously".

199 *Quantitative assessments of honeybee colony's response to an artificial vibrational pulse*, Bencsik et al, *Scientific Reports* (2024). Eight hives of British bees. If you are comfortable with technical details and graphs about frequencies etc, there are other interesting details in this paper.

200 *Long-term trends in the honeybee 'whooping signal' revealed by automated detection*, Ramsey, Bencsik, Newton (2017) doi.org/10.1371/journal.pone.0181736

201 Thanks to John Woods for this observation

Unconfirmed reports

Tickner Edwardes describes the following sounds:

Shrill snappish background tones during the honey season. He may have been referring to stop-signal beeps in crowded hives.

"Deep sobbing outcry" increasing hour by hour - drones being rounded up and forced out (*Lore of the Honeybee*)

Wavering intermittent hum - weak or starving stock (*Lore of the Honeybee*)

Weight, warmth, touch

Hefting means to tip one side of a hive to gauge its weight. This warns you if the hive is suspiciously light.

Honey is very dense. A thriving hive has 15+kg of stores going into winter. That's like a sack of coal - you **will** know if the hive feels suspiciously light during dearths.

If you're unsure whether a hive is empty, say in a dearth, tip opposite sides to gauge the difference; one end is often fuller. If you're still not sure - it's starving.

Hefting isn't really possible on tall, thin Warrés or heavy super-insulated hives. For example with 2 inch thick wood walls it is difficult to tell, with your senses, if one end of a TBH is heavier.

Hefting can fool you: Increased weight *doesn't necessarily mean the bees have plenty of honey.* Larvae and nectar are heavy! And sometimes in dearths, or Spring, bees will gather water to dilute honey stores (they can't digest pure honey!). This can make the hive heavier even though they are not gathering nectar. They are burning reserves.

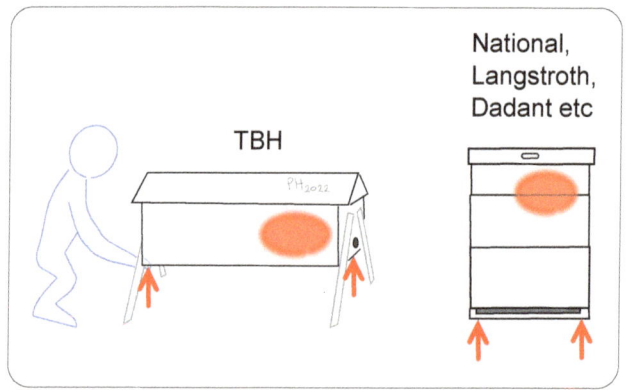

Hefting opposite ends to estimate weight and location of stores (red). Straightforward with wide, shallow hives.

Hefting a tall, narrow Warré is futile: you can't discriminate weight distribution. And it may topple. But you can tip the top box a few mm to gauge stores - most honey is in the top box in vertical hives.

Luggage scales *can be used for precision measurements:* interesting, but tricky and not vital. You weigh both sides, add the results and divide by 2 to get an average. A **windowed** Warré box full of empty comb may weigh 5kg.

It's possible - but expensive - to monitor weight continuously, electronically. If you have such monitoring, you'll see jumps in weight when swarms leave; daily cycles as foragers leave and return; weight gradually decreasing over winter, etc.

Warmth - your hand can often feel where the bee cluster is through a top cloth or thin roof. If you are about to open a hive, this informs you where the mass of bees are, or to put it another way, on cold days - where the most defensive ones are likely to be ('stop chilling our brood').

It's not 100% reliable as sometimes the bees are deep down in the hive and the top cloth is cool. But if there is *no* warm spot it alerts you that the queen *may* have laying problems.

Winter monitoring

Heat from winter cluster, or bees raising brood, can be felt through top cloth

Finger in entrance can feel warm, moist breath of living colony flowing past

Humidity on window pane - inadequate insulation, improve immediately, even if it's just duct taping bubble wrap around the hive

Loud buzzing in winter - fanning because entrance is blocked by e.g. dead bees - hook these out

Winter clusters hum - they are vibrating wing muscles to keep warm, it sounds like the normal flying tone, but their wings are decoupled so the sound is very quiet. It is only just discernable at 5°C with a stethoscope, but gets louder as temperature falls. It indicates the bees are OK.

Ice or dark stain on landing board - just the bees' breath condensing. They're alive.

Lots of paralysed bees outside hives on a fine day in e.g. January - the bees have been tempted out by sun, then got too cold (<8C) to get back into the hive. They can be swept into a box, brought indoors to warm up and released back by the hives.

Pettigrew records this as a particular issue when the bees forage on e.g. heather so in addition to being cold, they are heavy. Although that would be in August, it can get suddenly cold in the hills.

Hive very active after a long period of bad weather - you sometimes see a hive use the first day of good weather to drag out accumulated dead bees; there are a fair number of bees doing orientation and cleansing flights.

A hive is inactive during late winter whilst its neighbour is busy - the hive has to be warm enough, and have enough light coming in the entrance, to trigger this behaviour - so you may see a colony in shade do nothing on a marginal day. Another cause of relative inactivity is because the queen has not begun laying early - a sign of good traits matched to local rhythms. Another cause of inactivity is the colony is in poor condition or dead. In any case, opening the hive will not solve the problem, it will simply chill any survivors. Leave it for now and watch.

If it snows:

- **Check entrances are not blocked, bees need to breathe.** Even a little snow can build up on a landing board, though the bees' own warm breath will usually melt it away. Larger colonies are more at risk of suffocation.

- In some countries with deep drifts of snow, the beekeepers have to wrap hives in insulation and create a tunnel to a point at roof level, or use upper entrances.

- A blanket of snow on hive roofs is a *good* sign, the hive is well insulated. Sometimes there is a slight dimple above the cluster. If all the snow melts, or there is a very thin patch in one place, add insulation, if only bubble wrap.

Bees need to breathe. Light snowfalls like this are melted away at the entrance by the colony's warmth - proof the colony is alive. Deep snowfalls can suffocate bees.

Scraps of comb at entrance - mouse in hive. It will kill the hive over winter if it remains in there.

Dead bees, nibbled / torn to pieces at entrance - shrew in hive. Shrews can enter via a 6mm gap so some mouse guard styles may not work, and all shrew species are protected in Britain. It will kill the colony over winter if it remains in there so chase it out.

Fist sized holes in hive wall / entrance hacked larger - woodpecker. This is a learned behaviour, not all do it. If they start, wrap the hives in chicken wire so birds can't get near the walls.

Hive smashed open - Badgers will rip hives open **if** the ground freezes and they can't dig up worms. Outside Britain, bears. Once they learn hives are full of delicious bees and larvae, they will persist all year - they **ignore stings.** Fencing required - electric for bears. Smarter ones dig under fences, and bears may use overhanging trees as ladders.

Points to ponder as you watch a hive

How does this colony differ from its neighbours? For example,
- Colouration, origin, size
- Foraging - different pollen colours / surge times
- Intermittent or continuous behaviour
- Temper
- Response to stimuli
- How many different noises distinguishable
- Is anything *missing* which is normally present?

Why might this colony be acting differently to its neighbours / yesterday?
- Swarm caught at different time of year
- What's growing / crawling below it
- Hive type, insulation level
- Different comb topology (amount, age, whether it reaches the entrance, twisted or straight)
- Position: Facing prevailing wind? In a damp spot?
- Angle of sunlight / shading, on hive body / entrance
- Humid day - affects electrostatics, body hairs, pollen adherence

Don't just do something, stand there

Just watch them for 10 mins. Don't glance and assume; procrastination filters your thoughts.

Try defocusing your eyes, watching them from a novel angle, and deep slow breathing in front of the hive. Slow down. Imagine what they are saying to each other, their different personalities, the way a group about a task has its own purposes. Take joy.

Inspection summary

At the end of your inspection, record:

- What you learned
- Future actions required (e.g. fit mouse guard)
- Stings, location, size of swelling, how long it took to fade away, unusual reactions (this is useful to review if you suspect you are becoming sensitised)

One should always be slow to draw conclusions. Compare observations over time before deciding.

Further reading:

At the Hive Entrance by H.Storch, available from Northern Bee Books. Ensure you are getting the English translation. Avoid Amazon's version, it's a poorly printed pirated reproduction. Although pdf copies can be downloaded for free, that defrauds the IBRA research charity of their commission.

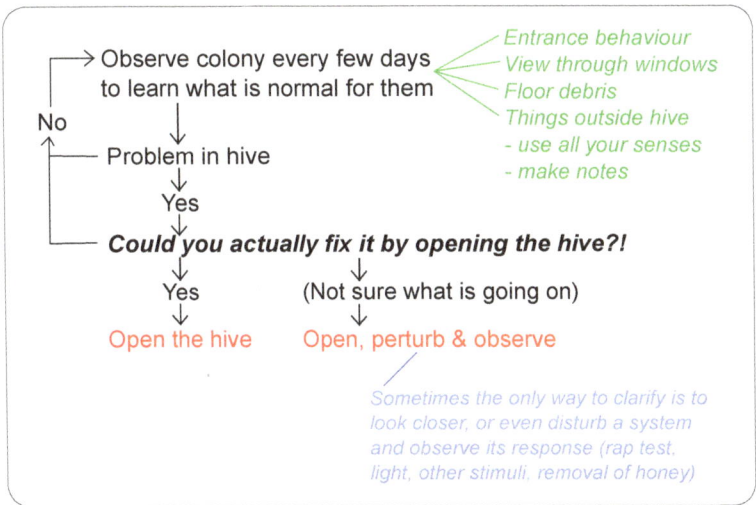

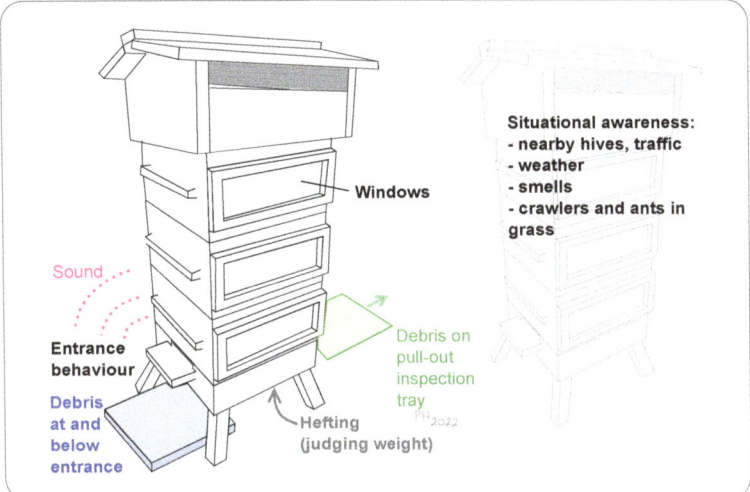

Review 8

There will be factors and patterns you spot which are not discussed above. Jot down some of your own ideas:

Chapter 11.2
Inspections: internal observations

The principle here is minimum interference. For a large scale commercial beekeeper, the priority is on speed (which amounts to the same - they can't waste time mollycoddling weak colonies).

Internal inspections should stop once a situation is clarified. You don't need to check everything; the longer the hive is open the more the bees are stressed.

An open hive inspection is like routine surgery - slightly risky for the patient, and weakens them. And remember, colonies don't behave normally while being smoked and taken apart.

Preparation

Weather: Avoid opening a hive if it is cold[202], windy or rain is coming – the bees will be very edgy, and more will be in the hive. If they are not flying & gathering pollen, it's too cold.

Time of day: best when foragers are out, reducing problems if the hive gets defensive. 10AM onwards. Bees are **much** calmer on hot days, because brood don't get chilled, but extreme heat + bee suits = stress for the human.

Protection: Cover ankles (tuck trousers into wellies) and wrists, these really hurt if stung. Wellies also prevent bees climbing up the inside of your trousers!

Neighbours: Warn them if performing a drastic intervention likely to rile the bees, like cutting into brood comb.

Move slowly and gently. Smells like stale sweat will alarm the bees. If you remain calm after making a mistake and riling up the bees, it will soothe them.

Plan out what you are going to do and have all tools nearby so the operation can be done as quickly as possible. Often taking notes is easiest using a voice recorder app and transcribing later - writing is tricky in sticky gloves! If you are opening several hives, open the edgy ones last as they will set off the others. Big colonies are generally more assertive.

Consider: If you *did* find something wrong, what could you do about it? Most 'corrections' are to maximise honey crop. Things move fast in the insect world – in one week a pest can multiply enormously. By the time something is visibly awry, it's often uncorrectable by a human – and the bees are already dealing with it.

Assume you will get a surprise – which may involve excited bees. It may help to imagine this beforehand, and visualise yourself handling it calmly and examining and learning and communicating calmness to the bees.

---

202 One baffling aspect of conventional beekeeping is the nonchalance about opening hives in winter – January, say – to apply oxalic acid (a miticide). Although the declared intention is to do this at a time when the hive is broodless, there *are* often some brood being nursed then which are at severe risk of being chilled, and burned by the acid – as every experienced beekeeper knows.

Just allow a little smoke to drift down - too much upsets the bees. Smoke your hands. A cover cloth across gaps helps retain nest warmth and scents.

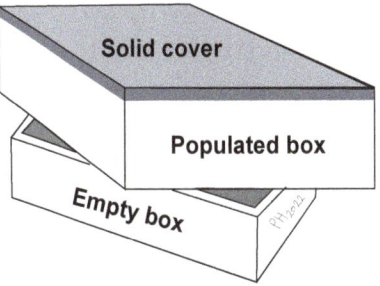

When splitting hives apart, cover boxes you are not inspecting with a wooden panel (cloth blows off!) and don't place bxes directly on the ground - young bees / queens wander off into grass and get lost! Instead, place boxes on an empty box (or inverted roof) at an angle, for just a few points of contact (less risk of crushing bees).

Procedure

Don protective clothing. Remove rings.

State out loud reason for opening hive. (Focuses your mind)[203].

Start with external observation. Looking through windows, the combs with a thick covering of bees are the brood combs and the bees will be more alert to interference with these.

Start by gently wafting your smoker around the hive entrance - not to panic the bees but just to dull their sense of smell. *Don't* squirt smoke directly in, it puts them on guard - a few wisps are more effective. Smoke your hands/gloves too.

After removing roof, feel for cluster heat so you know where the nest is before you expose it. (This gentle touching operation also calms you and thus the bees.)

Use tea towel or wooden board to hold in nest atmosphere and heat.

Talk - quietly - to the bees. Many beekeepers do this, it keeps **me** calm and by treating them as people, I automatically treat them with respect. A loud voice disturbs bees.

Only open the hive as much as needed, to determine something.

If you don't know what to do, do nothing. It rarely makes it worse.

Hold combs **over open hive** (queens can drop off)

Put combs back in same order & orientation

If you stand in front, you'll get covered in returning bees.

Warré hive inspections: always have a replacement top cloth prepared.

I sometimes take photos to augment my notes as these can be studied more closely later, and I always take photos when I see something unusual.

---

203 "Self training," "teaching others" and "Spring health check" are perfectly valid reasons. "6th routine check this year" is questionable.

When closing a hive up avoid squashing any bees if possible because any alarm pheromone released will start a lot of trouble. A smoker can help drive bees at this point.

Warning: colonies are often unsettled and defensive for a day after a thorough inspection.

Open hive inspections should be brief, and you often get distracted, so on the next page is a checklist of key things to look for.

Typical open hive toolkit:

A   smoker

B   lighter

C   dilute peppermint spray to mask sting scent

D   epipen

E   antihistamine pills

F   magnifying glass

G   queen clip

H   notebook + pen

I   sample jar

J   tweezers

K   duct tape

L   tea towel (top cloth)

M   water

N   knife

O   hive tools

P   feather / bee brush

Q   comb knife

Not shown: protective clothing; mobile phone / camera

And consider if today's task requires specialised items, such as:

- Spare bars / frames to replace the ones you're harvesting
- Box to hold harvested comb
- Bucket of water + cloth to wipe your sticky hands
- Inspection mirror
- Waterproof paper / notebook
- Thin wire to cut Warré boxes apart

Part 2 Chapter 11.2 Inspections: internal observations

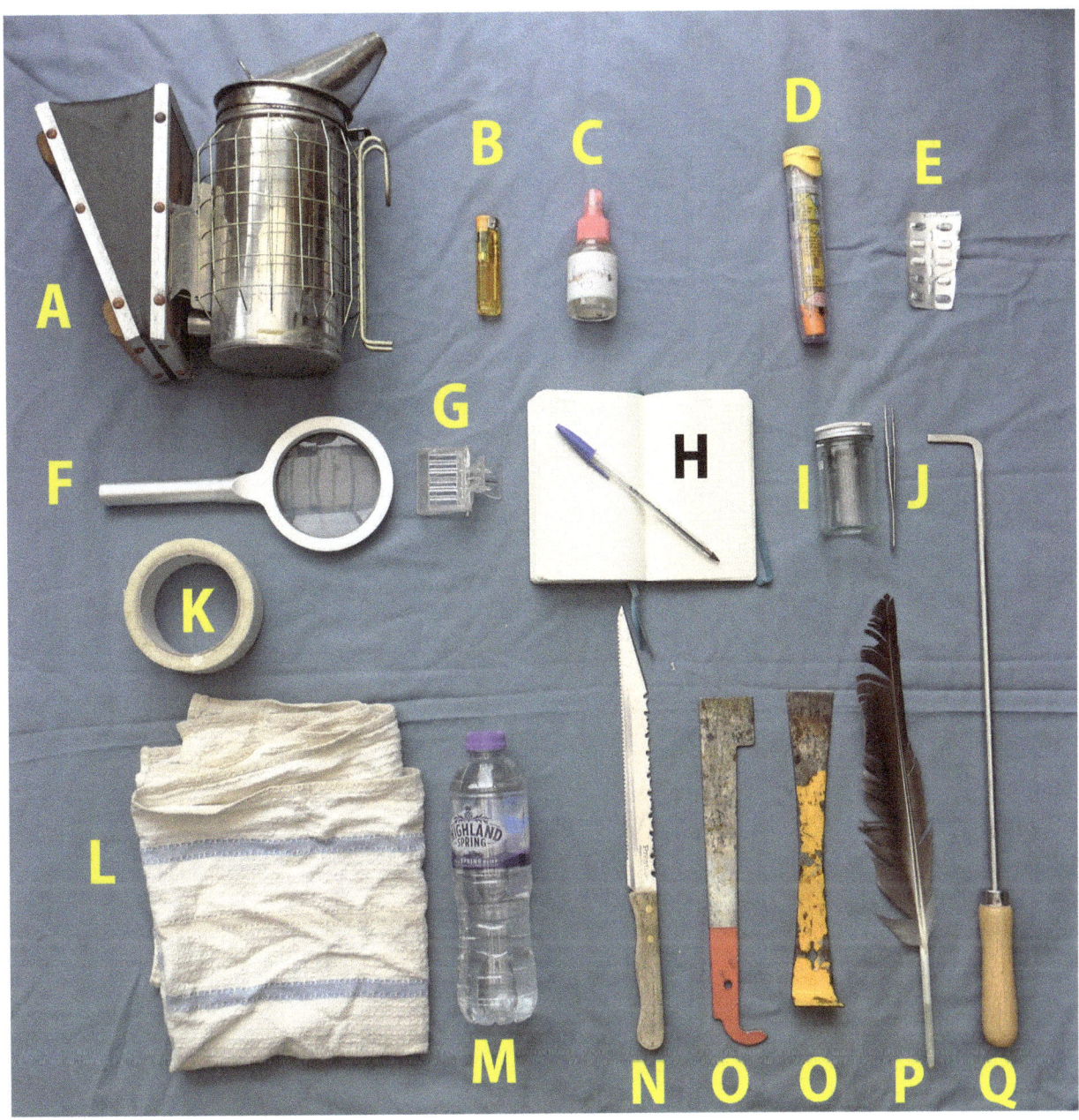

On the next page is a form you can photocopy and write on to record internal hive inspections. Just write over the grey text on your copies.

Aide memoire: hive internal inspection checklist

Apiary: Hive:

First, perform external observations before opening a hive. Then:

Date	
Time of day, weather	
Purpose of intervention	**State your reason out loud.** Never open a hive unless you have a clear reason to, although this may simply be "No reason to suspect anything awry, but have not inspected for ages and need to improve my skills". That's perfectly valid. **What do you think opening the hive will achieve?**
In / under roof	Woodlice love damp, indicating a leak
Stores	How many combs hold honey (does hive need feeding). Is there a shortage of pollen (normally stored in modest quantities in arc above brood).
Manipulations performed	Increasing free space in brood area to avoid crowding and premature swarming; correcting misaligned comb; splits...
Brood pattern	Solid patches or arcs of brood of the same age indicate a healthy queen. Spotty patterns indicate marginal fertility or possibly disease. Photos useful.
Brood visible - Eggs - Larvae - Capped workers - Capped drones - No. brood combs	How many combs of brood? If capped brood are present you know the queen was laying 16 days ago. If uncapped larvae are present she was laying in the last 8 days. And if eggs are present, she was OK 3 days ago. Personally, my eyes can't make out eggs, so I rely on photos which I study later. If there are **multiple eggs in a lot of cells**, and just see drones being raised, maybe lots of small drones, you may have drone laying workers. Useful mnemonic: look for BIAS - Brood In All Stages
Brood abnormalities, signs of disease	Healthy larvae look like **"curly pearlies"**. They should be **segmented**, and **lie curled in a "C" shape**. Do you see **discoloured or sunken capped cells, foul smells, larvse rotting / liquidising in their cells, yellow droppings inside hive**. Open a couple of **sealed** cells for signs of melted brood (AFB). Deformed adults may be due to chilled brood.
Queen(s)	Did you see the queen? Y/N. Also worth recording: colour of body, number of queens(!), any marking on her, where found in the hive, behaviour.
Adults	Are numbers exceptionally low / high? What fraction are drones? Temperament? Behaviour - bearding? Runny? Lethargic? Any signs of disease, e.g. wing damage; lethargy; yellow spotting (faeces) on comb / entrance?
Queen cells	Number, position, appearance
Pests	Wax moth ("cobwebs" across comb), beetles, wing damage. The roofspace can host harmless earwigs, bumblebees and solitary bees. Wasps in a hive are a significant problem.
Feeding (kg sugar)	Record how many kilograms of sugar you fed, not litres of syrup (as syrup varies in concentration). This should be a very rare event.
Medications / treatments	eg shook swarm, sugar dusting (this is a major section for conventional beekeepers; if you apply medicines there is a legal requirement to record it)
Honey harvested	(This is a major section for conventional beeks)
Miscellaneous internal observations	e.g. "hive has swarmed", "very clean floor" (can imply a hygienic colony), "lots of propolis", "comb not aligned to bars". If colony has died, take photos and record everything to help diagnose and learn. Take dead bees and floor debris and freeze them should something occur to you later.
Before closing the hive...	Did you achieve the stated aim at the beginning of this list? Have you taken all the photos you want to?
What I learned	

Conventional training just teaches one type of inspection, the full-on open hive type, and emphasises using inspection forms too - though their records emphasize factors like feeding, queen marking (age) and treatments applied.

However, most inspections should be non-invasive. As early as the 1940s, Storch commented (p.41-42) that excessive fiddling in the brood nest often upsets the bees resulting in listlessness, the queen and workers seem dispirited and stop laying, gathering nectar etc for a few hours. Modern researchers note the massive loss of heat and pause for repair work after beekeepers inspect. This is a big deal. You can lose several hours' honey production once a week during peak nectar flow.

The most significant information from an internal inspection is usually the condition of the brood, not the adults. The adults indicate the immediate state of colony health, but brood tell you what the hive will be like in 3 weeks.

What is a sensible level of intervention?

(1) Depends how much honey you want[204].

(2) Inspect more often when learning – to know what is "normal".

(3) TBHs need more intervention than Warrés (this is a good thing for learners).

(4) If you suspect disease, check the brood nest combs. Use a camera to record what you find.

What I advocate is that when you start beekeeping, you look inside hives several times a year, eventually tailing off to necessary swarm-minimisation measures where you simply make room by adding boxes or mving bars (depending on hive type). You need to know what healthy comb looks like, and lose any fear of handling bees, so when things do go wrong you can handle it.

And in Top Bar Hives, you need to ensure the combs are built straight so the hive is inspectable.

Inspections: summary

Unless you are manipulating your colonies to suppress brood breaks and maximise honey yield, 95% of the time you can monitor your colonies adequately through external observations[205].

"A hive is like a sterile isolation ward. Every time you open one it is bad for the inhabitants." - Simon Kellam

Top tip: leave them alone, they'll generally sort it out. 'Correcting a problem' prematurely[206] can hinder them - the bees are already fixing it.

---

204   "Probably one of the chief reasons why the enthusiastic beginner rarely gets honey is that he cannot let the bees alone. He is always disturbing them and looking to see how they are getting on. It is much as if one dug up a plant every day to see how its roots were progressing" - R.O.B. Manley, *Honey Farming*.

205   A lot of people confuse opening a hive every 8 days in swarm season (to check for swarm preparations - resulting in approximately 18 hive openings per year) with full-on health & disease checks. 18 checks a year may be appropriate for bee farmers tuning production colonies, *but they only need a few seconds to evaluate each frame*, they don't disrupt their colonies significantly while they flash-process signature patterns.

206   Example: **I saw what I thought was a drone laying hive, which "inevitably" dies, spontaneously recover in my 4th year of beekeeping. I had been wrong in my diagnosis. Possible explanation:** sometimes workers start raising a new queen (supersedure), and while "queenless" they permit drones to enter. If this happens in August, all the drones being excluded elsewhere end up in this hive, until they are queen-right again when they exclude drones like the other hives.

My notes contain several instances where on opening the hive **I decided to do nothing.** Sometimes this is the best option!

There is no need to inspect more than a couple of brood combs, *once you have a few years' experience* and know what is normal.

I stopped most internal inspections after -

- I opened a hive, caused great disruption and learned **nothing** I didn't already know from external observation.

- I realised most of the problems I spot are not correctable.

Review 9

Jot down 5 things you would look for in an open hive inspection:

1.

2.

3.

4.

5.

Chapter 11.3
Inspections: reading comb

Many subtleties. The 'skeleton' of the superorganism.

*This chapter has reference photos of **healthy foundationless** comb. This is what your bees are **trying** to build.*

Overall structure

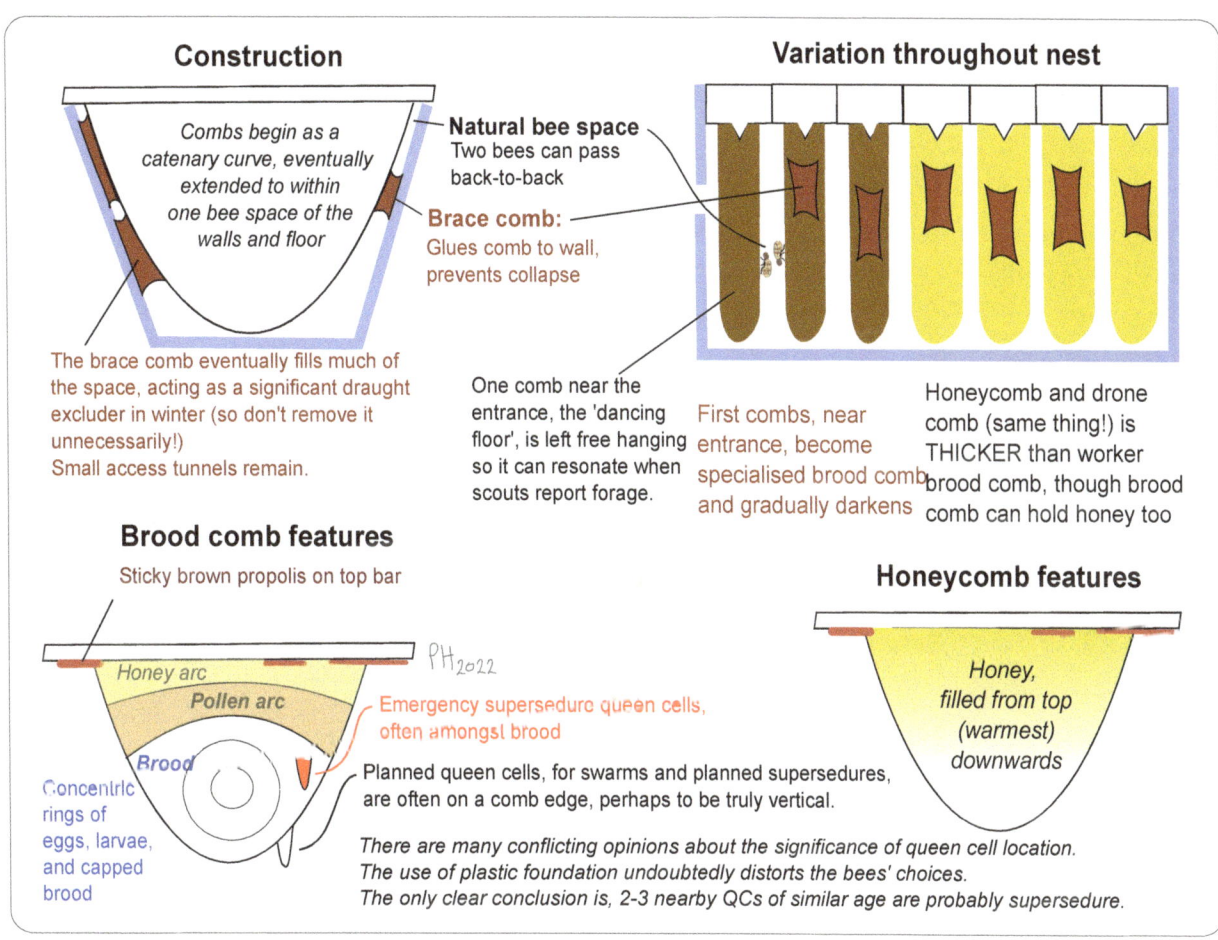

Natural comb varies according to its function. Foundation massively distorts nests.
The European Black Bee (Amm) stores a much thicker honey arc above brood than is shown here.

Generally speaking, cross (bridge) comb and brace comb happens primarily with **honey**comb, but will *eventually* build up in brood areas.

Bees aim for a constant *bee space* between combs, allowing 2 bees to pass back-to-back. Brood comb thickness varies with race - my bees like a 38mm bar spacing; North American bees prefer 35mm "Hoffman" spacing; African bees 32mm; Asian bees 29mm. Many humans impose a constant frame spacing throughout a hive but actually, whilst bees like the inter-comb gap to be constant - and the honeycombs can be very thick. See the top right image, "variation throughout nest"? The bees aim for constant spaces between the combs, but the constant bar width means the bees have to choose between their preferred (thick) honeycomb or preferred bee (passing) space between combs. Sometimes, beekeepers place frames a bit further apart to allow thicker combs.

New comb

New comb starts out white - pure wax, soft and flexible (delicate!). Over its first few weeks, the bees add a frosting of propolis round the rims of the cells, seen here as the reddish-brown mesh, which adds stiffness - forming a composite material stronger than either.

Over time, *brood* comb darkens as it incorporates fragments of silk coccoons, making an even stronger, leathery metamaterial. Honeycomb tends to be weaker, heavier (full of honey) and needs careful handling to avoid collapse.

Ladder comb

Bees will build vertical bridge ("ladder") comb upwards to cross gaps higher than themselves. This can glue frames in, or even fill large voids like empty supers added above frames.

Early stage propolis frosting.

Soon a stiff, crystalline lattice of propolis reinforces outer edges.

Very new (white) comb lying on old, dark comb. Bottom left: bee bread (part filled cells). Bottom right: capped honey (rippling ribs are added to old honey).

Old comb

Comb gets reworked and repaired, but after many years degrades and begins crumbling - "rotten comb".

At this point the bees chew away the comb above and let it fall, whereupon wax moths completely devour it, making room for fresh comb.

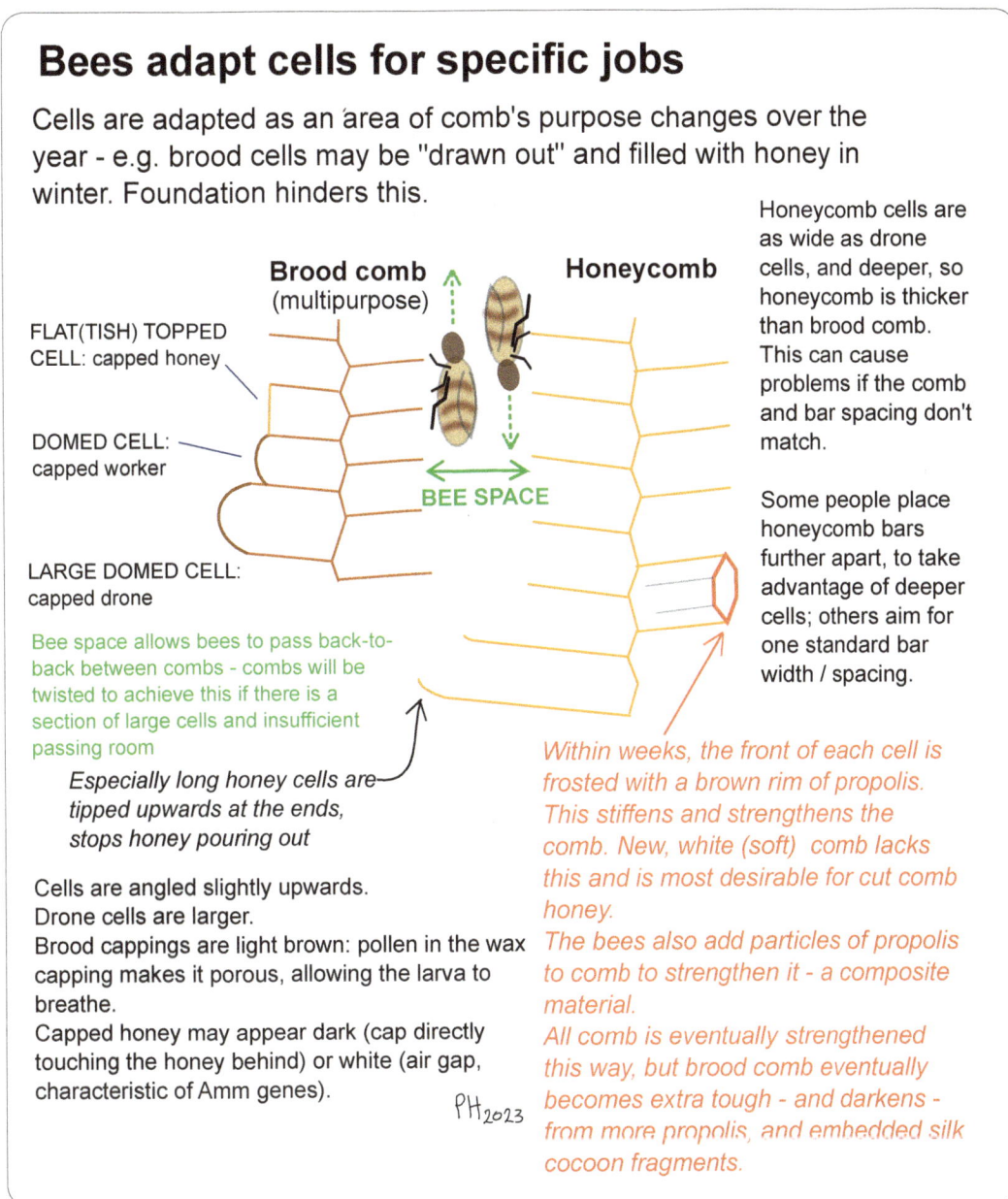

Many conventional beekeepers urge you to change old, dark comb as it is a "resevoir of disease". It's certainly a resevoir of chemicals like miticides and antibiotics if you use those, but intensive bee farmers tend to overlook that bees have evolved to live in crowded conditions and have evolved hygiene practises – nests are kept sterile. If comb causes a problem, they tear it down and build anew: they continually remodel natural comb.

So I leave it. But it depends on your circumstances. Rob Keller (napavalleybeeco.com) is surrounded by wine grape growers who saturate everything in fungicides. He does have to change old comb, despite being a treatment free beekeeper. Another major factor is no doubt whether you keep moving your hives to crops.

Simmins claimed[207] *the older the combs are the more protection afforded in winter.* Using a thermal camera, I confirmed old brood comb is a slightly better insulator.

---

207   *A Modern Bee Farm*, (1887) p.28

20 year old wild nest in wall cavity. Healthy, but house was being demolished. Entrance was at top left. Comb is in a corner for extra support and is almost 2 feet deep. Underneath was a mass of spider and wax moth webs and crumbs of digested comb. Image © 2017 Will Hanrott

From the same nest. Surface furrows made by wax moth; polished cell bottoms reflect light. Image © 2017 Will Hanrott

Old comb on foundation

With natural comb, bees remove old, rotten ones by chewing along the top honey arc and letting the stuff below collapse onto the floor, where wax moths devour it.

Foundation, particularly wired or plastic foundation, can't be bitten through and this image shows how bees have chewed through cocoons to remove individual cells. The ragged, papery cocoon edges protrude - quite different to wax cell walls. The trigger wasn't clear, possiby the cells grew too small, there was insufficient bee space to the next comb or maybe they just wanted to build drone comb (but it was October).

Brood cells chewed down to foundation. Image © 2023 Michael Cummings

Old honeycomb

Honeycomb left in a hive gradually acquires thick wiggly ridges and thicker wax caps. The honey is still perfectly edible.

Honey, I shrunk the larvae

Are old cells smaller, resulting in smaller bees?

This oft-repeated "Facebook fact" is a half-story.

In old **brood** comb, the holes in the cells can become *rounded*, not hexagonal as the corners gradually fill with propolis and cocoons.

As you can see here, the right hand side's large, drone cells have been gradually infilled. The lower right ones have ended up the same internal diameter as the worker cells... which have **not** shrunk. And unreduced drone cells can be seen top right! Perhaps the bees wanted more workers?

Cell size varies across a comb anyway. And if you've ever rendered black worker brood comb for wax, you'll get almost no wax. It's almost as if wax was removed as silk was added - it appears bees only permit internal diameter to shrink for a reason.

Whenever I see 'small' workers at my hives, I photograph them and invariably find it was a trick of the light / angle of view / their variant colouration / posture.

However, even if I haven't seen it myself, there **is** evidence that this stunting effect is a real phenomenon. A recent paper showed the effect using C-T scanning to look at pupae in old brood comb, and W Herrod-Hempsall definitely observed it. And you can definitely get short drones if your combs are not spaced far enough apart for full length drone cells.

Small cell theory

This school of thought holds that unmanaged bees build slightly smaller cells than those forced to build on foundation, and this contributes to varroa resistance.

Comb from a deadout. Smooth emergency queen cell in middle, scattered capped drones plus a patch of them - probably a mix of drone laying queen and laying workers - and a couple of dead larvae peeping out of cells.

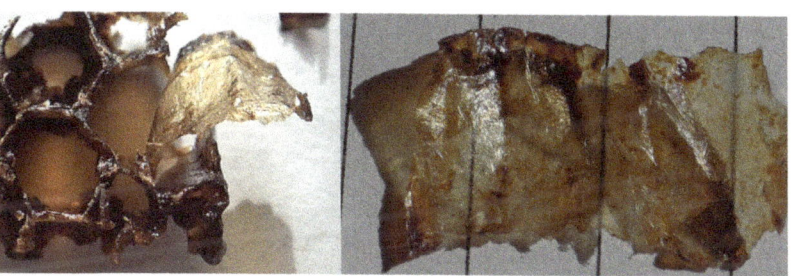

Dissecting 2-3 year old brood comb. There's essentially no wax, internal diameter is same as fresh cells'. Layers of tough silk cocoons can be peeled off, thin enough to see lines on paper through, and smeared with sterile brown propolis. Structurally very different to honeycomb and much stronger.

Worker bees of variable size, from W Herrod-Hempsall's 1930 volume *Bee Keeping New & Old*. He attributed this to some being raised in reduced cells.

In short, I think it's based on false premises[208], and cell size varies across natural comb anyway - there's no "one perfect size" of cell or bee.

Brood comb - reference photos

Queens return to lay in brood cells once they are empty, which often results in rings like this where older, capped brood are near hatching at the edges and a new batch of younger uncapped "pearly curlies" (healthy larvae) can be seen in the centre.

Classic brood comb from a Top Bar Hive

Sometimes short bees are simply headless

Rings are not universal though, as seen in the right hand image below. This features a mix of worker cells and larger drone cells, making it wiggly as thickness varies. Honey is in a band along the top, brood below (drones at bottom right).

Useful rule of thumb: Once hatched, one comb of brood will make three combs of bees.

Typical brood frame from a hive using foundation, giving a very regular array of pure workers. A ring of brood surrounds empty, hatched cells. The lines of cells are horizontal - natural comb has no consistent orientation, the cells don't always point upwards.

Capped drone brood, often described as "bullet shaped", appear to be popping out of the comb at bottom right, with capped worker brood almost flush to the face of the comb around them. Honey at top. A queen, spanning 4 cells, is visible just to the left of the middle.

Larvae should be "white, bright, and curled up tight". Segments should be clearly visible.

---

208 Dee Lusby (Arizona, resistantbees.com) noticed a correlation between health and natural comb, but misinterpreted historical measurements of cell size - which to be fair are pretty tricky to interpret - and concluded foundation was too large. **Further reading:** https://elgon.es/naturalcellsize.html - very readable article by Marcus Nilsson; *Treatment-Free Beekeeping*, David Heaf pages 80-83, summarised at www.dheaf.plus.com/warrebeekeeping/do_small_cells_help_bees_cope_with_varroa.pdf; *Natural Bee Husbandry*, issue 16, page 10.

Brood patterns

Spotty or not?

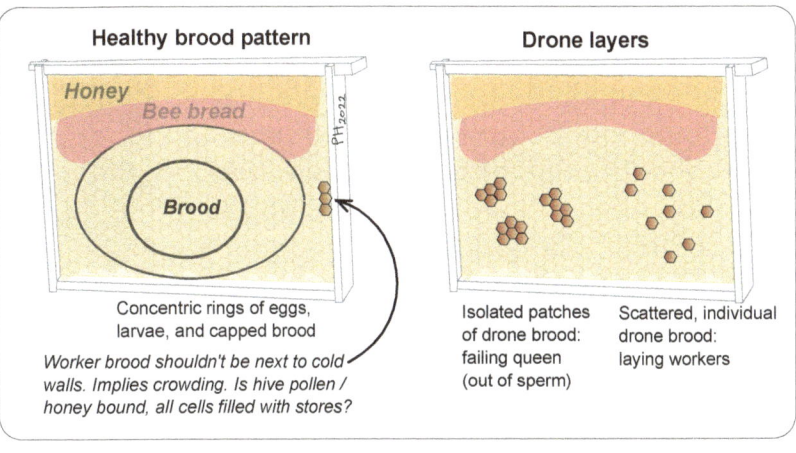

- At least 1 in 20 cells should be empty, to provide anchoring points for heater bees who regulate brood temperature; and as a sign of hygienic behaviour (highly desirable!). *It is not a sign of a failing queen, especially if you only see it on one comb.*

- Are those cells really empty? Or do they contain tiny larvae? (Different ages scattered irregularly on a comb is called *shot brood,* and causes panic in some beekeepers who assume the queen is failing.) The queen has run out of space to lay and is searching for odd cells to lay in.

- Any non-solid-brood pattern disruption will ripple through to the end of the year. A scattered pattern could be due to something as simple as honey binding, i.e. all available cells were filled with nectar during a big flow and the queen had to search around for any free cells to lay. If there are solid patches of brood *anywhere* the queen's probably fine.

Spotty brood ALL OVER a comb full of brood is a danger flag - that may well be foul brood, whereas what many call "spotty brood pattern" is just a queen laying an appropriate number of eggs for the current forage.

Brood comb: close-ups

Larvae in various stages of development.
Image © Rory Wills / Guelph HBRC

Green cells are empty, you're seeing grass through the comb

Brood cappings - not just "biscuit coloured"

Cap colouring is due to pollen being mixed with wax to make it porous (so larvae can breathe). The capping colour can thus vary, different shades of brown right up to a near black in some countries.

189

- partly because the comb below itself varies in colour (repeated brood rearing in an area darkens comb - the right hand cells here are in the centre of a comb);
- and underneath the left hand cells are young, **pearly white** larvae whereas the older, **dark** brood on the right are nearly mature, and their colour is visible through the thin wax. This is easily confirmed by uncapping a couple of cells, and note there are many holes round the older brood where sisters have hatched. **Despite the dark colour, there is no disease problem here.**

Distinguishing worker and drone cells

Here are some healthy worker brood. The curvature of the capping varies a bit - you'll see they're flatter in the next photo.

In summer, 15-20% of the brood comb should be drone comb. Don't try to remove it - it will simply waste resources (honey!) because they'll rebuild it. Using foundation to force worker cells results in the bees desperately trying to raise drones anyway, as shown in the next two pictures.

Here's *natural* comb showing a mix of worker and drone cells:

Close up of worker brood cells showing curved top.

Bullet shaped capped drone cells at top; capped worker cells below protrude much less. Brood cappings are rough (porous) unlike capped honey, and unlike most honey, brood patches have occasional empty cells. This hive uses foundation, forcing the queen to lay drones in worker sized cells. These will be small drones, less likely to mate. The drone cell sizes are uneven. They'll also be short - insufficient headroom between worker brood combs.

This photo is not taken at an angle, the right hand (drone) area has larger (wider and longer) cells which bulge out.

In the same hive, the bees have resorted to building drones cells at the bottom of a frame. Note the two queen cups, pointing straight down. Drone cells are (usually) almost horizontal.

Variable width combs like this mean the next comb will be built parallel to it - at an angle to bars, making the hive uninspectable - unless you have adequate spacing, clear comb guide edges, and ruthlessly correct skewed comb. Once you have some straight ones, you can move them apart and use them as comb guides.

This comb also features runways through it, allowing access to stores without going near cold winter walls; and some heavily distorted cells (top right) where the builders have had to correct for earlier planning errors!

Polished cells

The workers clean cells ready for eggs. Their inner walls are shiny, as if containing nectar; this is particularly easy to see if the comb is old and dark.

Purple eyed larvae looking out at you from cells

These are late stage larvae (ie, not white) which have been uncapped for a few hours which, for unknown reasons, expels the mites[209]. The uncapping by nurse bees appears to be triggered by a smell, possibly the varroa feeding wound going gangrenous.

Queen cells

Appearance

Development of a queen cell (QC)

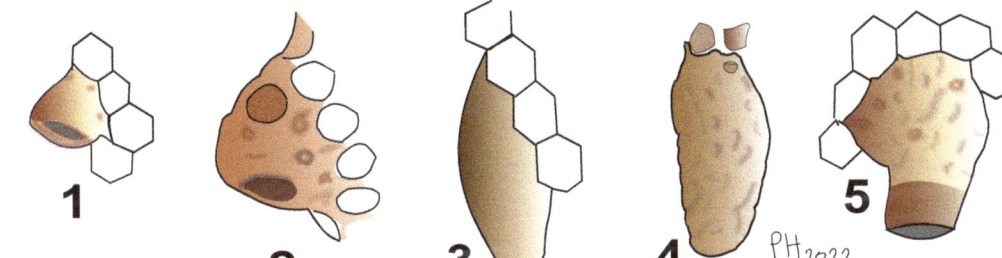
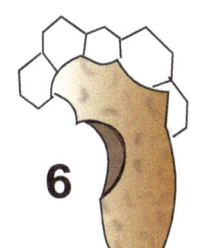

1. *Queen cup*, aka *practise cell*. Often seen - not significant. Possibly readied for emergencies.
2. Advanced enough to receive egg.
3. Capped, **smooth** so recently capped.
4. Dimples added over first few days; heavily ridged by day 5. *If no dimples after 3 days, it is dead, or a drone inside* (e.g. drone laying hives desperately trying to raise a queen).

   (Not shown) Final day - workers remove end cap: brown coccoon is visible (a *ripe* QC).
5. Queen has hatched. Smooth cocoon protrudes[210] (propolised, brown).
6. This queen was stung to death in her cocoon by a rival. Workers removed the body.

---

209   Heard this from Tom Seeley; Ron Hoskins confirmed uncapping is at the purple eye stage.

210   Huber observed that unlike worker and drone cocoons, queen cocoons only cover the front half of her body... which permits other queens to sting this unarmoured section! *Huber on Bees* (1841), the English translation of Huber's *Natural History of the Honey Bee*, p.94

QCs sometimes get resealed after the queen has emerged - I've heard of workers crawling in to clean them being entombed by their sisters! There are also reports of queens developing *wrong way up* in cells and dying because they can't chew their way out; and drones in QCs means the colony instinctively tried raising a new queen after the mother ran out of sperm.

Position on comb; planned or emergency?

Bees want space below the QC so it can hang vertically, and be nice and long. This may be in the middle of foundationless comb (see picture). Large vertical ones are - usually! - *planned* swarm or supersedure cells.

A QC crammed in *at an angle,* or with insufficient space, is an emergency one.

A modified worker cell is an emergency QC (so *usually* in the middle of comb[211]).

Emergency QCs are smaller.

Numbers

1-3 QCs probably planned supersedure. These will be the same age.

5+ QCs, probably swarm cells; these tend to differ in age by a few days[212].

Wax is not conserved when making queen cells. They're not tight packed hexagons, and the entire surface is porous to air - queens get unlimited oxygen.

Planned or emergency supersedure: QC in middle of comb, with space created below. Typically 1-3 such cells for a leisurely, planned supersedure. Note dimpling on its surface. This is natural comb from a Warre.

Properly hatched QC at edge of comb, with remnants of cocoon protuding. Cells below were removed by bees so it could hang vertically.

Capped queen cell above queen cup

---

211   Just because a QC is not at the edge of a comb does not mean it is an emergency cell. Queens are raised at the **edges** of brood nests at slightly lower temperatures than workers, and that does not necessarily mean the bottom edge.

212   I say 5+ because different sources give different figures; "6-10 across 2-3 combs", "12-20" etc. African bees may raise hundreds of queens! Big roomy hives with many bees raise more queens (*Bee Keeping New & Old* p.442). Roger Patterson has inspected thousands of colonies and breeds from those with his desired traits; and says higher numbers correlates with high swarminess. He also confirms Beowulf Cooper's "Peak Queen Number" observation that a given colony will always produce about the same number, i.e.15 whether it is swarming or making emergency queens.

Problematic queen cells

Dislodged larvae can also cause banana shaped queen cells, which are not usually viable.

Eggs

Don't get too hung up about eggs. I get by without seeing them - my eyes aren't good enough. If you see larvae, the queen was laying within the last few days.

After 3 days, eggs hatch and a larva drops into milky food (royal jelly). The amount of liquid food in the cell varies over the year - in dearths there may be little.

Huber and Burnens noted[213] fragments of silk at the bottom of cells in brood comb can be very shiny, giving the illusion of liquid if the sun shines directly in.

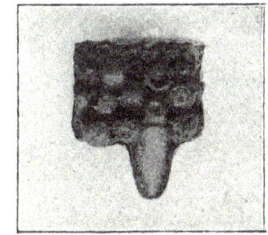 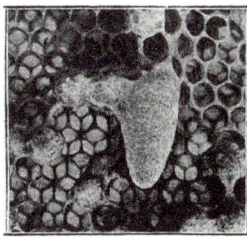

Smooth Queen Cells containing drones ("King Cells"). Source: *Bee Keeping New & Old* (1930), Herrod Hempsall

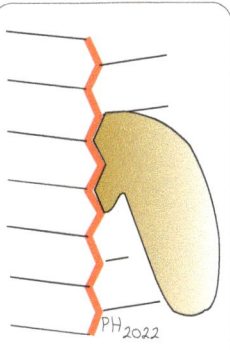

Bees cannot chew through plastic foundation so emergency QCs on foundation do not always hang vertical. Opinion is divided on whether this gives smaller queens.

Elongated

Non viable - rarely hatch. Often drones, torn down by workers. Female ones occasionally found, thought to occur when larva slips out of pool of food at top of cell.

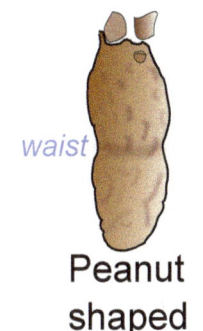

waist

Peanut shaped

Rarely hatch. If they do hatch, poor quality queen

There are four tiny white eggs in this photo. They're quite tricky to see!

Interpreting eggs

Front view   *Cross-sectional view*

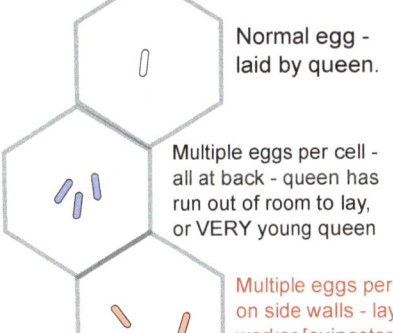

Normal egg - laid by queen.

The queen glues an egg to the centre of the back wall, at almost right angles

Multiple eggs per cell - all at back - queen has run out of room to lay, or VERY young queen

The egg often droops (but this is not universal): eggs in this orientation are about 3 days old, just about to become larvae

Multiple eggs per cell - on side walls - laying worker [ovipostor can't reach back of cell]

---

213 I saw this summarised in *Bee Matters and Beemasters* p.14, but can't find the source reference in Huber's own work.

Very new comb has round cells, The bees then soften it with body heat and surface tension pulls it into hexagons, like bubbles. The ends of brood cells eventually get rounded again by accumulated wax, cocoons etc.

Honeycomb

Thicker than brood comb. People sometimes put fewer frames, spaced further apart, in supers so more volume is honey, though this makes supers significantly heavier to lift.

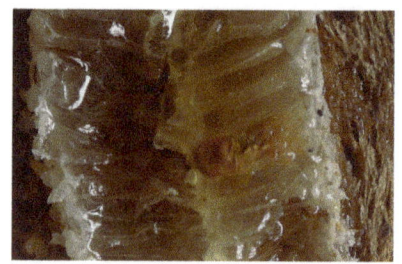

Cross-section of honeycomb (on toast). The cells are angled up, and occasional plugs of bee bread add flavour.

Capped honey. The central patch appears white as it has an air gap between the honey and wax cap - a characteristic of Amm and Carniolan bees (Brother Adam noted Amm uses a domed cap where Carniolans use a flat one). Other races place caps directly against the honey so the cell appears dark. I.e. this hive has a mix of drone fathers. A further effect visible is that as this is from a Warre hive, the comb was used for brood so is generally dark apart from the top arc.

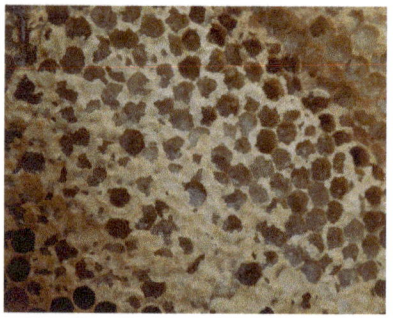

Honey and brood are usually uncapped with neat, round holes, including the temporary uncapping/recapping behaviour of some hygienic bees. Ragged, irregular, off-centre holes in caps indicate robbing.

Honey comb can be much thicker than brood comb. This striking cross-sectional drawing of natural comb in a skep shows the difference well - deep honey cells at the top, brood below.

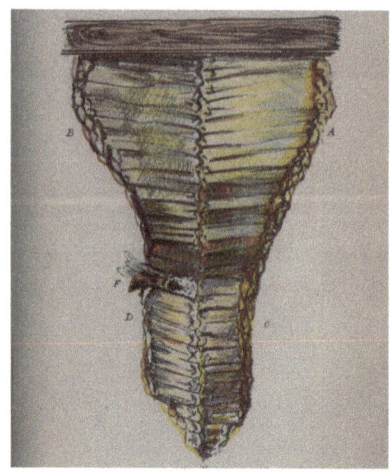

The drawing was by W. Munn, for Bevan's 3rd edition of *The Honeybee, Its Natural History & Management* (1870). Bevan stated the comb was about an inch thick at D-C. Eggs and larvae are at D, C. The nurse bee is 'lubricating' a larva (we would say feeding) at F.

Crystallisation

When honey crystallises, excess water is expelled from the crystals into the fluid between them. As this fluid rises above 21% water, yeasts activate and ferment it, causing bubbles and pushing plugs of honey out of the cells.

Pollen and bee bread

- Both are found in sunken cells, sometimes dark. A good sign - vital stores.

- A variety of colours (of both) indicates a healthy varied diet.

- Pollen has a dry, powdery appearance, like flour. If you knock a comb, raw pollen may fall out. Bees convert it to bee bread ASAP. Raw pollen is eaten by tiny pollen mites, picked up by foragers visiting flowers; these mites convert pollen to dust.

- Bee bread glistens - it has a cap of honey, not wax. It is predigested, and mixed with honey - this keeps longer than raw pollen (different pH). It's softer than pollen and can be forked out of a cell. Sometimes sold for human consumption, fresh bee bread should have a semi sweet lemony flavour.

- Some liken bee bread to Bee Cheese!

- By the end of winter, unused bee bread can go mouldy and is ejected from the hive as cylindrical pellets.

- Pollen is only stored in worker sized cells.

- Amm will also pack pollen *under* brood.

Not mould: crystallised (ivy) honey. Bees have sucked out fluid leaving inedible crystals, sometimes forming a crust at the top of the cell.

Ivy honey, beginning to ferment. The raised caps in the centre of the left hand image are a sign of capped honey fermenting

A wax moth larva trail (top right to middle left). It has chewed along the top of cells, eating bee bread, then pupated in a mass of webs. Characteristic small black tubular faeces. You sometimes see a line of uncapped brood ("bald brood") due to moths.

Below the white capped honey lie cells of bee bread. Its colour depends on the pollen origin, but it has a glistening layer of honey on its surface, different to unprocessed pollen, which has a dry, powdery appearance.

This is from a winter deadout. On the left is bee bread - some still preserved (glistening), other cells dull (mouldy). Towards the right the white crystals in cells are crystallised honey, probably ivy. Enough crystallised honey can act as a wall, blocking a brood nest from expanding in Spring. (There is also moth damage but we already discussed that)

Left to right: capped honey; nectar (glistening); bee breads formed from different coloured pollens.

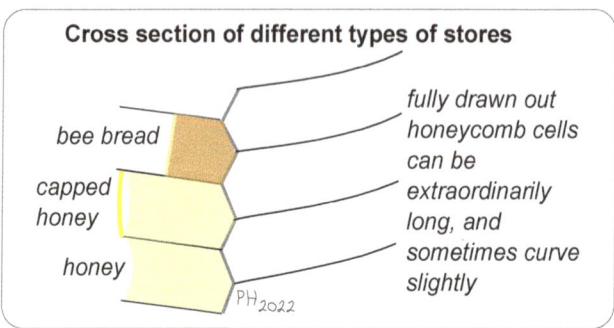

Same piece of comb with bee bread from different angles. Bee bread can look dull, like pollen but pull the comb apart and you see it's not a powder, it's sticky granules, and it doesn't fall out of the comb like dust if you tap it. On the right hand side of the cross-sectional view you can see layers of different colours, derived from different pollens.

Bee bread doesn't fill the whole cell, and is covered by a layer of honey. Colour is highly varied.

Mould and white stuff

If you just post a blurry photo to a forum and ask what it is you'll be assured it is [some nonsense]. People leap to conclusions when seeing white stuff on / in combs. Does it dissolve in water (sugar)? Or float, or melt in hot water (wax)? There are many types of white stuff in hives.

Not mould: wax moth webs. Outside the cells, and note black droppings on right.

Bee bread (centre: mouldy and white; right: unspoiled) from a winter deadout.
Image © 2024 Simon Morton

On seeing white stuff, scrape it away to judge its texture and see what's underneath. Very educational. Often not what you think.

If you see crumbly white wax resembling mould, it is probably just *bloom* - a condition where oils in the wax emerge on the surface. You may have seen it on candles. Google "beeswax bloom" for images.

In my experience, mould is *always* above bee bread, and doesn't signify anything other than "this is a cold, damp hive". Open honey ferments and bee bodies decay. However, chalk and stone brood do sprout out of larvae (I just haven't seen that personally).

Left long enough, things go mouldy when cold and damp. Some mould is normal in damp winter deadouts. Very occasionally, you see adult bee bodies in damp deadouts supporting *external* mould growth. A winter deadout with a damp, cold floor may have rotting bees on the floor but I've yet to see mouldy ones on a floor.

Walls and floors can go black with mildew. If feasible, this should be scrubbed off and scorched with a blowtorch or heat gun.

Sugar syrup will form a white mould in days without antifungal additives, which tend to be pungent and mask pheromone signalling. This is not an issue if the bees empty a feeder in a day or so.

The antifungal propolis which coats walls stops working below 10°C, so in winter black mould can form on walls away from the warm cluster. The bees *can* chew / lick this off, but it's not good for them. This darkening of the internal walls over a year or two is quite a good way of gauging insulation.

See also

Chalk Brood images in chapter 14.2.

Various moulds on bee bread from a deadout. The white mould in some cells is at a constant level below the cell entrance - it's on the flat surface of bee bread. Not a larval infection.

One box's walls are darker from mould (not scorching or shadows). It was a colder box, due to different material (pine not cedar) and thickness.

Examples

Mouldy bee bread from a deadout. The left hand image is the same comb, showing a cross-section, revealing the layered structure of a bee bread cell (topped by mould), with different colours of pollen; honey beginning to granulate; ultra thin silk walls coloured brown by propolis. An intriguing feature is the cross section shows sharp cornered, hexagonal cells - despite the walls being silk darkened by propolis, implying they've raised many brood: such cells would normally be infilled to a more cylindrical shape by remnant wax and silk.

Examples (continued)

Some finely matured moulds on bee bread in a damp winter deadout. Photo © 2023 Liz Robinson

This initially looks like mould growing on dead bees. Actually it is growing **through** them from bee bread below. Mouldy adults are discussed further in chapter 14.1, 'fungal infections'.

A fine selection of moulds on bee bread from a winter deadout. Photo © 2023 Liz Robinson

Entombed creatures

Problems that can't be moved are sometimes simply encased in propolis. This may be because they're too large, like a dead mouse; but people have also reported entombed brood – a response of unhygienic bees that can't groom; and varroa, wax moth and small hive beetles being corralled and sealed in to crevices.

The image shows dead bees on the floor of a hive. This heap of bodies was next to the entrance, and there were more outside that had been hauled out in the normal manner, but the colony chose this method of disposal for these ones. Possibly they were already stuck together.

Entombed bee bread

Bee bread is sometimes covered in a very thin layer of propolis, and thereafter left alone. On analysis, bee bread sealed away like this is found to contain high levels of fungicide.

This phenomenon was first remarked upon[214] by vanEngelsdorp in 2009, who found it seemed associated with colony mortality and re-used waxed comb.

The thin propolis layer over the entombed bee bread in this photograph appears grey due to lighting effects. Image © 2024 Michael Palmer

---

214   vanEngelsdorp et al, "Entombed Pollen", a new condition in honeybee colonies ssociated with the risk of colony mortality (2009) DOI: 10.1016/j.jip.2009.03.008

Further reading: pictures of healthy comb and various disorders - www.nationalbeeunit.com/gallery

Summary

Combs are essentially highly customised 3D-printed rooms made of earwax.

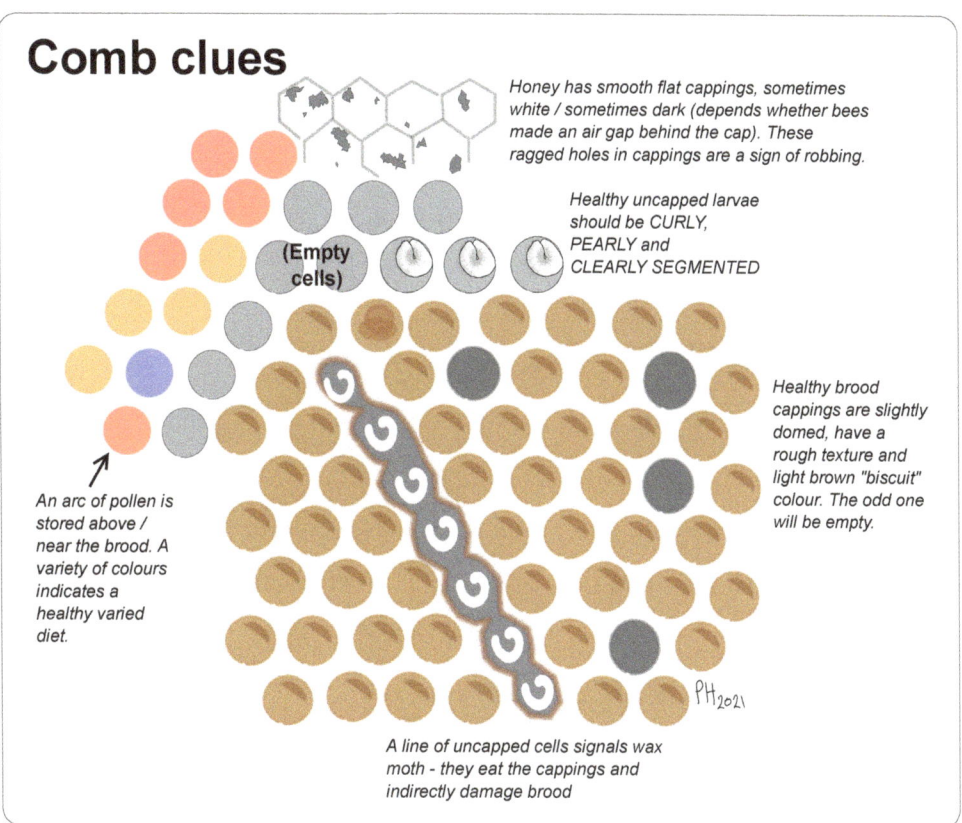

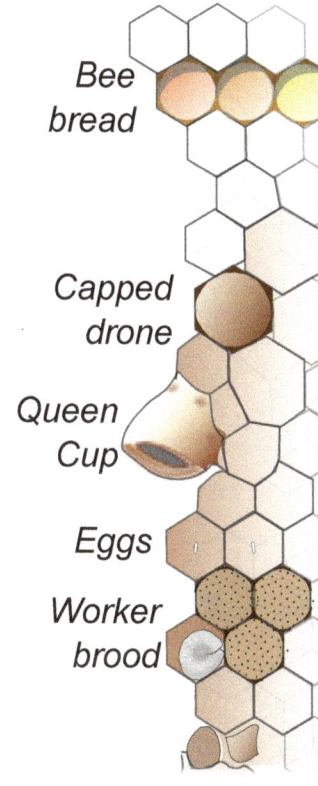

Review 10

How would you tell the age of a colony?

Examine, really examine, an unpopulated comb from a hive, the more complex the better. Take a few minutes. Use a magnifying glass. Write down the story it tells you:

Chapter 12
Managing TBHs, Warrés, skeps, logs & trees, and standard framed hives

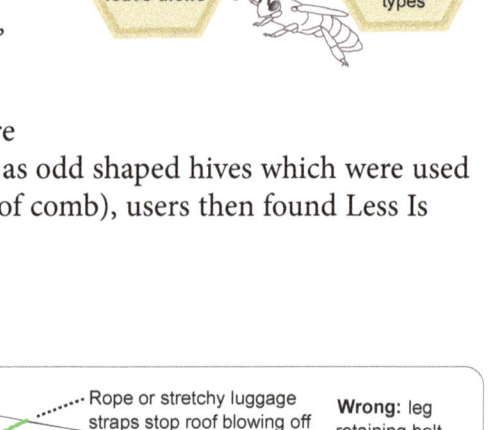

- TBHs need a LITTLE comb management
- Honey is harvested 'little and often' from a TBH
- Warres and skeps are typically harvested once a year
- Warres and skeps are almost totally leave-alone
- These principles can be extended to similar hive types

Einstein and his teaching assistant were preparing the annual exams for his students. "Professor" exclaimed the assistant, "question 1 is the same as a question you set last year!" "Yes" said Einstein, "but the answer has changed."

TBH usage in particular has evolved rapidly since they were introduced here around 2005. Originally promoted almost as odd shaped hives which were used conventionally (regular inspections, significant movement of comb), users then found Less Is More, i.e. the bees do better the less you meddle.

TBHs

Over the 14 years I've been using horizontal TBHs (hTBHs), I've seen many experiments in design and conflicting advice.

Unlike Warré hives, hTBHs require some degree of comb management. This provides a very useful and positive training experience.

They feature simplicity (fewer tools and accessories, no boxes to store), and no foundation: the comb is pure uncontaminated beeswax of known provenance, and allows the bees to decide the optimum cell-size mix. HTBHs' shape retains heat better than the high surface area, high volume Nationals. Additional equipment is simple and low cost (no queen excluder, chemicals). Wax moth has no empty boxes to infest.

Rope or stretchy luggage straps stop roof blowing off

Wrong: leg retaining bolt pointing in - impossible to remove legs to move hive with bees inside

Correct: nut and screw thread on outside - leg can be removed, hive may then fit in

If the legs are too thin, the whole hive will wobble when you open it

Some makes have features that can be improved. Wobbly hives save lives, or something. Anyway wobbling can, sometimes, tip bees over the edge into grumpiness. And you may not think you'll ever have to move your hives… yes. Yes, one day you will. With bees inside.

Further useful features are: there is less disturbance to the bees when taking honey, because the nest is not exposed. This, and the low intervention regime means the bees are less defensive. And, you only need to lift one comb at a time (~3kg). However, the fragile combs **can not be handled as casually as frames.**

A downside is size. Two Warrés can fit in the footprint of one hTBH. However the wide, splayed feet make them pretty immune to being knocked over by animals; they're robust for remote locations and they are unlikely to be stolen.

Design considerations and variants

Several commercial models are available. I avoid flat roofs (maintenance issues, no air void so less insulated and don't act as a sunshield). Self-built ones allow a no-compromise design with exactly the features you want; the cheapest I know of cost £90, using scrap wood. I confess I bought mine, then later retrofitted insulation.

Any modifications need to be done **before** populating with bees. Prioritise insulation – walls *at least* 25mm thick and plenty of top insulation. Cedar is best (40% better insulator, doesn't rot easily, much lighter than most woods). Pine doesn't last as well, and warps. Some UK suppliers make hives with walls just one pine plank thick - you can feel the warmth of the bees through these in winter - this is too thin, and you may see comb collapse from overheating in summer.

Rectangular Tanzanian style TBHs accept frames but are colder, thus less suited to Britain. Kenyans seem to be ignored by woodpeckers.

Warmth matters because large draughty volumes with open mesh floors hinder parasite control behaviours – the bees have to spend too much effort regulating temperature. Being cold blooded, bees are vulnerable to fungi, and the best way to avoid mould is to avoid condensation on cold spots in a hive's humid atmosphere. This is why solid floors are best - and bees will eventually seal mesh ones with propolis.

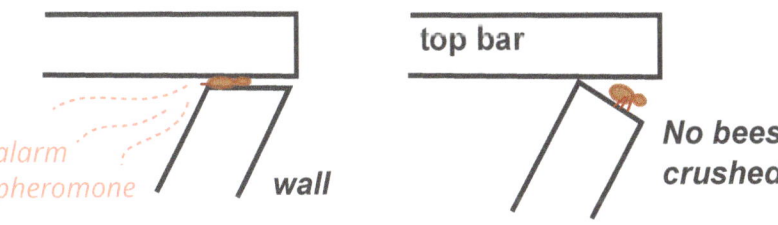

TBHs excel in windy conditions - their splayed legs make them very stable.

Les Crowder advises the top edge of a TBH should be a point (bevelled) so bees move off the ridge when you place a top bar back in place. This makes a huge difference in crushing bees / hive reaction.

Normal operation

After all manipulations etc, all bars are replaced so the bees have a solid roof of top bars, with no gaps.

(Some hives use thinner bars with a top cloth covered by a quilt, but the principle remains of allowing only one entrance point - or wasps get in.)

Viewing windows: A side window is only a marginal extra cost, but enormously useful.

Entrances can be at the ends or sides. My experience is end entrances encourage more orderly comb building along bars, but side entrances mimic natural nests better (comb is edge on to entrance) and both positions work. You want two sets of three 1" diameter entrance holes (at opposite ends of the hive) which can be blocked with corks; each colony needs one set of entrances. Two sets allows flexibility e.g. to split or merge two colonies in one hive. Entrance tunnels probably

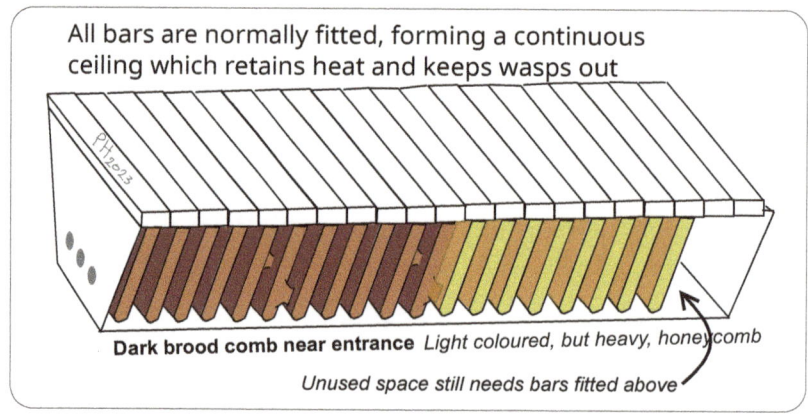

All bars are normally fitted, forming a continuous ceiling which retains heat and keeps wasps out
PH 2023
Dark brood comb near entrance
Light coloured, but heavy, honeycomb
Unused space still needs bars fitted above

I Mature TBH colony. Brood is generally nearest the entrance.

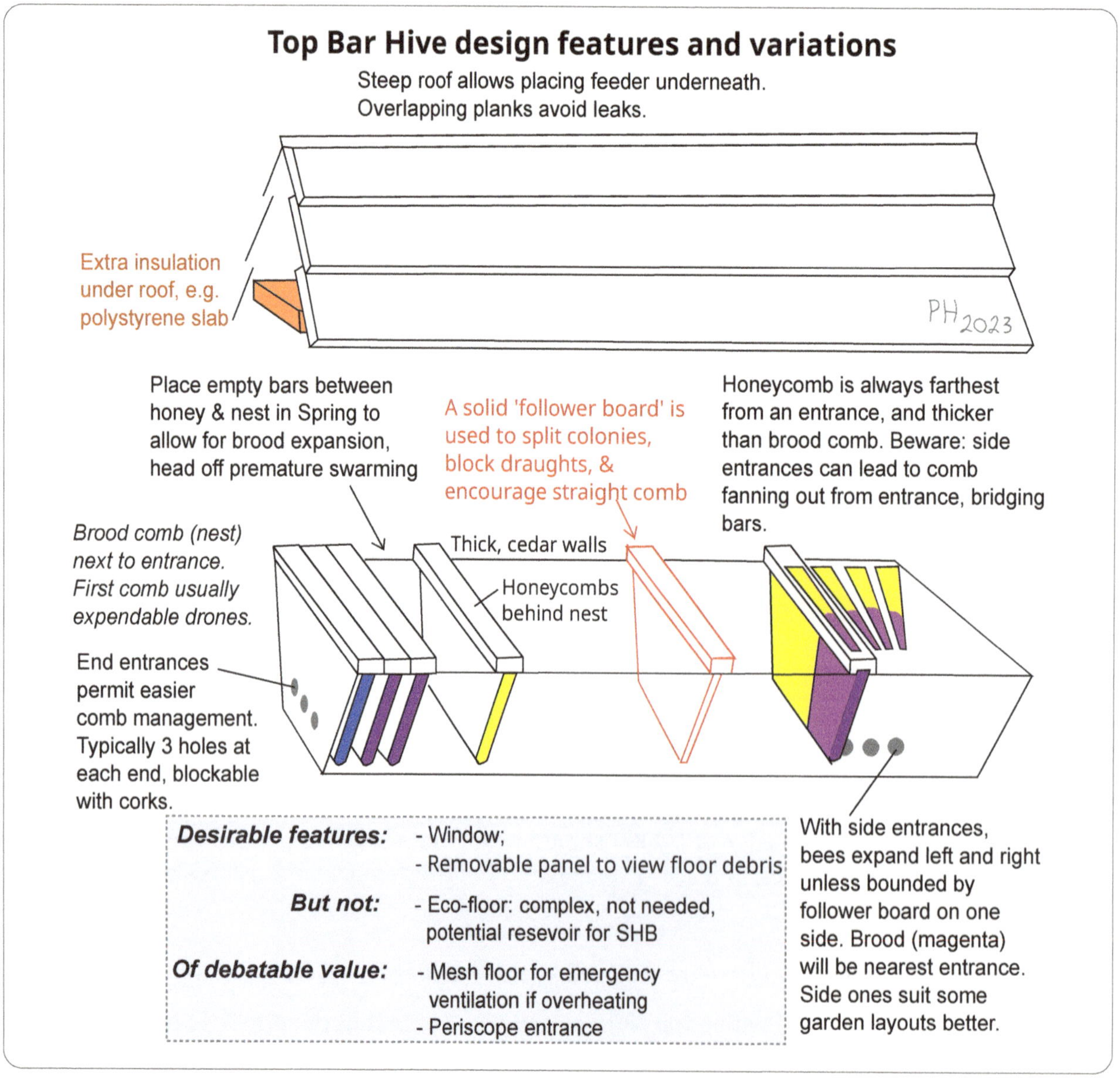

Top Bar Hive design features and variations
Steep roof allows placing feeder underneath.
Overlapping planks avoid leaks.
Extra insulation under roof, e.g. polystyrene slab
PH 2023
Place empty bars between honey & nest in Spring to allow for brood expansion, head off premature swarming
A solid 'follower board' is used to split colonies, block draughts, & encourage straight comb
Honeycomb is always farthest from an entrance, and thicker than brood comb. Beware: side entrances can lead to comb fanning out from entrance, bridging bars.
Brood comb (nest) next to entrance. First comb usually expendable drones.
Thick, cedar walls
Honeycombs behind nest
End entrances permit easier comb management. Typically 3 holes at each end, blockable with corks.
Desirable features:
- Window;
- Removable panel to view floor debris
But not:
- Eco-floor: complex, not needed, potential resevoir for SHB
Of debatable value:
- Mesh floor for emergency ventilation if overheating
- Periscope entrance
With side entrances, bees expand left and right unless bounded by follower board on one side. Brood (magenta) will be nearest entrance. Side ones suit some garden layouts better.

deter wasps a bit, but add complexity and make fitting mouse guards tricky. I suspect they make ventilating the hive harder work for the bees so I limit mine to about 5cm long in a block of wood (which still deters wasps and mouse guards can still be pinned in place). A landing board makes observing the bees easier.

Hive length: HTBHs tend to be 3' 6" or 4 feet long. Both work, but long ones with legs attached can be *just* too large to fit in a car. Assume you will have to move it one day and consider how you would remove the legs while bees are in occupation. 3 feet is too short, the hive will get crowded.

Mesh floors are not needed - if the hive is well insulated, and shaded after noon, you don't need extra ventilation to avoid comb collapse in heatwaves. As bees cool a hive by evaporation, a fully uncovered mesh can prevent them cooling on really hot days by disrupting the air flow. African hives are all solid floored!

In cold parts of England, TBHs with mesh floors have all their honey stored near the ceiling, away from the cold mesh. My TBHs originally came with mesh floors and I used to open these, to avoid comb collapse, for about 2-4 weeks in the hottest part of the year, but this has not been necessary since nearby trees grew and shaded them in the afternoon. I now use thick solid floors.

Removable floor inspection panels allow examination of floor debris for non invasive checking of clues like varroa mite levels. However, like mesh floors they add complexity and like mesh floors, they are a potential ingress route for pests like ants. So I no longer advocate them, I am now more focused on a warm, insulated floor.

Bar design

Semi frames: Combs deeper than 12" require an edge support to prevent collapse of soft, warm, new honeycomb holding 3kg of stores. Adding edge bars at 30° is tricky, but Helen Nunn cracked this by cutting slots in the bars, with a mitre box and gluing craft (popsicle) sticks in them. A simplified diagram:

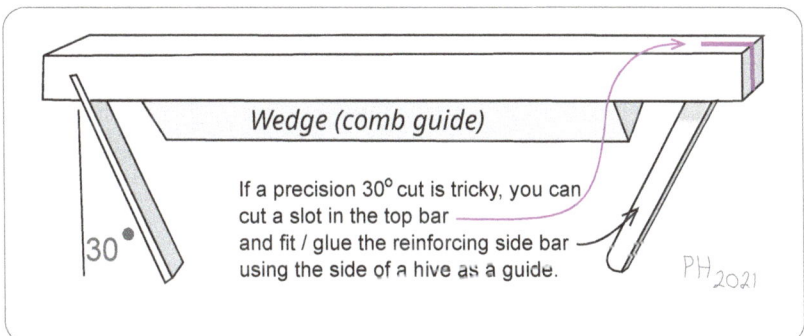

The side bars don't need to go all the way down. They just need to carry a significant part of the weight.

Helen actually went further than this, here's a photo of her bar design. She made thin bars with gaps between to discourage comb veering off the bars. The bar spacing is by wooden pegs at alternate ends of the bars (so you can rotate the combs 180° in the hive if desired), and the gaps are covered by a cloth, which has an insulating quilt above it (an idea from Warré hives). Result: completely straight comb, every time.

Helen Nunn's elegant bar design. Craft sticks in slots and gaps between (thin) bars. This hive has an extra section (an eke) above the bars. Image © 2016 Helen Nunn

Width is critical. Optimum width to avoid overlapping combs is debated and everyone I asked seems to use slightly different widths! If you prefer uniform bar widths, **38mm throughout** seems to work in Britain[215].

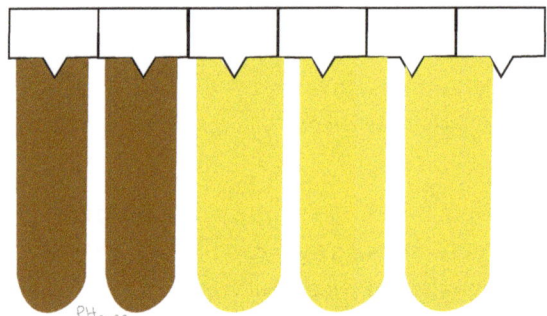

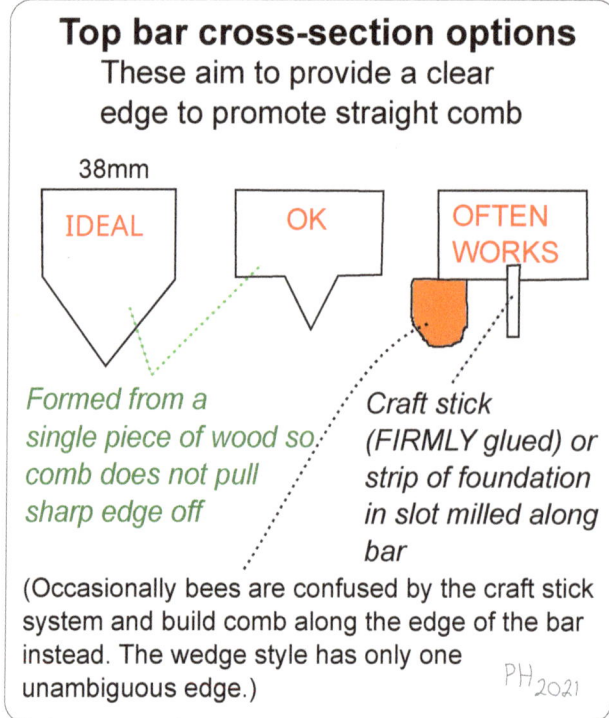

Worker brood comb (brown) is narrower than drone comb or honeycomb (yellow). If bar and comb width mismatch too much, combs are built out of step with the bars above, across 2 bars! This makes opening hives very difficult. 38mm usually works as a compromise bar width in Britain.

Don't just nail a triangular guide in place, it can pull out leading to comb collapse and a sticky mess.

You could get very fussy. Bees almost always build the nest next to the entrance, and the first, coolest comb is often a wider drone comb - and the brood nest expands and contracts over the year. Whilst some people use multiple bar widths, I found this complicates things, especially if you cycle bars from brood nest to honey area over a couple of years.

Do NOT use thin shims to vary bar width, it adds more complexity and solves no problems.

Length is best the same as your country's dominant hive type, i.e. 17" for British Nationals, 19" for American Langstroths, to permit some degree of interchangeability. 16" has been trialled and Backyardhive.com (USA) use 18". Shorter *may* give less cross-combing and less wavy comb; longer comb is heavier and more fragile.

In conclusion, following some sticky comb collapses, I now use semi frames for the heavy honeycombs but find that as brood comb is lighter, and tougher they can be unreinforced bars. All my bars use a lower sharp edge as a comb guide.

Using the hive

Populating your hive

HTBHs should be populated with a local swarm, ideally from a wild colony or non treatment beekeeper which has already proven its ability to survive without human management.

---

215 Comb spacing is usually 30-34mm for African bees (except *A m capensis*, 28mm), and 29mm for Asian (Apis cerana). Most American hives use 35mm (1 3/8") but some use the "Dadant" spacing of 38mm (1.5"). As explained above, in Britain (and Europe) 38mm is a safer choice with foundationless hives.
   Charitable donations of hives to Africa with the wrong spaced frames led to lots of cross combing until the importance of comb spacing was grasped.

Don't buy bees. Nucleus frames don't fit in [Kenyan] hTBHs. Commercial queens will interbreed with locals and can turn nasty after a year. The maximum-bees approach of a migratory honey farmer with queens selected for permanent superfecundity is counterproductive for static hives, which experience forage breaks. The resulting brood breaks suppress mites. More bees aren't always better, and may starve if not fed during dearths.

The downside of a swarm is that it may be a cast with a virgin queen, and if she fails to mate it may dwindle and die over a couple of months. But if honey is not your priority this is simply an interesting learning experience, and you repopulate next year.

Suitable races for horizontal hives - Russians believe[216] Italians and Amm fare best in vertical hives as they have a strong tendency to store honey above the brood nest, whilst Caucasian, Carpathian and Carnica are OK in horizontals. Horizontals are, of course, quite common in tropical areas which have different races.

Management

So, here's the thing. I find standard advice is wrong. It comes down to

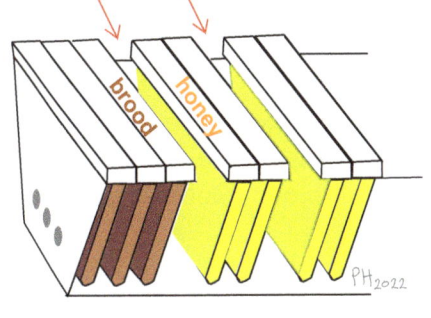

- **Move comb around** for various reasons - I find this makes no difference

- **Feed** - this often just causes problems unless you are honey farming. See *Feeding* in chapter 13.

Moving bars of straight comb apart gives no option but to build more straight comb between them (you add an empty bar between ones moved apart, and close the gap up tight). Follower boards can be used too but need moving every few days or the bees may run out of space.

Let's drill down into some comb management dogma inherited from conventional beekeeping, and what we actually observe in treatment-free beekeeping with survivor genetics.

"Old comb[217] accumulates toxins and must be cycled out of a hive". But many beekeepers consider old, black brood comb a valuable asset for a healthy hive; wild nests are full of it[218]. Chemical-free management means the foundation is not accumulating miticides, and our experience indicates there is no link to disease. To restate this, if the hive is healthy, why would you remove something the bees have taken years to create? They do remove comb if they do not like it.

"Unmanaged hives become uninspectable" **This is true, but** *"less is more"*. Bees gradually add brace and bridge comb. But every opening is a risk (damage the queen, chill brood, destroy the delicate pheromone balances, train them to recognise you as a threat etc) so I advocate **minimal** intervention:

---

216   According to GregB on Beesource. It may explain why some colonies produce combs completely filled with honey, whilst others only fill the top third and pefer a long brood nest - but this could be due to insulation or ventilation effects.

217   I like Les Crowder's criterion - "if you can no longer see light through it when held up to the sky". He's in America where crops are sprayed more intensively; I *never* replace old comb.

218   E.g. Moses Quimby, *Mysteries of Beekeeping Explained* (1853), p.19: paraphrasing the section "Old breeding cells will last a long time" he says: they will last 6-8 years; the bees do not get smaller as some think; he has 2 stocks with 12 year old comb; he has heard of 20 year old comb. He estimated removing old brood comb lost 10 to 25 pounds of honey per year. Black comb is tough and full of antiseptic propolis.

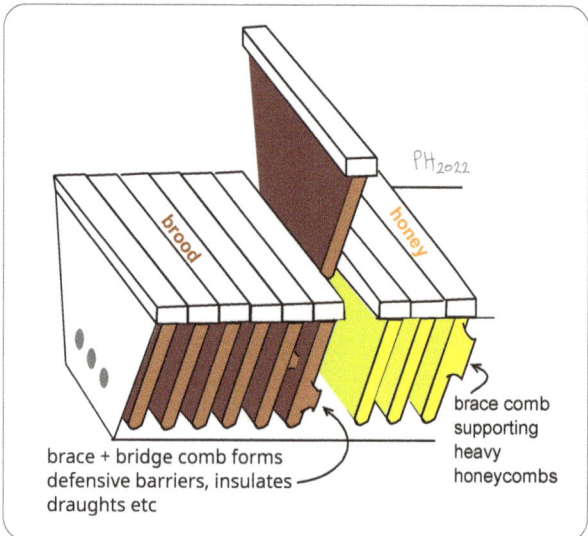

brace + bridge comb forms defensive barriers, insulates draughts etc

brace comb supporting heavy honeycombs

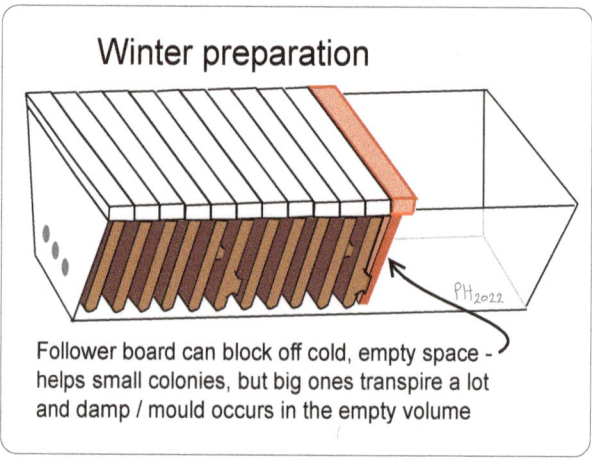

Follower board can block off cold, empty space – helps small colonies, but big ones transpire a lot and damp / mould occurs in the empty volume

Don't fill the empty void with crumpled newspaper. It will go mouldy when humidity condenses there.

Once the brood area's comb is straight there's no point inspecting more than the first 1-2 brood combs for disease, larvae etc

- A balance must be struck between the bees' needs, and *your* need to be familiar with how to handle combs, keep the hive inspectable, and know what healthy comb looks like.

- Once a swarm is in a hive I check it to confirm the **last built** comb is straight and true on its bar; if it is straight there is no need to move anything. Depending on weather, I do this about 4-9 days after the swarm is installed, any earlier and they may abscond; if later then correcting comb may require cutting too much comb. (Chapter 13 discusses *sorting out cross combed hives*.)

- I then check its expansion every 1-2 weeks and if they have problems building straight comb, I move bars apart as in the first diagram.

- Once a "straight and true" nest has been established, I leave it alone. It **is** inspectable if needed, but pulling it apart is pointless and disruptive. I learn all I need to from the end 1-2 combs.

Winter preparation: less is more

Some books advise moving honeycombs nearer the brood area in Autumn, to avoid winter isolation starvation. No one actually does this *in Britain* and the bees are fine – they'll reorganise and optimise the stores layout for you, usually intially clustering near the entrance.

However, it may be necessary to move combs to make space in some circumstances. For example, Canadian hives often fill with OSR honey. This sets rock hard in their winter and apparently bees can starve on it. So some judgement is needed here.

In winter, an hTBH's volume can be reduced to conserve heat by moving the follower board(s).

Swarm management

There is a widespread belief that if you do not give the brood nest room to expand in Spring, the bees will get blocked in by walls of honey and make many small, weak swarms which are unlikely to survive and weaken the mother colony.

But in the last few years I've found my TBH colonies have no problem expanding the brood nest in Spring to fill 50-65% of the hive, even if I do nothing. So I do nothing.

I am not aware of excessive swarms, in fact they seem to be swarming much less after a few years of commercial bees' prolific genes being deselected by tough love. About one swarm per year per mature hive now.

I do take a little honey after the first year, which probably helps. I don't see entire combs of crystallised honey (ivy, OSR etc) or pollen blocking expansion - though a nearby TBH user did find some hives died one cold spring when combs of frozen ivy honey led to isolation starvation.

Some hTBH users practise pre-emptive splits with the follower board, but I believe swarming is healthier (brood break for both halves, no comb transfer --> no horizontal disease transmission).

Opening TBHs

If you suspect something is awry because of unusual external observations, you may decide to open the hive.

It is usually easiest to start at the honey end. You need to cut away bits of brace & bridge comb connecting the main comb to walls / other combs [chapter 13 describes *Removing TBH comb: using a comb knife*]. You need to consider these factors:

- Always hold combs vertically.
- If you don't cut the comb away from the wall, it may tear off the bar and stay in the hive when you pull the bar up.
- But fully loaded (3kg), **new, warm honey**comb is easy to break - if you cut its brace comb adding support from the walls, it *may* collapse in **really hot** weather, unless you have semi-frames with reinforcing side-bars. As wax approaches our blood temperature it gets very soft.

Once you reach brood comb, it is stronger (and darker) because of reinforcing propolis, silk etc. And it's not loaded with 3kg of honey. But gentleness is still needed to avoid upsetting the nurse bees.

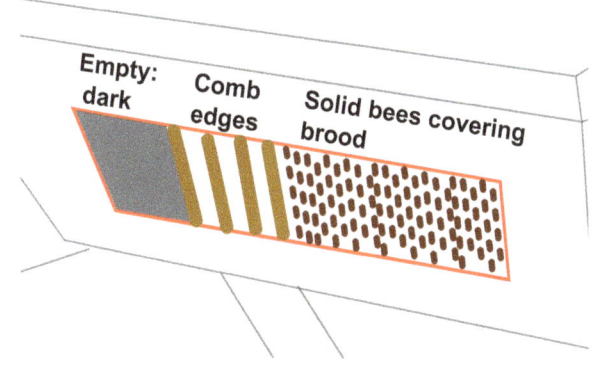

A window instantly tells you the brood nest location. Or you can feel for heat above the bars. With care, honeycomb can be manipulated without upsetting nurse bees; guards are at the entrance.

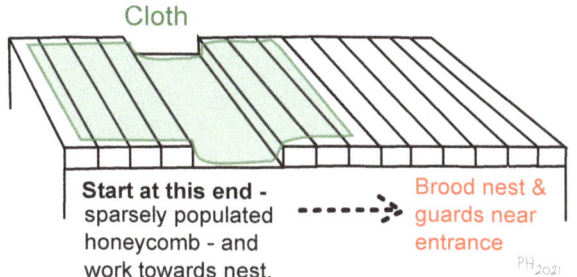

Place a cloth, like a tea towel, over the "working gap" to reduce light getting in, and bees / smells / warmth escaping. Sometimes people attach a rod at each end to weight it down in wind.

Typical top bar brood comb. Honey at top, capped brood below, pollen (bee bread) between them. Reinforcing craft sticks at sides. Image (c) 2017 Helen Nunn

Repairing fallen comb

If a comb breaks off, brood comb can be reattached with large hairclips, but a better method is the Rescue Bar developed by thegardenacademy.com

Honeycomb is heavier and more fragile. Don't try to rehang it unless it's very light. Leave it in the hive for the bees to rescue stores, then remove 1-2 weeks later.

Honey

Removing combs stimulates the bees to build and forage more, but detracts from other behaviours like varroa control. I take only 1-3 combs when the hive is heavy with stores. I don't take honey in a colony's first year.

As the combs lack wires, cut comb honey is easy to produce.

Feeding

If you do want to feed, it's a bit fiddly with TBHs. You generally need to open the hive, requiring a full suit, particularly when *refilling* feeders (because there will now be a bunch of bees all over the feeder).

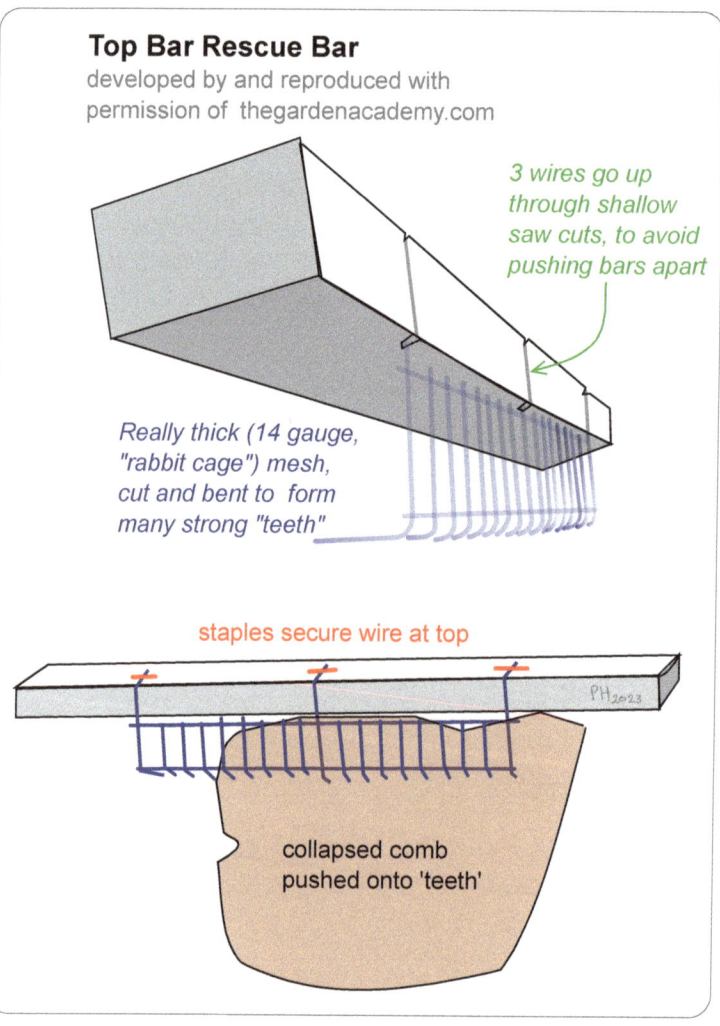

I use a couple of frame feeders which hang vertically in place of top bars.

You *can* pour syrup into a comb held horizontally, but it tends to run off rather than into cells. If you have spare crystallised honey, it can be laid on a [solid] floor.

A top feeder above the bars avoids stings, but requires a steeply angled roof, or an eke, to provide space for the feeder, and a modified bar (with a blockable hole) to allow bees controlled access.

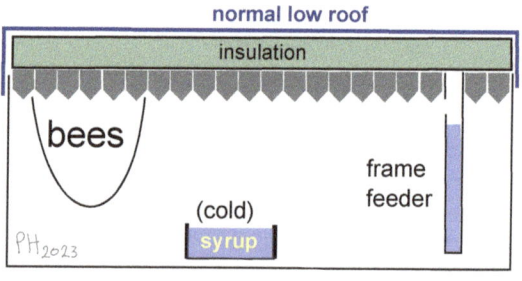

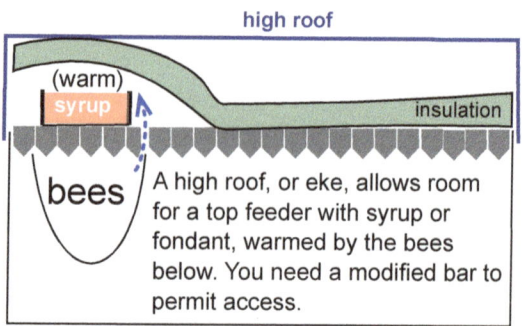

See chapter 13, *Key Skills, feeding* for more details on feeding TBHs.

Further reading

Balanced Beekeeping II, Phil Chandler (southern England)

Top Bar Beekeeping, Crowder and Harrell (US centric)

Warrés

Developed by the French abbot Warré, a simplified leave-alone "People's Hive" suitable for time-poor users, with just one size of box and no need for accessories like foundation or queen excluders. Boxes are added as the colony expands, there is one internal Spring inspection and usually just one harvest in Spring or Autumn.

Features

- **No frames or foundation:** bees build own comb and build *down* from the top
- **No queen excluder**
- Designed for simplicity: **minimal intervention and comb manipulation,** typically just one inspection and harvest a year
- More heavy lifting – add new boxes at *base (nadiring)* – but a full box of honey is just 25 lbs
- Well insulated, lower stress for the bees
- **Quilt box:**- a smaller box at the top of the stack stuffed with breathable insulation
- Lower internal volume / weight *per box* than framed hives, but as there is no queen excluder the brood can spread over multiple boxes, allowing for monster colonies in high-forage areas. Even a 2 box Warré has a larger potential brood area than a standard Langstroth 8 or 10 frame hive.

Windowed Warre with entrance raised above ground, luggage strap to secure it in high winds, and mouse guard fitted.

Honey is harvested by the *box*, not by taking individual combs. Near the end of a good season the top box is full of honey and it is removed as a unit, the comb - by now often attached to the sides of the box - is cut out as a batch and the honey harvested, usually by simply crushing and straining, because the combs are not on foundation.

You usually add boxes at the *bottom* of the stack (so, some heavy lifting involved) and harvest top boxes, so by the time you harvest a box its comb may be 2-3 years old and it has been brood comb. This gives the honey a lot of pollen and flavour, but it is dark comb with little wax to harvest.

Warre brood comb.

This hive is gaining popularity in the Anglosphere since David Heaf translated Warré's 1948 book *Beekeeping for All* in 2007 and wrote an updated how-to manual *Beekeeping with the Warré Hive* in 2013.

Preparing your Warré hive

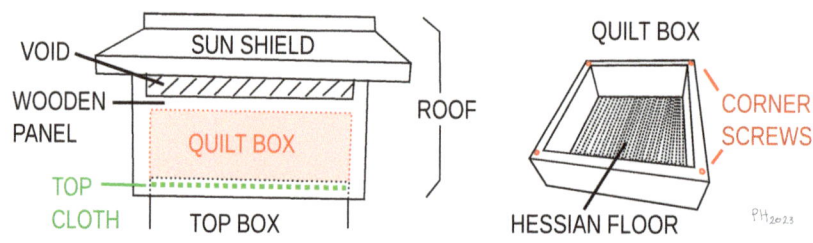

The **quilt box** has no wooden floor, you need to add a fabric base to stop its insulation falling through. Hessian is normally stapled to its base as a humidity-permeable membrane. I've also come across canvas and calico, and wire mesh.

> The wooden panel must not lay flat on the quilt box or humidity will be trapped in a damp quilt. The quilt must 'breathe'. Thus the corner screws. Alternatively, make a hole in the wooden panel covered with mouse-proof metal mesh.

Quilt filling: usually wool or wood shavings. Sawdust and *compressed* wool are not permeable enough to humidity - large colonies produce lots of water (eg honey processing) and too-dense quilts result in excessive condensation in the hive in Autumn, leading to soggy quilts and mould on floor.

Wood shavings seem consistently OK, so is straw (but don't use hay). Wool must be from organically raised sheep: most wool is treated with lethal insecticides.

Some people use internal baskets so the quilt can be changed easily.

Between the quilt box and the topmost box is the **top cloth**, an extra square of cloth, e.g. canvas. It is important this material has no protruding hairs as these trigger the bees to nibble it away, and bees can get stuck to hairy material and die. So if using hessian (rough sackcloth), it is stiffened ("sized") with e.g. beeswax melted on with a hairdryer, or flour-and-water paste. This is literally a consumable item; even sized it will only delay them about a year. If they chew through into the quilt, a cascade of sawdust can drop into the hive. Leave a few unwaxed patches for humidity to seep through. Hessian cut out of old potato sacks may contain insecticide: use untreated material.

The quilt box is filled with wool or wood shavings and acts as

- an insulator
- a humidity buffer (stabiliser)

but there is a potential problem – a damp, mouldy quilt - if there is no escape route for humidity except back into the hive. You see the wooden panel above the quilt box in the diagram? It presses flat against the quilt box forming a seal. So you need to allow slight ventilation by either

- drilling a few small holes in that wooden panel, or
- raise the panel 4 - 6 mm by add screws to each upper corner of the quilt box *(see above diagram)* or
- both

Warnings: if you make too big a gap, mice can squeeze in which would be Bad. If they are cedar hives then use stainless steel screws (brass reacts with cedar) and, the screws should not be right at the corners or they will hit the constructional screws which hold the quilt box together.

Improved bars to avoid the 'False Floor' ('Stuck Box') effect

Warrés sometimes swarm even when there are empty boxes below. The bees seem to mistake the bars in the box below them as a solid floor, and rather than build in the next box down they swarm.

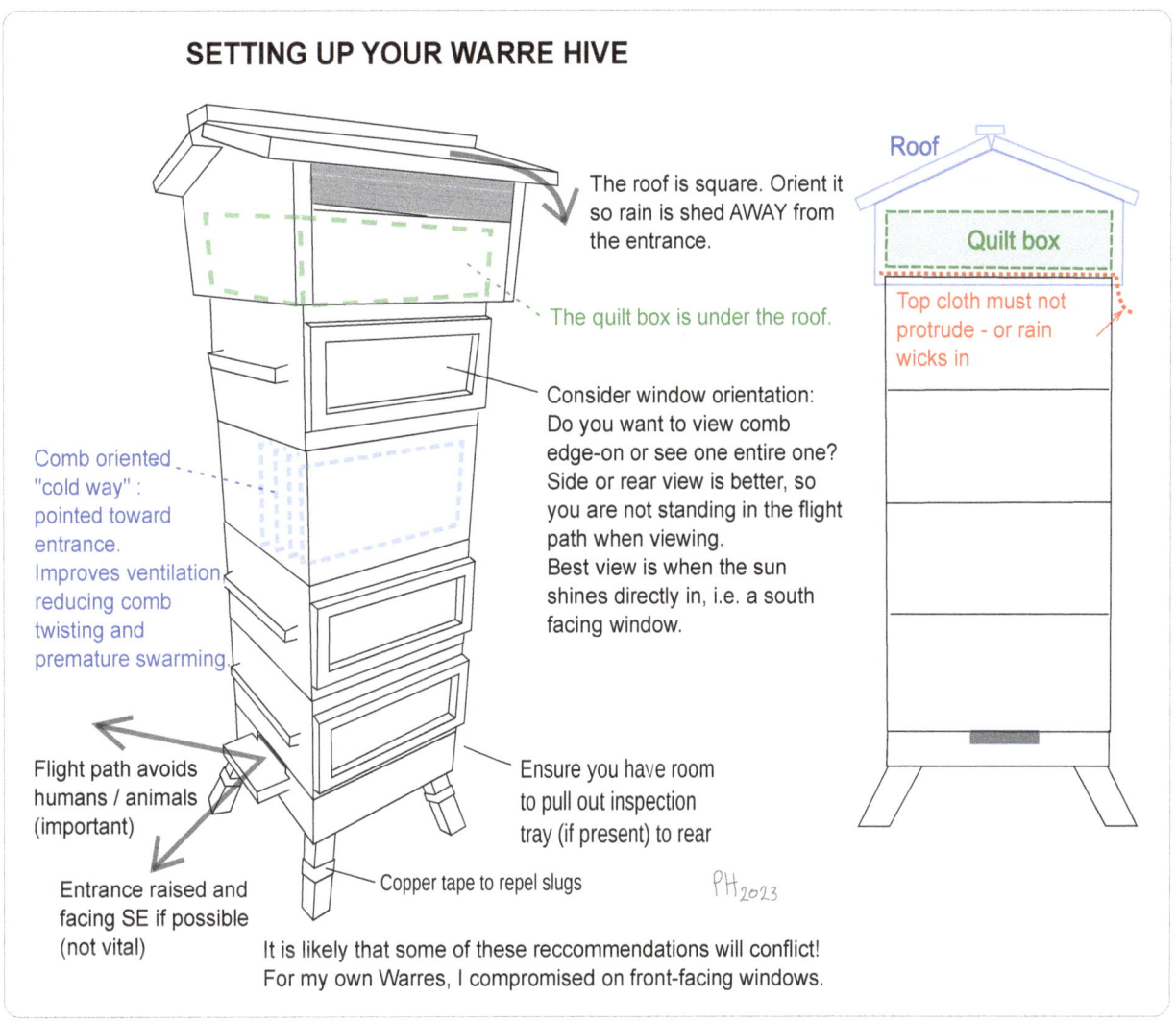

| Warrés are basically chimneys with a duvet on top.

- I definitely see fewer swarms, bigger colonies (= more honey crop) and less twisted comb with bars oriented cold way (pointing towards the entrance).
- Gareth John has found that using very thin Warré bars stops the False Floor effect and, a really deep, sharp lower edge inhibits cross combing.

This indicates to me that Warrés, which have solid floors, can suffer ventilation issues if you orient bars warm way; and perhaps the bars are too wide and restrict airflow. So I now use 19mm wide ones. Any narrower, and the gap between bars is big enough for the bees to build comb from the box above right through the gap, which I can assure you is a sticky disaster when you try to separate the boxes. Learn From My Fails!

Semi frames

You can use the normal Warré bars – flat bars of wood – and add a ridge of wax as a comb guide, and they work, though you sometimes get a lot of twisted combs.

I make fairly elaborate bars: it takes a lot of time and effort so you need a good reason to.

When a hive was knocked over in a storm, combs on these semi frames were fine, but many unreinforced combs were snapped off their bars by the impact, so I strongly reccommend these.

Their sharp lower edge (comb guide) is made with a router.

Management

Introducing swarms: use a 2 box hive for a small swarm, 3 boxes for a large one. Remove all bars except the top box's ones for about 4 days, *then replace them*. The reasons are (1) this ensures the bees have to start in the top box! (2) Big swarms sometimes abscond if they feel constrained - they prefer one big uninterrupted cavity.

Leave them alone for a few days so they don't feel disturbed and abscond. After 4-5 days you can be sure they have committed. At this point you **must** replace the bars in the other boxes or there is a danger the bees build comb down through >1 box, which will cause significant management problems later.

This diagram is necessarily simplified. You keep adding boxes as the bees build into the bottom box. By the end of summer I have had hives 5 boxes high, and I have seen pictures of towering Warrés in exceptional forage areas 7 or more boxes high - which present weight and safety challenges!

Warre semi frame

My current semi frame design, evolved from ones shown me by Gareth John and Brian Fiddian. Start with a bar of wood 315x19x24mm. The critical dimensions are in red. Ensure bars fit loosely (perhaps shorter than 315mm) because you don't want to find they won't fit in the middle of an open hive operation!

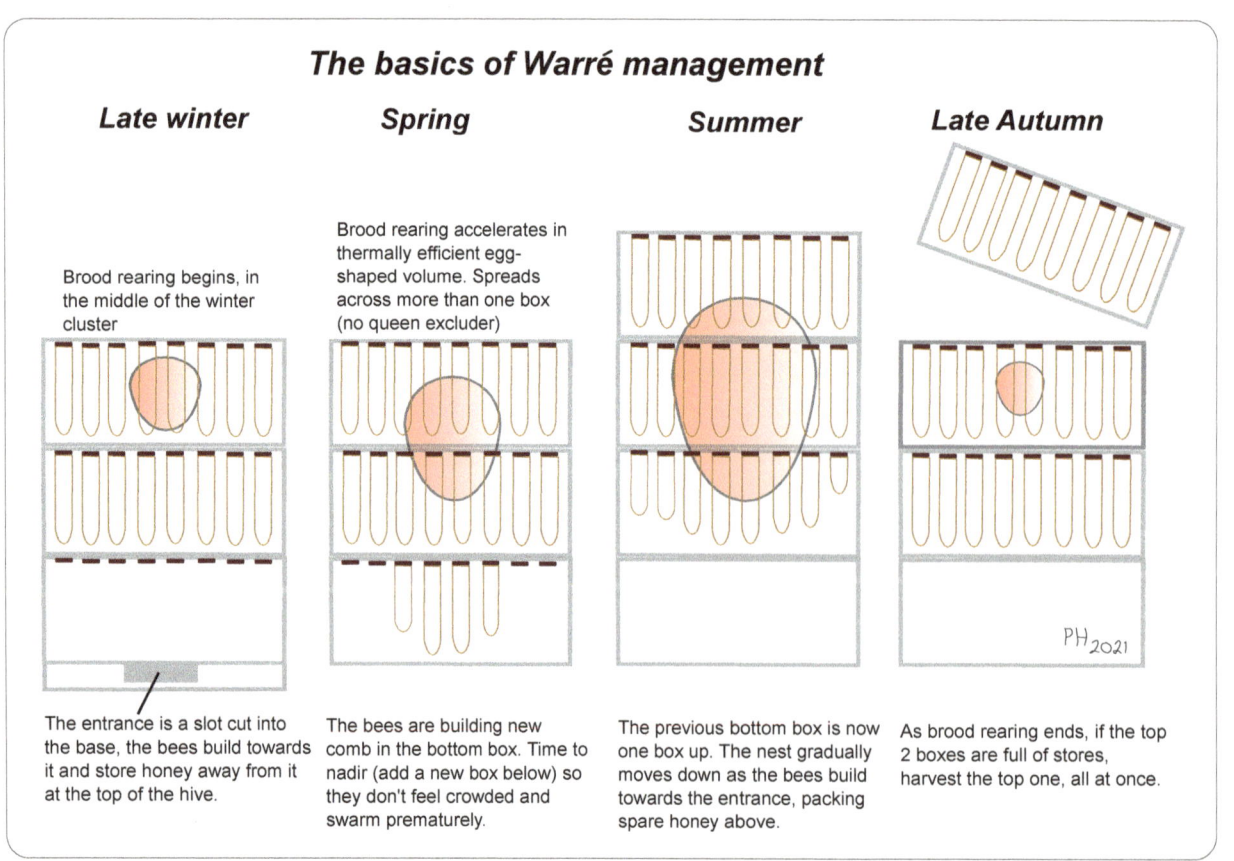

The basics of Warré management

Late winter: Brood rearing begins, in the middle of the winter cluster. The entrance is a slot cut into the base, the bees build towards it and store honey away from it at the top of the hive.

Spring: Brood rearing accelerates in thermally efficient egg-shaped volume. Spreads across more than one box (no queen excluder). The bees are building new comb in the bottom box. Time to nadir (add a new box below) so they don't feel crowded and swarm prematurely.

Summer: The previous bottom box is now one box up. The nest gradually moves down as the bees build towards the entrance, packing spare honey above.

Late Autumn: As brood rearing ends, if the top 2 boxes are full of stores, harvest the top one, all at once.

Separating boxes

- Insert a hive tool between boxes and give it a sharp rap to crack the propolis seal.
- If the boxes still seem connected after breaking the edge seal, comb is built right down to the lower box's top bars, and the boxes need cutting apart with cheesewire[219]

Nadiring - you don't need to lift all the boxes at once!

I can just about lift 2 full boxes at once, so temporarily split the hive into subsections during this operation. Use a sharp knife or wire to cut through comb if it joins boxes together.

Some people have made clever lifts and hoists. The most elegant (easiest to make) I have seen, was simply positioning the hives under the A-frame of an unused child's swing, allowing the beekeeper to use block and tackle to lift the hive.

Warré variations - what features to get?

Windows: Yes. You definitely want windows on every box, unless you have large numbers of hives (as they're expensive). The windows should be **glass,** not perspex - eventually you will harvest the box and need to clean propolis aggressively off the inside of the window, and you will damage perspex. But there are various window designs and ideally you want **thick insulation** over them so they don't get too cold – **this avoids condensation issues.**

At least one American Warré manufacturer found multiple problems with windows. Different climate? Less insulation?

Floors: There are various floor types offered. My recommendations are:

- The hive works best with a solid floor - mesh just creates problems
- I would go for a hive stand, if possible; my hives have splayed legs built into the stand to make the towers more stable.
- A pull-out inspection tray is useful, but if you don't use it often it can get jammed shut.
- Avoid sumps / eco-floors (overcomplicated, dubious benefits).

In countries with Small Hive Beetle, people sometimes integrate a beetle trap into the base.

Alternative entrances: rather than one entrance at the bottom, folk have tried one or two small round entrances in each box. This means the bees don't necessarily pack honey in the top boxes, so you need to be quite an experienced user as the management is then non-standard.

Super-insulation: I would say the minimum acceptable wall thickness is 25mm of cedar (more for other woods). I've seen octagonal Warrés, but these need bars of varying lengths; and double walled Warrés; these were very tricky to make. The makers did not build more. I gather their conclusion was these were a lot of effort to build, and the bees didn't fare noticeably

Window cover is edge on to viewer - my thumb gives an idea of how thick the foamed polyurethane insulation against the window is. NB: the comb in that top box was all edge-on to the window a few weeks earlier, but then I added a 5th box and they rearranged ventilation!

---

219 An un-wound bicycle brake cable makes an excellent cheesewire - tip from Mike Pighill. Bill Anderson recommends placing coins at corners after prising boxes apart slightly, creating a gap the cheesewire can move through.

better in our relatively mild climate. In Canada, Hubert Pilon (45.3° north) experimented with assorted insulation and settled on simply using 38mm thick cedar walls and a windbreak comprising mirrorised plastic bubblewrap, which proved the most durable wrap, the most effective at keeping the hive walls dry, cheap and easy to handle and store. (My own British Warrés have 25mm cedar walls, and I don't wrap them in winter.) Hubert tells me, "I let snow cover the entrance without problem. The colony can breathe through the snow and the winter is so cold that bees do not exit the hive until warmer days (in which days the snow will have partially melted, clearing part of the entrance). I have even seen hives completely covered by snow!"

Different box sizes: the critical dimension for most Warrés is that the internal cavity is 300 x 300 mm. You can make boxes any height, but deep ones weigh more so most people use 210 mm, or sometimes 100 mm if experimenting with supers.

In Australia, different standard lumber sizes mean they use 308 x 308 x 240 mm.

Supering rather than nadiring: Although the Warré is designed to be nadired, you can *super* it. This requires some finesse as Warrés don't use queen excluders:

- You have to time addition of the super carefully. It must be during a strong nectar flow.

- It needs a band of honey above the top box's brood to act as a natural queen excluder. Some people use a box of honey from lower down the stack for this, I don't have such strong flows here.

- It requires a strong colony and warm weather.

- Too soon (April?) and the bees will simply extend the brood nest up there. Too late (mid May) and the bees will ignore the super, leaving it empty all year!

- The successful supering I have seen, used shallow boxes, about 100mm high, and frames with a bottom bar. Otherwise the comb will be attached to the top bars of the box below.

I have yet to master supering in Warrés.

Further reading on Warré supering: warre.biobees.com/supering.htm

Warres modified for Canadian winters, manufactured by rebelbees.ca. The unusually thick wood is visible in the roof; the hive is wrapped in mirrorised bubble wrap. Photo © Hubert Pilon.

Experiment with empty box (21cm deep) as a super: after 3 months the bees had only built this comb UP from the bars below.

Experiment using box of empty comb from the bottom of the stack. Partially successful, they filled it with a mix of honey and nectar by October - I left it on the hive for winter.

Honey harvesting

The key is to avoid brood combs. Nurse bees won't leave them, and they will defend them. You need to process unpopulated boxes inside your house: cutting out combs in the garden will take ages and attract clouds of robber bees.

Get every bee out of harvested boxes and into the hive - young ones can't fly.

Locating the brood is simplified if you have windows, otherwise you will have to tip boxes back to look. Weight is also an indicator: tipping a box just a couple of mm gives you an idea if honey is in there - it's **very** heavy[220]! You can heft just the top box, it's easier to lift and judge than the whole stack.

The brood nest is covered in nurse bees. The brood nest doesn't generally go right to the walls, instead expanding downwards and leaving a honeycomb as an insulator next to the walls. Some people harvest this end comb if they just want one, knowing it will be brood-free. If you use a prolific queen in a hive with a restricted laying area (queen excluder) you may be more used to seeing brood right up to the edges of hives, and thus a lot of chilled brood.

(1) Basic harvesting technique

Initially we are trying to assess whether there is a broodless box we can harvest:

OPENING AND HARVESTING A WARRE HIVE

1. Remove roof and quilt box. Expose one bar at a time. Allow SOME smoke to waft under the top cloth - to dull bees' sense of smell and drive them down a little so you can gauge how many are present. You are NOT trying to clear the box of bees at this point.
Only expose as many bars as needed to ascertain bee numbers. Replace top cloth and smooth it down, so when this box is separated, bees are only exposed to light at the bottom of the box..

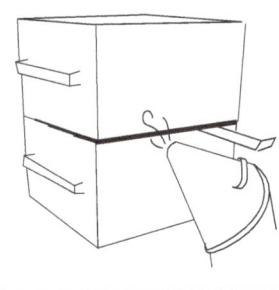

2. Crack boxes apart (break propolis seal) with hive tool. If combs are attached to lower box, use cheesewire to separate - ease the boxes apart just a bit with the hive tool to prevent the wire snagging on the wood; consider wedging 4 coins between boxes, one at each corner.
Smoke into this gap to move bees away from gap, then shift the top box forward 1cm so you can tip the box back safely..

---

220 Or you can use luggage scales; a full Warré box of honey is about 12kg, depending on wood type.

Smoking the bees out:

4A - straightforward way. BUT if the box is all honey, and the hive is crowded, smoking the box here risks the bees being driven down and pouring out the entrance.

4B - better way

4C - best way if you lack a clearer board: place the box (without top cloth) on its edge against a leg of the hive and smoke the bees out there. The box should be in contact with the hive as most of the bees will be non-flying: they need a path to walk along.

A chemical repellent called Bee Quick is popular in America. Its effect is so strong I cannot recommend it on a Warré stack as it may contaminate the comb below and disrupt long term behaviour.

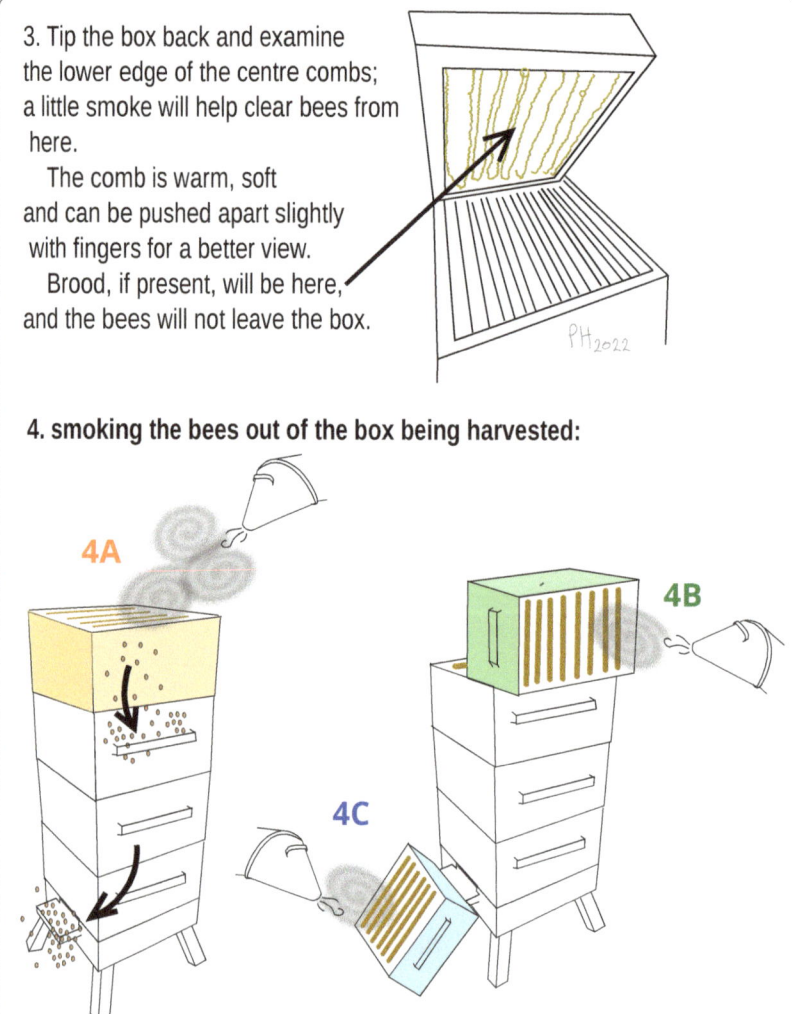

3. Tip the box back and examine the lower edge of the centre combs; a little smoke will help clear bees from here.

The comb is warm, soft and can be pushed apart slightly with fingers for a better view.

Brood, if present, will be here, and the bees will not leave the box.

4. smoking the bees out of the box being harvested:

Clearer board in use (2 boxes down - top box is semi empty, an experiment in supering). LCD thermometers in windows.

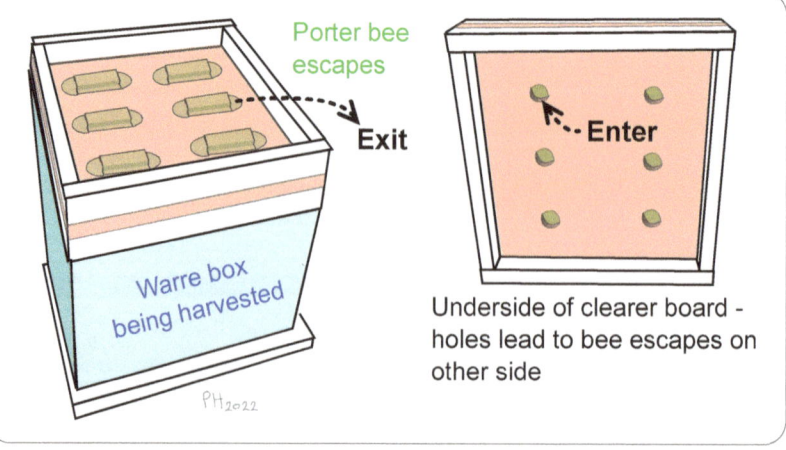

A home made clearer board: plywood (red) with rims added round sides; holes drilled through to Porter bee escapes. These are low profile and work well, as do rhombus escapes. Doesn't need to be fancy - can be made of cardboard! It can be flipped upside down to allow bees to go either way: it is the other way up in the hive clearing photo.

(2) Easier harvesting - with windows and a clearer board

This avoids having to clear hundreds of agitated bees from the combs.

After determining which boxes are (probably) just honey, bees can be cleared in a few hours by placing a *clearer board* between the honey and brood boxes; you return later and take the honey box(es) with minimal disruption. If using a clearer board, smoke is usually not needed to clear the harvested boxes - they'll be pretty much empty of bees. (However, when you lift them off the hive, you will need a little smoke in to the top of the hive to avoid crushing them as you put the roof back.)

Optional confirmation technique: LCD thermometers behind windows (zebra striped objects in photo) give a comparison of box temperatures. If the top boxes remain at the same temperature as the lower ones they have not cleared - nurse bees are keeping brood warm in there. If they're really cleared, they'll be about 10-12C lower.

Alternative harvesting technique off the stack - if you only want to open the hive once, boxes to be harvested are placed on a flat board **near** the hive (honey won't leak out - it's thick and forms a seal) and the clearer board is placed on top of the harvested board, flipped over, to stop bees getting in (see diagram). Do this very near the hive, with a clear path between harvested box and home hive so young, non flying bees have an easy walk home.

The honey is most easily processed quickly, while the combs are still warm.

Fun fact: Other creatures sometimes set up home in Warré roofs. David Heaf occasionally finds bats.

Curious convergences: the Japanese or "pile box" hive was developed independently during the Edo period (1586 – 1911 AD) and is eerily similar in design and management. It is still used by some hobbyists managing their local bee, *Apis cerana*. And in England, Edward Bevan published a similar design in 1827. Function defines form.

Further reading:

Natural Beekeeping with the Warré Hive, by David Heaf

Google groups, w*arrebeekeeping* mailing list

https://warre.biobees.com

Bees' nectar management

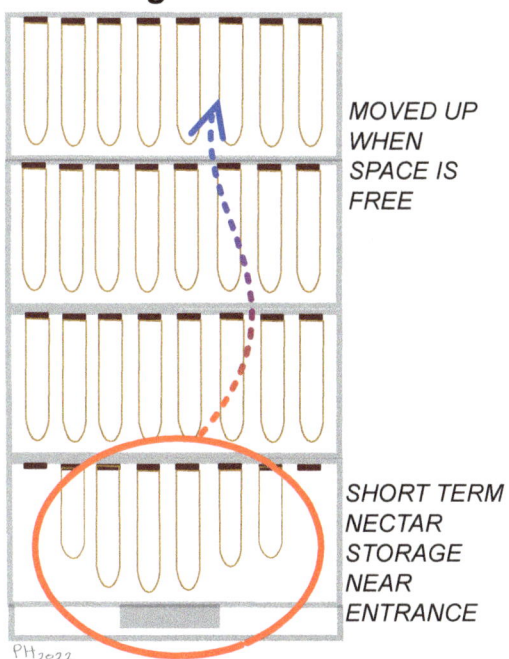

This is why you sometimes find apparently empty, unused combs at the bottom of a stack. It can even be a daily cycle, nectar moved up into supers during the night. Queen excluders inhibit this.

Skeps

Skeps are woven baskets. Rare these days, they were common in Britain up to around 1900.

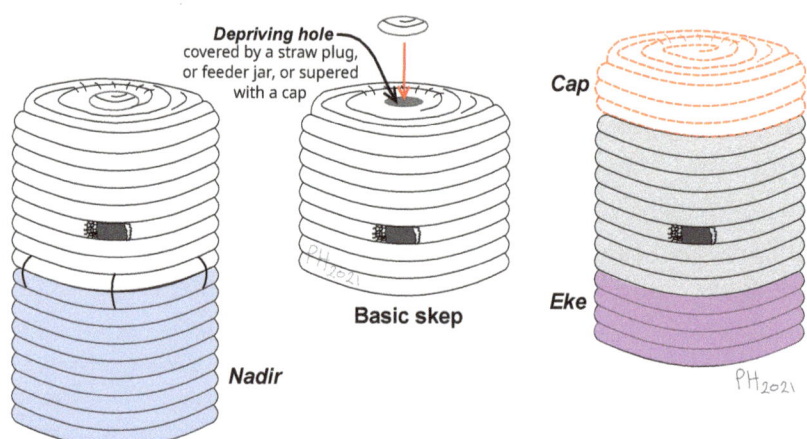

The walls should be over an inch thick and the test of a good skep is said to be that you can stand on it. However, that is tricky to weave, so these days many are thinner with added external insulation, because they don't need to be as rugged as a production hive.

Skeps use neither top bars nor frames. The bees attach the comb directly to the walls. To help support the comb, a couple of skewers, called *spleets* or *driving irons* or *cross sticks* or simply *sticks* are pushed through the walls.

Professional skeppists used stackable, flat topped ones with a hole in the top - a "reducing skep" - which gave them many management options. A **nadir** (skep with blocked entrance) can be added below to allow the brood nest to expand down and discourage swarming. **Supers** (above) and **ekes** (shallow nadirs, just hollow rings below) accumulate honey - eke comb is brood comb and its honey must be strained; supers yield less honey but it is pure, clean honeycomb which can be sold as cut sections, more valuable. Tall stacks of honey laden basketwork can be unstable and may need stitching together.

Because they are difficult to inspect for disease, most US states banned them by 1998. If you absolutely have to remove the comb for inspection[221], it can be cut out and then pinned back in place by re-skewering it with spleets. Remember that comb has an "up" and "down" (the cells are angled up). I don't think you can sterilise a skep.

Skeps are one of the best insulated hive types, and bees love them. Skeps have a reputation for swarming a lot due to limited room for colonies to grow, but you *can* expand them by adding sections above and below. The terms *supering* and *nadiring* predate framed hives: skeppists were very sophisticated. For example they used drone comb as a starter / guide comb in supers so queens didn't lay in them, giving pure honey[222]. They knew supers gave less honey overall than an eke, but it was pure honeycomb which didn't need straining.

They are populated with swarms, so when skeps were common these were greeted with joy - the opposite of how many beekeepers are trained now.

Pettigrew surveyed British honey harvests and concluded skep size was a key factor. In his *Handy Book of Bees* (1875) he recommends a steep sided, flattish topped type 16" or 18" diameter, 12" deep, 4" hole with a cap at top. Use the smaller diameter for small or late swarms. Illustration from his book.

---

221 The only reasons I can imagine for removing skep comb are to harvest it, or because there is an outbreak of notifiable disease in the area requiring inspection.

222 Pettigrew, p.79. Be sure to put the comb the right way up or honey leaks out!

The bees are allowed to do their own thing until honey is harvested in Autumn, by which time they are full of comb fixed to the walls by the bees.

Skep beekeeping was for *Amm*, a less prolific bee than other breeds which suits harsh, marginal environments[223]. Historical writers noted the inside of skeps became heavily propolised, and modern British users remark on this too.

A handful of enthusiasts like Chris Park in Britain (acorneducation.com) keep the skills alive.

Germany's Luneberg Heath skeppists operated at large scale (hundreds of hives) and use many sophisticated techniques, like *Pöttscher* - cute tiny skeps the size of a teapot, used as mini mating nucs and queen banks[224]. The last such beekeeper is probably Rob van Hernen, running 600 skeps in 2017. His skeps are shallower than the British tradition, so he doesn't need supporting spleets; he has some rectangular skeps, and uses queen excluders in some.

Different countries use different materials for low tech hives. You can still find rectangular and cylindrical cork hives in Portugal - in Britain this marvellous insulator is an expensive import. The photo shows one from a traditional apiary in Portugal; these hives were in full sun (~40°C in summer). Internally, they use sticks to reinforce comb and in the apiaries Martyn examined most didn't even bother with entrance holes: the bees just came and went via cracks in the walls!

You find similar simple hives around the world. An interesting management variation is suggested by the United Nations FAO[225] for traditional cylindrical Chinese hives - after harvesting honey from one end, the hive is turned *upside down* to encourage bees to rebuild into the now empty section.

Traditional Portugese cork hive with stone slab above cork roof; some are simply upight cylinders. Photograph © Martyn Townsend, 2023.

Preparation before introducing a swarm

a) Shelter

Even with inch thick thick walls, they are not weatherproof in the British climate so are often sealed on the outside with wattle and daub, or *clomb* (clay and straw) then sheltered in a variety of ways. British cottages sometimes had *bee boles* - recesses for skeps - built into their walls. If using dung, it must be from an organic farm - most livestock is doused with insecticides and their dung will kill the bees.

b) Inside must be smooth

Several historical sources stress you must smooth the inside of the hive - cut / sand down protruding straw etc - I'm not sure why but imagine bees tear wings on sharp protruberances.

---

223 Thomas Wildman, in *A Treatise on the Management of Bees* (1770) counted 8,000 – 10,000 bees in small hives and 18,000 in large ones. This is considerably fewer than beekeepers assume is normal now. However he also estimates swarms as 10,000 – 12,000 which is inconsistent, and I think illustrates the difficulty of counting bees. Estimating by weight is particularly tricky as workers can double their weight when stuffed. Despite the small populations, yields per colony were very high by our standards, because there was so much forage then.

224 A *queen bank* is where you store spare queens from e.g. merging two casts. One or more queens is stored with a few hundred bees to keep her alive until needed. Tiny hives have issues - they can run out of food very quickly; colonies in mini nucs may expand right out of the hive, abscond etc.

225 teca.apps.fao.org/teca/en/technologies/10135

Skep shelter at Sturt Farm, Hampshire. The skeps are about 40cm above ground, and the surfaces are sealed with dried cow dung.

A conical "hackle" or "skep bonnet" of hazel twigs forms a weather shield for this Butler style skep. Photo © 2024 Chris Park.

Chris Park opening a "WBC hive" to reveal a skep inside. This skep is supered with a second one - the lower skep has a hole in the top - and the cross sticks are clearly visible.

Square skep with alternative entrance design, and second entrance half way up, in bee bole (wall recess) at Chris Park's apiary.

c) Spleets and comb guides

Spleets - wooden skewers, basically - are poked through the walls,, typically 1/3 and 2/3 down the skep, to reinforce the comb.

Make sure part of them remains poking out externally, because you will need to pull them out when you eventually harvest the hive.

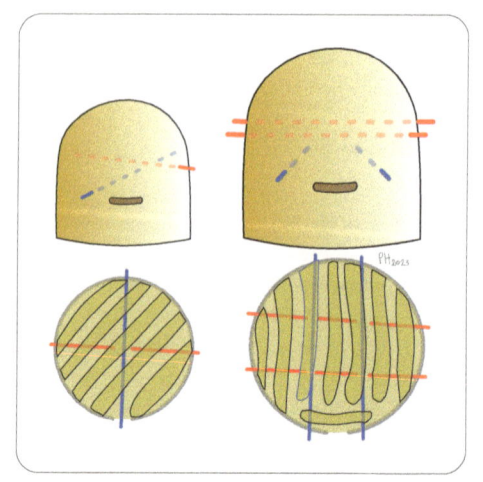

Spleets reinforce the comb. Orientation with respect to comb is not critical, but a larger skep, with more comb, will need more; large skeps may need 5-6.

Ideally, spleets are oriented at right angles to the comb, but you can't really predict how the bees will build it. Chris Park finds in Britain it's *usually* cold way, i.e. comb pointing towards the entrance, though if there is a large entrance the bees may build a small curtain of comb immediately behind it.

Some books talk of using comb guides - a small piece of wood with a wax starter strip, or piece of comb melted to it, wedged into the top of the skep. I'm not sure anyone bothers with that.

Small, stationary skeps handled gently don't necessarily need cross-sticks.

Skep walls were traditionally very thick.

d) Floors and stands

Pettigrew suggests floors are 1" thick (any thicker and their weight becomes an issue), and cut in pairs from deal (fir) boards in one piece which includes a landing board. His template is shown as **A** in this diagram.

You can cut a rebate into the floor to form an entrance under the rim of the skep, so the skep needs no door in its side.

Skeps need to be raised above ground level. Historically, 3 legged stools were often used; I've generally seen them at about waist height. The floor of the skep is thus wood; some people used rush mats.

Thomas Wildman, in *Management of Bees* (1770) warns that stone floors can be very cold; oak is colder than fir; and you can lose bees in winter if they walk onto a cold floor and freeze - so I would choose a wood that feels warm to the touch.

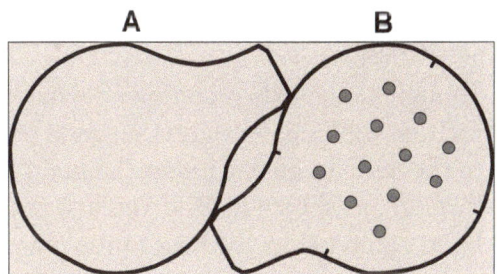

Design B shows notches for string to make moving skep easier, and holes for feeding bees from below. Floors are normally solid.

Wildman advised to put barley chaff (straw) and soot below skeps to deter slugs, snails & ants; snails can get into skeps and cause havoc; it also deters millipedes and woodlice which can enter in cold weather.[226]

---

226   He dismisses spiders and ants as minor pests.

A skep on a tree section. The vertical grain of the wood wicks away condensation.

Pettigrew style floors at Chris Park's apiary

e) Moving skeps

In my limited experience, adding 4 small notches as shown in **B** will be invaluable when moving skeps, when you tie them to the baseboard with string or cord: it will slip away without these. Pettigrew recommends you also use three 2-inch nails through the bottom of the basket into the floor-board, after placing thin coins between the basket and floor for ventilation. It is very important to leave the entrance and crown hole of the skep open, but covered with insect mesh, so the bees can breathe.

Quimby mentions the following trick[227]: turn the hive upside down when moving them. That way the combs' weight rests on their top, which is fixed, and they are much less likely to snap off.

Bagster suggests raising the skep off its stand for a few hours before moving it, so the bees are IN the skep not on the thing it stands on.

f) Feeding

Though some texts recommend what we would now call entrance feeders, these are notorious for inciting robbing. Pettigrew suggests you put food, like left-over honeycomb you are feeding back to the bees, in an eke under the main hive and use a special floorboard with holes for bees to go through, as shown in **B**. If you just put the food in an eke *without* a board between, the bees may build comb below the base of the main hive.

An opening in the top permits use of a contact feeder jar.

g) Ongoing management, and inspections

Skeps are very "leave-alone beekeeping". A swarm is introduced, and fills the skep with comb in a

---

227 Quimby was actually writing in the early 1800s about wooden hives just before frames were invented, rather than skeps, but the principle applies to any fixed-comb hive. His experience was with long bumpy trips on country roads using carts with no suspension. I think the skeps were generally wrapped in a sheet for transport. Interestingly, he remarks (p.188) that straw hive users didn't seem to have condensation problems in cold weather, though Chris Park has lost a couple to damp over winter.

month[228]. You then add a super. Chris Park advises that the key is to keep the holes between the baskets under 3" (7.5cm) in diameter. The queen then - usually - stays in the lower skep *without needing a queen excluder*[229] and you get pure honeycomb in the super. He does not harvest from the lower, brood, skep, leaving the honey in it for the bees to overwinter.

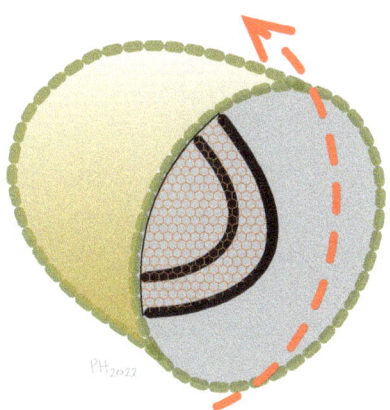

Inspection: Filipe Salbany pushes the fixed combs slightly apart, looking for evidence of brood; he is also feeling for warmth. (Conclusion: colony probably dying after swarming.) This skep is sheltered in a cork lined National shell, which is screwed together to protect against badgers.

To inspect, turn skeps up in a line parallel with the combs, so that none fall to one side. Bees naturally move **up** when disturbed, so unlike opening a framed hive, where they all pour out the top during inspections, bees almost ignore bottom inspectors.

Inspection is via external clues, or turning the skep upside down (you can wedge it between your legs, or in a big bucket).

According to Wildman, as soon as you up-end a skep, and give it some taps on the sides and bottom, the queen immediately appears, then retires. This behaviour can be used to catch her. Beware stings from workers, though.

Winter management

Straw was sometimes stuffed round skeps in bee boles for insulation. Pettigrew reckoned you couldn't have too much winter insulation.

Pettigrew describes the use of mouse guards. You need to block a skep's door so it has a 1" wide, 1/4" high gap surrounded by solid wood. You'll need to make these by hand.

Bagster[230] describes skeppists moving their skeps to the north, shadowed side of their house during winter. We discuss elsewhere how a good hard freeze is a good paasite break. However, Bevan says this tempted them out when too cold-but-bright. I think this is something which will depend on your climate and race of bees.

---

228   This assumes it is a reasonable sized swarm, the queen mates, there is forage etc - all the usual caveats!

229   I believe another reason she stays below is that queens don't usually cross honey ("I've come to the end of the brood nest, turn back"), and there will be a band of honey at the top of the brood comb in the lower skep. But in a poor year the honey band may be depleted so the queen wanders on.

230   *The Management of Bees, with a description of the Ladies Safety Hive* (1865) chapter 8

h) Harvesting

Cutting out outer combs - the simplest technique: the inner combs are brood, so just take the 4 outermost [honey]combs from an occupied skep.

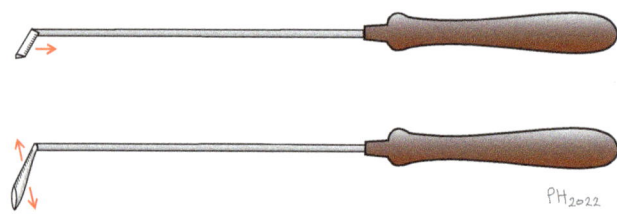

Drumming is particularly important for skep management (driving bees out for splits, harvesting etc) and is described in Chapter 13. Using heavy smoke taints honey.

Herrod-Hempsall describes using **skep knives** to harvest comb. One with a sharp edge pointing towards the handle cuts along the sides of the skep, then one with sharp edges pointing sideways severs them from the roof.

Throwing the bees[231], is another technique to get bees out of a hive - you can literally throw the entire colony out of a skep onto a sheet, or into an upturned hive, by thrusting the skep down over the receptacle and stopping it with a jerk. The bees are dislodged out. Modern beekeepers do this with single frames. Bevan says you may need to do it several times, and warns - the skep combs must be reinforced with crossed sticks first!

Bumping is another harvesting technique described by Bevan.

"After first intimidating the bees[232], another way to get them out, is to invert the hive and give it one or two sharp 'bumps' on the ground, at the edge of the crown on the side parallel to the combs. If carefully done the combs break away from the sides and top of the hive much cleaner than they can be taken out by any other way. [Remember a skep has slightly flexible sides.] *Brush the bees off into an empty skep, and transfer the combs as desired."*

Diagram showing exact point to 'bump' a skep, from Webster's Book of Beekeeping 4th Ed (published c. 1910).

Webster describes bumping in more detail, suggesting first smoking the bees and stopping up the entrance and drumming the hive to cause a commotion, leave for 2-3 minutes, remember to pull reinforcing sticks out first (good point!), turn the skep upside down and drive (smoke) bees away from the top edges of the combs. Then bump the hive in the exact position shown in the engraving, this breaks the attachments to walls and roof, and pull out the combs one by one, brushing the bees back in.

Whilst Webster's diagram is helpful, I believe you will get far fewer stings by first drumming the bees into another skep and then bumping an unoccupied skep. He actually warns against

---

231 Bevan writes of skeps, throwing and bumping in *The Honey Bee, its Natural History, Physiology and Management* (1887).
232 'Intimidating' = smoking heavily. Victorians were brutal towards their bees.

bumping exceptionally well-laden honeycombs without first diving out some bees, and warns that all bees should be driven out of deep skeps as their combs are prone to snapping in the middle, especially new combs on warm days.

i) Why did skeps fall out of favour?

Primarily, honey yields - you can process a hundred framed supers faster than a hundred skeps. Framed hives with interchangeable modules of a standard size are simply much more convenient for large scale operations.

Victorian promoters of wooden hives claimed they had far fewer insects and spiders preying on the bees, and fewer wax moth problems. Pettigrew, a skep user writing in 1875, considered wax moth a major problem[233]: one solution was to move bees to clean skeps every 3 years (queens typically lasted 3 years then).

Skeps were appropriate when wax was a premium product - it was in massive demand for bright, odourless candles. But as framed hives emerged, in the mid 1800s, mass manufactured paraffin wax candles were developed. By the end of the 19th century alternative lighting technologies had replaced most of the market for candles.

heritagecrafts.org.uk mentions British skep beekeeping waned after WW1, when a government restocking scheme only gave a subsidy for bees in boxes, not skeps.

j) Fact check: did skep users select for swarminess?

Short answer: some did, many inadvertently did the opposite!

Long answer: Larger scale skeppists rely on lots of swarms to rapidly populate many skeps at the beginning of the season. The Luneberg Heath skeppists northern Germany, where beekeepers may have hundreds of colonies by the end of the year, then combine weaker ones so they have empty skeps next Spring. So they select for swarminess. But the kind of beekeeper who routinely killed his heaviest colonies was selecting *against* swarminess (see next Fact Check).

Books written in the 1800s seem to imply you *could* get several swarms from one hive in one year - but sometimes not for several years. (It's difficult to say for sure as they only mention these things in passing - they weren't aware it would be interesting now.) The bees obviously tuned their response to that year's forage opportunities. There seems to have been a jump in swarminess when the British government imported a lot of Dutch heath bees in the 1920s[234] to replace losses from the Isle of Wight disease; and another when people began introducing Carniolan stock to Britain. Both these strains are dark bees, and partially informed beekeepers then assumed that Amm bees must be swarmy - whilst simultaneously arguing that all the Amm had been wiped out!

k) Fact check: did skep users routinely kill their bees as some claim?

Short answer: It's not the hive type that determines if people kill their bees, it's the beekeeper.

Long answer: there are elements of truth in this statement, at least in England up to about 1910.

---

233 Recently (2020) John Haverson reported wax moth only get out of control in colonies that are not queen-right... he notes that though the moths can chew into wood, they don't seem to damage straw skeps. David Heaf suggests this is due to the high silica content of straw.

234 Mentioned by Brother Adam reference reqd and I also read on p.128 of Bee Matters and Beemasters, 2nd Ed by Herbert Mace that John Charles Bee-Mason 'imported 1,000 326 Dutch skeps 1914, sending 1,000 swarms from Bures to all parts of the British Isles'. [Bee-Mason made films like "The Bee Master" in 1913 viewable online.] In fairness, Herrod-Hempsall, originally a critic of Dutch bees changed his mind saying "they don't seem to swarm much in cooler Scotland" (*Bee Keeping New & Old* (1930), p.575-6)

Many skep users would kill the lightest and heaviest colonies with burning sulfur, and overwinter the medium ones - the reasoning being that the weak would not survive winter and the heavy ones had the most honey to harvest. An ex-skep user looking back on this[235] reflected that he didn't know why they did this - it would have been simple to drum the bees out of one skep into another without killing them. They'd just always done it.

It is notable that Simmins, in *A Modern Bee Farm* (1887) never discusses skep users sulfuring (killing) the bees, which supports modern skep users' assertion that that was only ever done by a few people. Pettigrew's *Handy Book of Bees* (1875) strongly urges against sulfuring, listing advantages of uniting with other stocks for winter. Alston, in *Skeps: Their History Making and Use* (1987, p.66) notes that by the early 18th century skeppists mainly used supering precisely to avoid killing bees. Robinson's *British Bee-Farming its Profits And Pleasures* (1880) has a chapter on 'How to avoid the Brimstone-pit', by drumming. Basically, every reference written by historical skep users advises against sulfuring.

Bevan wrote in 1870 that *unlike Europeans,* British skeppists were ignorant of advanced techniques to maximise harvests like migratory beekeeping, and simply sulfured their bees[236].

William Herrod-Hempsall writes[237] of how he and his brother learned to drive bees and in the 1890s, did this for Nottinghamshire skep users who were delighted their bees didn't have to be killed for harvesting. The boys used the bees to populate their own hives.

However, this casual killing wasn't limited to skep users. Moses Quimby wrote[238] in 1853 of people with early *wooden box hives* sulfuring bees.

Some Canadian Langstroth beekeepers routinely killed bees, as late as the 1980s[239] because it wasn't worth the trouble of keeping them alive over winter, with typical losses of 40%. It was cheaper and easier to simply take all a hive's honey, then restock with mail-order Californian bees in Spring. This practice died out when packages of bees tripled in cost[240], but cyanide is still licensed for killing bees in Canada (google "*cyanide beekeeping*") although Ron Miksha, a former Canadian bee farmer & inspector tells me no one has killed colonies there since at least 1990.

---

235 Tickner Edwardes, *The Bee Master of Warrilow* (1920) Chapter XXX, "The Bee-Burners": "Thus a complete reversal of the doctrine of survival of the fittest was brought about". He points out that skep beekeeping relies on swarms, and those are most likely to emerge from the heaviest hives because those chose not to swarm this year. Another point against sulfuring bees is that it taints the honey! Edwardes does not mention this, but beekeepers in the 1800s probably assumed there was an endless supply of swarms, they were very common.

236 *Bevan on the Honey Bee*, revised & illustrated by WA Munn, 3rd Ed, p.137

237 *Bee-Keeping New and Old* (1930), p.3-4. Routine killing of bees by skeppists is also advised against by W B Webster in his Book of Beekeeping, 4th Ed p.83-84, written around 1910; he argues it's easy to drive them into a new skep and have More Bees.

238 *Mysteries of Beekeeping Explained,* p.168. Quimby lived in New York State. He points out they were throwing money away.

239 You can still (2020) find a document titled *Beekeeping for Beginners* on the Alberta government website which states "Once the honey supers have been removed, the bees in the colony can be killed if you plan to buy package bees in the spring." (https://open.alberta.ca/dataset/2819821, page 2 of document number Agdex 616-23). The topic is discussed online here: www.quora.com/Is-it-true-that-some-beekeepers-kill-their-bees-over-winter-and-if-so-why

240 When varroa reached the USA in 1987 the Canadian government closed the land border to imports (people still smuggled them over) and surviving beekeepers switched to importing costlier packages from New Zealand and Australia, except in Manitoba where they now raise their own bees one year to use them the next. Cyanide use within living memory confirmed by Paul Kelly, Guelph University, Ontario in Zoom talk to CBKA 9/11/2022

There is, however, some truth to the claim. Apicide was widespread in Europe after the fall of the Roman Empire. Reading Latin texts, scholars of the 15th and 16th centuries realized that beekeeping was much more profitable in Roman times and that bees were never sacrificed to get their products[241]. As the 17th & 18th centuries progressed, people imported and adapted ideas from Crete, Greece and elsewhere with the aim of preserving the bees and making harvests easier and larger, resulting in modern framed hives.

l) Variations

Before skeps, the Anglo-Saxons used the *alveary*, a cone woven of wicker rather than straw. They are weatherproofed with a layer of e.g. clay & straw, sometimes with added lime, sand or gravel to resist the gnawing of mice. Jane Denby describes hers (pictured) as a "swarm machine", presumably as the narrow cone constricts brood.

I've seen a skep with multiple entrances at different heights.

m) Further information

Books

Skeps: Their history making and use by Frank Alston

Make your own skep, Rev E Nobbs, being updated by Chris Park and republished by BIBBA

Handy Book of Bees, Pettigrew (1875) – my primary source – large scale Scottish skeppist

Online

www.degoederaat.nl/dir/english/ - Rob van Hernen's website (runs 600 skeps) - videos of key operations like shaking bees from one skep into another.

janesbees.ie - general info, runs courses, sells skeps

https://heritagecrafts.org.uk/bee-skep-making/ - background info

Chris Park runs skepping courses - beesfordevelopment.org/bee-involved/courses, and @chrischarlespark on Facebook / Instagram

bijodivers.nl, boombijen.nl (Netherlands)

Alveary in Jane Denby's garden, with weather shield.

---

241 UN report *Good beekeeping practices for sustainable apiculture*, ISSN 1810-0708 p. 25 . The surviving Roman literature was a vague about the exact methods used, but the Enlightenment scholars now knew it was possible.

Log & tree hives

Log hives are basically hollow logs, generally with the ends capped with discs of e.g. wood. The more basic ones have comb fixed directly to walls and ceiling, and are inherently low-intervention and harvested by cutting comb out; so direct inspection of comb isn't done, which frustrates some bee inspectors. Other styles use top bars or occasionally frames.

The *Zeidler* style, cutting cavities directly into trees, is being reintroduced to Eastern Europe from its last practitioners in Russia.

An intrinsic advantage is that they are pretty near what European bees evolved for - typically a long, 9" diameter cavity with very thick walls. The enormous thermal mass of the trunk provides a very stable environment and the rough walls trigger propolis coating - some people are now roughening the inside of normal hives to promote this.

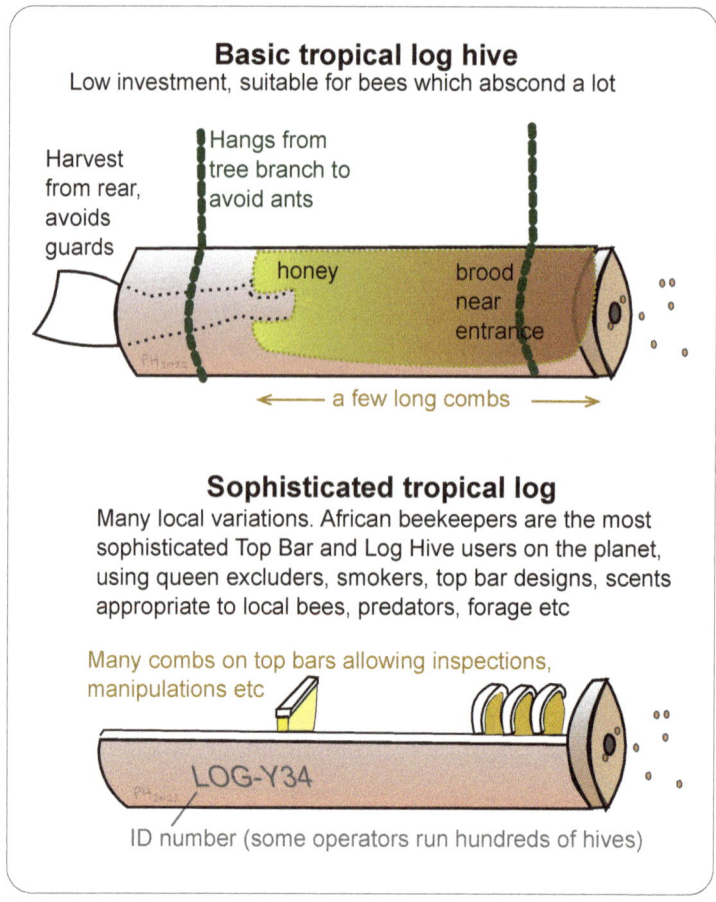

African hives may be 10m - 30m high and it's too hot for bee suits, so there is considerable skill involved.

Ethiopian log hive users smear the inside of the cavity with cow dung - a natural disinfectant, apparently - and once dry they smoke the inside with local herbs to make the cavity smell more attractive. The end discs may be made of woven straw. In some cases, harvesting is by forcing all the bees out with smoke, the bees then move to another cavity; but it's more common for the beekeepers to try and harvest non-destructively.

Russian interest in traditional log hives is reviving; there are production apiaries of 50-100 log hives in the Urals, some larger, though honey from these costs 10-20 times normal.

In the Himalayas, *Apis cerana* in forested areas is traditionally kept in thin walled log hives[242]. The inside is scorched to kill pests and remove smells as *A. cerana* is quite sensitive to smells and doesn't like painted hives. Hives are often positioned on the walls of houses, raised, and sheltered by the eaves. The bees are essentially unmanaged and allowed to swarm at will; at the end of the rainy season, capped honey is ready to harvest. Because there are so many swarms, some hives are harvested destructively (i.e. all comb is taken: the locals enjoy eating the brood) but if they are mounted horizontally, as *"sleeping hives"* the beekeeper can sometimes manage to

---

242   This info is drawn from an article on the courses run by *Elevated Honey Co* in Shangri-La, Tibet: www.permaculturenews.org/2019/06/24/a-look-at-traditional-asian-beekeeping-methods

harvest some comb whilst leaving a viable colony. Note the similarities with skep beekeeping on forage-rich heaths - the "business model" works because the bee and colony numbers explode in forage-rich environments, limited only by the number of available cavities not swarms; so beekeepers can harvest sustainably.

Ornate Latvian log hive with crafter Maris Kirnis. As a production hive, it has a huge internal cavity and large rear hatch - a child would fit in. The entrance design features thin slots which hornets cannot enter, and no shelf so sick bees fall straight out. This one is freestanding, but most are tree mounted.

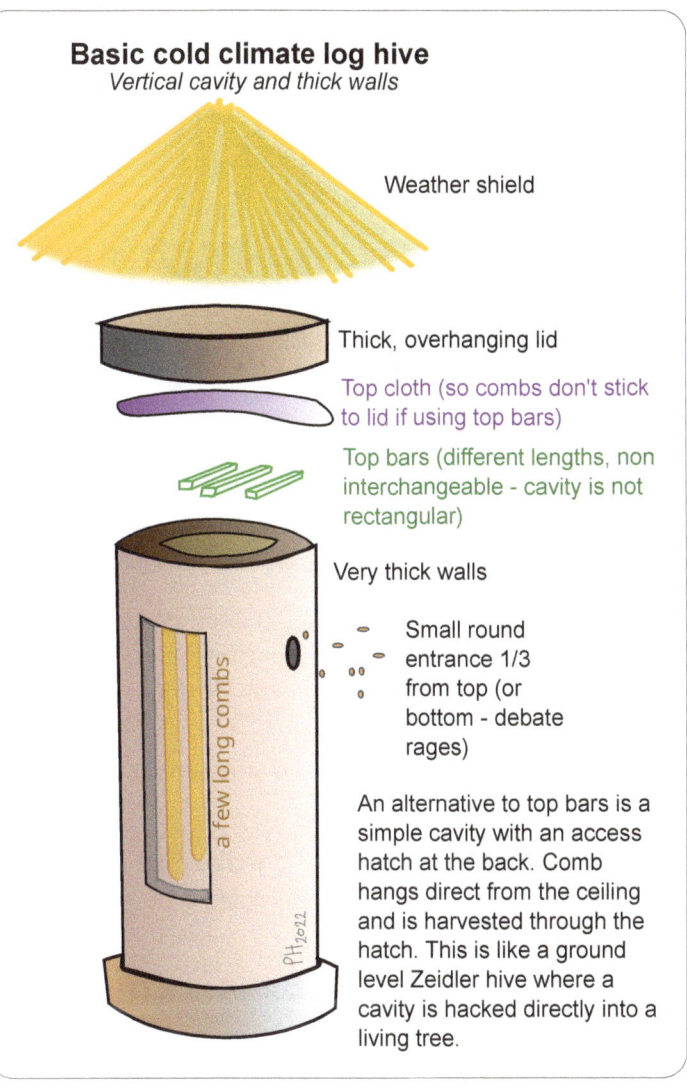

In vertical (cold weather) logs intended for harvesting, it is standard to access comb from the side. Top bars are uncommon in these as they add complexity. Some vertical logs raised above ground level and intended only for conservation, have a simple bottom hatch which allows some inspection.

This style works in areas of forest and mixed crops, but in monoculture crop areas, *A.cerana* can't follow its frequent swarming pattern.

Simulating hollow trees

There is a general shortage of really big old trees with ideal cavities, or wide enough to cut a Zeidler-style cavity into, so a hybrid approach is to elevate a log hive on long legs, or mount it high up to provide fit-and-leave-alone bee homes. A number of relatively light, often double walled designs have been used, but these are tricky and expensive to make so most people try that a few times, then converge on a simple hollowed out log, desite the extra difficulty of hauling

Log hive apiary in Oman. In the Mediterranean and Middle East, horizontal clay and pottery tubes are also a traditional hive. Photo © 2020 Jane Denby

Freedom Hive up tree in Oxford

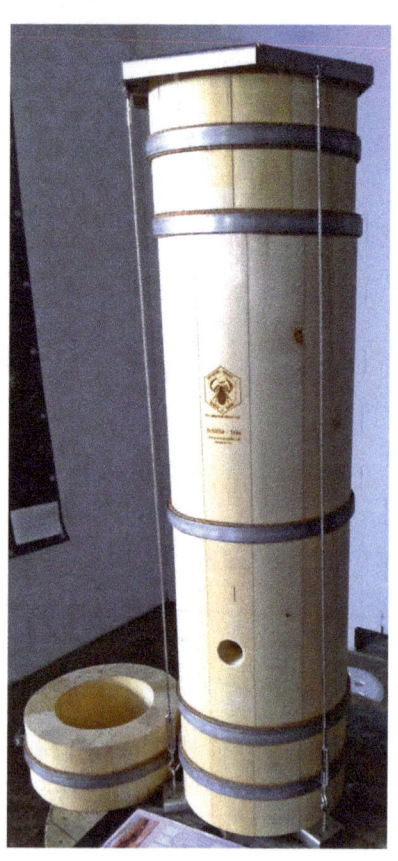

Schiffer hive

such a heavy object up high - several people have broken hips etc doing this.

A hive strapped high up a tree is well suited to the far north's forests - the entrance never gets blocked by snow, and they are a lot trickier for bears to spot.

This style is becoming quietly fashionable. Here's some of the innovators:

- Matt Somerville and John Haverson (beekindhives.com, UK)
- Simon Kellam (justbeeecohives.com, UK)
- Torben Schiffer (beenature-project.com, Germany)
- Mick Verspiuj (boomtreebees.com, Ireland)
- Mārcis Bauze-Krastiņš (the-cirgale-society.business.site, Latvia)
- Willi Herzog (freethebees.ch, Switzerland)
- Ivan Pigarev (https://return-of-the-bees.com , Russia)
- Michael Thiele et alia (apisarborea.org, USA)
- Piotr Pilasiewicz (bartnictwo.com , Poland)
- bijodivers.nl, boombijen.nl (Netherlands)

Often with removable bases rather than side access panels, these are generally easier to observe than actual tree hollows and users have reported many interesting behaviours, and refined their designs over the years (for example noting bees prefer round cavities). They're popular in forests, and semi public spaces as people can get near and feel connected, but safe as they are below the bees' flight path.

Further reading

The Tree Beekeeping Field Guide, Jonathan Powell - EPUB eBook available from the Natural Beekeeping Trust

The Arboreal Apiculture Salon, freelivingbees.com/the-salon

Key search term: **Zeidler** beekeeping

Running a standard framed hive on low intervention principles

Assumption: you have a handful of static hives run for fun not profit.

Such hives *can* be used by low-intervention / natural beekeepers; many have run them treatment-free (i.e. no chemical miticides) for decades. You don't have to use foundation: natural comb is possible on frames. We would usually add lots of insulation[243]. Although they are designed as honey factories, you don't need to run them like sweatshops.

Frames should be of the type giving bottom bee space, not a gap at the top. Bees naturally hang comb from the ceiling to avoid heat loss.

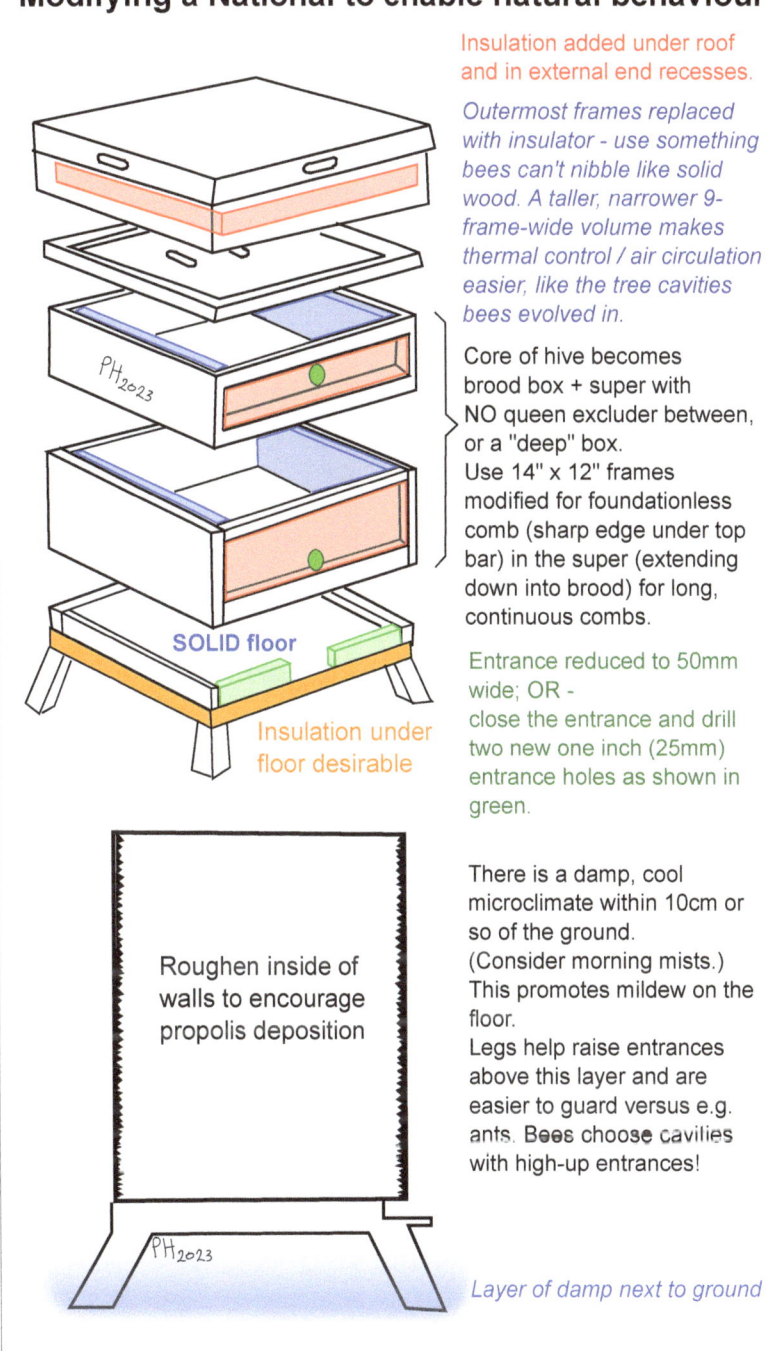

Modifying a National to enable natural behaviour

Insulation added under roof and in external end recesses.

Outermost frames replaced with insulator - use something bees can't nibble like solid wood. A taller, narrower 9-frame-wide volume makes thermal control / air circulation easier, like the tree cavities bees evolved in.

Core of hive becomes brood box + super with NO queen excluder between, or a "deep" box.
Use 14" x 12" frames modified for foundationless comb (sharp edge under top bar) in the super (extending down into brood) for long, continuous combs.

SOLID floor

Entrance reduced to 50mm wide; OR -
close the entrance and drill two new one inch (25mm) entrance holes as shown in green.

Insulation under floor desirable

Roughen inside of walls to encourage propolis deposition

There is a damp, cool microclimate within 10cm or so of the ground. (Consider morning mists.) This promotes mildew on the floor.
Legs help raise entrances above this layer and are easier to guard versus e.g. ants. Bees choose cavities with high-up entrances!

Layer of damp next to ground

In Britain, use 38mm centre-to-centre spacing, not 35mm "Hoffman". If the bees can't pass freely back-to-back the combs begin getting bridged and merged.

---

243 *Insulation gives bees options,* freeing resources to address other issues. Horses need 30% less feed over winter if they wear a coat! Derek Mitchell's work has shaken up how people view insulation and is discussed elsewhere; there is other peer reviewed bee research, such as a study of 43 Langstroth colonies in Illinois showing lower winter food consumption and improved winter survival when wrapped with extra insulation. *Honey bee hive covers reduce food consumption and colony mortality during overwintering* (PlosOne, 2022) St Clair, Beach, Dolezal - Caniolan and Italian bees.

The whole point of low intervention is **not** to open and disrupt the nest every week, but the point of framed hives is to enable inspection when you need to. So as long as the comb is straight, allow them to adhere comb to the walls. They'll be easy to cut away *if you have to inspect* - just leave the nest (brood comb) alone. However if you want to peer at or harvest honeycomb, that's less disruptive. You should be able to tell a lot by examining entrance behaviour and debris.

Don't keep moving combs around, between hives etc. Don't practice pre-emptive swarm control, don't requeen with non local queens (they won't be varroa resistant) - ideally, populate with a swarm from a local unmanaged colony; and stop feeding the bees except in extended unusual weather.

Don't use a queen excluder. It distorts behaviours and prevents the bees locating the brood in the warmest part of the hive, the top. If you really want to use supers, a QE is often unecessary - *usually* the queen won't cross the honey arc above the brood. Some people don't use supers at all, but take honey frames from the outer walls of the hive, because brood are raised in the warm centre.

Consider a perspex crown board. This is essentially a retrofittable window allowing you to look under the roof and see, for example, if supers are crowded (a trigger for swarming). And consider the skeppist technique of tipping hives back and looking at combs from below, for eg swarm cells, rather than opening the hive and pulling frames out.

This photo shows two Nationals converted by Gareth John. Gaps have been infilled and walls and floor thickened. A hessian top cloth, rather than a solid crown board, under an insulating quilt retains heat and nest scents. Not visible is an insulating air gap between the original walls and the added end panels. Gareth has also chosen to provide two 1" diameter entrances, more defensible than a slot. Bees were using both entrances when I took the photo.

Chapter 13
Tips and techniques: key skills: processes and pitfalls

Feeding (honey substitutes, pollen & supplements); smoker use; using comb knives; harvesting wax; correcting wonky, collapsed and cross comb; shook swarms; drumming; uniting (merging) colonies; splits; requeening; finding queens; moving hives; cut-outs; trap-outs; growdowns; dealing with dying colonies; killing bees

> "A surgeon knows how to operate. A good surgeon knows when to operate. A really good surgeon knows when not to operate."

Medical mantra (Dr Roger Kneebone)

Feeding

It is worth pausing to consider *why* you feed, because it is easy to lock colonies into a cycle of dependency. Feeding causes breeding.

Static hives in gardens will experience dearths. Encouraging laying all year, with no brood breaks, leads to rapid starvation in dearths - and excessive varroa.

I find feeding superfluous, barring emergency feeding of a starving swarm. The bees adjust to circumstance. It is fascinating to see a colony establish in its first year, behaving very frugally, and flourish the next – I am content to wait for a honey surplus. However, if you live in an area without mature trees and bushes, your bees may need help getting established and building up reserves, particularly if they are a late swarm. If they struggle in their second year, try different bees.

If you do decide to feed - be aware there are some gotchas. Syrup can get mixed with honey, giving a flavourless crop. Human made syrup, fondant, and most pollen substitutes are basically junk food. If you really need to feed, consider transferring combs from another hive.

We'll deal with feeding sugar syrup / fondant first, then a few words on pollen substitutes at the end of this section.

Question: if you have to routinely feed your bees – there are probably too many stripping the area – what happens to other pollinators?

Feeding: usual objective

To increase honey production by stimulate laying[244]; then, to prevent resultant starvation due to too many mouths during dearths - because feeding disrupts the colony's self-tuning to local forage.

Other reasons to feed

As explained in chapter 9 swarms should **not** normally be fed.

Splits and *shook swarms* can struggle. Helping bees after you've deliberately weakened them does not propagate anti-survival traits, so feeding is appropriate.

Feeding is also a good idea if you only have a few hives and are at risk of losing all your bees. Those of us with many hives can afford to let the unfit die - up to a point. Even I feed sometimes.

You may also have specific local issues, such as:

North America's canola (OSR) honey rapidly crystallises in cool Langstroth hives, becoming inedibly rock hard in winter.

In coniferous forest, the major forage is **honeydew**. Another problematic harvest is **heather honey**. If this is all the bees have to overwinter on, it can be fatal for **non-local** bees over winter (local *Amm* bees are adapted for heather; and can tolerate honeydew if they have some floral honey).

Feeding: methods

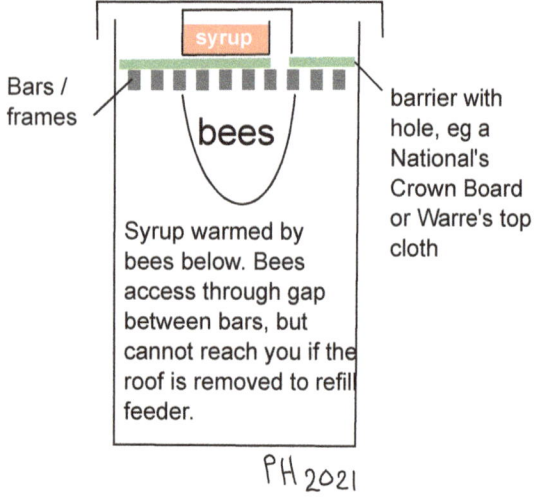

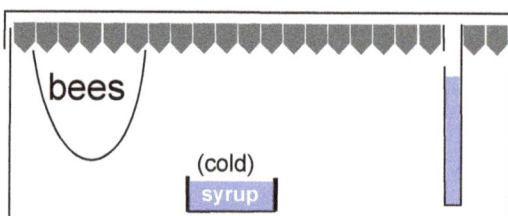

Many horizontal hives have low roofs with no room for a feeder above the bees, so the feed cannot be warmed by them. Besides, standard TBH bars have no access gaps between them, though you can modify bars.

Options include a TBH frame feeder (difficult to source or make) and a tub of syrup. Refilling either results in bees defending their new food source and you are NOT isolated from them so wear protective gear!

This is one issue driving people to develop new styes of horizontal hive.

You can use honey from the same apiary, this can be as simple as swapping frames between hives. **Don't use honey from another apiary (risk of pathogen transmission).** Honey is the ideal pH for bees' gut bacteria and contains micro-nutrients including essential amino acids, vitamins,

---

244   Some beeks stimulate laying to have enough bees for the earliest spring **pollination** contracts and nectar flows. It's just a different emphasis, for commercial migratory hives.

enzymes, phenols, flavinoids, minerals, lipids, etc. Sugar contains none of these. Bees fed on honey live longer.

How quickly do bees take up syrup?

Extraordinarily rapidly. 1 - 2kg of sugar fed as strong syrup in under 24 hours.

Why are my bees ignoring syrup?

There is a nectar flow on. **Bees prefer nectar.**

Cold weather. Below about 10C it can't be evaporated into "honey". This gives yeasts the opportunity to ferment your syrup.

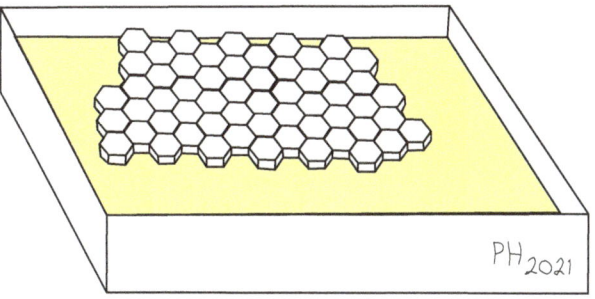

Some bees always drown in syrup so many people favour contact (mesh) feeders. Empty comb makes a handy raft when feeding syrup, or you can pour some directly into empty comb.

If they can't smell it they may not know it's there.
Adding a scent like crushed herb stems or a drop of lemongrass oil helps; or dribble a little trail of syrup from the feeder to the cluster.

They have filled all their comb (they are "congested" or "honey bound"). Unable to lay brood, this can trigger unnecessary swarming.

Feeding: potential pitfalls

Syrup is **surprisingly bad** for bees - see health downsides below.

You create a **cycle of breeding** at wrong times[245], resulting in dependency and *a lot more work for the beekeeper.*

Overfeeding results in tasteless honey (dilutes the fragrant nectar). Some beeks dye their syrup green to avoid harvesting it as honey.

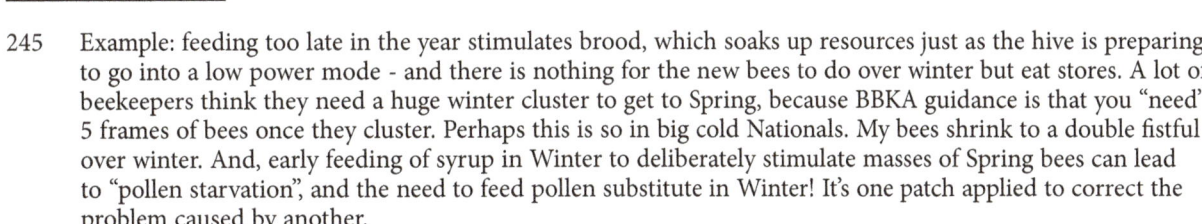

Robbers and pests: spilled syrup / feeding during the day can attract robbers. Feed in the **evening** when flying has stopped, and **reduce entrances.** Spilled syrup can attract ants and slugs.

Bees can drown in syrup. If you have a simple feeder like an open topped tub, put rafts in it – for example pieces of straw, so the bees can grab something if they fall in.

Flypaper effect (1): Strong syrup can act like glue, sticking bees to floors, and crystallizing on wings and hairs, causing difficulties moving.

Flypaper effect (2): Fondant can absorb humidity and become a sticky mass trapping bees for a slow death.

Stings: feeders placed inside hives are defended by bees – see diagram - TBHs are tricky to feed.

Late feeding saps stamina as winter bees must work hard to turn it into honey. Leads to weak

---

245   Example: feeding too late in the year stimulates brood, which soaks up resources just as the hive is preparing to go into a low power mode - and there is nothing for the new bees to do over winter but eat stores. A lot of beekeepers think they need a huge winter cluster to get to Spring, because BBKA guidance is that you "need" 5 frames of bees once they cluster. Perhaps this is so in big cold Nationals. My bees shrink to a double fistful over winter. And, early feeding of syrup in Winter to deliberately stimulate masses of Spring bees can lead to "pollen starvation", and the need to feed pollen substitute in Winter! It's one patch applied to correct the problem caused by another.

spring growth and, sometimes, nosema[246].

Late feeding can stimulate laying if it emulates a nectar flow, so should be done in a few big feeds.

Feeding can stimulate swarms at weird times of year.

Excessive Autumn feeding following a "starvation warning" from the NBU can lead to the bees filling every cell, leaving no room for the queen to lay winter bees. The colony then dies over winter despite plenty of stores.

Wax made from sugar syrup is more brittle than combs made from honey[247]. Foundation users don't see this but it makes foundationless combs trickier to move.

| Foraging on... blackberry juice

Avoid [*beet*] **sugar grown in the UK** - following Brexit it is sometimes treated with neonics, whereas imported *cane* sugar tends not to be.

Health downsides of syrup

There are peer reviewed hard science research papers[248] on the effects of common syrup recipes, which shock even me. The effects can be grouped as *nutrition, HMF and acidity* -

<u>Nutritional deficiencies</u>

Sugar syrup and the American alternative, High Fuctose Corn Syrup (HFCS) depress immune responses[249], probably due to lack of micronutrients.

Syrup can ferment or go mouldy if bees don't use it within a day or two. Beekeepers sometimes add thymol to it to stop this, but it is extremely pungent and disrupts pheromone signalling in a colony (thymol based mite treatments made my bees very grumpy). I don't recommend it! It is only relevant for large scale beekeepers making many gallons of syrup per batch - they can't visit out-apiaries every day, so use large feeders holding several days' syrup[250]. I only give my hives about 1kg of sugar per feed, which they can store in a day.

<u>HMF (hydroxymethylfurfural)</u>

Created by heating sugars[251], too much is toxic to bees.

Overheating syrup (to dissolve the sugar) increases concentration - dramatically - reducing bees' lifespan – significantly. **Boiling** syrup **massively** increases the HMF levels. Many recipes say boil your home made syrup for 10 minutes or more! Luckily this effect is pretty unnoticeable if you warm it to just 50C to dissolve the sugar. *[Frizzera paper referenced below]*.

---

246   Storch, *At the Hive Entrance*, p.55.

247   Pettigrew's Handy Book of Bees (1875), ps. 31 & 48; confirmed by AskJustBees on beesource.com who adds comb made from honey smells different and is probably a better swarm lure (2024)

248   *Possible side effects of sugar supplementary nutrition on honey bee health*, Frizzera et al (Italian; Apidologie Jan 2020) DOI 10.1007/s13592-020-00745-6 ; also *Impact of different feed on intestine health of honey bees*, Mirjanik, Gajger, Mladenovik, Kozaric (Bosnia and Hertegovina; Apimondia, 2013)

249   *Diet-dependent gene expression in honey bees: honey vs. sucrose or high fructose corn syrup*, Wheeler & Robinson, Scientific Reports 2014 DOI: 10.1038/srep05726

250   Thymol also helps vs nosema, but nosema is not generally a problem for low-intensity beekeepers.

251   Formed in cooking through the Maillard reaction and caramelisation. No known toxicity to humans.

High fructose corn syrup (HFCS) has relatively high levels of HMF, elevated further if it is heated.

HMF levels can rise during storage [of large batches] if exposed to warmth or light.

Acidity

Lemon juice or other acid is often recommended to speed 'inversion' of the white sugar (sucrose) to the glucose / fructose mix found in honey. Unfortunately it is very easy to overshoot and make the mix *too* acidic[252]. But there's no need for acid as the bees' enzyme sucrase (invertase) does the same job.

Adding lemon juice often lowers pH to 2 to 3, way below honey's normal 3.4 – 6.1 range.(Average 3.9. Sugar syrup is typically pH6.) The acidity varies with factors like how ripe the lemons are so you cannot control the pH. *[Frizzera paper referenced below]*.

Some beekeepers enter an Autumn feeding frenzy

Acidity increases production of HMF during heating.

pH affects useful gut microbiota[253].

If you do decide to add acid, use vitamin C (ascorbic acid) which is found in nectar and check pH is near 3.9 with e.g. pH strips.

Example

Mirjanic et al[254] found the following impact on worker lifespan after 3 years' tests:

Diet	Average life of bees (days)	Comments
Honey	27	
Sugar syrup	22	Oldest bees are best foragers!
Acid inverted syrup	12	The acid seems to damage the bees' intestines.

In separate and surprising research, A. Martin Ewert and colleagues found adding a low dose of thymol to syrup was extraordinarily bad[255].

---

252   Some people use pH strips to check their syrup is in the honey region, 3.2 to 4.5. Water varies in its mineral content (hard, soft etc) and the correct recipe for adjusting pH will vary by region.

253   Randy Oliver says one reason given for adding acid is 'to kill nosema in the gut'. He found research testing this. It doesn't. scientificbeekeeping.com/the-nosema-problem-part-6-treatment, "Acids in syrup"

254   *Impact of different feed on intestine health of honey bees,* Mirjanic, Gajger, Mladenovic & Kozaric (2012). Testing a wide range of diets, they found adding brewer's yeast or wort to *acid inverted* syrup somewhat improved lifespan, but adding the same things to *enzymatically inverted* syrup, or honey, *reduced* lifespan further. I.e. the same additives - opposite effects! So be wary of recipes in books - yes, including this one. This is why major laboratories like OxfordBeeLab.com are doing a lot of proper, systematic research on nutrition for commercial beekeeping.

255   *Effects of ingested essential oils and propolis extracts on honey bee health and gut microbiota,* Allyson Martin Ewert et alia (2023), doi.org/10.1093/jisesa/iead087 - their control group's lifespan was 16 days, which fell to 9 days when fed low thymol doses. They mention this effect has been seen in other studies and may be due to thymol inhibiting the enzyme acetylcholinesterase, which is how some pesticides kill insects. It's not clear why high doses don't do this, or why their control group's lifespan was only 16 days - possibly related to the highly artificial caging conditions.

If you still decide to feed...: recipes

Standard feeding advice from Beekeepers' Associations is "worst case" advice - bees shouldn't routinely require regular replenishment.

Situation	Standard advice	Problem with standard advice	My recommendation
Spring	Weak syrup – to simulate nectar (1kg sugar per litre of water - use refined, white sugar NOT brown)	This assumes they are starving as they come out of winter, or you want to deliberately stimulate early laying to maximise honey crop. More mouths to feed early in the year risks starvation in odd weather. But the rate limiting factor for raising early bees is number of nurse bees, sometimes pollen, not usually nectar.	Let poorly adapted colonies perish. Conventional advice about avoiding brown sugar is correct, it contains potentially lethal impurities for bees.
Dearths	Weak syrup	A dearth induced brood break helps control varroa. Weak syrup simulates nectar and may stimulate inappropriate laying.	I never feed weak syrup. For emergency feeding in really extreme conditions (if hive is starving) I may use strong syrup.
Emergency feeding (bees starving)	Strong syrup (2kg/litre) - so they can get their energy back as fast as possible	Not normal. Tends to happen with queens selected to lay continuously (Buckfasts etc)	If they are starving because of beekeeper action like a split, rather than being poorly adapted stock, feed strong syrup. Do not warm above 50C whilst preparing, and do not add thymol. Add 10% honey from same apiary if available, or a weak scent and trace nutrients from crushed herbs and nettles. If adding acid, check pH is 3.2 to 4.5.
Autumn	Strong syrup (2kg/litre) - so they can build up 20kg of 'honey' before winter	Assumption: you took most of their honey stores. Assumption: all hives need 20kg winter stores. Adding thymol to syrup as a preservative is smelly, disrupts colony pheromone signalling and detection of damaged brood (hygienic behaviour). Feeding stimulates late laying. They go into winter with far more mouths to feed than they should, and by spring the winter bees are part worn out by raising brood from too early on. Not all fondants are the same – do not use ones with fats in them such as palm oil: bee feeding fondant must be just sugar and water. Make your own or buy it from a specialist beekeeping supplier.	Let poorly adapted colonies perish
Winter	Fondant if emergency feeding is needed, because they need at least 10C to process syrup to honey. (Some Americans suggest dry sugar crystals, not sure they work well in our climate). Fondant must be placed above the nest to keep it warm and accessible, which is tricky in e.g. TBHs. Fondant has no smell so may be ignored if not directly above the nest. Fondant does not stimulate inappropriate laying.		

Bottom line: syrup is just calories, not food.

Pollen substitutes and wonder supplements

Tellingly, bees ignore pollen substitutes if they find enough pollen elsewhere; and they don't store them, they consume them immediately to make food for larvae. (Cheap pollen substitutes are typically just soy or pea flour [protein] mixed with sugar, yeast and a little oil.)

You shouldn't need supplements in Britain / Europe unless you're a bee farmer who seeks to accelerate spring growth to pollinate early crops like orchards. Beyond bees' native range (Europe, Africa, Asia) you could conceivably hit nutritional deficiencies in local pollens.

Don't feed pollen substitutes in Autumn / Winter. Colonies judge when to make winter bees by a lack of pollen[256]. As discussed in chapter 14, pollen patties boost Small Hive Beetle numbers, and if you buy *real* pollen bear in mind it can vector American Foul Brood.

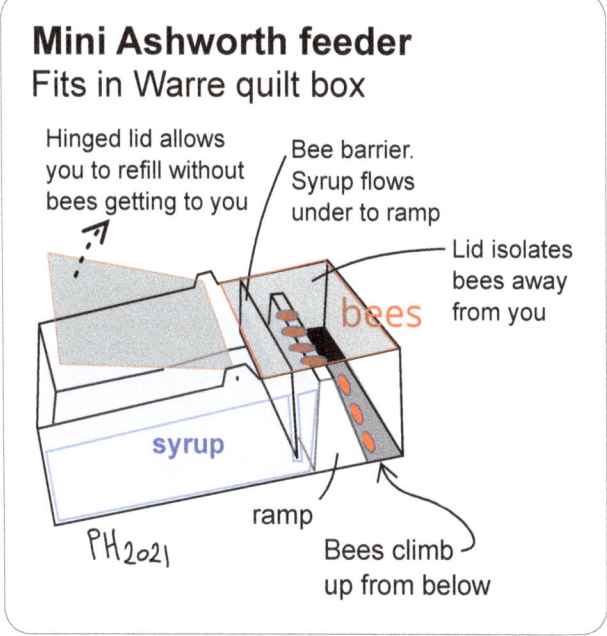

| How I feed Warrés.

Feeding supplements - something to think on: Marketeers love selling mysterious products their customers don't understand and can't tell if they are working: complex financial products, supplements, extended guarantees, miticides. A customer who is afraid is not analytic - and if your bees survive, you tend to attribute that to the reassuringly expensive product you bought. Next time you see a claim about "proven to help" have a look at the actual proof they quote on their website... if you can find it. No one ever reads these papers! They are generally a one-year study on a handful of hives, and often by a party with a connection to the seller; or take place in apiaries with hundreds of hives denuding an area of pollen. The sign of honest science is some ambiguous and negative results[257]. Many supplements are repurposed animal ones, with no consideration for bees' physiology[258].

Smoker use

Tricky to keep going; can be disruptive if overused. But they have their uses.

Don't use blown air. If you watch videos of blowers used to clear bees, they work OK whilst the airstream is on the bees, but don't deter the bees from surging back to the same spot as soon as

---

256 Mattila, H.R., Otis, G.W., 2007. *Dwindling Pollen Resources Trigger the Transition to Broodless Populations of Long-lived Honey Bees Each Autumn.* Ecological Entomology 32:496-505

257 Gregor Mendel is famous for discovering the laws of genetic heredity by crossing peas and showing traits like height and flower colour depended on parents, that some traits were dominant and others recessive, etc. There is some suspicion that he manipulated his published results (confirmation bias) as his F2 pea ratios were **exactly** 3:1. As it happens, he initially used bees but could not control their breeding, and the crosses were aggressive, so he got rid of them.

258 UN FAO report *Good beekeeping practices for sustainable apiculture*, ISSN 1810-0708 p.55

the draught disappears. So it takes a long time to clear them from a box, whereas *smoke has a more deterrent effect and gives you a few seconds* to move boxes etc before they return.

Don't use a water spray. People tried these a lot in the early 2010's, and the idea keeps resurfacing because the internet never forgets dumb ideas. They don't work well at the best of times and they won't move bees which are defending something. Used excessively the bees get soaked, cold and huddle in dispirited clumps to get warm, and again, don't move.

They may work better in hot countries. I've seen reports from America that spraying sugar water distracts 'hot' hives.

There are two main uses for a smoker:

1. Masking your own smell;
2. Driving bees from A to B, for example to clear them from the top of a hive before you add a box, which could squash some.

The common errors in using smoke are:

Using hot smoke: let the smoker settle to a steady state after lighting it. A fuel which smoulders is ideal – one which goes up in flames is not. If it burns your hand, it's too hot: add green grass to the top to cool it.

Using too much smoke: just let a little drift in the entrance of the hive a minute or so before you open it, *don't* squirt puffs directly into the entrance. This numbs the guards' sense of smell. If you pump clouds of smoke into the hive, you panic the colony whose instinct is to gorge themselves on honey and prepare to abscond, because there is a fire nearby. The exception to this is when you need to drive the bees off combs, but this can be overdone - use too much and they panic and rise up in an agitated cloud. **When harvesting honey, excessive smoke can result in the honey tasting of smoke.**

Early black & white photographs show beekeepers almost universally lacked protection: most used at most a simple veil. However some beekeepers have always been in a hurry, and the more industrially-minded 19th century authors were obsessed with heavy smoking, revelling in its use in "intimidating," "subjugating" and "subduing" the bees.

BJ Sherriff's invention of the full-coverage "fencing veil" suit in 1966 triggered a step change in the handling of bees.

Wrong fuel - unpleasant (acrid or mildly toxic) fumes. Ideally, use natural products like dried grass, pine needles, cotton, dried flower heads. I use some cardboard or paper in the mix, to get the main fuel smouldering (cardboard rarely has glue in it these days). Some people add dried citrus peels, lavender, rosemary or similar fragrant stuff to the mix, which may be a good idea, but bees use pheromones to communicate so don't overdo this. Petroleum products are poisonous to bees.

Smoking starving colonies stimulates them to gorge. It is better to feed first, inspect later. Much calmer.

Smoke yourself before opening a hive. Use a good few puffs on areas of you which are likely to smell – armpits, gloves, face (close your eyes!).

I only use intense smoke occasionally, if brushing the bees away with a goose feather doesn't work. (However, this generally indicates they are nursing brood, so consider whether you really ought to.) Heavy smoking actually upsets bees, disrupting the hive and making them defensive for some hours afterwards.

If you're having problems lighting a smoker, say because it is a very high humidity day and the fuel is damp, beeswax makes a good ignition fuel. Many people (me included) resort to a blowtorch to get the smoker going.

Many African smokers are extra large, as there is a belief there that you need copious smoke to handle their bees - this just ensures every bee is in a huge cloud around you - what you really need is a little smoke for a long time (patience)[259].

Smoke moves bees from brood comb straightforwardly, but once they get to open honey cells they stop and stick their heads in.

Smoke is useful during dearths, masking the open hive's smell from robbers.

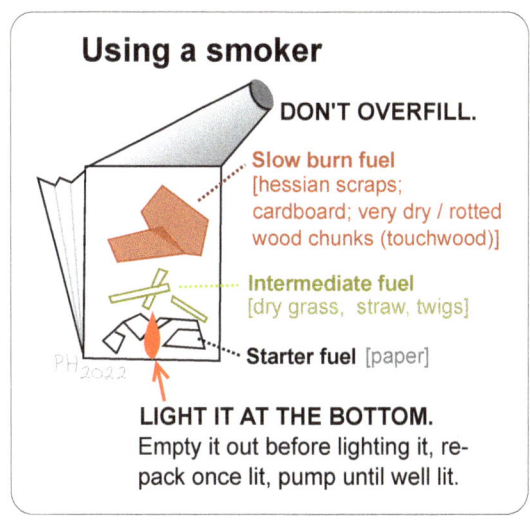

Smoke doesn't affect honey yields, but clumsy inspections do. Use as little smoke as possible around supers, ideally none, as it can be tasted in the harvested honey.

Removing top bar comb: using a comb knife

If comb is just secured to a top bar, it can snap off really easily if it experiences lateral pressure – so **always hold comb vertically.**

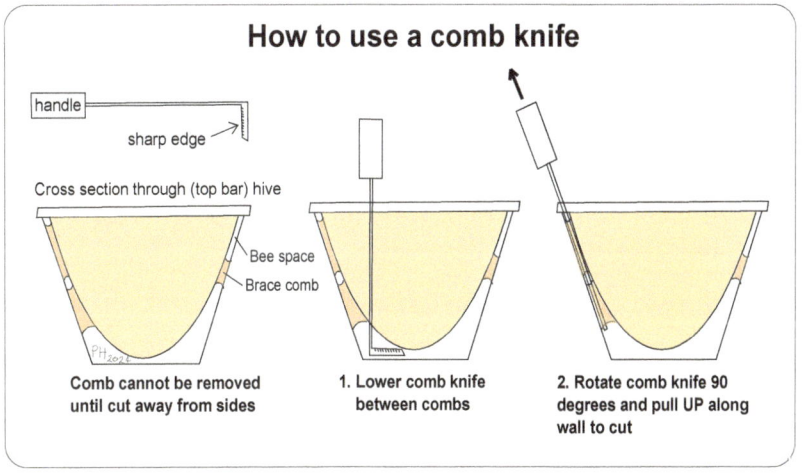

When comb is glued onto hive walls by brace comb, don't use a normal, flat bladed knife to cut through. You will almost certainly apply so much side pressure that the comb collapses. Instead, use a *comb knife,* an L-shaped blade which can be slipped between combs, rotated and pulled straight up to cut through comb.

Comb knives are normally used with Top Bar Hives, but occasionally people want to inspect a single Warré comb. This is trickier, because the Warré hive was designed assuming you will only ever want to remove an entire box. You need to add some tape and a mark to the knife so you can tell when it is at the correct depth as its blade disappears into a box, and so you know which way the blade is pointing.

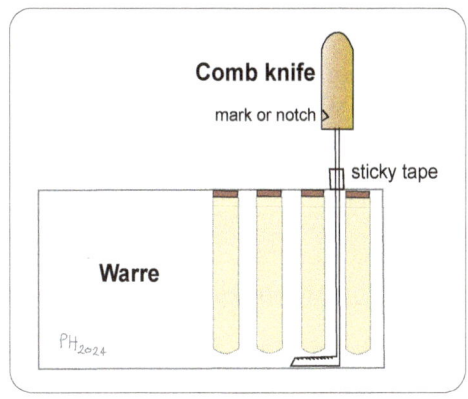

---

259 Personal conversation 2021 with Filipe Salbany, who has extensive African experience.

Harvesting wax

Usual method in Britain: melt comb in very hot water and strain out any grungy bits. E.g: wrap combs in an old stocking, weigh down so it doesn't float, and melt in water. Wax gets through and floats.

In warmer climates: a solar wax melter is easier, though there's a risk of overheating and discolouring the wax. Use old glass - modern stuff often blocks heat and UV.

Beeswax melts around 63C[260]. But it is very time consuming and you get surprisingly little wax for the effort.

After extracting honey, you are left with comb which is sticky and some contains coccoon fragments, propolis, bits of bees, etc. If you're using a Warré hive, where all comb was used for brood, the wax will be difficult to process and the result will not be very pure - I don't bother processing Warré combs, I just throw them away! My top bar hives give lovely pure honeycombs which are easy to process into pure wax.[261]

I have had a bit of success with a solar wax melter, dripping through a muslin bag. You can google various designs on the Net. However I found, in my climate, it only got hot enough to melt wax about 2 days a year! A consideration here is that beeswax discolours above 85C. Apparently a really hot greenouse sometimes works.

Top tips:

Beeswax vapour has a low flash (ignition) point, just 205 C, similar to motor oil - so melting over an open flame, like a gas cooker, is a Really Bad Idea.

Before melting in water, **cut out old dark (brood) comb.** It is mainly coccoon and almost no wax. It will take longer to process this than the white honeycomb, and discolour all the other wax, reducing its value.

Also discard also any comb filled with **pollen.**

Dedicate a large old pan and seive (or muslin bag) to this task. Cleaning them afterwards is a nightmare, you cannot use them again for cooking.

When cleaning afterwards – remnant wax will attract wax moths - you will need boiling water to shift wax from the seive. Do **not**, repeat **not,** pour it down the kitchen sink as it will block your drain. Instead do this operation outside on e.g. your drive.

Small bain-maries don't work too well, because they do not transfer heat quickly enough to keep much wax melted.

These days what I do is:

- Boil water in an old pan on an electric hob
- Dump in comb
- Repeatedly strain out "bits"
- Remove from heat. The remaining wax solidifies as a disk floating on the water.
- Frantically clean hob, floor etc before my wife sees the mess.

---

260 At nomal British temperatures a big lump of cast wax is like a rock, and more brittle than a comb - but will become much more plastic as it warms in your hand.

261 This is one reason Top Bar and log Hives are so popular in Africa. You get 2 products, and wax is easy to transport. Pesticide-free African wax is a premium product, in high demand from the cosmetics industry.

Comments:

Producing pure wax is a lot of trouble. These days I keep a little for waxing bars and chuck the rest out. Sometimes I remember to give some to the local blacksmith: apparently it is the best material for coating their tools.

Keep stored wax well wrapped / boxed to avoid attracting wax moths.

The slower wax sets, the more solid and bubble-free it is.

Beeswax doesn't burn – it melts and its vapour burns. It needs a wick to turn it into a candle; wicks are surprisingly sophisticated. Propolis in the wax looks grainy but adds an incense-like smell. However, most people prefer pure white candles (bright, smokeless, no smell), which may require repeated melting to purify.

Historically, the aim was to remove impurities and get the purest, whitest wax possible, as it sold for the highest price. Starting with virgin (white) comb was best. In medieval times people made as much (some sources say more) money from wax, than honey, because churches needed huge numbers of beeswax candles. It is still a major income source for e.g. African beekeepers, because their wax is pesticide-free and highly sought by cosmetics companies. It had, and has, other industrial uses like lubrication, watertight seals, the 'lost wax' casting process etc.

Purifying it was a fraught, hours-long process involving boiling water over fires, and straining through cloth bags. They tended to do the process on a larger scale and the other way round - boil the water vigorously to get a froth of pure wax, which they skimmed off, rather than strain *out* the impurities. Temperature control was tricky and beeswax has a dangerously low flashpoint (204 C). Be grateful for electric hobs.

Correcting comb problems

Wiggly comb hampers manipulations. But fiddling with comb on a bar can result in irreversible damage, and comb collapse. So first consider if you really need to correct the issue.

In particular - if you have wiggly comb in a Warré hive, don't worry about it, unless you live in a country where inspectable frames are compulsory. You don't really need to remove Warré combs until you harvest a box for honey (you can view them from below if you want to inspect).

If however you see comb in a horizontal TBH begin to veer off the bars, you need to be pretty ruthless about correcting it. Because once one comb goes awry, others will follow in a ripple-on effect, leading to an uninspectable hive (all bars welded together). Sometimes you can cut one end of a **new** comb away from the "wrong" bar with a hot knife, and kind of melt it onto the "right" one. New comb is pretty soft when warm.

If the comb snaps off when you try this, lean it against a wall inside the hive so the bees can strip its stores, and remove it a week later before they start building really weird comb around it.

If comb has fallen off a top bar, it is very tricky to reattach. Although theoretically you can suspend the comb from the bar with string, good luck tying knots when you are wearing gloves! The gloves will be sticky, because snapped comb oozes honey, and there will be a lot of upset bees everywhere.

Curving across bars

Comb collapse

Rubber bands

Bottom of frame supports comb

Replacing comb in a frame is simple - just pop it back in and restrain it with a couple of rubber bands. The bees will rebuild connective wax to the frame.

But a comb in a frame, or even a semi frame, is unlikely to collapse. What causes collapses is usually when the hive gets hot in sun, which softens wax, and there is a good nectar flow so the bees load 3kg of nectar into a new, soft honeycomb. I have come across this a couple of times in TBHs, which is why I try to place them where they get shade from noon onwards.

Sorting out cross combed hives

People sometimes leave hives unmanaged for too long, and they fill up with an unmanageable mess of cross comb.

My advice is, leave them until early Spring when the bee numbers are few, and they have contracted to a nest. This is your opportunity to go in and brutally cut out comb from the other end of the hive, forming a cavity where you can ensure if they *do* build comb, it's straight. Leave the cross-combed **brood** area: cutting this risks too much damage / killing the queen, and if they've lasted years there, why disturb what works.

Example: TBH

Replace bars above empty space with ones with a clear comb guide (a sharp edge, or strip of foundation) to promote straight comb.

Irregular comb can happen in any hive type

Cross combing tends to happen when the nectar flow is intermittent and comb building stops and restarts.

And free built comb will occur in any volume where the bees can build without clear guides. Even framed hives with foundation can go awry if the frame spacing is wrong; a classic error is to forget to put new frames in a hive in Spring. The bees build awkward brace comb every which way in the free space, forming an uninspectable mess.

Shook swarms

Objective

A drastic manipulation where the adult bees are transferred to a new hive, abandoning brood infested by e.g. varroa, chalk brood, EFB etc. It does not always literally mean *shaking* the bees.

Summary

Set up a clean empty hive, ideally in the same location as the original one so returning foragers don't get confused

If the purpose is to purge a disease, you start the adults afresh with a new, empty hive and **no comb**. They rebuild from scratch.

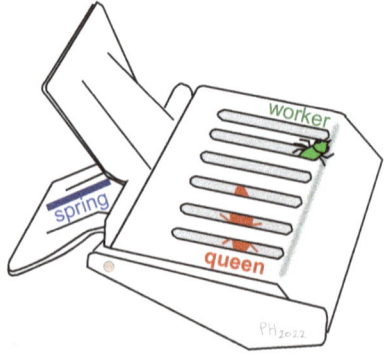

Queen clip. Queen is too large to exit through slots.

Transfer the queen and adult bees. You may decide to catch the queen in a queen clip first which can be put in your pocket, where it is warm and dark and out of harm's way. Gently **brush** or **shake** or **drum** the bees off the old brood comb, which is then discarded. Release the queen into her new hive.

Alternatively, the "**pioneering**" method is to smoke the bees down out of the boxes and then take away the brood comb.

Loss of brood in Spring is a bump in the road, not a car crash, as the queen is at maximum laying rate. The earlier you do it the more time the colony has to rebuild and grow strong that year.

What can go wrong

Don't do this late in the year. You are trying to emulate a natural swarm (which usually occurs in May/June). By August, a shook swarm stresses the colony. The ideal time is April, when the colony and queen are healthy and the brood losses relatively low.

If you leave the colony with NO comb, they have no options. (And I suspect if they are sufficiently upset they could even abscond. No stores or brood left to defend.) If you now start feeding them, for example their own honey, to make up for lost stores - where will they put it? Feeding and colony development stalls.

Young bees are weak flyers, have never been out of the hive, may get lost in grass. They really need a sloping ramp in.

There is a risk the now broodless and agitated colony will abscond.

Fortunate happenstance: This is a great opportunity to do a really thorough inspection and learn from the old comb.

Drumming

Objective

An alternative to brushing or shaking bees from one box to a new one. Bees are surprisingly calm when drummed. The downside is it takes a long time (~10 minutes, some sources say 30) - they do not want to leave brood / stores.

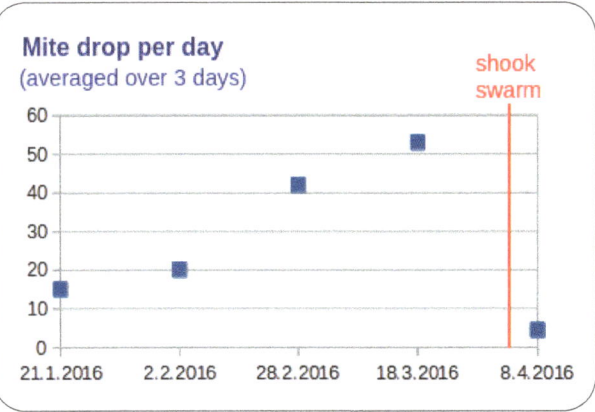

Shook swarm reset varroa the one time I had problems using local bees

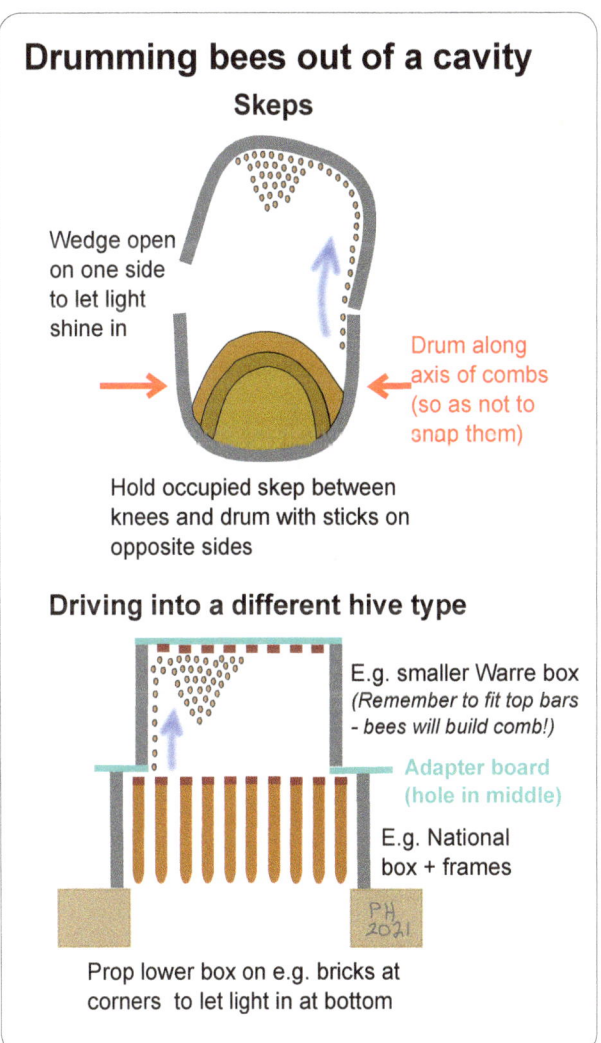

Summary

You use a **little** smoke from below to encourage the bees to start moving[262], and drum sticks on opposite sides of the **lower** box. (Hands would be rather numb after 30 minutes on a wooden box.) They migrate **up** and away from light.

The rhythm is fast - about 120 beats per minute, constant, and strong. Imagine saying *dumdumdumdum* constantly.

The first few seconds are chaos. After 4-5 minutes the bees start to flow upwards into the emtpy box. By 10 minutes, most bees and the queen should be clustered on bars in the upper box. Confirm by peeling back a corner of the top cloth.

Once the majority are in, place the new box of bees on top of its hive (in the same position as the old hive). The stubborn remainers can be shaken or brushed out of the old box onto a board in front of the new hive.

If there are several boxes, start with the one with brood and queen in it. It's possible to just drum the combs you wish to inspect or harvest.

The new hive will need feeding as it has been deprived of stores.

The adapter board has a hole cut in it to fit the top box, and could simply be cardboard.

Pettigrew, a skep beekeeper described a slightly different technique (which sounds easier) where the upper skep is not tilted - no light gets in but the 2 skeps are mated mouth-to-mouth and a sheet tied round the join so no bees can escape. He adds that bees are reluctant to move in cold weather, so pre-warm the recipient skep, and if necessary pour some syrup onto the bees - they slurp it avidly and their temperature and activity rises. If the skep is light (under 30 lbs) it is faster to shake the bees out into a skep below them, held between your legs.

With comb in this orientation, some bees have no path upwards.

What can go wrong

Some people say it works well, others poorly. I suspect comb orientation matters, i.e. the bees need a clear path from *all* combs to the upper box. It probably works better in skeps, with their slightly flexible woven walls.

William Herrod-Hempsall noted[263] that drumming doesn't drive every last bee out; that the bees will gorge on honey before leaving; and it only works if a skep is completely full of comb.

Robinson advises[264] that if a swarming hum is heard while driving, all is going well; if the bees are quiet they have not been "terrified or smoked sufficiently".

---

262   Herbert Mace mentions that the purpose of the smoke is to induce them to fill up with honey [implied: compliant]. *Bee Matters and Beemasters*, 2nd Ed (1931) p.43

263   *Bee Keeping New & Old* (1930), p.753-5

264   *British Bee-Farming, its Profits and Pleasures* (1880), p.107

Uniting (aka merging, combining)

A very simple procedure, but one must be careful not to trigger fighting between two colonies. Conflict is prevented by controlling the **scent**, or **speed** of introduction.

Good reasons to amalgamate colonies:

- Small casts often dwindle and die.
- To unify a queenless colony with a queen-right one. Creates one strong colony and you know the genetics are OK.

Questionable reasons:

- One large colony will make more honey, and deal with threats and parasites better than two small struggling ones.
- Two merged colonies consume far fewer winter stores in total[265], and are more likely to survive through Spring.
- Creation of 'packages' of bees.
- 'Equalising' hives.

Worth knowing:

- Colonies recognise nestmates by scent.
- Only princesses fight. Mature queens tolerate other mature queens. But their workers will fight if they perceive a threat to their colony. Young nurse bees are more timid.
- Colonies naturally merge! Workers from failing colonies decamp over a few weeks to nearby related ones and relocate stores there ("silent robbing").
- You can remove one queen before the merge, to requeen a queenless colony.

Method:

Casts can simply be thrown in front of an occupied hive so they can march in. Sounds crude - but generally works[266] - but pay **careful attention** to "what can go wrong" below. Just let the bees choose their preferred princess.

For vertical hives: the *newspaper method* is simple, and works well. Place boxes of bees from hive A on top of the boxes of hive B, forming one continuous cavity. But between them, you put an air-permeable barrier, typically a sheet of newspaper with small holes or tears in it. This allows the upper boxes to breathe, and for the scents of the two colonies to mingle. They can't immediately reach each other because the newspaper is in the way, so they can't fight. The bees will eat through the newspaper, but it takes a while, by which time they all smell the same and think they are the same colony. The bees will remove the newspaper.

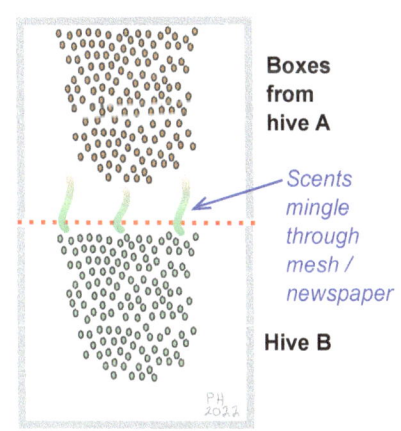

| Newspaper method

---

265   Quimby, *Mysteries of Beekeeping Explained* (1853), p.169-170.

266   Gareth John tells me he does this for up to 10 days after the first cast was hived and has never seen a problem. But he would hesitate with **prime swarms** as they have a much greater sense of identity.

For horizontal hives: the *divider board* serves the same function as newspaper. It's not airtight. You must remove it later. Alternatively use a newspaper screen. But if the access between the two colonies is small, like a 2 inch hole covered in newspaper, there is a danger they will fight - both see the other as invading their nest - so you need to ensure they are next to each other (not at opposite ends of the hive) and there is a large porous surface between them. A masking scent is advisable.

What can go wrong

(1) Disease / pest transfer - Merging a random swarm of unknown origin into one of your hives is risky. If it is from a commercial source, it may be carrying mites, disease and poor genetics / hygiene practises. Consider not merging swarms of commercial origin.

(2) The newspaper method may be inadvisable in hot weather - bees need ventilation.

(3) Spraying bees with sugar water - some old sources recommended this[267], or a scented syrup. This can result in a miserable mass of bees stuck together. They die, and it can trigger robbing from nearby hives.

(4) Fighting risk - Sometimes walking casts in to hives ends in fighting. This is *very* rare, but by examining these failures we learn a lot. In short, if the incoming bees arrive suddenly and can cluster together as a separate mass inside the hive, they may retain a strong sense of group identity, feel threatened and trouble starts. They need to be thoroughly mixed in to a crowd of strangers, whereupon they act submissive.

Merging swarms: risk reduction measures

It's all about convincing the bees they have nothing to defend.

(a) by Controlling speed of mixing

(i) very rapid mixing - packages are created by carefully removing **all** queens, then the workers are shaken (mixed) into one box. Abruptly finding themselves in a crowd of strangers, with nothing to defend, the workers are confused but not defensive.

(ii) slow mixing - small entrances help - the incomers can't barge in as a mob. Ideally, the entrance should have occupied comb directly behind it, not a volume where the incomers can assemble - for a vertical hive you may want to block the normal entrance and pull an upper box forward.

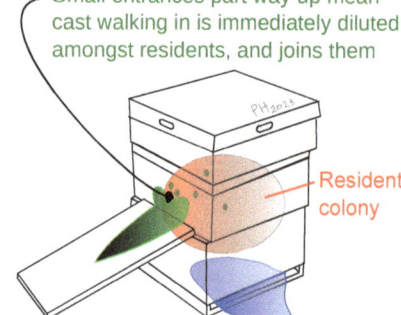

Small entrances part way up mean cast walking in is immediately diluted amongst residents, and joins them

Resident colony

Entering via wide base entrance allows incomers to gather in one mass, remote from populated brood nest combs - they gather as one unit, then fight

Small entrances next to residents reduces risks

Merge while foragers are flying - *not* **the evening**[268] - to ensure incomers encounter foragers at the entrance. (This is the opposite advice to hiving swarms in unoccupied hives, where hiving in the evening is advisable to reduce their likeliehood of absconding.)

---

267 e.g. A. Pettigrew, *Handy Book of Bees* (1875), p.134.

268 Wildman (*Management of Bees*, 1770) recommended an evening merge "so regicide is over before one queen can flee with half the bees". Bevan (1870) also mentions evening merges being recommended by some earlier writers. Herrod-Hempsall argued for evening merges to ensure all bees were present to be covered with the pea flour he used as a masking scent. But after reading widely, I think **scents** and **daytime merging** are the best strategy to avoid fights. The residents will have fewer foragers present, who tend to be the most trigger-happy guards.

(b) Masking scents

A masking scent totally prevents fighting[269]. Uniting different races is tricky and definitely requires a one. However, if you're doing a newspaper merge, it's slow enough there is no need for added scent. If you use a scent, it should be weak, like **2 drops of essential oil in water.**

<u>Advanced merging - obscure points</u>

Be aware that a small swarm is not always a cast. It's possible to get tiny Primes if they leave late, after casts have already taken most of the flyers! Primes have more sense of cohesion than casts. They can still be united via the newspaper method, though.

Ideally, you want to merge two swarms the same day both are captured. If not possible, Quimby advised, don't march a small swarm into a large one's hive - instead, march the **large** one into the **small** one's hive, so the newcomers are somewhat subdued.

When merging casts, most people just let the bees choose which princess they want. Swarming bees are full of food, thus good tempered, and their natural excitement makes them easy to mingle. But some advocate caging one queen to protect her. Good luck finding and removing every unmarked queen in a swarm!

For established colonies, drumming bees out of one hive into another is said to be very effective as they are "demoralised" by the process and don't fight, but the small print in such discussions assumes you've removed one queen.

Webster recommended[270] simply shaking swarms together within 24 hours of swarming.

Splits

A colony is split into 2 or more colonies. It's straightforward.

Always ask: does this help the bees?

Objective

- Increase stocks.
- Prevent swarming (trick bees into thinking they've already swarmed - an *artificial swarm*).

(Occasionally...)

- To raise a new queen.
- To add a queen of your choice to the queenless half. Obviously the wisdom of this is debatable if she's non local.
- With the intention of recombining the two colonies into one big one, for more honey, once the impulse to swarm has passed.
- Varroa control – a Spring split causes a brood break, and avoids one huge colony which is a better breeding ground for mites.
- To calm down a large, defensive colony.

---

269   I was opposed to using scents because *colonies communicate via pheromones.* What just happened? Are we queen-right? Are we under attack? I changed my mind after a couple of routine merges ended in carnage. I am very dubious about using air freshener; whilst it is *said* to work well, its proponents are commercially motivated and prioritise convenience. It would be my last resort - see Wikipedia's entry on Air Freshener and check the Toxicity section.

270   *The Book of Bee-Keeping*, 4th Ed, p.54 W.B. Webster

Methods

Walk-away splits (simplest) - Basically you need to split the combs in a strong hive between 2 hives. Each colony gets about half the brood and stores. The queenless half will raise an emergency queen if it has eggs. You do it in Spring or Summer so the new queen has drones to mate with, at a time of day when most foragers are out. You want a warm sunny day as brood will be exposed.

This isn't a very effective method of swarm control, since the colony with the original queen still has her, and Brood In All Stages (BIAS), and often continue making queen cells. And sometimes the other half builds QCs and swarms too.

You don't absolutely need to know which half has the original queen if you are simply splitting to create a new colony, *unless* you are doing this because you spotted capped queen cells and are trying to head off swarming.

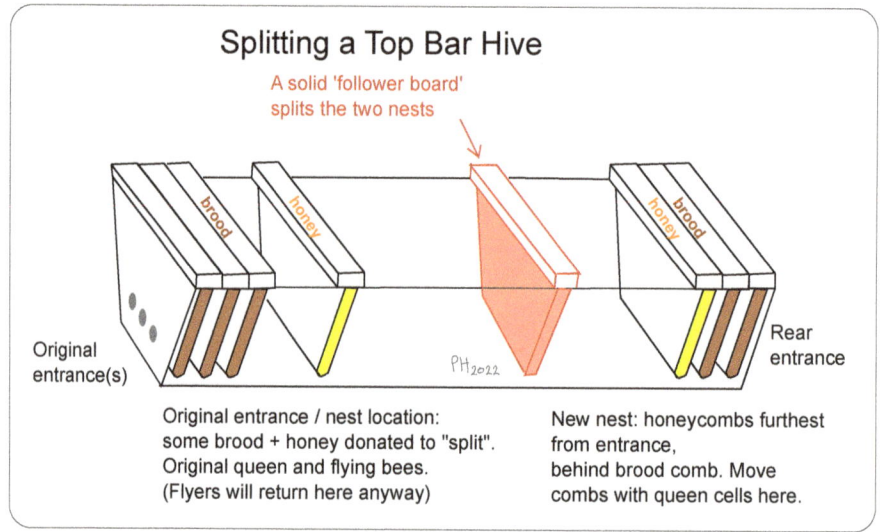

Ideally, you will end up with 2 hives in two locations (which may only be a metre apart). You want the original queen, with nurse bees and brood, in the new position and the queenless colony in the old position [for TBHs it may be easiest to rotate the hive to achieve this]. The queenless one will be reinforced by returning foragers whilst the queen-right colony will not suffer an interruption in laying. Thus each side has an advantage.

If you haven't inspected your brood nest for a while and you have to cut through cross combed brood comb, splitting the hive will **really upset the bees**, so warn your neighbours. Consider a preliminary inspection and mark bars with queen cells so the actual split can be done quickly.

Don't inspect for the first two weeks as the disturbed bees may kill their princesses. The two halves will be relatively weak colonies so reduce their entrances. After two weeks check in case they need feeding in a forage dearth. Splitting a hive in two usually results in less honey that year.

Variations

Most of these basically separate the queen & flyers from the brood & nurse bees. They have a higher success rate than walk-away splits.

Pagden, Heddon - split the colony in two. The brood / nurse bees are in one, a metre from the original spot. The queen goes in an empty hive in the original position and *flying bees will return to this hive at the original position*, automatically separating the flyers and nurse bees for you. The nurse bees raise new queens in the displaced, queenless hive. The age distribution of the bees with the queen is older than for a natural swarm, but like big casts, they usually survive.

Demaree - split a hive in two sections *vertically* (no need for 2nd hive) separated by queen excluder. Reduced queen pheromone in the separated brood area leads workers there to raise more queens - so you can get lots of swarms if done wrong. Hive must be visited every 5-7 days to release new drones trapped behind queen excluder. Usually a pre-emptive swarm control procedure.

Snelgrove, Horsley - vertical splits like Demaree, but use special boards with movable parts rather than queen excluders. Horsley refined the Snelgrove method for out-apiaries where frequent visits are less feasible.

Shook swarm - you can shake the flying bees and queen into a new box (typically with no comb). This is about as close as you can get to a real swarm.

Taranov - similar to shook swarm but with a higher proportion of young bees, and fewer mature foragers. Mimics a natural swarm.

If you have no queen cells but still want to make a split: at the risk of over-complicating things, you can aid the queenless side by cutting some cells away below cells with eggs. This gives the workers room to extend the eggs' cells into larger emergency queen cells. It does sometimes work, but the success rate is low.

Potential pitfalls

Timing: Note that splits pretty much **have** to be done in swarm season - the new queen will need to mate. If bees are already destroying drones (indicating a shortage of forage) it is a Bad Time to do this. (Splits often boot out drones, it's a stressful time and they lack spare resources.) Another reason to split before August is otherwise they won't have enough young bees to overwinter.

Splitting a colony into more than 2 will result in smaller colonies less likely to survive.

Colonies don't split in nature! There are usually plenty of swarms (shared round a group of beekeepers) and no need for splits. Repeatedly forcing bees to raise emergency queens eventually weakens them[271]. Natural swarming is also healthier than splits, of course, because a swarm leaves parasites behind, and **both mother & daughter colony experience a brood break which reduces parasite load.** (Discussed further in chapter 14.1 as Vertical and Horizontal transmission of pests.)

An interesting downside is that the queenless part of a split tends to raise *lots* of queens and then often swarms, leaving it quite weak. This was noted by Winston in a series of studies[272]. A related problem is that if you split a colony as soon as you see queen cells, you select for swarmier bees.

Requeening / forced supersedure

Objective

Give a colony a new queen because -

- the old one has died or stopped laying;
- or the bees have become defensive.

(Also, conventional beekeepers occasionally change queens in response to disease[273]. Inexperienced ones also tend to over-obsess about brood laying patterns, assuming a patchy pattern means the queen is failing.)

---

271   The largest, healthiest, most fertile queens are raised from eggs workers choose, in proper queen cells, rather than emergency queens in hastily converted worker cells. *High-quality queens produce high-quality offspring queens*, Yu, Shi, Zeng, Yan & Wu (2022), doi.org/10.3390/insects13050486 . Tellingly, emergency queens are often rapidly superseded. And repeatedly raising queens from what is essentially the same colony is a rapid path to inbreeding and loss of valuable rare alleles.

272   A very readable analysis at: theapiarist.org/doing-the-splits. Total number of colonies studied was only 8.

273   The belief is that certain lines are "resistant" to foul brood, nosema etc. Since nosema and foul brood seem to only be seen in intensively managed colonies under nutritional, etc stress (see chapters on diseases), it's not clear that some queens really are more resistant than others. But perhaps some lines are less inbred and retain a few of the wild colonies' hygienic traits.

Method

- Don't rush in and replace the queen.
- She may be fine and you have misinterpreted a brood break caused by a forage dearth.
- The bees may already be raising a new queen, which preserves the local / survivor genetics you have established.
- Temper changes can be due to a stress you don't know about, like an animal attack or forage dearth, and the bees are fine again a couple of weeks later.

If you decide the colony does need a new queen, then consider, instead of buying a queen -

- allowing the colony to die (natural selection)
- merging a swarm into the hive
- merging this colony with a queen-right one
- killing the queen so they raise a new one - see *Finding Queens* below
- transferring a comb of brood from another hive.

Thus you retain local, survivor genes.

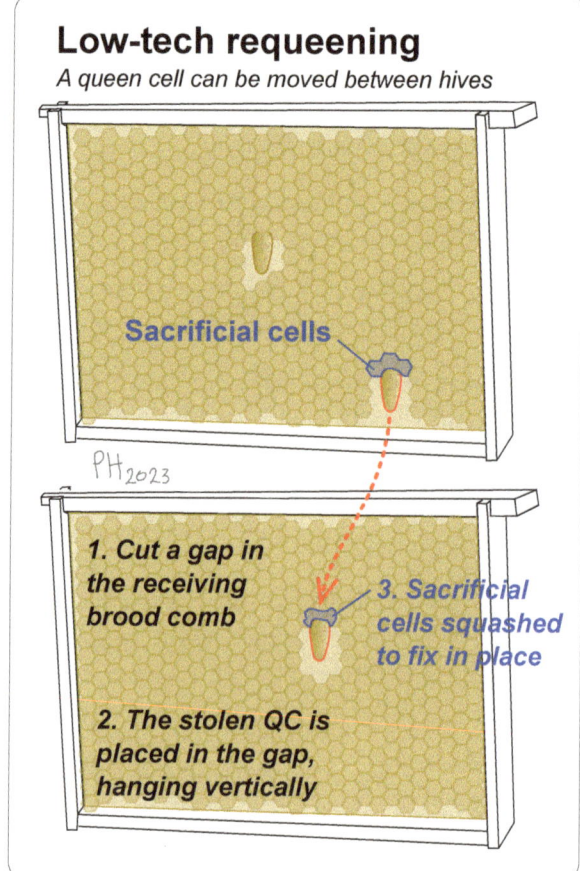

If the hive has become drone laying, and you really want to requeen, then

Case 1: Drone laying queen [small patches of drone brood in worker sized cells] - add comb with a capped Q cell (unless it's too late for a new queen to mate). You need to carefully check the comb for its queen so you don't leave the doner hive queenless![274] The sides of the foreign queen cell needs protecting with a short tube - the end is tough enough to resist stings.

Case 2: Drone laying workers [scattered individual drone brood in worker sized cells] - introduce a new queen, together with healthy brood from a healthy colony. The new queen is generally in a queen cage to protect her for a few days.

What can go wrong

- No drones to mate with (wrong time of year)
- The bees may kill the new queen because...
  ◊ They already have a queen
  ◊ They won't accept a queen until they have no chance of raising a new one from eggs - ie for 3 days after losing their original queen

---

274 Not a good idea to brush the bees off, too much risk of damage to a queen; or shake the bees off, which risks damaging the queen cell's larva / pupa

- ◊ she doesn't smell like them yet - which is why new queens are often introduced in cages with candy plugs, which take the bees a few days to nibble through.
- ◊ You were wrong, they're not queenless.
- ◊ They *think* they're queen-right (typical of drone laying workers).
- ◊ They are already raising new queens (i.e. their own family), which is why conventional beekeepers *pinch out* (euphemism for *kill*) capped queen cells if they want to requeen with a known line.
- ◊ They just do, sometimes[275],[276].

And, the queen may attack the workers![277] – If she bites your hand, to avoid her being killed ensure she is hungry, she will then beg for food first and everyone chills out. And clean your hand, to ensure she does not smell of you.

Temper follows the queen, a calm queen calms the colony, but it can take a couple of days for her pheromones to percolate through a large hive.

Queen cell protector

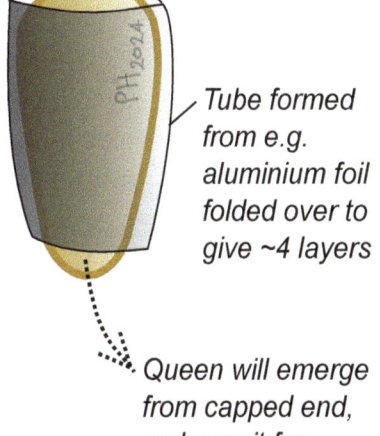

Tube formed from e.g. aluminium foil folded over to give ~4 layers

Queen will emerge from capped end, so leave it free

A tube round the queen cell stops workers and rival queens stinging through it. They don't sting through the thick end cap. You can buy plastic ones or roll your own.

Finding queens

Why?

I see queens very rarely - I don't see a need to.

Opening hives and handling frames is always slightly risky[278]. You can just rap the hive to confirm they are queen right (if the bees buzz then immediately settle down, you know she is OK).

Of course you *can* open a hive to look for brood (or look through windows) to confirm she's laying. Seeing eggs standing on their end usually means she laid them within the last day; on their sides, within 3 days. But lack of eggs is ambiguous; she might simply be on a brood break due to swarming or a nectar dearth.

But if you **really** want to find one bee in the crowd, here's how. We'll assume she's unmarked and just looks like a pointy drone, though sometimes she stands out due to distinctive colouring.

Always ask: does this help the bees?

---

275   Someone asked how to introduce a virgin queen on a forum. "Just run her in [to the hive entrance]" advised 2 people with 26,652 and 22,051 posts. But "I just did this with 12 virgin queens I had raised" said another with a mere 7,498 posts. "I lost 3 of the 12 virgins. It isn't completely reliable." Just because someone posts a lot, and speaks with great certainty, is not a reliable indicator of good advice.

276   www.sussex.ac.uk/lasi/research/beekeeping - if you smoke a hive heavily you can introduce a new queen immediately, with no losses. Beekeepers continue using cages and stuff because... tradition?

277   *Beekeeping New & Old* p. 679 , Herrod Hempsall

278   Leading to *occasional* mishaps like inadvertently killing queens, or dropping them in the grass; training the bees to recognise you as a threat; enabling disease; attracting pests, etc. Looking for a queen is time consuming and means the hive is open longer, stressing the bees. Virgins, reently mated qs and recently introduced Qs may be balled and killed if you disturb the brood area too much. Never open a hive without a good reason.

Methods:

Use minimum smoke and disruption, so she is calm and not hiding, and the workers are not milling around in agitation.

Her job is laying eggs: she will **usually** be in the brood nest, the combs thickest with bees or to put it another way, the combs with eggs.

Within the brood nest, she is unlikely to be on sealed brood, she's most likely on areas with eggs, younng larvae and polished cells (just cleaned for egg laying - very shiny inner walls, particularly easy to see on old, dark comb). This generally narrows the search to 2-4 combs. But in runny colonies she could be anywhere.

Keep an eye out for a spare heir. 5-15% of hives have 2+ queens. Queen colouration can change dramatically with age, so whilst one may stand out with distinctive colouring, the other just blends in with the workers.

Finding an unmarked queen in a bad tempered hive can be challenging. One trick is to temporarily split the hive in two: after a couple of hours the section with the queen should be a lot calmer than the other half. This makes the task more manageable.

If queens are alarmed they flee for a dark place. So hold the comb up to face the sun, then rotate it so the dark side is in sun. Then watch the edges of the comb for a large bee moving rapidly to the dark side. I find this the most effective technique.

Defocus your eyes and look for a ripple of odd behaviour on a comb.

Virgin queens are more or less ignored by other bees so you **won't** see a ripple of odd behaviour around her.

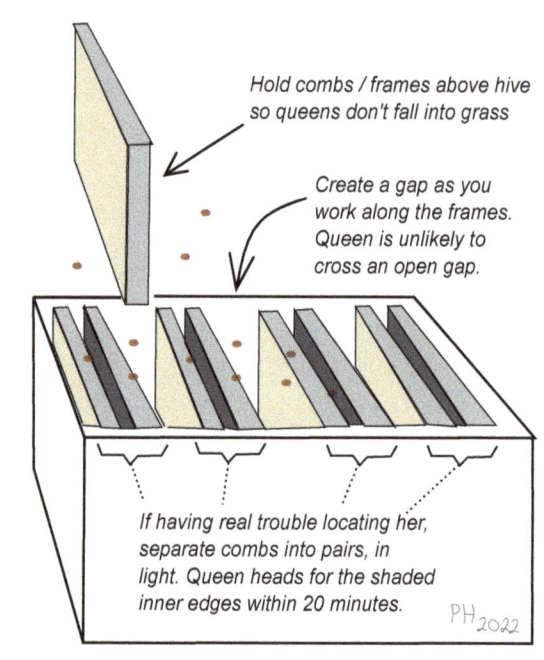

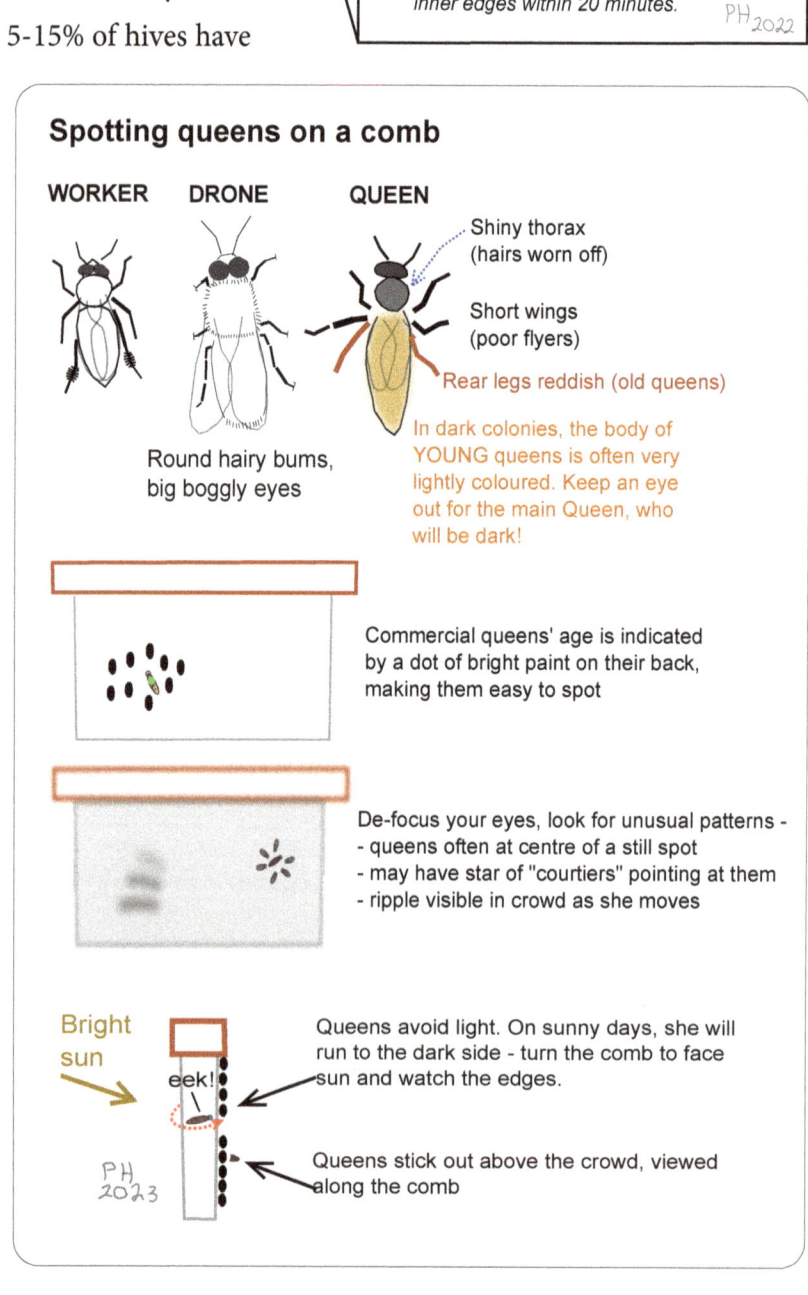

Skeps can be turned upside down. If you then and tap on the sides and bottom, the queen immediately appears (Wildman).

If you watch a swarm cluster, queens pop up to the surface every 5 minutes.

Large scale bee farmers sometimes use a queen excluder at the bottom of a box and seive the bees through them to find queens rapidly but brutally – drones and queens are left in the box.

In large hives, you can often narrow down which box she is in by separating the boxes. Often the noisier, more agitated box, making a "dull roar" lacks the queen.

Moving hives

There are three general aspects to consider here:

Safety for the driver (seal the hives)

Safety for the bees (comb collapse, ventilation)

How far are you moving the hives

So, plan for the worst and mull the following specifics:

Some hives are heavy or bulky. Do you need a second person in a bee suit to help?

Block entrances **at night** when all the bees are in.

For very long moves start in the early morning to be sure there is daylight to set up the hive at the far end.

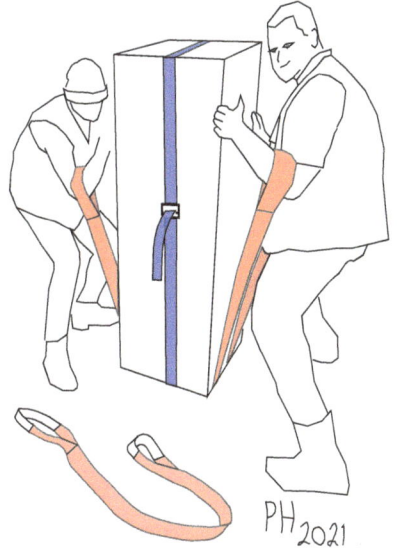

Workmen using a pair of **lifting slings** (red) around their forearms to lift tall, heavy objects. Man on right: good posture, straight back, lifting with legs. Man on left: bad posture. Ratchet strap (blue) locks boxes together.

Major moves are best done in winter (e.g. November to January) when there are fewer bees and they are naturally less active. Fewer things to go wrong.

Hives on 2 wheel trailers will bounce everywhere, it is a recipe for trouble. Hives inside a car or van benefit from a much smoother ride, and you're aware if there is a bee leak – escapees don't simply get whipped away by the wind and lost.

Consider bees escaping during the move: Examine how the bees enter the hive before moving it. Some old hives have warped floors, etc and the bees come in though the rear or roof! If unrepairable, use duct tape to temporarily seal gaps they may get out of. The most problematic hive here is perhaps the WBC where there is an internal box. Some friends reported that wrapping their entire WBC in a gauzy gardening fleece used for winter sheltering of plants worked to keep the bees in, it is sold in huge rolls and they can breathe through it. **Take duct tape and plasticine in the car to block gaps.**

Just to be clear – block the entrance!

If driving a hive inside a car or van, keep a veil near you in case some bees get out. Or there were some *underneath* the hive you didn't spot! Driving in a full bee suit with hood up isn't a good idea as your visibility is low. From my experience driving with swarms in the car, a couple always leak out, but are mainly interested in getting back in to the comfort of the queen. However if a *lot* get out beware, because of course they are not a swarm – they have a hive to protect and it is being vibrated and moved.

Consider a tarpaulin below the hive if there's a possibility that honey might leak out.

If the hive is dependent on a bottom mesh for ventilation, *don't block it* by putting the base directly on a seat.

For tall hives, you may have to break them down into a couple of half size hives to fit in your car, or simply due to how much you can lift.

Lock hives together with luggage straps so their roofs don't come off, boxes slide around etc.

Ventilation is crucial for moves longer than a few minutes: the bees get agitated by car vibration and use oxygen. Also with an entire colony in there, they get hot, and if you are moving during summer you can easily get comb melting. Simply putting a mesh over the entrance is unlikely to be sufficient, especially on hot days. You need to make a special ventilation device. This could be as simple as adding an empty box on top with insect mesh sandwiched between it and the top box. Or, an

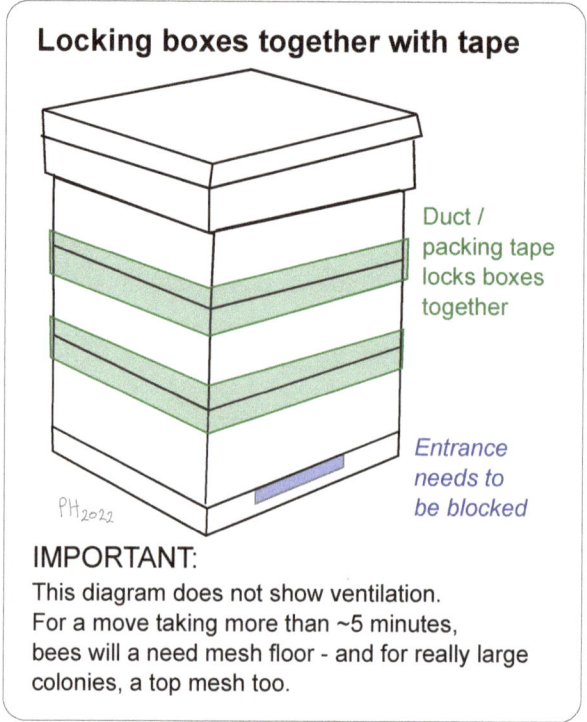

open mesh floor. You can buy a temporary mesh roof (a *travel screen*) for framed hives. If your car has air conditioning, use it. For a 20 minute drive they will probably have enough air in the hive without extra ventilation.

Comb collapse in hot weather: a big problem if the comb is simply attached to a top bar. (I now add side bars to make "semi frames.") Comb is softer in warm weather and likely to be heavily loaded with honey and brood. Uncapped warm honey is runnier and can seep onto bees. If the queen is squashed the colony will die; if the entrance is blocked by fallen comb the bees can suffocate. BUT if a hive has not been opened for years the combs will be locked together and to the walls with brace comb. This is ideal! Don't inspect the hive before moving it! This is not such an issue with framed hives - their combs are supported on 4 sides and very robust when moved, especially if they use wires embedded in the comb.

Comb collapse in ultra cold weather: comb gets brittle and can break; propolis loses its stickiness. I have heard of this in frameless hives in American winters (-20C). If you agitate the bees an hour before the move, the heat they generate will warm the comb, making it more robust.

Comb orientation is important: as a car accelerates and brakes you want the axis of acceleration to be along the combs so they don't snap off. Even framed hives have potential issues with crushing bees during sudden stops if the frames are oriented wrong.

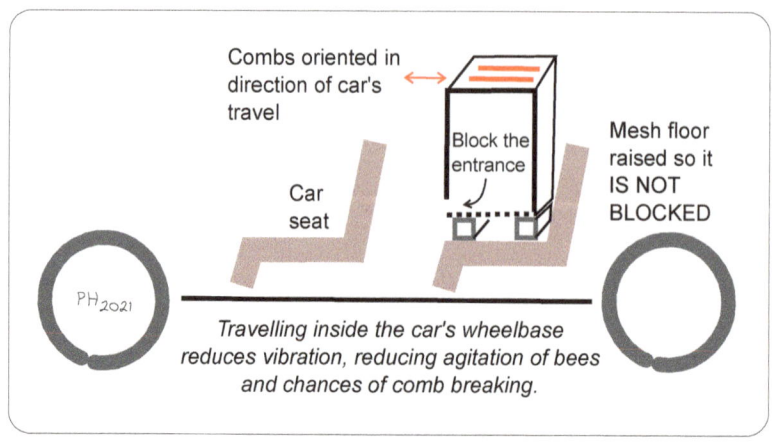

| Moving hives by car.

Long journeys: bees get thirsty, a gentle water spray through the mess every hour or so helps them. Don't overdo it – if they get wet they overheat trying to dry off.

On arrival: remove the entrance block. The bees will pour out and orient, but will be anxious and on alert so wear at least a veil.

You may well find a few bees show up back where the hive **was.** These are foragers who got trapped out at night. Leave a small cardboard box where the hive was so they can gather in it - they will gather in the box as the evening cools and you can brush them gently on to the landing board at the new location. For short moves (under 3 miles) this may be necessary for 3 successive nights as bees can be slow to learn the hive has moved within their memorised map.

Tip: clearing bees from the top boxes (supers) of a hive before the move, allows you to remove those boxes and gives a much shorter, lighter stack to move.

<u>The '3 foot / 3 mile' rule:</u>

Either move a hive less than 3 feet or over 3 miles. Distances in between cause problems.

I would add:

If you move them more than 3 feet, force them to pause as they fly out so they think "where am I?" and re-map.

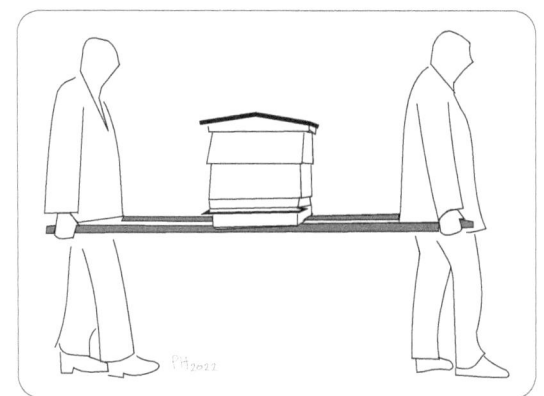

"Stretcher" method with 2 poles

Reason: Bees remember where their hive is with great accuracy[279]. If you move the entrance more than 3 feet (1 metre) returning foragers get confused and may not find it, ending up clustering at the last known location. (They have not evolved to live in trees that move!)

If the new position is within 3 miles of the old position, they recognise landmarks and navigate back to the old one.

This is a rough rule of thumb. If you move a single hive say 30 metres in an open field, it can still be seen and the returning bees may well find it. They will however get confused if there is visual clutter like other hives around, or you move their home to the other side of a landmark.

There are a number of tricks you can use here to reset their internal map.

When moving a hive, say, across a garden, you can do it in 3 foot steps, one per day.

If the hive suddenly jumps to a new position more than 3 miles from "home" they seem to realise and re-orient their internal map. So for a mile move, first move it temporarily 5 miles away, then back a few days later.

You can also place a maze of grass and sticks in front of the hive entrance. This forces any bees exiting the hive to stop and think. Realising something is amiss, they re-orient. This is my usual method for short moves, it works well.

Bees will re-orientate after 3 days confined inside.

---

279 Albert Muller has observed that if you move a hive forward 50cm, returning foragers bounce off the front – they are on a memorised descent path and not expecting a wall there! Drones however have no problem finding the entrance, presumably because they are not particular about whch hive they enter.

Drones know a wider area than foragers. If you move a hive 3 or 4 miles, a considerable number of drones return to its previous location[280]. But they migrate between hives, so don't worry about them.

Migratory beekeepers claim they lose few bees this way despite frequent moves: they just give returning bees an hour or two to realise their hives are now on that truck over there (the bees follow the smell of their mother hive). I don't believe this. I always see a few dozen left-behinds. I think migratory beekeepers accept orphaning some bees as a cost of doing business; their queens will replace 100 bees in an hour. Quimby remarks that merely moving hives around an apiary can lose a disastrous number of bees.[281]

To avoid new comb being twisted and cross-combed, try and orient the hive in the same direction if possible (i.e. if the entrance faced south, maintain that orientation at the new location). That will help avoid twisted / cross comb, because bees tend to continue building in the direction they settled on in the first location – this seems to be linked to a magnetic sense.

Cut-outs: removing established wild nests

Sometimes people ask you to move a nest from a hollow tree, a chimney or wall cavity. It's messy, sweaty and it needs to be done quickly once you start, to avoid chilling brood. *Don't get taken advantage of. People want to offload their problem onto you.* Ceiling cavities are particularly taxing: working for a long time with your arms above your head. *It is OK to change your mind and decline to move them after assessing the work involved.*

Wild comb cut and tied into a Warre frame. Rubber bands are often used. A full frame, with a lower edge to stop comb slipping out is essential. Photo (c) Will Hanrott 2017.

(American) YouTube videos often portray cut-outs as straightforward, with nests behind wood and plasterboard, but British houses feature brick cavity walls and almost inaccessible eaves, and the process is protracted and traumatic. Only about half the colonies moved from British houses survive. You may not find or inadvertently kill the queen. So if someone asks for bees to be removed from a house here, I ask why. And is the house Listed [282].

My sole experience here involved sledgehammers, crowbars etc. We inflicted significant damage, only viable because the house was due to be demolished a few days later.

Easy alternatives to consider first: Wildman recorded that you can drum bees out of a tree hollow. There are also products like Bee Quick and Honey-B-Gone which have a

---

280   Pettigrew, *Handy Book of Bees* (1875)
281   *Mysteries of Beekeeping Explained*, 1853, p.79. He is disparaging a solution some people at the time proposed for robbing, to move hives around an apiary to confuse the robbers. It did more harm than good.
282   In Britain, some old "Listed" houses cannot be altered or repaired without council approval and specialist contractors.

marzipan-like smell which bees flee from (a *'forced abscond'*); and Dettol (the concentrated stuff, not the weak surface cleaning spray version).

Less obvious equipment list:

- Small hive, with full frames, to put the comb in. A useful innovation by Shân and Clive Hudson is to staple plastic Netlon mesh with 1" holes to one side, leaving a frame sized flap of Netlon below ready to wrap round and fix to protruding pins on the top bar.
- 'Bee vac' – modified vacuum cleaner operating at very low pressure. Use as a last resort - danger of harming bees, queen. Google for examples.
- Container for honeycomb.
- Sheet to place on floor so dropped bees can be seen.
- Tupperware box for floor debris (varroa etc) to examine / learn from later

Procedure

Prepare by watching YouTube videos on "honeybee cut-out."

If done carefully, the bees are surprisingly calm, especially if you open from the side away from the entrance where the guards are.

Block all but one entrance.

Prioritise catching the queen. If a queen is left behind, the colony will probably reboot. When you open the nest up, the queen is going to run for the darkest crevice she can find. Prioritise finding her and getting her in a queen clip. Don't assume there is just one queen.

Lay a frame with Netlon mesh horizontally and put pieces of comb in it, then wrap and secure the flap of mesh round tautly, e.g. round the lugs at the end of the top bar. The bees will knit the comb back together.

The brood will be near the entrance. Once you have brood comb in the recipient hive, transferred bees will stay in it - workers' instinct is to cover and protect the brood.

Leave one stub of comb in place and leave the flying bees to settle back there for an hour (as you sort the hive and have tea), so you can make a final vac collection from that stub before clearing up.

You won't get *all* the bees so, place a closeable box in the position of the original nest – with a bit of their comb - and returning foragers should gather in it. Ideally you move them in the evening to the new site. (Some people advise sealing the entrance after you've removed the bees - this will trap some in there.)

Sometimes people drill a hole into the honey area and pump smoke or a smell in to drive bees out. Another trick is to use a thermal camera to locate the brood nest.

Throughout this, examine the nest and take loads of pictures of the comb, its floor debris and varroa (or lack thereof), brood patterns and even the walls. These will be interesting to lots of people and help settle arguments on how bees live in the wild. Consider uploading the photos / videos to freelivingbees.com . We designated one of our group as the photographer to leave others' hands free.

Remember it's already stressful for the bees, and excess smoke causes bees to panic and run into crevices. Often the smoker sitting on the floor below the cut out with the smoke it produces drifting gently up is enough to mask alarm scents.

Trap-outs (very tricky)

Some sources suggest a *trap-out* for nests you cannot access. This involves a lot of work and time for questionable results.

The idea is to divert returning foragers from the original nest into another hive like cavity. Eventually the original bees run out of food and abscond. This might take 3-5 days for a new swarm or months for an established nest.

These are used by people who wish to bleed off workers to reinforce other colonies. I have not seen proof anyone has ever got a complete colony, including queen, out of a wall this way - though they may have starved her to death.

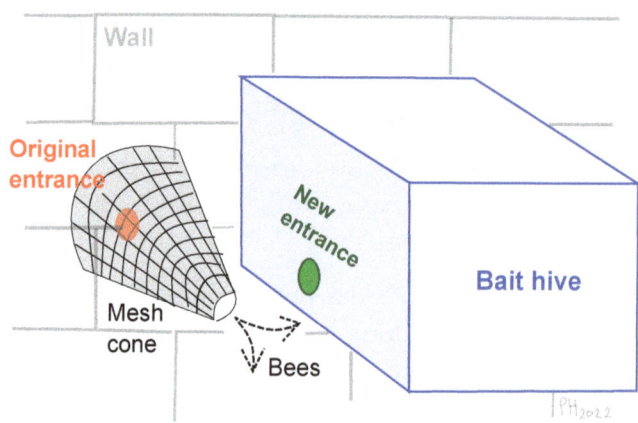

The trap-out principle. Bees can fly out of the original entrance but cannot figure out how to get back in (the cone is typically 2 feet long), so join the bait hive instead. The new entrance is as close to the original as possible.

Further reading: A Guide to the Safe Removal of Honey Bee Colonies from Buildings, Stewart & Roberts, northernbeebooks.co.uk (primarily British buildings).

Growdowns

Purpose

Moving an established colony from one hive type to another hive type without destroying comb.

Method

You place the box with the established colony above the hive you wish it to grow down into, cutting a hole to match the smaller box out of a wooden panel.

Discussion

It is important to give the bees only one route out, at the bottom of the stack. If there is an upper entrance, they will ignore the lower boxes and stay in the top one for at least a year.

It's a slow process and you may wish to consider drumming instead and write off the comb you cannot transfer manually. It might go faster if you start during the Spring flow.

Dying colonies and deadouts

Emotional impact; dealing with loss

Everyone loses colonies, and it's upsetting, particularly your first one; but due to the amorphous nature of bee colonies, they live on in the area through their drones, swarms and sister colonies. Look forward to repopulating the hive with a swarm in Spring, and focus on what you can learn.

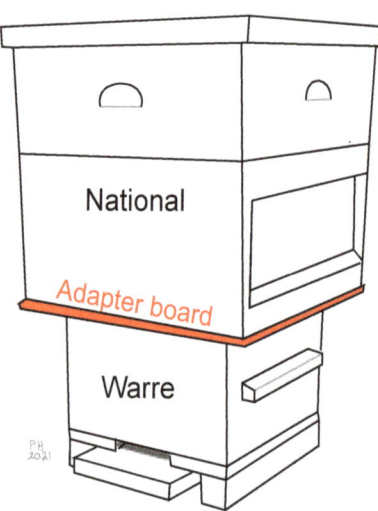

A successful growdown by Luzia Barclay. The colony, which was never fed, took almost 3 years to grow down sufficiently for the National to be removed and in retrospect would have gone faster if there was only a National brood box in place. Luzia used windows to monitor comb growth and added (nadired) a Warre box before finally harvesting the National boxes. Diagram based on photo from Luzia.

Rather than worry about hive losses, remember your bees are part of a "distributed superapiary" where the local **population** persists but **individual nests** come and go. These colonies are continually swapping genetics; whereas colonies requeened with identical genetics never adapt.

Wild bees continuously swarm and repopulate cavities and the number of nests varies year on year, but bounces back rapidly after bad years. These days, 7 years is old for a discrete colony[283], but if an established colony dies I smile and say "they live on", especially if they split in two (swarmed).

I find examining the *deadout* (dead hive) to find out *why* they died helps. Usually it is just queen failure, with a last burst of drone production - a natural and peaceful end. If it's something else, you need to know anyway.

In the end, individual colonies will die. But you can give them a happy home, and learn.

Options

Pragmatically, you can spend a lot of time trying to prop up failing colonies but if they are not queen-right, it is rare for them to survive. The main options are:

Often, the colony will supersede a failing queen if you just **leave it alone.**

If it's a swarm you caught and it acts queenless, just merge it with another colony.

If it is hopelessly not-queen-right[284], you can merge survivors with another colony if they are disease-free, harvest honey from the failed hive and set it up to attract a swarm.

Most low-intervention beekeepers just leave the colony alone and watch it dwindle - or reboot. You learn most from struggling colonies[285]. If it dies, you can take your time examing comb and deducing why. Chapter 15 covers post mortems in detail.

At the end of the season, the best options are allowing them to die and harvesting; or uniting with another colony (two or more weak ones may get through winter if combined).

Honey farmers' perspective

Conventional beekeeping views an unproductive hive as a problem. So standard advice is to re-queen immediately, which is another form of colony death, and unless you rear your own queens, introduces poorly adapted non-local genes to the area. Before rushing into this, remember, *there will always be more swarms next year.*

Another oft-touted option is to transfer in eggs (a comb of brood) from another hive. So drone laying workers can raise an emergency queen. However, this generally needs to be done 3 times over 3 weeks before they stop drone laying and raise a queen, you end up with two weak hives instead of one strong one for your trouble. There's no point giving them eggs if, by the time a new queen is ready to mate (3 weeks) there are no drones to mate with (September). Another factor is, are there enough bees left to raise a new queen? What ages are they? They need to survive long enough to raise not just her, but nurse her first generation to maturity - another 3 weeks.

---

283 Wildman, writing in 1770 and an exceptionally accurate observationist, recorded that colonies (not queens) live 9-10 years (one known definitely 30 yrs by Mr de Reaumur) but 10 years was generally considered the upper limit. *A Treatise on the Management of Bees*, p.68. My own colonies tend to last 2-4 years then die of queen failure, possibly due to a general lack of drones in this area, or persistent background levels of pesticides, but I collect plenty of related swarms from the neighbourhood to repopulate empty hives. Swarms take more time to collect than commercial beekeepers can spare, which is one reason they buy queens.

284 If the bees think they're queen-right, like drone-laying workers, the "rap test" won't be reliable. You have to open the hive to confirm.

285 This is how Gareth John discovered some colonies just go into a non laying "rest mode" for many months.

Some beekeepers are very quick to requeen despite no compelling evidence of queen failure. They get paranoid the moment they see a patchy brood pattern - yet the next frame may have a perfect pattern. And remember, a queen will likely stop laying during a forage dearth.

Killing bees

Obviously no one ever wants to kill bees, but one day you may need to kill a colony because, for example, of aggression or disease. This is a really last resort.

The usual technique is a soapy water spray, which can be easily washed out of a hive. It suffocates the bees it covers. Don't use insecticide. It will permanently poison the hive.

The following poisons all require the hive to be sealed, including any mesh floor -

Petrol and deisel fumes are very poisonous to bees, but would be massively dangerous if killing a wild colony in a wall or chimney. Beware lit smokers. Petrol melts poly hives.

Superglue fumes (isocyanates) are intensely toxic to bees.

CO_2 (dry ice or a fire extinguisher) used to be used in Canada but was de-listed by the authorities there, perhaps because it's quite dangerous to humans too (inadvertent suffocation, burns).

People used to use SO_2 (burning sulfur) but it will contaminate remaining stores - the honey will taste bad.

Isopropyl alcohol or ether are sometimes used. These are said to be quick, but that's only true if used at high doses, so use an entire bottle. Extreme fire hazard.

In Britain, it's illegal to poison honeybees because once they are dead, their contaminated stores will remain, and will cause a ripple-on circle of death as other creatures raid those.[286]

In Britain, honeybees are not a protected species, but most other bees are.

---

[286] Pest controllers have to remove the comb and block entrances. Cambridge Environmental Services were fined £1500 for not doing this and killing a nearby hive (23 Feb 2010, Ely Magistrates, HSE vs Russell Calne - RSPB Legal Eagle, June 2010 issue no. 61)

Part 2 Chapter 13 Tips and techniques: key skills: processes and pitfalls

Part 3: Beyond the basics - health, science and further reflections

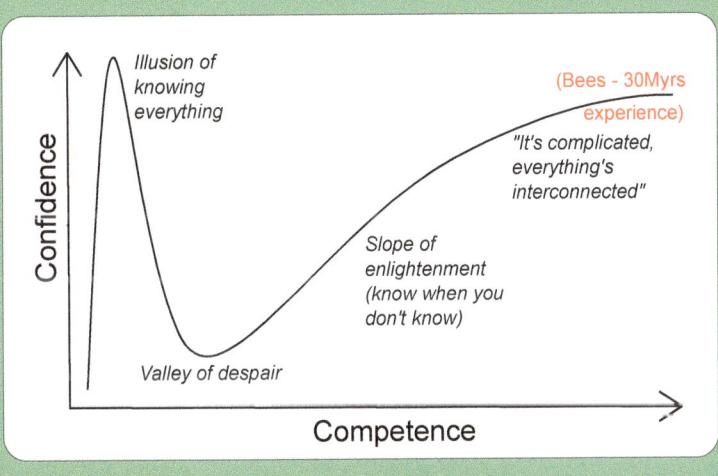

The Dunning-Kruger effect: novices tend to overestimate how much they know, then as they grasp the subtleties of a subject they recalibrate their self-image.

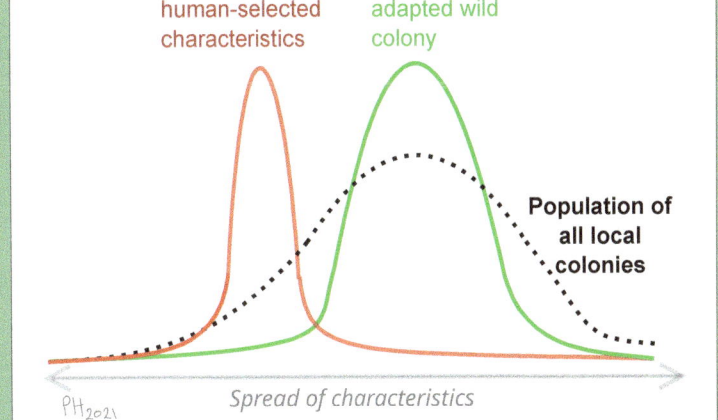

Commercial bees are relatively inbred, and if imported from another area are poorly adapted to local challenges.

Chapter 14
Pests, diseases and disorders

14.1: Pests, diseases and disorders - overview

14.2: Details of specific pests and diseases, rarely required

Chapter 14.1
Pests, diseases and disorders - overview

Some diseases and pests are *Notifiable*, which means you are legally required to report them to a Bee Inspector before they spread[287]. In Britain at the time of writing these are European Foul Brood (EFB) and American Foul Brood (AFB) – two diseases which spread rapidly and took many years to essentially eradicate. We expect some invasive pests to arrive eventually - the Asian Hornet, Small Hive Beetle (SHB), and the Tropilaelaps mite.

If worried it is one of these, call a bee inspector. It is a free expert consultancy service, they are vastly experienced and you always learn something from them.

Bee diseases don't affect humans (apart from the rare Stonebrood), but the digestive systems of infants below 12 months cannot neutralise botulism whose spores are sometimes present in honey (harmless to adult humans). So never feed honey to infants.

Disinfecting: If you find a hive has died with *any* serious disease, eg SHB or Tropilaelaps, then it is advisable to **sterilise the hive** before re-use. Burn all the comb, and scorch the inside (use a hot air gun until the wood smokes - safer than a blowtorch). Ask a bee inspector for advice on sterilising because some advice, like boiling an entire hive in caustic soda for 20 minutes, is not very practical. 1 part household bleach (sodium hypochlorite) diluted with 10 parts **cold** water kills everything after 20 minutes, is not too hazardous and can be washed off.

Then **sterilise your tools (including smoker), clothes, gloves and boots.** Don't put your bee suit's hood in the washing machine, it will tear the hood's veil – that will need hand washing. Do up zips before washing a bee suit. A little washing soda added to the detergent will help remove propolis.

> Soda is an ambiguous term:
>
> *Washing soda* is sodium carbonate. Strong alkali, USED FOR TOOLS AND WELLINGTON BOOTS ONLY. It's not clear if it is effective at sterilising EFB in wooden parts, it definitely doesn't kill AFB.
>
> DO NOT USE *baking soda* (sodium bicarbonate, much weaker alkali)
>
> AVOID *caustic soda* (sodium hydroxide) - far too caustic and dangerous, and problems disposing of the waste.

American Foulbrood forms incredibly robust spores which can survive boiling water and last for years – this is why you should never feed bees imported honey. Paradoxically, it's both a really virulent disease which spreads rapidly, but in some cases it seems very difficult to deliberately infect a hive! It seems to depend on how stressed the hive is: more on this below. On learning of AFB in a hive, the National Bee Unit will send inspectors to every known hive within 5km to check them. The Bee Inspector may require you to burn frames from an AFB hive in front of him – even kill the bees and burn the entire hive. This is one reason to join the BBKA (the insurance covers replacement). Yes, this is draconian, but it is why Britain and Europe have almost no AFB.

Interestingly, bees seem to have relatively few genes associated with an immune system. Instead, they enhance their weak personal immune systems with colony level behaviours: grooming; ejecting damaged larvae; individuals leave the nest if they are ill; antiseptic propolis is gathered and spread around; brood breaks to match forage. Some of these traits are deliberately bred out by breeders, and pesticides such as neonicotinoids have been shown to disrupt many colony level behaviours.

Key insights - diseases and pests

Medicines: it is quite striking that as of 2024 there are no medicines which **cure** bee diseases. All treatments amount to either:

Misleading symptoms: crawler with K wing, which eventually folded forward as if dislocated; but tracheal mites ruled out by dissection. Ragged wings like an old bee, but not so old it has lost all its hairs. Hive happy and thriving. A 40 year experience beek explained to me he sees similar damage from **foraging on borage**, which has prickly leaves, which was in bloom.

- Let the bees sort it out
- Help the bees sort it out (shook swarm, acaricides etc give them a breathing space - but all treatments have some collateral damage)
- Kill / burn them all and restock from healthy ones (scorched earth policy)
- Mask the symptoms
- Improve nutrition

The healthy bees used to restock losses... are simply ones raised in a low-stress environment with a varied diet, and not inbred to the point where they've lost key immune functions.

Which may shed some light on why left-alone colonies thrive compared to intensively managed ones.

However, there are some actual cures being looked into - see 'cutting edge research' below.

Fungal infections can be a major risk in cool climates, because insects are not warm blooded. These thrive in cold, damp conditions like a hive in winter or bees struggling to keep brood warm in Spring. It's often said that what kills bees is damp, not cold.

Propolis is good. It's antiseptic, antifungal and may even inhibit varroa. Its effectiveness at infection control is at least as great as soap's on human health. Rough walls stimulate bees to lay it down: tree cavities and skeps accumulate thick layers. But sticky gummy stuff slows down busy bee farmers, so they propagate bee strains bred for low propolis use, and like smooth walled hives whose walls get less propolised. Curiously, the Asian honeybee *A cerana* doesn't use propolis.

You're often looking at multiple, misleading symptoms. Once a colony is weakened by one thing, like poor diet or chilled brood, it is more susceptible to secondary problems. As most training stresses a handful of extreme risks, beekeepers often have these in mind and jump to the most dramatic conclusion from what they see, and treat inappropriately - stressing the colony further.

Examples of misleading or multiple symptoms:

- Perforated cells don't always mean varroosis. The fundamental issue may be starvation, or a mild brood disease, or chilled brood requiring ejection.

- Once brood die for whatever reason, they rot, going brown, then black and slimy, and they stink. This resembles foulbrood at first glance but, rotting is from opportunistic secondary infections post mortem - it's not not what killed them. *Example:* stimulative feeding in Spring can lead to too many brood for the nurse bees to keep warm in a cold snap. Dead brood are ejected and rot or, if there are too few nurse bees, they may even rot in the cells.

- Nutritional stress (no jelly in cells) can lead to brood dying with ambiguous symptoms.

Don't give a colony a huge area to patrol which they can't use (i.e, too much drawn comb in the hive). It means they can't guard it effectively and parasites can boom.

Unlike humans, the whole organism (colony) is continuously regenerating, and rapidly replaces failing parts. Individuals are undifferentiated, like plant cells. This creates a natural robustness to pests and disease! They can replace queens, sometimes they have overlapping queens; and their high staff turnover rate makes it particularly straightforward to fire & replace unsatisfactory brood and workers. They can even abscond and set up a new nest elsewhere. Beekeepers use this inefficiently themselves when they restock from healthy colonies, but if you are not intent on maximising production, why close off the colony's inbuilt options?

Example: workers only live a few weeks and, to the superorganism, are expendable and replaceable - like your red blood or skin cells. If they feel unwell, their instinct is to leave the colony - taking any pathogen load with them[288]. The queen's laying rate adjusts to the shortage of workers (she can pump out 1500 eggs a day if needed). The infection is flushed out by a flood of healthy bees *providing* the colony has a healthy queen and enough workers / stores to cover laying surges. Usually, large colonies are more resilient than small ones.

Things move fast in the insect world. By the time you notice a problem, they are already dealing with it. What more can you do to help? Usually nothing.

To put this another way, most of the activity of beekeepers who think they are "managing" or "helping" bees overcome an infection is generally disruptive or too late.

Varroa treatments: these mites have so far become resistant to three families of chemicals, and I don't think the manufacturers claim 100% kill rate for anything. They simply *reduce* mite levels, whilst generally damaging the bees.

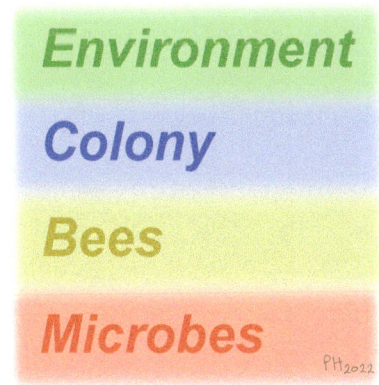

You mainly interact at the colony level. There are many aspects you cannot control.

---

288   In effect, evolution has tuned the workers to wear out just as their bodies saturate with pathogens. Perhaps a worker who lived too long would be an infection vector? Consider also how drones are used as expendable varroa bait and spare queens are killed. Bee life and death is ruthlessly efficient.

Bees can be vectors - for plant diseases and fungal spores[289]. In Europe, there are restrictions on moving hives in areas with Fire Blight, *Erwinia amylovora*.

Perhaps gut microbiota are the key to novel disease resistance: bees can't evolve quickly versus a **new** threat like a novel insecticide – that needs new genes and they are too reliant on social hygiene – though they **can** rapidly reactivate dormant (recessive) behaviours vs mites. But microbes can evolve incredibly quickly, and local gut biota seem key to why some strains of bee can handle honeydew and rhododendron. And recently researchers have realised animals' gut *virome* is of huge significance, keeping the microbes in balance, and contributing useful genes.

Brood ejection should be routine, not remarkable. Even ants kill diseased brood[290] (in their case they spray the pupae with antiseptic poison to kill pathogens). Breeders over-focused on other traits and are now scrambling to reintroduce hygiene traits they'd inadvertently de-emphasized.

Hives can be too clean. Medical researchers suspect autoimmune issues are often linked to insufficient exposure to low levels of pathogens[291]. Sterilising hives removes challenges and immune systems aren't trained.

Many diseases are always present at low, asymptomatic levels (almost as if they are commensal..?), and not a problem unless they get out of balance[292].

But many diseases are not. One can argue all diseases are omnipresent at low levels and symptoms indicated a stressed colony. There is a bit of a grey area here - is one bacterium or virus an infection?

---

289   *Survival and probability of transmission of plant pathogenic fungi through the digestive tract of honey bee workers.* Parish, Scott, Correll & Hogendoorn (Apidologie 2019)
290   *Destructive disinfection of infected brood prevents systemic disease spread in ant colonies,* Pull et alia (2018) doi.org/10.7554/eLife.32073.002
291   Guy Thompson discusses this further in *Old Friends,* in *Natural Bee Husbandry* No.24 Summer 2022
292   Many bacteria use *quorum sensing* - once their population reaches a critical density, they switch on a different set of genes and change their activity. Some drugs disrupt these signals, which is called *quorum quenching.*

Example: EFB.	Counterexamples: AFB and Chalk Brood.
It is possible for EFB to be present at such low levels it is symptomless ("subclinical"). You hear different stories here, and it is easy to argue that whether it is "omnipresent at subclinical levels" or not just depends on your threshold definition. But actually, it seems what's really going on depends on what country you live in: **In Europe**, a study[293] indicated EFB is indeed omnipresent at very low levels and "infected" hives are simply those where the bees have been pre-weakened by other factors. **In England and Wales**, the NBU did extensive research[294], checking a huge sample of 4600 apiaries with hypersensitive, very accurate PCR tests (partly to ascertain how accurate their new Lateral Flow Devices were). They found EFB is **definitely absent** in symptomless apiaries[295]. (Of interest: They also found that the Lateral Flow Devices were highly accurate in indicating EFB when it had reached clinical levels, but they need to be used properly, i.e. the inspector checks an obviously discoloured larva. If you check a larva with subclinical levels the LFD probably won't pick it up, and it can only test so many larvae.) **In America**, EFB is inconsistently controlled across states, resulting in continual re-importation to cleansed areas by migratory beekeepers.	Les Crowder recounts an encounter in his book Top-Bar Beekeeping, during his time as a bee inspector in New Mexico. He met the retired Dr Lyle who had researched the AFB bacterium and realised it feeds on already putrefying flesh, but is not pathogenic - it does not cause the disease, it is a symptom of already-dying brood. Dr Lyle and Crowder agreed the problem was associated with old, black comb - it can accumulate pathogens[296] - and strains of unhygienic bees. Dr Lyle challenged him to deliberately infect a healthy colony with new comb. Crowder had the opportunity to transfer a comb of AFB infested, dying brood into a small healthy hive and found the AFB was rapidly cleared up. He retried and it was cleared up again. This is intriguingly similar to Roger Patterson's attempts[297] to deliberately infect colonies with Chalk Brood by swapping queens, transferring brood from an infected hive etc. He found it was impossible to infect healthy colonies, it seemed to be a property of particular hives, not bees (not genetics).

293   *Bacterial community associated with worker honeybees (Apis mellifera) affected by European foulbrood*, Erban, Ledvinka et al (2017) - Czechia - DOI 10.7717/peerj.3816 . They found a (more sensitive) technique called HTS (High Throughput Illumina Sequencing) detected EFB even when PCR could not, and EFB was present in all hives in an apiary where some hives showed symptoms. They concluded there are low levels of local strains of EFB endemic to an area ("enzootic").

294   Summary on NBU website: EFB_-_The_National_Bee_Unit_Perspective_2.pdf

295   I discussed this with Professor Giles Budge, who was chief scientist at the NBU when they did these tests. Was it possible, I asked, that local bees or their gut microbiota were co-adapted to local strains of EFB which was omnipresent, but could lack resistance to non-local strains? And surely the European study, which used HTS (High Throughput Illumina Sequencing) could detect lower levels than PCR? (That's why the Europeans used HTS.) Budge replied [email 16/2/2023]: "Mp is **definitely absent** in many UK apiaries and it is **unlikely due to antagonistic microbiota**, it is more likely because of an effective control policy restricting spread. **Our tests are *far more sensitive* than conventional PCR**. It is difficult to compare findings in non-uk territories with England and Wales, because we have historically very different control policies." [The E&W policy is targeted destruction, not scorched earth.]

296   Elsewhere in this book I assert old, black comb is a Good Thing. In general, yes, but I guess you can have too much of a good thing and, obviously, it depends exactly what is in it. Comb in hives treated with miticides gives the queen long term exposure to chemicals known to affect long term fertility and health, but comb from a completely treatment free source is probably good for many years unless contaminated by nearby pesticide use. Crowder (TBH expert, 50 years' experience) reckons the time to discard it is, when you hold it up to the sky and can't see light through it - and although he sells fresh wax from his treatment free hives at a premium, he assumes black comb is best discarded. He's in Texas which may have more agricultural pesticide use than Britain.

297   *Beekeeping: challenge what you are told* p.213

What can we conclude from these contradictory examples?

- Tentatively - one does seem to need a threshold level of pathogen to present clinical symptoms;
- England & Wales have done a particularly good job at eliminating foulbroods;
- But some colonies seem to have behaviours or environments which promote resistance.
- This is **empirical** evidence by massively experienced beekeepers (Crowder: ex bee inspector / bee farmer in southern USA; Patterson: 70 years' experience in UK). This is a different form of evidence to a peer reviewed scientific paper under controlled conditions, but one thing I learned by talking to COLOSS scientists is that many only have experience across a narrow range of hive types, bees etc - their PhDs etc place other demands on their time and they crave datasets. **Both types of evidence are useful. One is narrow and reproducible, one is broad but less controlled.**

Many small shocks works well. Just as bee colonies collapse from being pushed too far, pests and parasites **need just the right conditions to thrive.** They're really quite delicate, specialised organisms. Many small factors = prevention, avoiding the need for a chemical hammer to reset numbers.

Breeders have weakened stocks massively, inadvertently selecting out disease and pest resistance traits which are the default in wild colonies.

We've moved a LOT of pests and diseases around the world. Not just imports: exports too. Asian honeybees now face novel challenges from wax moths, tracheal mites[298] and sacbrood; African bees have been gifted AFB, EFB, varroa and nosema ceranae[299].

Swarms spread "good" diseases. Pests and disease can spread *horizontally*, like when you split a hive, or *vertically*, when a child-swarm is birthed by a mother colony. There is

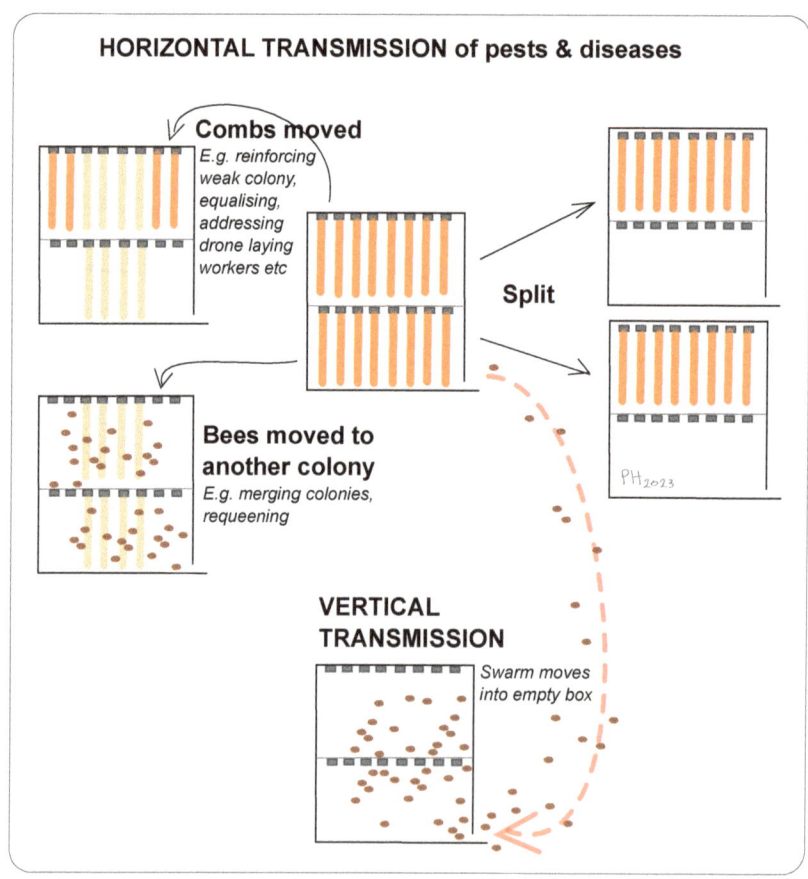

Vertical transmission occurs between mother & child: rapidly lethal parasites destroy their own transmission path. But **many intensive beekeeping techniques promote lethality of pests and disease**, by manually spreading them **horizontally**, and then nullifying any penalties for the diseases by propping up failing colonies.

---

298   Scottish BKA video lecture on traditional Japanese beekeeping, Jan 2022
299   *Honeybee health in Africa - a review* (2015), Pirk, Strauss, Yusaf, Démares, Human - Pretoria DOI: 10.1007/s13592-015-0406-6

no selection pressure on the parasite in horizontal transmission (it's basically deliberate cross-contamination), but if it is too lethal then host colonies cannot survive long enough to generate swarms. Vertical tansmission is a **massive** selection pressure for parasites to reduce in virulence and co-evolve with their host[300] - and a big reason to use **swarms, not splits.**

Time of year affects some diseases. For example in early Spring, there are lots of larvae and relatively few nurse bees. The nurse bees may not be able to keep up with cleaning / ejecting if EFB breaks out.

Poor nutrition weakens bees. They are much more resistant to disease with access to multiple pollen types and plentiful nectar / honey.

Many syndromes are symptoms of OVERmanagement. Low-intervention beekeepers simply don't see much disease: there isn't that much in hives the bees run themselves, populated by swarms from tough survivor colonies. Pests and diseases are a symptom of a stressed hive.

The bottom line is, if you suspect disease, shout for advice and if it's a notifiable one, contact a bee inspector.

Key insights - beekeepers

Forums' business oriented pragmatism: Feedback to queries tends to devolve to *'Miticide, Regicide, Bung Junk Food Inside'*. This *Pareto Principle* approach 'solves' the bulk of *symptoms,* but not root causes.

Most diseases are caused by beekeepers: Many beekeepers assume it is normal to continually fight disease and pests in a hive(!!!), without grasping it is their procedures promoting the problems. Densely packed hives promote disease. Brood factories are also mite factories.

Beekeepers leap to conclusions. Most diseases occur when colonies are pushed to their limits and can't cope with one more stressor. Most symptoms actually have multifactorial causes. As soon as the beekeeper notices one thing wrong they patch the symptom (requeen, feed etc) and continue as before. When a cause is announced with great certainty on ambiguous symptoms, it's a sign of false confidence (Dunning-Kruger effect). Not everything is mites!

Do treatments cause problems? Michael Bush suggests that medicines and fungicides disrupt bees' gut microbiota, which reduces the digestibility of pollen and may weaken them. He develops this idea at www.bushfarms.com/beesccd.htm. I would add that sterilising a hive with miticides is bound to reduce resilience: a hive is a mutualistic superorganism of several species, not just bees, and some 'problem' lifeforms are naturally present at a low, non-symptomatic level and perform a function. Stonebrood (*Aspergillus fumigatus*) displaces the worse nosema. Chalkbrood (*Ascosphaera apis*) protects versus EFB. Wax moth devour and sterilise rotting comb. Basically, complex webs of life have more correction mechanisms.

---

300   Sweden burns any hive infected with AFB - but not wild colonies, because even if those test positive for AFB **it is not symptomatic.** *Vertical transmission of American foulbrood (Paenibacillus larvae) in honey bees (Apis mellifera)* (2006) Fries, Lindström & Korpela   https://pubmed.ncbi.nih.gov/16420974

It's good to cycle between strategies, to avoid parasite buildup. Parasites tend to have fewer options than the larger brained bees working in the continuously-adapting social network of a colony. Beekeepers run colonies the same way each year...

Disease spreads from managed hives to wild nests, not the other way round. Factors involved are:

- Moving hives around a country to follow crops
- Buying queens and entire colonies from remote, sometimes foreign, suppliers
- Re-using comb after spinning (same comb not guaranteed to go in same hive)
- Commercial pressures resulting in hasty work and poor hygiene
- Dense apiaries (aids pathogen spread)
- Frequent inspections, removing propolis, suppressing swarming, and stimulating laying all disrupt natural pest control behaviours
- Requeening with purchased queens inhibits bees adapting to local pests.

Wild / feral bee colony health

I'm talking here of free living colonies which have survived a year or more unmanaged. By this point they are no longer *feral* ['escaped from domestic stock'], but *wild* - they have un-learned domestic behaviours - and as we'll see in chapter 16, after ~2 years, their genetics converges to the local wild genome.

The dominant 'conventional' trope prioritises money. Authorities promote intensive management and have a reputational and financial stake in being Right.

But.

- You keep hearing anecdotes about hives being abandoned and surviving for years until the property owner calls in a beekeeper, who finds them flourishing.

Wild nests **never** show the yellow spots & streaks of dysentery below the entrance. Just lots of dark, antiseptic propolis.

- Roger Patterson has removed several hundred British wild colonies. He's *never* seen foulbrood; occasionally mild cases of chalk or sac brood, which are relatively common in managed colonies.[301] As he says, winter kills weak wild colonies.

- Couston mentions cutting out several dozen feral colonies in Scotland and none had any brood disease, even when it was present in the area[302].

---

[301] BIBBA webinar 1/9/2020 on "challenge assumptions!"

[302] *Fly with the Beeman* (1989), p.59, 72, 76-77. Couston was a Scottish Specialist Advisor (equivalent to Bee Inspector), Markets' Convenor of the Scottish BKA, and chairman of the Bee Farmers' Association. Given this tended to be in small village apiaries, some still using skeps, the dark bees were probably pretty pure Amm. However he also writes that at that time, Scottish beekeepers were not required to destroy colonies with AFB. It was very noticeable that some stocks of the local dark bees in hives were tolerant of AFB, showing very little cell perforation in their first season of infection. Some owners just tolerated this as a low level chronic infection. Yet the wild colonies were free of AFB. Also of interest, but nothing to do with disease, he mentions that no wild nests had more than 10kg of honey - always full of dust and debris - reports of wild colonies accumulating vast stores are hugely overstated (in that area). He managed to save 5 and breed from them and though healthy, they were swarmy and produced little excess honey.

- Karl Colyer specialises in cut-outs in England and has never seen foulbrood (chalkbrood is common and he sees some bald brood)[303].

- Bailey analysed results from an international survey covering several hundred wild nests which were examined by experienced beekeepers or scientists,[304] and found almost no disease in them. This was particularly striking where he had data on nearby managed apiaries, some of which had very high incidences of foulbrood, nosema, and some had tracheal mites.

Related to this, there are lots of unmanaged wild colonies around, if you only look up. There are ~8 in my village. They're usually around roof height, in trees, walls or roof spaces. Although their existence or viability is denied by many conventional beekeepers, I monitor the local ones and can confirm they do not die out over winter / get repopulated by swarms escaping from hives. They tend to be long lived – one was continuously occupied for 19 years.

In the 2000's, iconoclasts like Phil Chandler in the UK, John Kefuss in France, Michael Bush, Les Crowder and Dee Lusby in the USA began asking: how do these colonies survive when so many managed colonies struggle with varroa mites, disease and other ailments? These beekeepers had decades of experience and had seen changes in beekeeping and landscape, so had sufficient self confidence and perspective to separate assumptions from fact. Pointing out the Emperor's New Clothes was not welcomed by the established beekeeping organisations, but gradually their views have gained traction.

The reasons *we know of* for wild colonies thriving are:

- Weak colonies are not propped up. High selection pressure.

- Bees' hierachy of needs understandably puts starvation as a higher, more immediate crisis than disease prevention. When a commercial colony's honey is taken, the entire workforce is mobilised for emergency foraging. Beekeepers conveniently interpret this as "bees like to work", but study of wild colonies[305] shows they really *don't* like to forage so hard that they die in 6 weeks (they live much longer in the wild): they use free time to maintain nest hygiene.

- Wild bees also tend to have much wider genetic variation than ones from a breeder - they haven't had useful traits inadvertently **de**selected or suppressed (like being able to stop laying in dearths).

- There is evidence that imported bees are more susceptible to local strains of disease[306].

- Propolis. Their nests are completely saturated with it[307].

- Wild nests are *well hidden,* unlike ground level hives, and have small scent plumes (small entrances, lower / less active population). Drifting and robbing is rare. (Drifting spreads disease in packed apiaries.)

- Bee colonies have *evolved to fail,* so old nests don't become pest resevoirs. Over the course of a few years, they're essentially a *migratory* species. The population ebbs and flows around an area, with *healthy* swarms budding off as *old* nests die out.

---

303 BIBBA webinar 17/1/2023 "Rescuing bees from buildings". He also mentioned he has yet to see an established colony which is not dark - yellow bees don't seem to survive unmanaged in Cheshire.

304 Bailey, L. *Wild honeybees & disease* (1958) Bee World 39, p.93-95. Even back in 1958 he remarks the likely reasons are: reduced colony density; swarming is not suppressed; old comb along with pathogens is digested by wax moth; not propping up weak colonies by feeding. He credits the late E B Wedmore with starting the investigation.

305 Torben Schiffer (personal communication). His book isn't available in English yet.

306 The massive COLOSS *genotype environmental interactions experiment.* 621 colonies across 11 countries over 3 years. The local bees had lower viral loads of eg acute bee paralysis virus and other viruses.

307 Torben Schiffer (beenature-project.com) has shown air from wild nests is sterile, whilst air from conventional hives seeds dramatic mould cultures on agar plates.

Historically, bee diseases were less common

A major British Victorian-era reference book was Bevan's *The Honey Bee, its Natural History, Physiology and Management*. Bevan was a doctor as well as a beekeeper, and extremely well read and connected. Here's his list of bee diseases from 1870, when skeps were still the standard hive. It's remarkably short by modern standards:

- Foulbrood *[no distinction between American and European]*
- Dysentery
- Vertigo *[poisoning or possibly a viral disease- bees act drunk, or lie on backs dying in severe cases]*
- Tumefaction of the Antennae *[antennae swell at ends, antennae and forehead go yellow, occurs in May, bees become languid and may die. I don't know what this is.]*
- Pestilence, aka Faux Couvain *[dead larvae, possibly chilled brood]*

And that's it. No mites, chalkbrood, stonebrood, pesticide poisoning, or widespread die-offs (excepting the winter of 1782-83, no one could ascertain why).

These days we might split these into subcategories but, if the treatment is the same, you could say - why distinguish?!

For treatments he recommended cleanliness and cordials (like sugared ale). His medical experience shines through in this simple, practical advice: "*The reader must now perceive the importance of feeding, and that the transition from health to langur and death is less frequently to be ascribed to disease than the want of the necessary means to continue the vital energy.*"

From the brevity of the chapter on disease, despite medicine being his "real life" profession, we can infer it was not a major distraction for beekeepers at the time.

Pettigrew, a Scottish skep beekeeper writing in 1875[308] lists only foul brood (common) and dysentery (rare, associated with damp, which he reckons is a flaw of wooden hives).

Erik Tihelka and Sofia Croppi researched the history of foulbrood[309], and point out:

- Before the 20th century there was no standard nomenclature for many diseases and descriptions are quite vague; many symptoms overlap.
- But you can find scattered surveys for colony loss rates and see they were, say, 8.1% between 1936 and 1941 in Bohemia - figures which seem very good today.
- When framed Dzierżon hives were introduced, German beekeepers had a phrase "Dzierzonstöcke, Faulbrutstöcke" (Dzierżon hives, foulbrood thrives) - reflected in the upsurge in the mysterious "Isle of Wight disease" which correlates with the change from skeps to framed hives in Britain[310].
- It seems likely the change from fixed comb skeps to movable frames allowed disease to thrive through deliberate movement of frames between hives, e.g. splits and re-use of frames after honey extraction. Swarms and fixed comb are intrinsically healthier.

---

308 *Handy Book of Bees*, 2nd edition, pages 117- 120 (the only pages on disease in this 200 page reference work)
309 *A brief history of foulbrood: from superstition to veterinary medicine and back*, Natural Bee Husbandry #16, Summer 2020
310 Also discussed in the Introduction.

Viruses and bacteria - general

- Viruses are much less persistent than bacterial infections. Viruses are basically RNA which isn't a very stable molecule and breaks down within hours in honey. In contrast, bacteria, like Foulbrood can remain viable on comb for years.

- Bacterial diseases only affect brood. Perhaps because they are warm, growing and surrounded by food. This is curious, as adults fly in sunlight, where you would expect ultraviolet to help sterilise viruses.

- Most bee viruses are RNA, not DNA based. CBPV has a unique structure (4 strands of RNA). RNA viruses mutate rapidly.

- Due to immense variation within each virus species, the boundary between viral species is fuzzy. Quite a few viruses are being reclassified as just local variants of the same thing. DWV comes in many flavours: - A,B,C; Egyptian Acute Paralysis Virus; Japanese Kakugo Virus; and China's VDV-1. In practice, names don't matter, symptoms determine your course of action - paralysis, deformed wings, sacbrood etc.

- *Some* bees (and other insects) seem able to suppress *some* viruses using RNAi (interference)[311].

- Bee viruses are predominantly spread by human activity.

- Varroa amplify many virus problems. Mite-resistant colonies have much lower viral loads than most. Acaricides will reduce, but not eliminate some viruses.

- Virus effects are suppressed by good quality pollen forage.

- Several (not all - check specific entries in next chapter) viruses are vectored and amplified by varroa, so miticides may help these ones; but as miticides interfere with hygienic behaviour I would only use them as a last resort.

- Crowding promotes virus transmission between adults. Roger Patterson points out[312] Hoffman frames space combs 35mm apart, whereas natural nests' comb spacing is 38mm in Britain - and as brood comb is always the same thickness, 38mm gives the bees 30% more headroom between combs.

- There are probably thousands of as-yet undiscovered viruses associated with honeybees. Most will be ignorable, for example, AmSV1 was discovered in 2023 but as yet, there's no clear indication it causes problems. (There's a correlation to queenlessness, but other causes weren't checked.)

- Researchers have detected occasional opportunistic infections by generalist bacteria and fungi with broad palates, such as *E. coli* and *penicillium*. Such pathogens infect many creatures and are pretty omnipresent in soil etc. These are too rare to cover here, and seem to be secondary infections rather than primary causes of death.

---

311 *Understanding the intimate relationship of the honeybee and its viral pathogens in order to tackle bee mortality from a different angle*, De Graaf, De Smet, Ravoet, Wenseleers. This is potentially major.

312 BBKA News, August 2019, p.265-266; also in Roger Paterson's book *Beekeeping: challenge what you are told*, p. 81 onwards; and Lazutin mentions the Russian standard spacing is 1.5" in *Keeping Bees with a Smile* p.90 footnote 31.

Fungal infections

Apart from the gut parasite nosema, these only infect brood.

As noted in Chapter 11.3, *Reading Comb - mould and other white stuff*, you don't generally see mouldy adult bodies. When you do - rarely - see what appears to be mouldy adult bodies, it generally turns out the mould is growing from something under them.

After saying this, I was given samples of adults with mould growing on them from a February deadout (see composite picture showing bees before and after disassembly). And yes, there is a dust-like surface coating of fungus. However the mould didn't infect living adults: the inside of this worker shows no sign of mycelia. I believe these bees died in a damp spot in the hive and a post mortem fungus, perhaps mildew grew on their surfaces. The drone shows slight ingress of mycelia but had probably died months before and is stilllargely hollow. Another clue is that all the mouldy bees' bodies were externally intact, suggesting it didn't penetrate the bodies.

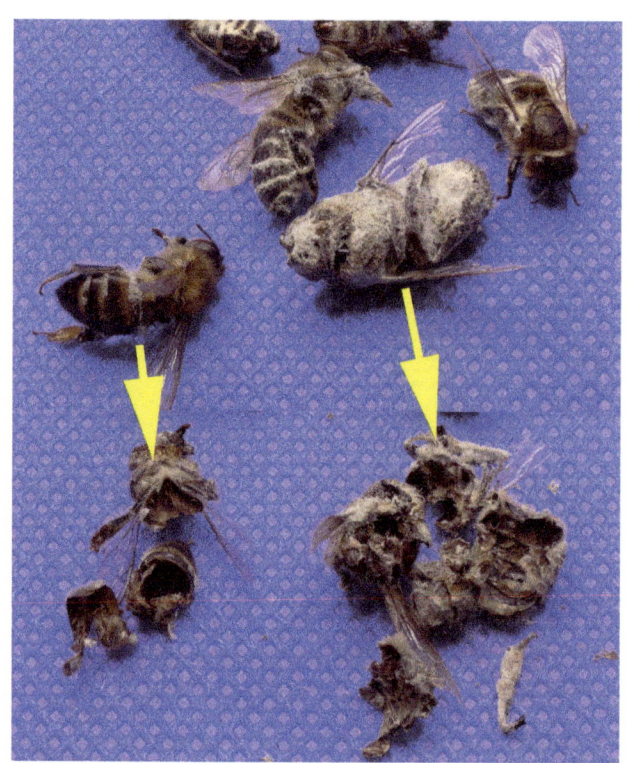

Surface mould on bees from a February deadout. The worker and drone were hollow when pulled apart.

Torben Schiffer in Germany has investigated tree cavity nests and noted they always have a heavily propolised comb right behind the entrance. His tests with Petri dishes show the air in wild nests to be sterile, whereas conventional hives' atmospheres cause massive mould growth.

Pioneering research

Exciting new treatments are continually being developed like DSRNA (Double Stranded RNA) versus varroa, exotic fungi versus AFB / Chalkbrood / Stonebrood[313], etc. Many, like RNA interference are technologies from human healthcare which turn out not to work in insects.

Such innovations have two huge hurdles to cross before they're released as a product -

- **Usability** - how do you actually apply them to a hive in such a way that enough bees, eggs, parasites etc are treated? Even the ones in capped brood cells?

- **Regulatory approval** - for a small market like Britain (outside the EU) it just isn't worth the manufacturer's while to pay for expensive tests and reviews to prove they're safe. The safety bar is very high because hives produce human food. For really novel treatments like bacteriophages versus foulbrood, the first stage is to agree with the regulators what *kind* of proof they need and *create the regulations*.

---

313   *Fungal microbiota isolated from native stingless bee species inhibited pathogens of Apis mellifera*, Tejerina et alia (2023) doi.org/10.1016/j.funbio.2023.07.003

Whether they work is irrelevant if they're unavailable. Most 'breakthroughs' vanish without trace.

A couple of novel treatments show promise. For example, Paul Stamets' mushroom extract for viral diseases (www.fungi.com). It seems unlikely to cause harm and there is peer-reviewed evidence that it helps – a lot - with DWV and LSV[314].

Others seem dubious. Dalan Animal Health promotes an oral vaccine for queens - but investors are wary after the Theranos "fake it 'til you make it" scandal. Dalan has one published peer reviewed evaluation[315] which has been soundbited to make it sound like a cure for AFB. It is not. It just makes the vaccinated bees more tolerant to it - and those bees are more likely to spread the bacteria to neighbouring hives. It would not be licensed in the UK, where we have under 100 cases of AFB a year.

Epidemiology: societal factors

Assumptions matter. Most of the world has a zero-tolerance attitude towards disease. America, however, has had to pragmatically live with a higher disease burden due to some unique challenges.

The primary issue is the extensive migratory beekeeping industry - nowhere else are millions of hives moved right across continents every year. The majority are taken to one area in spring to pollinate almonds, an ideal way to transfer pests and pathogens. Another factor is inconsistent regulations across all its states. Illinois burns AFB hives, but until recently many states permitted the use of OTC to suppress (not cure!) AFB symptoms. America also seems to have a higher proportion of people who believe inconvenient laws don't apply to them.

A pattern emerges as you read more about bee diseases: Americans assume they are common and widespread! Whereas in Britain a lack of AFB symptoms really does mean zero AFB spores, in American research papers you see statements showing they think a few hundred spores per bee is normal background level[316]: what they're interested in is, is it symptomatic yet. Again, in Britain, when testing for nosema with a microscope, one spore is considered a positive result. Whereas in America, under 5 million nosema spores per bee is considered economically viable.

So you will hear conflicting statements, even among research scientists, about whether "all colonies have asymptomatic traces of most diseases" versus "healthy colonies have **zero** foulbrood, etc spores / bacilli". Once again, it all depends what your starting assumptions are.

Another cultural difference skewing perceptions of "normal" is the accepted norms of how to treat animals. The Animal Protection Index, api.worldanimalprotection.org , is a good indicator of a culture's attitude. America achieves the lowest grade, D.

---

314 *Extracts of Polypore Mushroom Mycelia Reduce Viruses in Honeybees*, Nature.com, 2018

315 http://doi.org/10.3389/fvets.2022.946237 - 60 hives at apiaries in **Austria and Spain**, double blind experiment. 28 - 30% extra larvae survived AFB infection. **But 30% is a poor reduction for a vaccine, especially a highly transmissable pathogen like AFB**. Any vaccine that doesn't result in high enough efficacy to produce herd immunity- meaning a lasting efficacy of >70% disease *prevention* (not just reduction in symptoms or lethality) has the potential to increase the spread of a disease and drive mutations. And AFB is a bacterial infection, not a virus: bacteria respond to vaccines very differently because they infect new hosts much more easily.

316 doi: 10.1186/s12898-020-00283-w quote: "*it has been shown that virulent spores are present even in asymptomatic hives, having an average of 158 spores per bee; however, an increase in spore loads of 30% (to about 228 spores per bee) can lead to a clinical outbreak of the AFB.*" This research was looking at 40 colonies in Maryland, USA. Another interesting point this paper raises is that when AFB levels are low (asymptomatic) nurse bees inadvertently *spread* the spores rather than act as agents of social immunity, because they're not ejecting infected brood.

Re-using comb

Conventional beekeepers consider clean drawn comb an asset, because it can be re-used to get more honey, faster (bees use a lot of honey to make wax).

This has led to practices like saving last year's comb after honey extraction and re-using it in random hives. This is a recognised risk, as nosema spores can last a year in wax, and AFB spores many years. Beekeepers are generally careful to only move frames within apiaries and if disease spreads, well, it's a cost of doing business. It's generally assumed that supers (honeycomb) are OK to move unless the hive had Foul Brood.

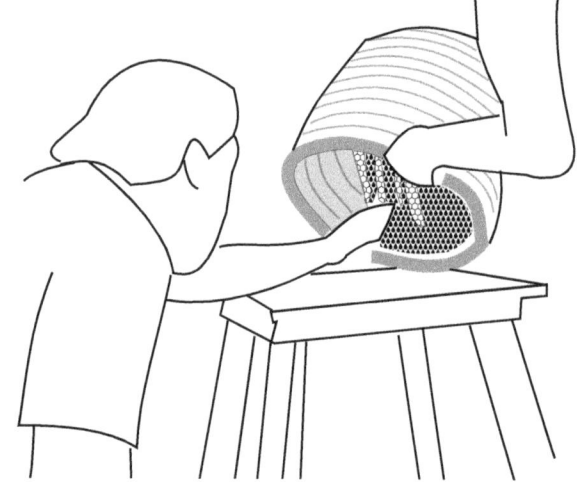
Fixed comb hives can be inspected

I occasionally re-use comb myself, as bait comb, if I believe it undiseased.

An interesting trend can be seen when you talk to enough beekeepers. In general, conventional ones believe you should discard old (brood) comb as it spreads disease. Leave-alone beekeepers prize old, dark comb pointing out that wild nests are full of it, and healthy.

Perhaps the problem with old comb is accumulated industrial waste: miticides, antibiotics etc?

Age and stress effects

Bees can exhibit odd stress related 'displacement' behaviours, just like humans under stress. "We can't express what we want to, so we'll do this." Aggression, propolis barriers and excessive queen cells probably indicate the bees feel constrained and unhappy. Conventional beekeepers often express surprise at how calm left-alone colonies are.

Most animals have age related symptoms like arthritis, cancer etc. With age, individual bees lose hairs and their wings get tattered; but do old colonies have particular issues? Their comb (skeleton analogue) accumulates toxins and structural issues, but unlike large animals this, like the component bees, is continually replaced. So the colony never suffers sensescence. However cavities decay and may fill with detritus and parasites, triggering 'end of life swarms' (aka absconding, and perhaps mislabelled CCD) after some years. And every time the colony supersedes, they effectively roll a dice - one day their new queen will fail to mate or be eaten on her mating flight. So individual colonies don't last forever.

Larger colonies, which means older ones if you populate hives with swarms, can be noticeably more assertive if approached. This may be because they can afford to lose guards, or simply because larger colonies contain more bees of all temperaments.

Hormesis: use it or lose it

Some toxins are a helpful tonic at low doses - a phenomenon known as *hormesis*. Likewise, doctors are aware that immune systems which are never challenged, atrophy.

There is a similar concept in psychology of *optimal anxiety* - you need a little stress to keep in mental shape. And in botany, *thigmomorphogenesis* refers to plants responding to buffeting by weather by growing stockier and stronger than ones in greenhouses.

There is an argument, therefore, that low levels of varroa mites etc are beneficial. You don't want hygiene habits to decay.

Some pathogens are always present at low, asymptomatic levels. Without varroa, DWV viruses per worker[317] are 10^4 and the bees are OK. Symptoms like deformed wings kick in when varroa infestations boost these to 10^{10} - a million times higher.

Are some pests commensal?

A number of people have pondered[318] whether – like termites which need help digesting cellulose – some "parasites" are actually symbiotes or, at least, commensal - which is to say, they themselves benefit but do no harm to the host. For example, wax moth larvae eat old brood comb, converting it to sterile crumbs.

Complacency

There's a difference between what's important and what's measurable, and sometimes people get fixated with metrics. Bee inspectors worry about hives with uninspectable comb, because that's their primary method of spotting disease.

But they worry for a reason. There **are** occasional instances of disease, and varroa, found in "natural" hives.

If foulbrood is in an area inspectors need to check **all** hives nearby. It's **vital** that your hives are inspectable - even if you prefer to monitor them non-invasively.

> ### Review 11
> There are parallels between factors affecting bee and human health. Which ones do you see in common?

---

317   Prof. David Evans, writing in BBKA News Feb 2023 p.53. But as discussed elsewhere, it depends on which pathogen you're talking about and which country, as some regulators take a zero-tolerance approach to the worst diseases and have essentially completely eliminated them.

318   The earliest example I know of is H J Wadey in a 1964 lecture.

Chapter 14.2
Pests, diseases and disorders - specifics

Here, we focus on how to identify diseases and pests, and your options if they occur.

Disclaimer

I have not seen most of these myself.

First steps in diagnosis

If you encounter something worrying send close-up photos and detailed observations to a **trusted expert**, like your mentor, who has first hand experience with disease - there's a difference between book learning and real field experience. But they can't diagnose without data, so explain the issue clearly.

Key point: sometimes it is appropriate to poke around in the brood nest. But it disrupts the bees - they may be trying to control varroa by keeping the nest warm; they may already be cleaning out cells infected with chalk brood. So always ask yourself:

> *"Why do I want to open the hive? What will it achieve?"*

If the answer is "I have to know if it is poisoning rather than a virus", fair enough. You may be able to help.

If you do find disease, sterilise your gloves, tools, boots, smoker etc.

The following chart lists readily observable external indicators. But many are ambiguous: colonies may not exhibit all symptoms of a disease if it has not progressed far. And they may have secondary infections.

After narrowing down possibilities, check how to distingush causes and what to do next in the next section's alphabetical listing.

Good records help diagnoses ("this only started a week ago").

Symptom chart

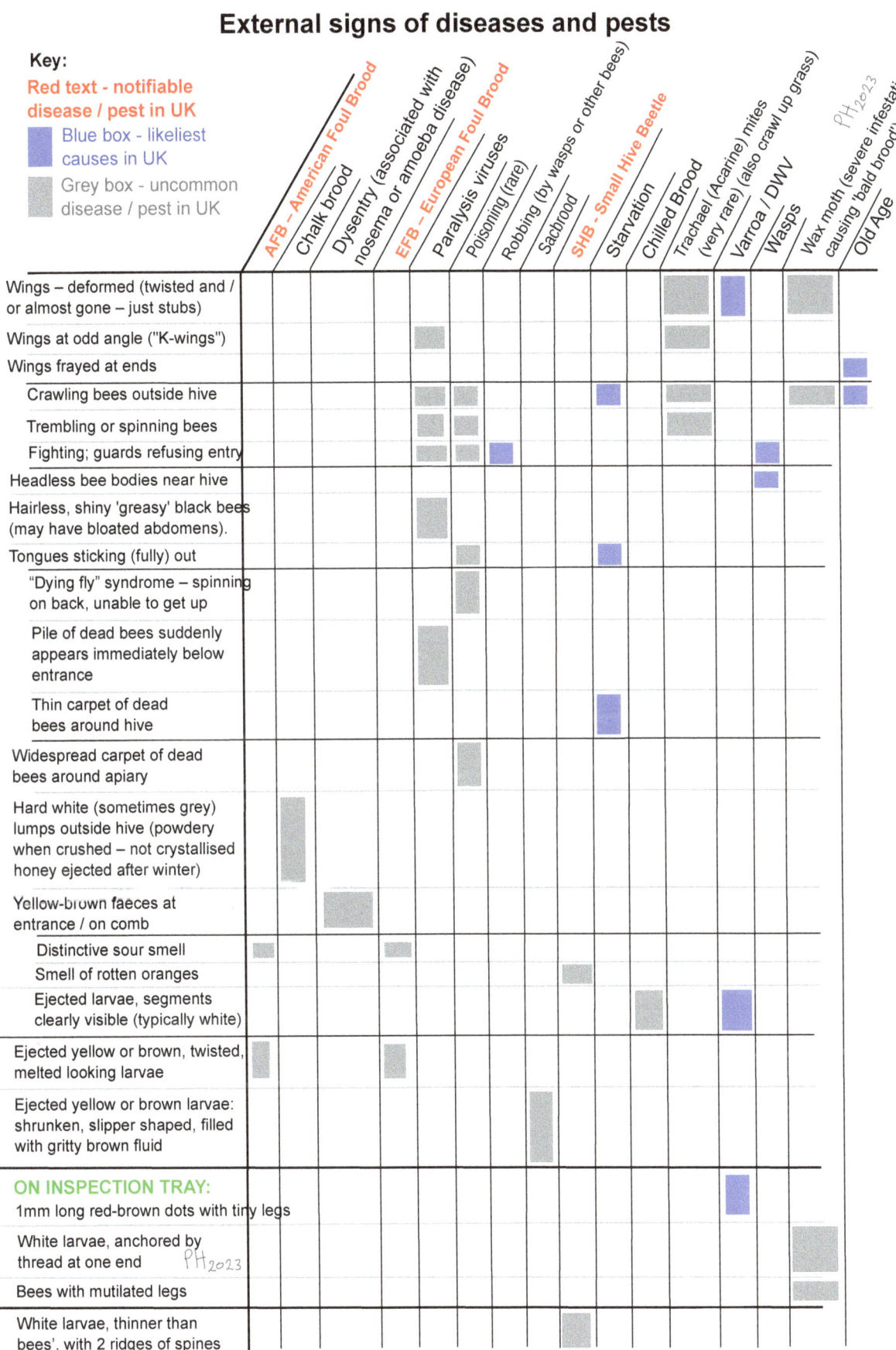

Pest identification

Varroa mites and Braula flies are both reddish-brown and about the same size, though braula tend to be seen *on* a bee's head whereas varroa cling *behind* it (actually, most varroa cling *under* the bee, and aren't visible looking at a frame). Pseudoscorpions are harmless to bees and eat mites.

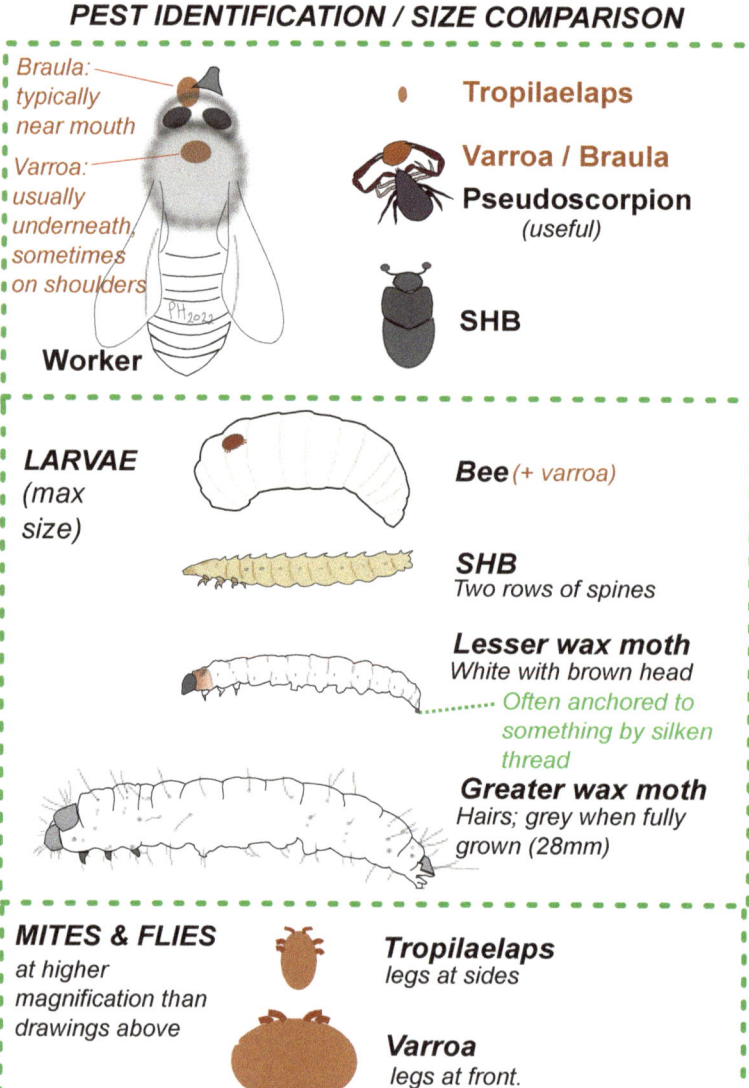

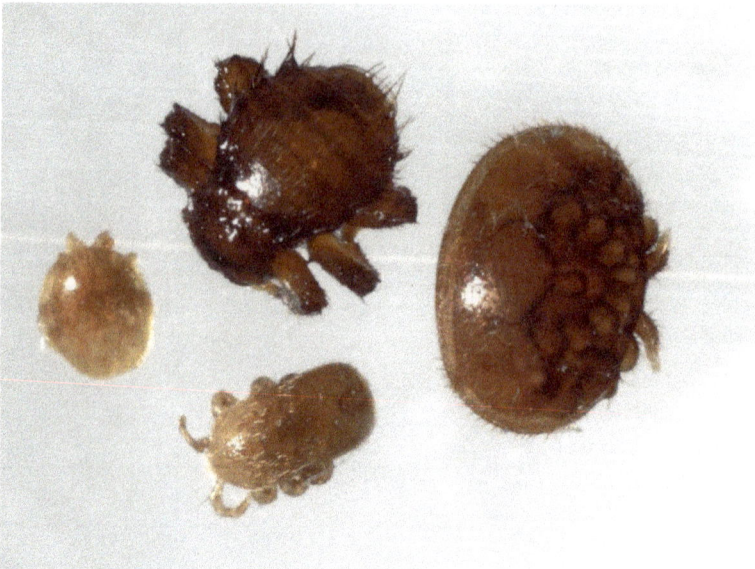

Top: braula fly (harmless). Right: varroa mite. Bottom: tropilaelaps. Left: pollen mite (harmless). Image Crown copyright, source: nationalbeeunit.com

Brood disorder identification

Varoosis, PMS
(PMS = Parasitic Mite Syndrome, very severe varroa infestation)

Sunken, discoloured caps. Perforations, from nurses uncapping and checking suspicious brood

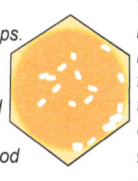

White crystals resembling sugar, inside empty cells and forming a crust round the rim, is mite droppings (frass, guanine). This indicates extreme infestation (PMS).

Hygienic colonies' nurse bees eject white larvae, often with shrivelled wings, typically at purple eye stage.

If out of control (PMS), larvae rot and discolour in the cells faster than nurses can eject them.

Chilled brood

Larvae may die in sealed or unsealed cells, turning shiny grey-black. Sunken caps, sometimes dark. nurses remove rotting larvae.

Chalkbrood

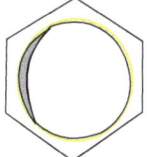  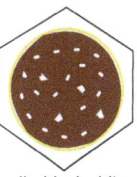 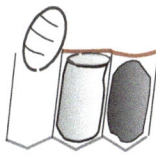

Pupae mummify first into **very hard** cylindrical white pellets, ejected by nurses - if they can keep up. They eventually turn grey-black - the spore forming stage.

Stonebrood
Very rare
Hazardous to humans

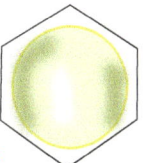 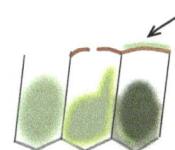

Cappings may be slightly green, and perforated.

Larvae are green-yellow (sometimes black) and mouldy. End up shrunken and hard.

Sacbrood

 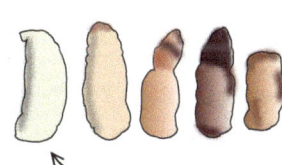

Larvae die in characteristic twisted posture in cell with tail poking out, rather than head, then darken

← Membraneous sac containing fluid

EFB
Notifiable

Foulbrood victims characteristically darken through beiges and magnolias to black

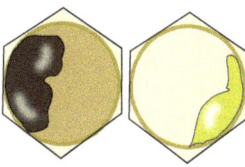  

Black or yellow melted twisted larvae stuck to side of cells

Dark "scale" (rubbery fragments of melted larvae) stuck to walls and back of cells Easily removed.

Dark, sunken cappings, some perforated

(c) Paul Honigmann 2023

AFB
Notifiable

   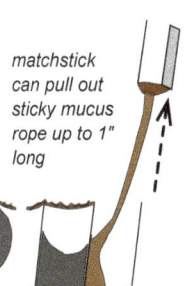

Dark melted larvae on bottom of cells. Can be dull white, yellow, brown

Dark "scale", difficult to remove

Dark, sunken cappings, many perforated

matchstick can pull out sticky mucus rope up to 1" long

Alphabetical listing of pests, diseases and other dangers

Acarine
See *Tracheal mites* and *Isle of Wight Disease*

Addled brood
Archaic, obsolete term for nonspecific brood problems, e.g. misshapen or discoloured.

AFB - American Foul Brood
See: Foul Brood, below

AmFV - Apis mellifera Filamentous Virus

Symptoms

White haemolymph in (squashed) bees

Discussion, treatment

Minimally harmful, but remarkable in two ways. It's huge for a virus - it's visible with a normal light microscope! - and it turns the haemolymph of bees milky white.

Amoeba disease (Amoebiasis)

Rare.

Symptoms

Similar to nosema - swollen abdomen and diarrhoea; quivering wings, inability to fly; faeces at hive entrance and on honeycomb.

Discussion

Caused by *Malpighamoeba mellificae*.

Distinguishable from nosema by microscopic examination of the bee gut. The amoeba is 3-15um wide, and round, whereas nosema is rice-shaped and about 3-6um long.

Spread through faeces - and thus by beekeepers transferring stores or combs between hives.

Affects adult bees in temperate zones but, curiously, not tropical or subtropical areas.

Affects bees mainly in Spring.

Treatment

Reduce other stresses eg damp, cold, pests, nutritional stress.

Colonies often recover in warmer drier weather.

Ants

British ants aren't much of a problem but I assume they steal nectar and honey, and they can certainly caused newly hived swarms to abscond; if I find a nest in a hive roof I brush it out.

Ants are strongly dependent on **scent trails.** If they are coming up hive legs scatter **cinnamon dust** on the ground around the hive to break up their scent trails. Don't put cinnamon *inside* the hive, it smells strongly and will disrupt pheromone signalling.

Don't use ant powder near the hive, it will kill the bees.

They avoid wood ash and you can sprinkle this below a hive if they persist, but be careful handling it - it forms sodium hydroxide (lye) when wet / rained on, so don't handle with bare hands or get the dust in your eyes. If it contacts you, wash with lots of water.

I have heard used coffee grounds or limestone powder round the hive legs are also effective. I found placing hive legs in water fairly ineffective as I kept forgetting to top up the water. Vaseline on hive legs got dusty in days and the ants just walked over it. They can enter through mesh floors. Keep nearby vegetation from touching the hive or the ants use it as a bridge. Bees can use strong fanning (wind pressure) to blow ants out of hives, and Japanese bees have been seen wing-slapping to flick ants out[319].

Ants are a major pest in the tropics: some African beekeepers hang hives from trees due to ants. Americans used to worry about fire ants, until Argentinian ants arrived and they realised what a *real* ant problem was. It doesn't matter how many AA you kill because they form near-infinite supercolonies and they *just keep coming* along those scent trails.

CBPV (Chronic Bee Paralysis Virus) can infect ants, which may therefore be a vector. Deformed Wing Virus infects Fire Ants, which have become established in Europe and raid bee nests.

Asian Hornet

Symptoms

If they are present in your area and you see lots of guard bees at the entrance, it is likely that Asian Hornets have been visiting.

Discussion

A specialist bee predator which was accidentally imported to France and spread rapidly. They have reached the Channel Isles and in 2023, 49 nests were destroyed across England. **This is the biggest threat to British bees we're currently facing,** much bigger than varroa, it is real and imminent and decimating colonies in France and Spain, mainly by disrupting foraging. It is vital that all nests are found and destroyed to prevent them establishing. If seen, photograph them and email the picture to alertnonnative@ceh.ac.uk but don't expect a rapid response, so then contact your local BKA and ask for help.

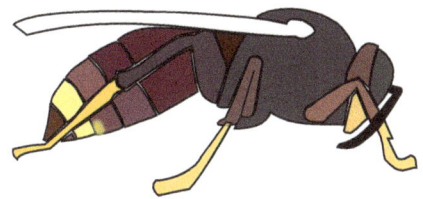

Asian hornet, *Vespa velutina*

European hornet, *Vespa crabro*

Honeybee

Often described as a "mostly black hornet with yellow legs", but like bees, colouring of both European and Asian hornets varies somewhat - descriptions overlap and I suspect some photos are mislabelled. Queens are different again. Behaviour is a better indicator.

Whereas common European hornets (see *Hornets* entry) will just take the occasional bee opportunistically, these ones will hawk (hover) outside a hive entrance, picking off returning foragers (heavily loaded bees can't dodge as easily). A side effect is they terrify the bees which huddle at the entrance or in the hive and refuse to go forage. Asian honeybees enter and leave their nests at speed but ours are easy prey.

---

319   Spangler and Taer (1970), doi.org/10.1155/1970/49131 and Seko, Morii and Sakamoto (2024), doi.org/m8q7

They established in France around 2004, and have been seen entering hives at 4-5°C to prey on dormant bee clusters. After an initial bad year, the local beekeepers found mass trapping of queens in Spring dramatically reduced numbers. One nest can eat 11.3kg of biomass / insects[320] so reducing nest numbers is critical. Crude traps catch some European hornets, but AH start flying 3-4 weeks earlier (March-April vs April-May).

But perhaps the situation is not as desperate as it seems. French natural beekeepers report that the hornets are most active in July and August, and their bees... just stay in and eat honey. Obviously this is a problem if you've harvested it all. Another potential defense is shimmering beards, a coordinated wave of motion across a mat of bees which makes them seem like one big, offputting animal to the hornet. This is common among Asian bees, and has been observed infrequently in wild European bees when European hornets pass the nest, so is probably a latent behaviour.

Warning!

A big AH nest is much nastier to deal with than the relatively chilled out European hornet's[321]. AH nests contain many contain *thousands* of hornets, they chase you, and are much more enthusiastic about stinging. They are tough, difficult to squash. Their stings are long enough to penetrate bee suits. Hornet stings are said to feel like "a red-hot nail hammered into you". And they usually start a secondary nest near to the first.

Treatment

Muzzles - i.e. wire meshes about 30cm from the entrance. The electrified "Lyre" type works well, but is expensive and possibly a hazard to small birds. The AH wingspan is about 40mm so it can't power-dive through a 25-30mm gap to grab a flying bee. For minimal interference to bee flight, you only need the vertical bars so some people use 30mm square mesh and snip out the horizontal bars, and John Woods has found all you need is vertical string bars.

Some French beekeepers trap an AH worker, stun it with a zapper, and put poison on its back – usually fipronil based, which kills the nest (probably illegal). **Don't do this!** Sometimes the hornets continue hunting, and if they touch bees at the entrance of a hive, that hive is dead.

French companies jadeprode.fr and gardapis. eu have developed innovative traps that only catch Asian Hornets, minimising bycatch.

Guard bees ball an Asian Hornet that got too close. These are European honeybees, not *Apis cerana*. Photo © Pelayo Martinez Beiro, Spain, 2020.

---

320 *Not just honeybees: predatory habits of Vespa velutina in France* (2021) Rome, Perrard, et al doi.org/10.1080/00379271.2020.1867005

321 Key word: **relatively**. The phrase "hornet's nest" has always neant "highly hazardous situaution". European hornet nests 'only' contain hundreds of hornets.

John Woods' simple but effective muzzle (cage) which encloses the entrance, formed from string stapled to wooden slats.
Bees learn to fly through the "bars" freely, but hornets hit string as they swoop on bees and cannot get them. The bees ignore the hornets and foraging is not disrupted.

An idea of mine - attach a mirror of mylar (weatherproof reflective plastic, available in rolls) to disrupt attack swoops.

As the UK is cooler than France it's hoped we won't see the really big nests. And as the UK is more densely populated, more nests should be spotted. More queens will die over winter. Their narrow gene pool (all from one queen) may hamper them. However, they will be a big problem and there is no magic cure. To reduce hives' 'visibility' (scent plume) to this pest, only open hives rarely.

Given a choice of one or two hives or a large apiary nearby, these hornets preferentially predate the larger target - one of the rare cases where a large commercial operation near you helps.

Whilst hawking outside a hive, they are said to be oblivious to humans nearby killing other hornets. Badminton rackets followed by a boot heel are the preferred weapon.

Although a vertical slit entrance is said to work against European hornets, it's irrelevant for AH. Asian Hornets prey mainly on slow, returning foragers in the air but sometimes warily approach the landing board, attempting to knock a bee off. If they misjudge this the honeybees may ball and kill them by a combination of cooking them, suffocation, and pumping bee venom in faster than they can remove the stings; 3-4 bee stings in the thorax will kill them in time[322]. This is a common response of Asian honeybees, Apis cerana; European honeybees don't yet display this behaviour often but it is a latent response in some colonies. They do after all already ball wasps; a colony was seen balling a European hornet in Germany in 2019, and the photo here shows the balling response to a Spanish Asian Hornet.

The press sometimes confuses Asian Hornets and others with similar names which are even larger and scarier, but haven't reached Europe yet.

Remarkably, the European honey buzzard, a specialist wasp and hornet predator, is beginning to prey on the huge Asian Hornet nests! It's very rare in Britain but it's hoped it will help in France. Unlike bee-eaters, it's not just picking off the odd hornet: it's breaking into their nests and ripping out chunks of brood comb to feed its young[323].

---

322 More info: *Lethality of Honey Bee Stings to Heavily Armored Hornets*, Gaoying Gu, Yichuan Meng, Ken Tan, Shihao Dong and James C. Nieh, Biology 2021, 10, 484. https://doi.org/10.3390/biolog

323 *Exploitation of the invasive Asian Hornet Vespa velutina by the European Honey Buzzard Pernis apivorus*, Macia, Menchetti, Corbella, Grajera & Vila (2019) in *Bird Study*

There's at least one enormous dragonfly which eats hornets[324]. Two French parasitoids, a conopid fly and a nematode, now attack AH but had little impact. Spanish bumblebees have an unusual "Stop, Drop and Roll" defense against AH[325], but they're much bigger than honeybees.

Further reading: John Woods has compiled more information here: https://groups.io/g/asianhornet-beehiveprotection

Badgers

Badgers ignore hives until a harsh winter freezes the ground so hard they cannot access earthworms. Once they discover bee larvae they are addicted - hive raiding is a learned behaviour. They also dig up wasp nests and eat their contents. They just ignore bee and wasp stings: they close their eyes, and the insects can't sting through their skin.

Badgers can literally smash their way into most hives; they generally topple a hive to get in it, so strapping / weighting them down is useful in badger country. Sometimes a really strong fence, often electrified is needed - but it needs to extend below ground as they're burrowers.

TBHs are intrinsically stable on their splayed legs, and seem badger-resistant.

The African honey badger or ratel - actually related to wolverines - is on another level. It has been known to fight off young lions, and enthusiastically devours African honey bee nests, commercial hives etc.

Bald Brood

See also: Wax Moth

If the bees are doing this deliberately, it's not a problem. If wax moth are doing it, it **is** a problem:

As wax moth larvae tunnel through combs, they typically create a **line** of uncapped cells, or irregular patterns, with a trail of characteristic webbing. If this uncaps sealed brood (pupae), they die. If the webbing covers unsealed larvae, those die too - either trapped in the cell, or if they do manage to emerge as adults, they are dessicated, and damaged by the webbing, with shrivelled wings or malformed legs.

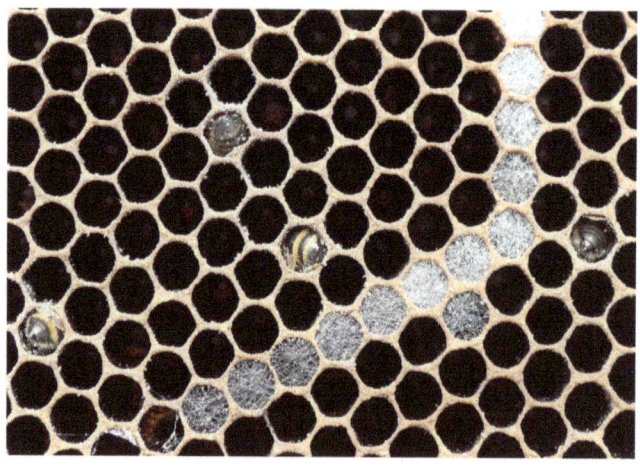

| Wax moth trail in a deadout (and some starved bees)

But if the nurse bees simply remove caps that isn't a problem. They may do this if the bee space between combs is too small - workers uncap cells to get by - and it can also be due to a genetic flaw in some queens. Or the bees may be *temporarily* uncapping to check pupae.

Grindrod & Martin at Salford University make the point that different groups of bees uncap and recap, so seeing **patches** of bald brood is good! It indicates these may be recapped shortly, don't

---

324 www.youtube.com/watch?v=EosniBoMuDY - I suspect the video shows an American dragonfly.
325 Quantifying the impact of an invasive hornet on Bombus Terrestris colonies, O'Shea-Whelley et al, Communications Biology 2023,

requeen! Bald brood from wax moth occurs in lines, or irregular patterns. (See images on their website varroaresistant.uk)

Bears

Although the last wild bear in Britain was hunted ~1000AD, they are a significant problem in some countries. Like badgers, they ignore stings and don't find a hive wall a significant barrier.

The usual defence is an electrified fence at maximum voltage. It needs wires on the *outside* of the posts, good grounding to a buried chicken wire mat; cut the grass below the fence, and moisten the earth in droughts. The fence must extend below ground as bears burrow.

The hives need to be far from the fence so the bear can't hook them over, and there must be no overhanging climbable trees.

The good news is that as bears are territorial, once you've trained the local bear this fence stings (perhaps with bacon grease or peanut butter in muslin on the wires) it will keep others away.

Beekeepers

Try this thought experiment:

Consider beekeepers as a parasite.

Badly adapted parasites stress their hosts to death.

I'll leave you to think on that.

BQCV - Black Queen Cell Virus

Symptoms

Queen pupae turn a "distinctive" yellow, then literally black, then die. Eventually the body is enclosed by a membrane, like sacbrood. Curiously, workers don't display this symptom, though their lifespan is greatly reduced.

Discussion, treatment

Spread rapidly from Australia and S Africa to almost every country now except Mongolia and Uganda. **Not observed in colonies where bees are allowed to select and raise queens themselves**[326](!), and tends to be a co-infection with nosema.

Polish researchers report it is very common and present in basically every apiary which has nosema (which seems to transmit it), and "is by far the most common cause of death in queen rearing apiaries". Bear in mind that such apiaries use "breeder nucs" to raise 20 or more queens simultaneously, and continuously, unlike a colony allowed to swarm or supersede naturally.

You generally need to rip open a queen cell to see this, so it's only checked for if there is no other obvious reason for a colony to dwindle.

Treatment

Requeen from healthy stock, remove potentially infected comb, reduce other stresses, i.e. no known cure. Bees could theoretically self-medicate with plant extracts to reduce viral load[327].

---

326 www.beeculture.com/bee-keepers-guide-to-honey-bee-viruses
327 Bay laurel (*Laurus nobilis*) as potential antiviral treatment in naturally BQCV infected honeybees, Aurori et alia, 2016

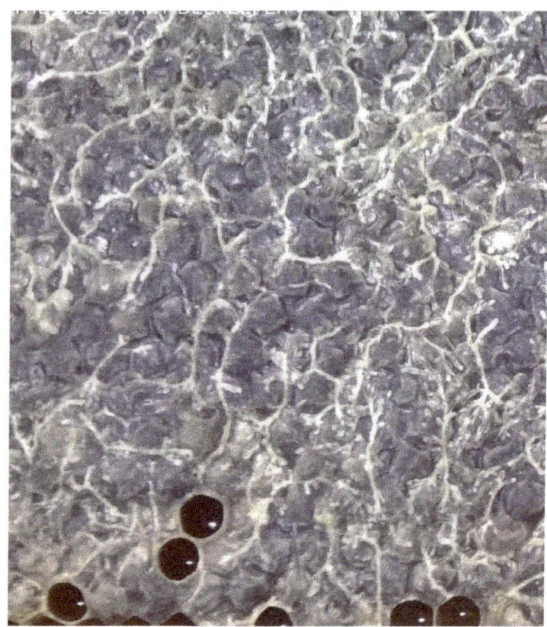

Honeycomb with characteristic Braula larvae tunnels below wax caps. Photo courtesy of Antipodes.

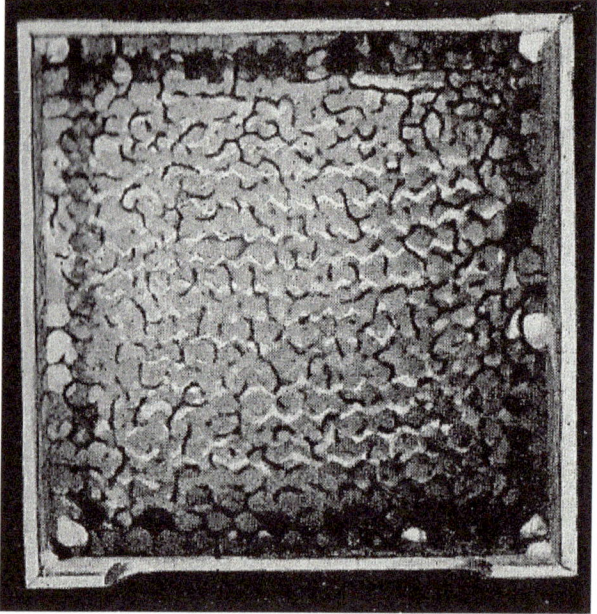

Don't confuse braula signs with these ripples **above** the surface, i.e. the normal reinforcing of honey cappings after a few weeks.

Braula flies

Symptoms

- About 1.6mm long and reddish brown, Braula look like varroa mites at first glance, but they have 6 long legs rather than 8 tiny stubs, and cluster round the bee's head, where they attach by gripping hairs. Braula are more spherical than varroa and look much lumpier on a sticky board.

- Easiest to spot on the queen. If the hive has braula there will be several on her head.

- Whilst it does not harm the bees it damages the *appearance* of comb honey because its larvae tunnel just under the cappings.

Braula on backs of bees in a badly infested hive. Photo courtesy of Antipodes.

Discussion

The bee louse, *Braula coeca*, is a tiny flightless fly that eats a mix of honey and saliva from bees' mouths - they can stimulate the bee to regurgitate food. That's why they cluster on the queen - she's fed frequently - I've seen a report of 20+ Braula being driven off a queen with tobacco smoke (Devon, 1980s). It hasn't been seen in Britain since we began treating with varroacides in the 1990s, so the photos are from 'Antipodes' (I don't know his real name) who lives in Tasmania, where they don't have varroa yet. Braula are widespread in Africa.

Thomas Wildman, in *Management of Bees* (1770), talks of a pin head sized red insect parasite which was a minor pest, though its numbers could build up in old hives. On p.238 he mentions red lice, they can jump 2" and badly infested hives can have 2-3 on each bee. I suspect these were *braula* flies.

CCD - Colony Collapse Disorder

Symptoms

Originally dubbed *disappearing disease:* workers suddenly abandon colony, leaving queen(!), significant honey and pollen, capped brood and a few nurse bees.

Tellingly, the remaining stores are not immediately eaten by other animals (wasps, SHB etc), implying contamination. Don't feed the left-over stores to other bees. Those colonies will not build up and eventually die.

Discussion

Appears to be multiple stresses - no single cause - and suddenly the bees cannot take any more.

Queens *can* fly, if they are slimmed down for a few days - so whatever the trigger is, it is sudden and extreme.

Associated with intensive migratory apiculture and pesticide use. Rare in UK. Europe had one bad year, USA repeatedly suffered from 2006-2012, triggering media panic about a worldwide "Beepocalypse". In the US, commercial beekeepers could claim compensation from the ELAP scheme, but only if the losses were from CCD; but surely they wouldn't misreport losses from other causes.

Self Professed Experts claimed it was due to one key factor - *Nosema ceranae* arrived in the US around 2009; others think it was Israeli acute paralysis virus (IAPV); many suspect miticides and neonics contributed. *No one knows.* The most convincing single-factor explanation I've see is the beekeepers at uWatch.co.uk who got fed up with high colony losses in hives next to frequently-sprayed fields, and had access to laboratory testing facilities at CEH. This showed that colonies which absconded in winter left behind stores with massive levels of glyphosphate. "We lost 31 out of 44 colonies to glyphosate winter before last and some samples of stores tested up to 540 x the legal limit."[328]

Michael Bush theorises that dousing the hive with antibiotics and miticides or even essential oils like thymol kills gut microbiota; suddenly the pollen is indigestible and there is malnutrition in the midst of abundant food. Similar to how antibiotics upset human stomachs[329].

Undiagnosed CCD?

It can be extraordinarily difficult getting a straight story from someone on e.g. a forum. Once they have an "answer" from an "expert" they often stop looking further, resulting in incomplete, ambiguous reports - exactly like the confusion surrounding Isle of Wight Disease. So here's some direct observation from a true expert:

Filipe Salbany tells of giving a Treatment Free queen of good lineage to some beek, who then had 11 'swarms' from that hive. Filipe investigated and found:

- They were the same colony repeatedly trying to abscond. The beek kept putting them back in the hive.
- The beek was treating them with oxalic acid 3x a month (why?!) at 3x the recc dose (!!). Filipe had stressed it was a TF lineage!
- The bees wouldn't go on the combs he'd treated.

---

328 Private communication, 14.6.2021 - lab results and discussion are in the *Thames Valley Pesticide Awareness Initiaitve Year 2* report - www.bee.watch

329 See NHS.uk, "side effects of antibiotics"

This made me reassess CCD - maybe abscondings are more prevalent than I thought (after all African bees do it regularly) and CCD really is Bad Beekeeping Syndrome, as some say. However, in CCD they leave the queen behind. Maybe simply because her wings are clipped?

Chalk Brood

Common, not usually serious

Symptoms

- Very hard lumps, white or grey, on floor / in front of hive. These are shrunken mummified brood. They resemble dried woodlice. *Note:* do not confuse with honeys like ivy which can set as hard white lumps of sugar. These tend to be shiny and would be present in honeycomb and dissolve in water. *Note:* the lumps are soft if wet, i.e. after rain.
- In brood comb, larvae covered with white fungus.

Discussion

A stressor but not usually the cause of colony death.

This fungus, *Ascosphaera apis,* tends to be symptomatic of a hive under stress, and brood being damp & chilled.

It thrives best slightly below brood temperature which is why it is more common -

- In poorly insulated hives
- When brood are suffering from Chilled Brood (see below)
- When hives are opened on cold days

Large colonies will raise the brood temperature to suppress it - a fever. Small colonies try, but lack the numbers[330].

Chalk brood mummies are the size of bee larvae and rock hard when dry. Their insides have a powdery appearance.

Brood on the outside of the nest are most susceptible. Larvae are initially covered by a white, fluffy fungus. The bees will control it if the colony is otherwise healthy (an example of hygienic behaviour, giving a pepperpot brood pattern; some gene lines are more susceptible than others). Larvae dry up, shrink and form hard pellets which the nurse bees eject; white ones are noticeable but later stage **dark grey ones spread the spores.** Spores can survive 15 years in wax etc. Although easily transmitted, don't worry about seeing just a couple of the hard white pellets, it is not usually a serious disease – like a cold to us. I have only twice seen a serious case in a low-intervention hive. (Shook swarming eliminated the chalk brood, but they didn't rebuild up enough to survive winter. I should have given them a smaller, better insulated hive and moved it out of deep shade.)

People sometimes confuse mouldy pollen with chalkbrood, but pollen crumbles easily between the fingers.

---

330 *Social benefits require a community: the influence of colony size on behavioral immunity in honey bees,* Bonoan et al (2020) doi.org/10.1007/s13592-020-00754-5

In severe cases the bees start to tear down the comb.

Chalk brood mummies (and lesser WM larva) among debris under a comb. Photo © 2022 Will Hanrott.

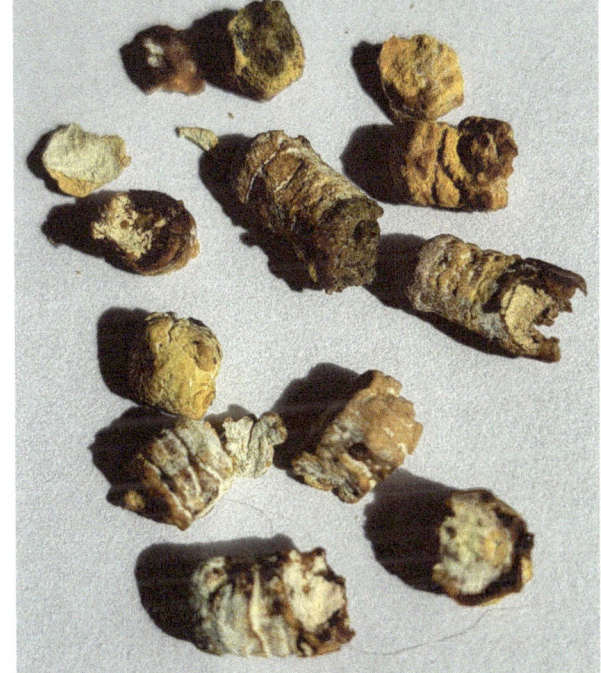

Chalk Brood is more common in cooler regions[331].

The key to chalkbrood resistance is hygienic bees, ejecting infected larvae. Reducing damp with sunshine, ventilation etc will help, but if your bees aren't hygienic it will persist.

See also: Stone Brood - much rarer fungal infection, also minor, sometimes a different colour.

---

331 *Identifying the climatic drivers of honey bee disease in England and Wales*, Rowland, Rushton, Shirley, Brown & Budge, Scientific Reports (2021). Also, in *Honeybees of the British Isles*, p.27, Beowulf Cooper mentions that *Amm* permit brood temperature to dip on cold nights. This correlates to continuous low levels of chalk brood, and he posits that this has been selected for as an acceptable collateral loss for low fuel use.

Chilled Brood

Due to weather / lack of insulation; not a disease

Symptoms

- Ejected larvae in front of hive - usually drones - after a sudden cold night.
- Larvae in combs may turn black, though I've only seen white ones ejected for this reason.
- Wing stubs[332]

Discussion, treatment

Chilled brood refers to larvae which are not warm enough during their development. This is at the edges of the combs, or nearest the entrance.

If they survive chilling, they do not live as long as they should - and just a few days knocked off a flying lifespan of 6 weeks can severely weaken a stretched colony. Their learning (memory) is also impaired.

Causes

- Stimulating (feeding) the colony to lay too many eggs in Spring. On a frosty night, the cluster contracts and cannot cover all the larvae with nurse bees.
- Splitting a hive, leaving insufficient adult bees to keep all the young warm.
- Checkerboarding.
- Overheating (e.g. Africa)[333].
- Pesticides (too few adults left to cover brood).

One solution is warmer hives.

'Checkerboarding', a swarm suppression technique almost guaranteed to chill brood in the British climate.

CWV - Cloudy Wing Virus

Wings look milky / cloudy

Discussion, treatment

Very little research. No treatment available. Possibly the new extract from fungi.com would help? Some strains of bees are said to be more resistant, and propolis is said to help.

Said to be common in colonies collapsing with a heavy varroa load.

---

332 Storch claims this can be due to chilled brood or poor nutrition - perhaps such wing stubs would not look ragged thus could be distinguished from DWV?

333 UN Food and Agriculture Organisation, teca.apps.fao.org/en/technologies/7332

Death's Head Hawk Moth

Symptoms

An enormous, 3 to 5 inch long, screeching moth in your hive.

Bees build propolis entrance barriers to block it, allowing just 1 or 2 bees through at once, but these are removed when not required as they impede traffic.

Discussion, treatment

There are three species across Africa & Europe. Their caterpillars live on plants not native to Britain but the adults eat honey - and emit a shrill cry when threatened by bees, mimicking a queen's piping, which seems to scare the bees away!

Very rare in Britain. Don't steal significant honey. Ignore / admire.

Famously featured in the film *Silence of the Lambs*.

Colouring varies somewhat but the skull marking and sheer size is striking.

DWV - Deformed Wing Virus

See also: Varroa

Symptoms

- Crumpled, twisted, and ragged wings, sometimes just stubs where wings should be, on crawling bees;
- Ejected larvae with ragged wings outside hive (hygienic behaviour by the nurse bees);
- Dwarf bees (shortened abdomens);
- Many dead mites on the bottom board (~1mm long, magnifying glass required, can resemble translucent white or rust-red coloured pollen blobs; you can put sticky – vaselined – paper on it for 3 days to get an average mite drop / day figure, but it is tedious). Not all hives have slide-out inspectable bottom boards.

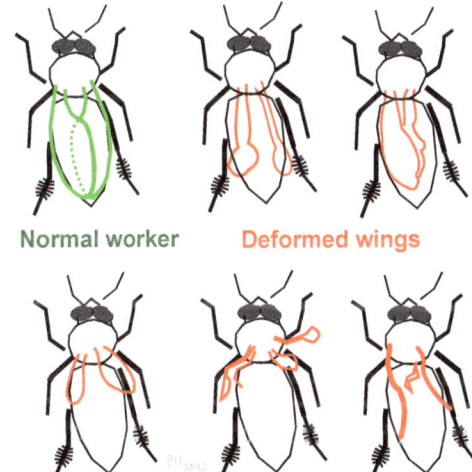

A worker's wings normally fold neatly over most of their back, with the smaller pair hidden below.

Discussion, treatment

This virus is normally present asymptomatically at very low levels in most bees. Varroa mites bite into a body part where DWV replicates and act as a vector, and amplify the levels hugely[334], and this causes distorted wings. Such bees can't fly and are a drain on the colony.

Damaged wings aren't always DWV. They can also result from age, hot smoke (melts wings!), or bees fighting. Guards bite robbers' wings, and workers bite drones' wings when expelling

---

334   Up to 1,000,000 times

them from hives at the end of mating season. However, this snips sections off - they don't twist and distort wings. Tickner Edwardes observed[335] that the last of the winter bees, old bald and tattered, dodder about slowly at an entrance in Spring while new workers zoom by – they can no longer fly, so guard.

DWV comes in at least 4 strains: A, B, C and D. When the first variant, DWV-B was dentified, in Ron Hoskins' varroa-resistant bees in Swindon - where he had not used miticides for 19 years, perhaps that is significant - type B was initially thought to be non-symptomatic (it isn't) and crowd out (block) type A[336] - but that's been disproved.

A 2019 study[337] found DWV-A was prevalent in America, DWV-B in England and Wales. But by 2022 the situation was changing - the more virulent DWV-B is rapidly becoming dominant worldwide, probably because DWV type B can also replicate in the varroa mite whilst DWV-A cannot.

Similar viruses recently reclassified as forms of DWV are found in Egypt, China, and Japan.

Recent American research[338] found that bees exhibiting Varroa Sensitive Hygiene cannibalised infected larvae, so the virus continued to propagate in the hive. I conclude their bees' hygienic behaviour is different to mine, who eject infested larvae. I don't see deformed wings in my bees.

Suggested action: none! (Natural selection.)

DWV can also cause body abnormalities. This bee has multiple deformities, most likely due to DWV (though there are other possible causes like chilled brood). Fur on the end of the body shows it has not been cut in two: it never had an abdomen. It's stuck down by some honey I spilled.

Dysentery

Often associated with nosema

<u>Symptoms</u>

- Brown spots and dribbles (faeces) at entrance *(healthy bee poo is yellow)*
- Spotting of faeces on combs in hive. If it has progressed this far, the hive may smell[339].
- In extreme cases, forms a thick crust which resembles proplis as it dries, but crumbly and not as shiny.

---

335 *The Bee Master of Warrilow* (1921, long before varroa) 3rd Ed. p.45-46

336 Wrong! Bee virus research is moving on rapidly and even while I was writing this book this result was shown to be untrue. See www.theapiarist.org/superinfection-exclusion/ written in 2021 by Prof David Evans, a virologist (utterly dismissive of natural varroa resistance, but an expert in his own field). He did find that superinfection exclusion operates to block *very similar* viruses but not ones as different as types A and B. People were very excited at the time as if it had been true, it would have helped explain the resilience of Ron's colonies in the presence of varroa. Search for "superinfection exclusion DWV" for further info on this.

337 https://doi.org/10.3390/v11050426 – multi-year study by some Big Names in bee science. Old samples show DWV-A was dominant in the UK in 2007 implying some kind of selection pressure for B.

338 Doi.org/10.1038/s41598-021-88649-y

339 Sources are inconsistent about what it smells like: "bad", "fresh bread", "fish"...

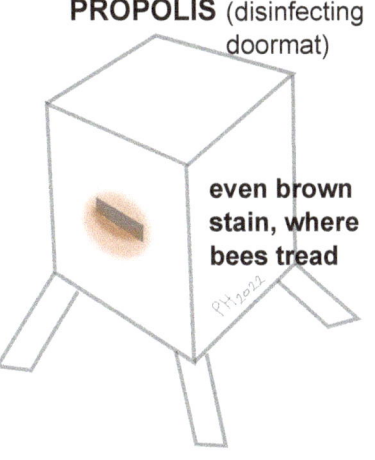

Discussion, treatment

It's actually diarrhoea: in humans, dysentery means blood among the faeces, but someone used the wrong term and it stuck.

Can be caused by:

- lengthy confinement from bad weather so bees cannot hold their bowels in and void inside the hive; especially if fed weak syrup (excess water accumulates in gut)[340];
- food fermenting[341];
- nosema infection (winter: *N. apis*; *N. ceranae* is active in summer but doesn't cause dysentery)
- (If it is a cold day in Spring, there may be nothing wrong. Nurse bees, raising brood eat a lot and may be emptying their bowels of normal **yellow** poo near the entrance on days when they cannot fly.)

Observations:

- Dysentery seems **strongly** associated with *colonies fed syrup*, which generally has a very different pH to honey.
- I've never seen or heard of a wild, unmanaged colony exhibiting dysentery.
- I've heard just a couple of reports (in 10+ years) of comb spotted with bee faeces among my low-intevention beekeeping circle, after many weeks of non flyable weather.

Racial differences

Cold-adapted races can "hold it in" much longer than Italian bees, and are more tolerant of the buildup of remnant solids ("ash") in dark honeys and honeydew. American beekeepers tend to use Italians... so replace honey with syrup over winter to avoid this problem, but as mentioned above this seems to have its own dysentery related issues!

---

340   scientificbeekeeping.com/the-causes-of-dysentery-in-honey-bees-part-2 - Randy Oliver summarising Erwin Alfonsus' 1935 paper, *The Cause of Dysentery in Honeybees*. Alfonsus found no correlation to solids (ash) in the diet.

341   This *may* be syrup fermenting, or *possibly* be due to honey crystallising. The crystallised honey theory is: the crystals express water and as the fluid between them exceeds 21% water, yeasts etc re-activate and ferment the honey. You will come across contradictory information on this, because bees can eat solid honey if they add water to it and eat it immediately. But they can't eat rystallised honey safely after it's separated into crystals-and-fluid and fermented.

So basically, if your hive smells of alcohol, expect your bees to get dysentery; if they're simply munching rock-hard crystallised stores, they should be fine. It probably varies with source - OSR is said to be safe, but ivy can definitely eventually ferment.

Thus, in 1870 Bevan agreed with Madame Vicat's belief that it was caused by feeding on honey that had become candied (crystallised) [*The Honeybee, its Natural History, Physiology and Management*, 1870, p.116.] Bevan and Vicat were Major Influencers of the time. They may not have known about the nosema microbe, but even back then they could see the root correlation, which appears to have been obscured by Much Ado About Nosema.

Amm don't have issues with dysentery even on heather. Writing in 1875 in Scotland, Pettigrew writes of taking his skeps to the heather and states dysentery is rare; he noticed a correlation with damp, blaming the early wooden hives. (He reckoned it could be cured by feeding with boiled sugar syrup, but as discussed under "feeding" in chapter 13, we now know boiling would generate high levels of HMF, which isn't great for bees.)

| A healthy propolis bloom at an entrance.

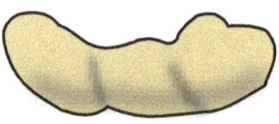 

| Healthy faeces is yellow, wrapped and held together by its peritrophic membrane, unless it falls a long way and splats.

| Bee faeces on car door

Another *Amm* example: Herbert Mace, writing in 1931, devotes two chapters of *Bee Matters and Beemasters* to D M MacDonald, a prominent Scottish beekeeper, mentioning p.122-4 MacDonald was very opposed to non native (yellow) bees, and he wintered his bees for 25 years on heather honey[342].

Modern beekeepers - using, most likely, other races - seem to mention more dysentery issues on heather, and some feed their bees syrup so the bees don't eat the heather honey, which upsets their tummies. Note the parallel to Pettigrew feeding syrup, but he didn't have to do it routinely.

Further reading: Search for *honeybee dysentery scientificbeekeeping*. Randy Oliver has untangled the contradictions and has a comprehensive, if complex, explanation and debunking of some myths.

Earwigs

Harmless to bees, but occasionally present in warm hive roofs.

EFB - European Foul Brood

See: Foul Brood, below

---

342   MacDonald's bees were seriously affected by Isle of Wight Disease, showing even local bees raised by experts aren't resistant to everything.

Exotics

Flies, exotic

For completeness, I should mention there are a number of flies which parasitise or predate on bees, but they are minor pests:

- Tachinid fly (Africa) - lays eggs on bee abdomen, larva burrows in
- *Senotaina tricuspis* - central and southern Europe & Ukraine. Lurks on hive roofs, ambushes bees and deposits flesh-eating larvae on ther backs which burrow in. The females can produce 700 larvae and it can be a serious pest. Not yet seen in the UK but climate change could allow it to live here.
- Bee conopid (wasp-like appearance; Africa) - lays eggs **in** bee abdomens
- Braula (widespread) - see separate entry
- Robber (Assassin) flies - thousands of species. Large, with a loud buzz, some even predate hornets. Attack flying bees, use venomous bite to dissolve insides, suck this out. A seasonal pest in some parts of the world.
- Dragonflies, for apiaries near freshwater
- Probably many others

Mammals, exotic

Bats have been seen flying above hives, if bees fly late enough.

Mites, exotic

See also: Tracheal mites, Tropilaelaps, Varroa

There are several obscure mites which aren't often mentioned, because they rarely cause a problem for bees and can be ignored. For example, various tiny **pollen mites** are often seen hitch-hiking on bumblebees, and occasionally on honeybee hive floors.

> Don't leap to conclusions: These initially seem to be braula as they are the right size and colour and clustered on the head, but the proportions are wrong - the legs are on the long sides. Adam Smith found these on bees crawling under his hive. They may be soil mites, Macrocheles muscaedomesticae; or Poecilochirus mites, which parasitise flies; but they're not a danger to the hive. Photos © Adam Smith, 2023

The **braula fly,** which has its own entry here, resembles a mite.

There's an online mite identification tool at https://idtools.org/bee_mite

Nematodes

Discussion

There are parasites like roundworms which affect other insects, typically by ambushing them and jumping from the ground, entering through an orifice and then eating them from inside. Surprisingly, they aren't mentioned in lists of bee pests, though researchers have shown they *can* infest bees by e.g. spraying bee larvae with tiny young flatworms. It's not clear what the symptoms would be[343]... but they seem worth considering as a possibility if nothing else seems to fit.

Wasps, exotic

Unusual wasps

There are thousands of species of wasp, many very specialist. One type of note to beekeepers is the **Bee Wolf,** fortunately solitary, which is a specialist honeybee predator and a minor pest in some areas. There are about 135 species worldwide and the main European one, *Philanthus triangulum* used to be rare in Britain but has dramatically expanded its range. They need sandy soil for their nest burrows.

Mutilla europaea (rare, obscure wasp)

Symptoms:

Healthy young bees lie dead in front of entrance. Colony otherwise healthy. Pile of dead bees below brood nest (inside hive), some still moving weakly.

Discussion:

A scent-disguising parasitic wasp, found in woodland - mainly parasitises bumblebees but, rarely, found in honeybee hives. The female wasp lays eggs in bee larvae, and the adult wasps eat the honey stores. Google for images and check through all the combs. The adult's size is limited by size of host cocoon, so in honeybee nests they will be smaller than honeybees.

Foulbrood - AFB, EFB

Notifiable diseases - you MUST -

- *contact a Bee Inspector*
- *self impose a lockdown (don't move hives, don't sell honey)*
- *sterilise hive tools, gloves, boots, smoker, suit etc harshly*
- *not move frames etc between hives*

The same but different: key points

- There are two types, **AFB** (American) and **EFB** (European).
- Their symptoms are broadly similar but the microorganisms behind them different, so their spreading mechanisms are different.

---

343   Googling *Steinernema carpocapsae*, widespread and used in biocontrol, I suspect you'd see discolouration and on dissecting the larva or insect, it would be full of tiny flatworms, like wriggly threads.

- EFB comes in many strains: speed of infection and lethality varies. This may lead to confusion with other disorders, such as chilled brood.

- Both foul broods are extremely infectious so it is vital to alert bee inspectors if you think you have either, so they can check nearby apiaries.

- AFB - need to destroy colony;

- EFB - colonies can recover eg via shook swarm, but the NBU reccomends destroying small / heavily infected ones.

- AFB has a rapid onset (kills larvae within 3 to 12 days depending on ERIC strain). Bee inspectors actually prefer to find AFB as it doesn't "hide" asymptomatically - unless you treat with OxyTetraCycline!

- Rotting, smelly bodies can arise in many ways, so the introduction of cheap, accurate Lateral Flow Devices for both has helped a lot.

See also: Neglected Drone Brood - looks similar until you spot that there are way too many drones in those hives.

Other symptoms

- These vary depending on how bad the infection is. If it's bad there will be rotting larvae at all stages, capped and uncapped, in many colours, for both types of foulbrood. It is rare that all the cells in a comb are affected.

- Pinhead sized dark brown spheres around the landing board[344] (scale).

- In hygienic colonies the brood are removed so all you see in the hive is a "pepperpot" spotty brood pattern ("failing queen"). *Note:* 1 in 20 empty cells is **healthy:** fewer holes means the queen is short of laying space.

- Ejected brood outside entrance - yellow, brown, or even melted-looking / unsegmented depending on how far the disease has progressed.

Characteristic sunken cappings, scattered brood pattern, discoloured larvae. Except - the bee inspector determined it was actually "chilled brood", rotted after a cold snap drove an overextended colony into a tighter cluster. Don't assume! Photo (c) Mike King 2022

European Foul Brood. Healthy looking larvae right next to melted looking and rotted ones; pepperpot laying pattern. (Also at least 3 varroa.)

---

344   Storch, *At the Hive Entrance*, p.51

Differences between AFB and EFB

AFB, *Paenibacillus larvae* kills by directly attacking the tissue of the larvae, creating sticky goo. Its spores can survive at least 75 years in honey and on woodwork, and survive boiling!

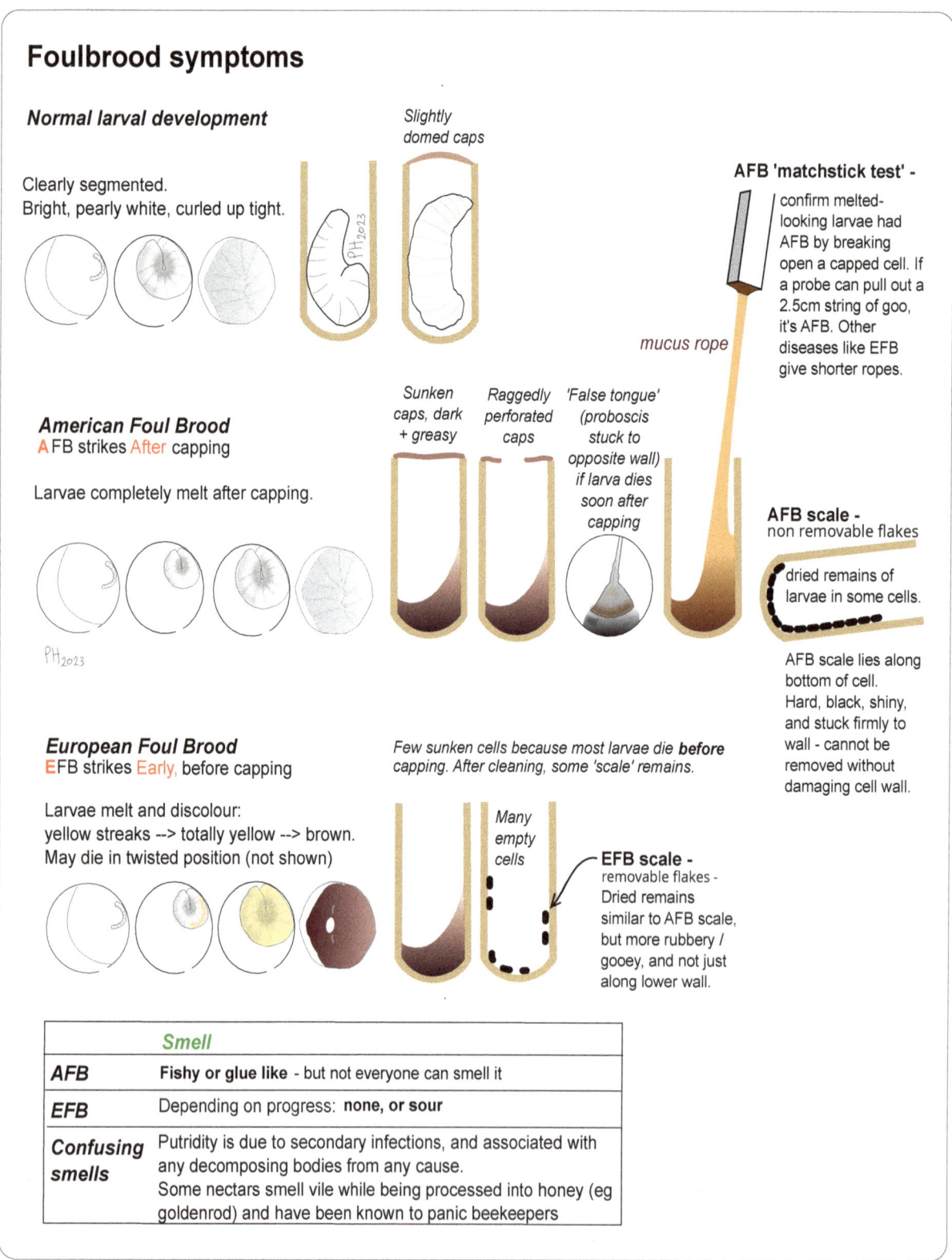

	Smell
AFB	Fishy or glue like - but not everyone can smell it
EFB	Depending on progress: **none, or sour**
Confusing smells	Putridity is due to secondary infections, and associated with any decomposing bodies from any cause. Some nectars smell vile while being processed into honey (eg goldenrod) and have been known to panic beekeepers

| Absence of "false tongue" does not mean abscence of AFB, it is rarely seen.

EFB, *Melissococcus plutonius (Mp)* grows in the gut and kills the larva indirectly by starvation, the larva then rots. It can survive in honey well over a year, so don't transfer honey frames between hives.

Definitive diagnosis

Many of the visual clues in the above diagrams overlap with other diseases; EFB has no consistent presentation, because discolouring, decay and smell depend on secondary infections[345]. Even (well developed) Chalk Brood, a relatively mild pathogen,

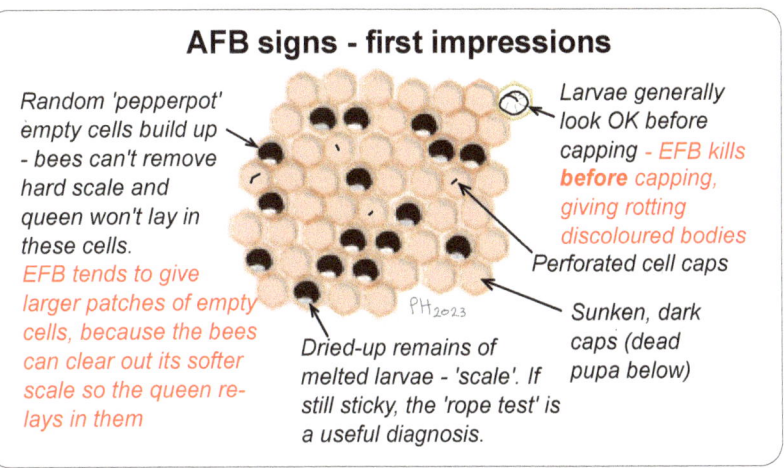

Heavy EFB infestations can also exhibit a few sunken and perforated cappings. The perforations look 'different in size and location' but my source couldn't describe exactly how

can resemble EFB's melted larvae. Brood cells with irregular, torn cappings (bees pulling bodies out) are also seen with e.g. sacbrood and advanced varroa infestation. (The holes of healthy cells *being* capped are round.) Cappings can look dark simply because the pupa below is nearly mature (dark eyes) or the cell walls are heavily propolised.

Bee inspectors have even seen hives with, simultaneously, both EFB *and* AFB.

However, the rope test and false tongue are pretty exclusive to AFB, and I haven't heard of scale associated with anything but foulbrood. So here's what to check next if you're suspicious:

'Scale'

The dried remains of pupa, sticking tightly to the cell wall and not human-removable without damaging the cell wall; the bees can't remove it. Dark and shiny, so easiest to see if you hold the comb up so light shines from behind you. EFB scale is soft, house bees remove it. AFB scale is hard, pretty impossible to remove and glows greenish-blue under blacklight (aka UV-A, or near ultraviolet). Google '*AFB scale image*' for pictures and '*Holst milk test*' if you want further confirmation, but lateral flow devices are better.

Rope test

Select a larva that looks infected but not dry (slight melted look). Use a toothpick or matchstick to swirl the contents of the cell and slowly withdraw it.

If the contents pull out up to an inch in length (2.5 centimeters) then snap back, the cell is most likely infected with AFB.

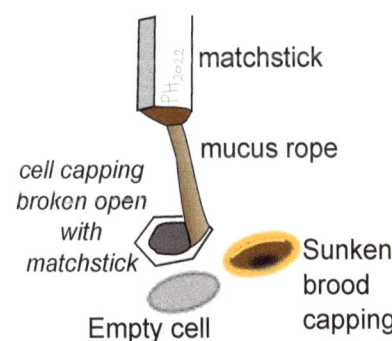

The "rope test". Seeing a suspicious cell capping, the beekeeper uses a thin stick and pulls a mucus-like string out, confirming the presence of foulbrood. However, a dark cell often simply has a dark cap - colour can vary depending on what pollen the bees gather.

---

345 And **EFB lineages' symptoms can vary slightly.** Variant ST39 rarely presents the "melted down" stage so is trickier to spot before it spreads to other hives. (Victoria Tomkies & Giles Budge, EFB article, BeeCraft October 2022)

An EFB "rope" will snap by the time it is pulled out just 1.5cm.

AFB bodies melt completely into a snot-like goo, EFB remains are not as sticky.

Lateral flow devices

Bee inspectors can now confirm the presence of EFB with a lateral flow device. These can be used instantly in the field and might miss 1 in 5 EFB infections (often because of sampling error like choosing the wrong larva) but they will basically never give false positives[346].

There is a separate lateral flow test for AFB. I don't know the false positive / negative rate but it's pretty good.

Both tests are available from vita-europe.com

Technique

- Levels of foulbrood DNA in floor debris and honey don't correlate well with infection, can give false positives from dead bacteria, and gum up the LFT device, so those aren't used.
- Instead the test is done on suspect mashed-up larvae.

In Scotland, if foulbrood is confirmed, inspectors take samples of larvae, and also nurse bees from brood combs: if foulbrood is present the nurses will be spreading it by trophyllaxis. These are checked in a lab for followup tests, e.g. are subclinical levels present? What genetic strain is it? Did a shook swarm work?

Using LFDs, inspectors therefore may not need to disassemble the nest and inspect *every brood cell* as they used to, but this is a matter of debate and negotiation.

Microscopic examination

Visible at x1000, but not a definitive test and technically difficult, requiring stains and oil drops. AFB spores are about 1.3x0.6um, EFB bacteria are about 1.0x0.5um. The organisms are so small they display Brownian motion, and the examination is best done on site because by the time a sample reaches a lab, similar microbes are feasting on the nutrients.

Preventing foulbrood spread

McLain observed that foulbrood spreads downwind[347], perhaps because bees tend to drift downwind.

The Roots[348] reccommend, if foul brood is present, to only open hives *in the evenings*, as the infection can be carried by a single bee travelling between hives. If you trigger robbing your entire apiary is at risk. A single brood frame or even drop of honey can transfer the infection.

If foulbrood kills a hive and you cannot sterilise it immediately, seal every crack so other colonies don't rob it.

---

346 From a Zoom talk by Prof. Giles Budge in 2021. "Sensitivity (true positive rate) = 0.81, Specificity (true negative rate) = 0.995" . Determined by comparing with lab based PCR tests which are very accurate.

347 Mentioned in passing in 1887 *Report on Experiments in Apiculture,* Washington (Google Books). At that time EFB and AFB were not distinguished.

348 *The ABC and XYZ of Bee Culture*, 1910 edition, p.138, Root & Root. They also observe on p.141 that "German black bees" (Amm?) are more susceptible to both AFB & EFB than Italians or Carniolans; and feeding or a strong nectar flow will suppress the symptoms... until the diseased stores are accessed again, at which point the disease restarts. However, Swiss research started in the mid 2010s for 8 years 8 yrs found no difference in racial susceptibility, and NBU research around 2011 found all patrilines are equally susceptible.

How EFB spreads between hives

This is not completely understood. It is unusual in having a very slow, asymptomatic development. This means it can spread efficiently[349] to neighbouring colonies, or via colony / queen sales, and indicates it has co-evolved with bees for a very long time. Once it hits a critical level colony collapse can occur rapidly.

In Britain, EFB is more common in areas of high rainfall, possibly related to poorer foraging conditions leading to brood cannibalism.[350] It is more common in areas with a high apiary density.

EFB may recur up to 2 years after infecting a hive, so be extra vigilant in that period.

Nearly all methods of spread are beekeeper mediated with the exception of swarming. Yet shook swarming is the preferred method of control for EFB! This has led to some very confused thinking with many beekeepers believing swarms are a major disease vector. (Whenever a disease breaks out in an apiary, the beekeeper is never to blame). This is undoubtedly linked to the advice from many Master Beekeepers and websites that the first thing to do with a swarm is... feed it. *No*. You force it to digest all the honey in its stomachs building new comb.

How AFB spreads between hives

Dr Mark Goodwin[351] in New Zealand did extensive research on AFB and concluded:

"It is quite difficult to infect a colony with AFB although some beekeepers seem to be very good at it. Under trial conditions to you need to feed about 5 million AFB spores per litre of sugar or honey to infect a colony."

Goodwin emphasises:

- Bee spread (by drifting) is much less common than beekeeper spread (by swapping brood between hives, moving infected hives, swapping supers and other hive parts, packing hives densely, and permitting robbing of hives that collapse due to AFB). He saw many examples where neighbouring beekeepers had very different incidences of AFB.
- Having hives in straight lines and all painted the same colour **increases drift.**
- **Feeding pollen is high risk.** The design of most pollen traps ensure that many of the AFB scales that bees remove from a hive end up in the pollen trap with the pollen.
- Swarms can very definitely carry AFB with them, or nest in infected cavities. **Don't hive swarms on drawn comb.**

In Britain, AFB is very rare, about 100 cases / year, and outbreaks are always imported to an area. This is evident looking at NBU maps showing how an area is clear for years, then there is a sudden outbreak. Importation can occur, for example, by bees licking an "empty" jar of imported honey some kind soul left out for them. **Never feed bees honey from another apiary.**

---

349 There are various EFB lineages and with improving DNA tech (*Single locus nucleotide sequencing* and soon, *whole genome sequencing*) the inspectors are beginning to be able to trace exactly *where* an EFB outbreak came from. So far they have eradicated 11 sequences from the UK. One seemed resistant to OTC. Some are suspected of being resistant to shook swarm. Some cause more severe symptoms - which makes them easier to spot and eradicate!

350 *Identifying the climatic drivers of honey bee disease in England and Wales,* Rowland, Rushton, Shirley, Brown & Budge, Scientific Reports (2021) analysing 317,000 apiary visits(!) by Inspectors over 10 years

351 Dr Goodwin seems to have died and his employer, HortResearch was folded into the New Zealand Ruakura Research Centre, but his work is still viewable at aucklandbeekeepersclub.org.nz/UFresource/HOW_BEEKEEPERS_SPREAD_AMERICAN_FOULBROOD_DISEASE-pss_(1).pdf and his book *Elimination of American Foulbrood Disease without the use of Drugs* is available from apinz.org.nz/shop/publications/afb-book/

Spreading mechanism within hives

AFB's spores are <u>in its scale</u>. Every time the nurses bees try to clean it off (which they can't) they pick up more spores and spread them. So AFB has evolved to kill its victims after capping, hidden away so they can melt into scale, and the more it kills the better it spreads.

EFB spreads differently within a hive. Bee larvae defecate only once, just before spinning their cocoon. The bacterium stays within their gut until this time. If the nurse bees eject a larva before this stage, they also eject most of the infective agent, so it's in EFB's interests to have evolved so its host survives to pupation. Once the bacterium is out of the larva it is spread very readily by nurse bees cleaning the faeces and, usually, the melted body of the victim.

It makes little difference to EFB if the larva dies before or after capping once it's defecated, and although most die before capping, in heavy infestations you'll find some capped melted bodies, resembling AFB.

(Remember - EFB is the **E**arly onset one - discoloured larvae before capping - whereas **AFB** discolouration tends to be **A**fter capping.)

Treatments - overview

Bee inspectors look at the severity of the infection and decide the best strategy.

Treatment	AFB	EFB
Fire (wooden hives)	Bees and entire hive and contents completely burned	Brood and honey comb, and scorch the hive walls (burn entirely for very bad cases)
Bleach (plastic hives)	Sodium hypochlorite solution	
"Shake and starve"	Originally used for AFB. Ineffective.	Surprisingly effective for EFB. The shook swarm breaks the breeding cycle and removes them from infected comb / hive.
OxyTetraCycline (a.k.a. OTC, Terramycin)	Merely suppresses symptoms. Disease recurs 22% of the time[352].	Not permitted for EFB in UK. Disease recurs 22% of the time.[353]
Requeening (combined with some of the above)	Only used in USA. No evidence requeening alone works, but it does sell queens.	

---

352 The 22.2% figure for AFB reoccurrence after OTC treatment is from *Adoption of partial shook swarm in the integrated control of American and European Foulbrood*, Mosca et al (2023), mdpi.com

353 The 22% figure for EFB recurrence after OTC treatment is from a Zoom lecture by Prof Giles Budge to Cambridgeshire BKA on Foul Brood in 2021. Also mentioned in *Development and Validation of a novel field test kit for EFB*, Tomkies et al (2009). Budge worked at the NBU for many years. The NBU gathers data on the recurrence of AFB in apiaries after using OTC, shook swarm and / or burning. See also *BeeCraft*, April 2020, p.8 where bee vet John Hill criticises some American states' use of OTC for both EFB and AFB, simply suppressing symptoms, followed by hives often being moved. "It is common for OTC to be used prophylactically twice a year on all colonies."

Treatments - details

Fire: AFB spores can persist 3mm inside wood, so scorching with a hot air gun is insufficient. Burning of woodenware, dead bees, wax, honey etc is done in a deep fire pit in presence of an inspector.

OxyTetraCycline (aka Terramycin): *does not cure AFB*. It is not an antibiotic - it simply stops the bacteria replicating, giving the colony a breathing space to recover. AFB recurs the next year in 22.2% of colonies treated with OTC Kills up to half the brood[354] and persists in honey for up to 32 weeks. Unsurprisingly, an antibiotic resistant strain of AFB is spreading rapidly in America.

Shake and starve (shook swarm): As early as 1875, Pettigrew recorded that swarms don't get foul brood in their first year. Until the 1960s[355], there was a now-banned method of dealing with AFB in Britain called "shake and starve". This was found to be ineffective for AFB, but the NBU has found it surprisingly effective for EFB - they found that it recurred next year in only 2% of shook swarmed colonies. Brood are sacrificed. You transfer the adults to a clean hive, and do not feed them for a few days to maximise the number of micro-organisms digested, and you force them to build new comb (or use undrawn foundation). All old comb is burned. You then feed them to help them prepare for winter. Even with feeding the colony is unlikely to survive if this is done later than July. The old hive is sterilised with flame or sodium hypochlorite.

Propolis: There are clear indications that this suppresses EFB[356] and perhaps AFB[357]

Hygienic bees: Some say that Foulbrood and chalkbrood are virtually non-existent in mite resistant bees[358]. The life cycle of AFB requires it to go through a vegetative and spore state. Spores are transferred to the brood by feeding. But spores are only formed after capping of an infected larvae, and it dies in the capped cell. Hygienic bees will remove and eject infected larvae far from a hive **before** capping, and dying. This breaks the foulbrood cycle, making it impossible for it to reproduce. *But...*

*Important - don't do this - **never** transfer combs from a hive with foulbrood symptoms to another hive, even if someone tells you it's OK because they're hygienic359. Sure, it's been done experimentally but by scientists and experienced bee inspectors. For you and me, it's a sure fire way to spread AFB and destroy our apiary.*

---

354 *Effects of European foulbrood* treatment *regime on oxytetracycline levels in honey extracted from treated honeybee colonies and toxicity to brood,* Thompson, Waite, Wilkins et al *Food Additives and Contaminants* 2005. Untreated (control) brood mortality was 13% after 17 days, versus 64% (powder OTC treatment) or 48% (liquid treatment).

355 Mentioned by Budge in the aforementioned Zoom lecture, and in a 1964 pamphlet I have by H J Wadey, editor of BeeCraft

356 *Antibacterial effects of propolis and brood comb extracts on the causative agent of European Foulbrood (Melissococcus plutonius) in honey bees (Apis mellifera)* - Murray, Kurkul, Mularo et al (2022) . A key point is that propolis does not disrupt the bee gut biota like OTC, or leave residues which prevent the sale of honey.

357 *In vitro evaluation of antimicrobial effect of propolis against Paenibacillus larvae genotypes in Turkey* Aygün Schiesser, Ö mür Gençay Ç elemli, Asl Ö zkrm, Nevin Keskin, Hacettepe University, Turkey. Schiesser has published at least one other paper on this and another on using rhododendron extracts versus AFB.

358 Backed up by research by Spivak & Reuter: Resistance to American foulbrood disease by honey bee colonies Apis mellifera bred for hygienic behavior (2001). https://hal.archives-ouvertes.fr/hal-00891903

359 Vaccines use a *weakened form* of a disease, not a massive dose of living bacteria.

Diet: Tihelka and Croppi[360] rediscovered several intriguing 19th century references to foraging on poplars and conifers helping against foulbrood, and hives spontaneously recovering from foul brood after being moved to willow forage. Likewise, blueberry pollination in North America tends to be associated with a rise in EFB, which is reversed by nutritional supplements[361] (blueberries are poor bee food and rain limits foraging). Bailey mentions[362] *"it is well known, for example, that European foul brood is rapidly suppressed during good nectar flows"*.

Other potential treatments for AFB

The University of Helsinki is developing the Dalan AH oral vaccine, discussed in the last chapter (I am concerned it simply helps *mask* symptoms without curing, like OTC), which is fed by workers to the queen who passes on vaccine particles to her eggs. Phage therapy is being investigated by Thomas Brady and Heather Hendrickson[363]. Lactic acid producing gut bacteria inhibit AFB in laboratory trials but not, it seems, in the field[364]. Silver nanoparticles derived from the camphor tree are being trialled.

Historical note

Up to about 1906 people didn't realise they were looking at two different diseases and could not figure out how Foul Brood spread, because experiments gave inconsistent results. Foul Brood, of whatever type, was rampant. We only got AFB under control in Britain with the foundation of the Bee Inspectorate in the 1940s and a draconian regime of burning infected colonies. It's still endemic in America. EFB can be slower to build up so more difficult to spot, and is still relatively common. There are usually less drastic options for dealing with EFB.

Honeydew Flow "Disease"

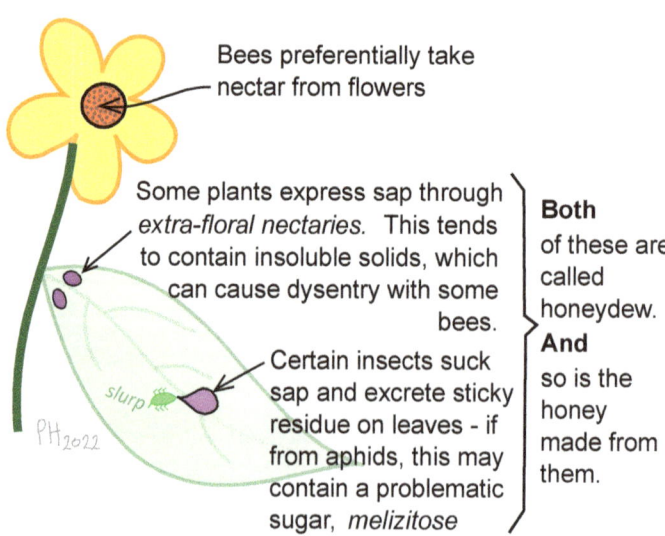

Symptoms:

Basically digestive problems when consuming honeydew. Bees forage on honeydew if nectar is unavailable.

- Swollen abdomens, abdomen tipping, impaired movement, twitching and terminal paralysis.
- Reduced lifespan of affected bees. Gets worse in winter, when hives can die in bad cases.
- Alternatively - almost oppositely - it can cause dysentery.

---

360   *A brief history of foulbrood: from superstition to veterinary medicine and back*, Natural Bee Husbandry #16, Summer 2020

361   *Occurance of Eurpean foulbrood disease (EFB) during blueberry pollination in the Vancouver BC area* Yeganehrad, Karim, Moarefi - good quality pollen based nutrition reversed EFB infections

362   Bailey, L *Wild honeybees & disease* (1958) Bee World 39, p.93-95.

363   Trending - google *phage therapy foulbrood bees* and you will now find many papers.

364   S. Lamei, *The effect of honeybee-specific lactic acid bacteria on american foulbrood disease of honeybees*, Swedish University of Agricultural Sciences (2018); Vásquez, Forsgren et al *Symbionts as Major Modulators of Insect Health: Lactic Acid Bacteria and Honeybees* (2012)

Discussion, prevention:

Seen in areas where bees feed on honeydew (extensive fir trees - Black Forest etc). I'm not aware of it in Britain.

It is only a problem *if you take all the floral honey* so there is nothing for them to mix it with[365].

Some texts claim honeydew will kill the bees if it remains in the hive; this is simplistic. If true, their habit of gathering it would be rapidly selected out. My impression is, local bees whose gut microbiome is adapted to honeydew are OK - certainly some writers claim Amm are OK with it, or at least the ash (solids) in it, and Amm live in northern forests. But import foreign bees, and you can expect problems, especially if you kill their microbiota with antibiotics[366].

Honeydew can refer to two different sources and they cause different symptoms. It is a sugary liquid exuded **either** directly from some plants' extra-floral nectaries, **or** excreted by sap sucking insects. It's mainly sugars, generally less nutritous than nectar[367], but the stuff directly from the plant contains insoluble solids ("ash") - those cause the dysentery[368] - whilst the stuff from sap sucking insects contains a problematic sugar, melezitose. Whatever the source, it makes a distinctive honey which is a big crop for some beekeepers.

The melezitose in honeydew from aphids is a complex sugar. Bees (more accurately, their stomach microbiota) can't digest this well and it builds up in the stomachs of foragers, reducing their lifespan[369] and causing bloated stomachs. In winter, the bees won't defecate in the hive so it builds up in them, and if the stores contain lots of melezitose, mortality increases over winter. Melezitose also makes stored honey set rock hard making it inedible over winter.

One American site[370] notes that bees foraging on honeydew can get exposed to pesticides farmers use against scale insects, and if they're collecting honeydew they're not getting nectar, so they need protein (patty cakes). To me that is more a sign of poor forage - the bees wouldn't go for honeydew if there was nectar around.

Hornets

See also Asian Hornets, and Wasps

European hornets, *Vespa crabro* are rare and not a significant bee pest in Britain. Italian and Cyprian bees come under more pressure from these and have developed partial defences like heat balling them, but these don't reach the temperatures necessary to cook Asian Hornets[371].

---

365 Personal communication, Ulrich Beckmann, a German beekeeper familiar with honeydew

366 This is a strong argument in favour of using local bees. As Rich Tetlow pointed out to me, bees get their gut microbes when they're fed by nurse bees. Imported colonies could have a mismatched gut microbiome to the local diet, and they're not going to get it by mating with local drones.

367 Quimby remarks bees ignore honeydew if nectar is available (p.51)

368 Note the parallel to heather honey - some beekeepers find their bees get dysentery overwintering on it due to its insoluble solids, but *Amm* is fine with it

369 Seeburger et al 2020, *The trisaccharide melezitose impacts honey bees and their intestinal microbiota*. https://doi.org/10.1371/journal.pone0230871 - a very readable and informative paper about honeydew.

370 beeinformed.org/2014/09/25/honeydew-a-mixed-blessing/

371 For more on how bee species defend against hornets, see section 4 of *Hornets and Honey Bees: A Coevolutionary Arms Race between Ancient Adaptations and New Invasive Threats* by Cappa, Cini, Bortolotti, Poidatz, Cervo (2021), doi.org/10.3390/insects12111037 . Interesting points: if Asian hornets are hawking, A. cerana speed up as they approach the hive and bullet into the entrance, whilst European bees get confused, slow down and are easier prey. One Asian bee species uses animal dung to repel a specific hornet - but only that particular hornet. The take home point is that European bees have more latent anti predator defense strategies than most people realise, but they are only expressed in a few colonies and are not sufficient to handle Asian Hornets yet - but *A. cerana* shows they *could* be.

Warning: hornets will happily defend their nest at night too. Their sting is long enough to penetrate bee suits and said to feel "like a red hot nail".

Isle of Wight Disease

Symptoms

Like a paralysis virus - bees unable to walk straight, or fly, trembling. K-wings. Beekeepers in e.g. 1912 also recorded dysentery and swollen abdomens.

Discussion

In the 1910s-1920s "Isle of Wight disease" was a big problem in Britain, but no one really knows what it was.

By 1931, perhaps earlier, beekeepers realised it was at least 2 things which overlapped, one which paralysed and killed bees fast, but tended to burn out and not spread unless people moved hives around; and another slow[372], so it could spread sneakily. By ~1930 bees had evolved resistance and the "disease" disappeared.

Timeline

Year	Accepted explanation at that time
1904	Nosema - and it is definitely present and responsible for some losses, but it wasn't universally present, and burns out too fast to spread as fast as IoW disease did.
1921	Microscopic tracheal (acarine) mite discovered. Widespread. Undoubtedly killed a lot of colonies. It gets the blame. But *tracheal mites do not cause paralysis*.
~1930	"IoW Disease" vanishes - bees now resistant
1930s+	Br Adam goes on intensive marketing campaign claiming IoW was conquered by a magic Buckfast hybrid and in the face of all evidence, all other bees are susceptible, and dead. This is authoritatively debunked at http://dave-cushman.net/bee/iowdisease.html - he interviewed people who lived through this - the local bees never died and the problem was actually confined to people who were still using skeppist techniques with the new framed hives, and using imported bees or hybrids.
1931	Herbert Mace reviews and concludes it was 2 overlapping factors: a strong association with nosema in the first years, and a longer term trachael mite (acarine) problem which remained until the first successful acaricide, Frow's Mixture was developed (which we would consider far too toxic and carcinogenic now).
1963	Bailey isolates CBPV, Chronic Bee Paralysis Virus, which fits the symptoms
2020s	Epidemiologists conclude it was probably multifactorial, but mostly due to a combination of CBPV, and increasingly intensive beekeeping, which promotes disease spread. Tracheal mites were discovered at the same time and as the New Scary Thing, got the blame for every instance of every disease by people who didn't look further once they had an idea, much like varroa mites are now.

---

372   Pages 109-114 of *Bee Matters and Bee Masters,* 2nd Ed (1931), Herbert Mace.

Treatment:

Whatever it was, it doesn't exist now. We would probably treat it as 2 different diseases now.

Note that bees evolved resistance to the Tracheal mite in just a few years. Now, about varroa...

Further reading:

- *The "Isle of Wight Disease" : The Origin and Significance of the Myth* by L. Bailey, published by the Central Association of Bee-Keepers 1963.
- Brother Adam witnessed it first hand, but had an interest in promoting his Buckfast bees. He argues it was indeed acarine mite - DOI 10.1080/000572X.1968.11097180
- *Village Bees – The native & near-native bees of Britain and Ireland*, Beowulf Cooper (1968) gives good arguments that IoW Disease did not wipe out even the majority of native bees.

Mice

(mainly a winter problem)

Symptoms:

- Mouse + nest in hive
- Distinctive droppings
- Smell of ammonia (urine) or mice
- They eat the body, leaving legs, wings and a few heads

Discussion:

This is why we fit mouse guards (see chapter 8, *winter preparation*).

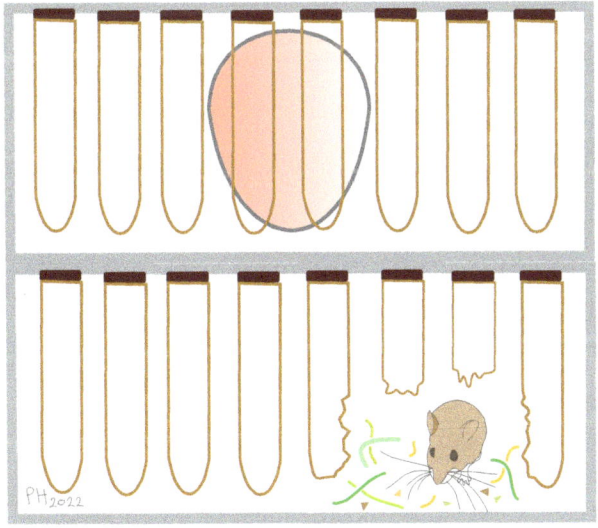

Mice nest among dried grass, comb fragments, shredded paper etc and prey on the winter cluster's peripheral bees.

Mouse nest in insulation above bars (isolated from bees)

Mouse droppings are dark brown and pointed at the ends.

A hive is a lovely warm, dry winter home with a buffet. Mice are insectivores, and one mouse can kill an entire colony over winter, when cold prevents bees breaking cluster to drive it out. They eat honey too.

The smell of an old rodent nest can make a swarm abscond months later, so clean and air hives which had mouse nests in.

Really well insulated hives (eg double walled) and tree nests do not need mouse guards because the bees don't need to cluster, they can defend their home all winter.

Resembling the image of old comb eaten down to foundation by bees in chapter 11.3, this mouse-nibbled foundationless comb around a mouse nest has bedding material in the chewed area, an uneven edge, and incidentally shows interesting distorted cells where it transitions from worker to drone brood comb.

Neglected Drone Brood

If there are too many drones due to a failing queen / drone laying workers etc, the remaining workers can't nurse them all, ventilate the hive properly etc. Drone brood die and rot (like EFB) then dry to form scale (like AFB) but you can tell it's Neglected Drone Brood because there are way too many drones.

Nosema (Vairimorpha)

See also: dysentery

See also: Amoeba Disease

I view this as an opportunistic infection indicating a deeper gut problem, like poor nutrition.

The US attitude to nosema casts insight into their beekeeping style. Over there, under 5 million nosema spores per bee is considered economically viable. But in Britain, one spore counts as a positive result[373].

Thymol, a strong antifungal, is effective against nosema. So some commercial beekeepers add it to their syrup. However, as we discuss under dysentery, the problem is really that they are feeding too much poor quality syrup - the nosema infection is a secondary symptom, not a cause of the dysentery. And a strong odour like thymol can swamp colony pheromone signalling.

There seem to be a lot of strident warnings that nosema is a dangerous disease which are not backed up when you actually read the literature. It's a mild gut problem, probably exacerbated by feeding junk food at the wrong pH to bees, barely kills colonies and can be ameliorated by feeding with probiotics[374] (friendly gut bacteria). However if it becomes severe it can halve lifespan because it impairs pollen digestion, and affected bees become unable to secrete royal jelly, and thus unable to feed brood.

---

373 This was mentioned in a BBKA microscopy course I attended in 2016.

374 *Deleterious interaction between honeybees and its microsporidian intracellular parasite N. ceranae was mitigated by administering...* Khoury, Rousseau, Lecoeur (2018), doi: 10.3389/fevo.2018.00058

Nosema is a single celled (microsporidian) fungal parasite (not a microsporidian protozoan, as stated in outdated texts) and comes in two types, *Nosema apis* and *Nosema ceranae*[375]. These tend to weaken, but not kill colonies and are most often seen in periods of long confinement when they cannot expel it from their guts (defecate). *N. apis* is seen in Winter and *N. ceranae* in Summer, so they affect colonies differently. Only *N. apis* causes the dysentery symptom, whilst *N. ceranae* results in dwindling as foragers die away from the hive. After infecting the gut, spores pass out in faeces.

Nosema tends to result in spotting on combs and at the entrance, rather than a full-on uncontrolled fecal flood. The combs must not be transferred to other hives; the spores remain infective for a year.

The British National Bee Unit did a random apiary survey in 2010/2011 and found 40% of colonies had *N. ceranae* and 46% had *N. apis*, but there was no strong correlation to poor colony health. This is not the case in Spain where it is strongly correlated. Caucasian bees (the grey mountain kind, adapted to a low humidity enviorornment and a short overwintering confinement) are said to be most vulnerable to nosema and *Amm* the most resilient.

Definitive diagnosis: if you mash up bees and look at the goo under a light microscope at x400, you can see the small rice-grain-like spores (round things the same size are most likely just air bubbles). And if you dissect to examine the gut, a healthy bee gut is reddish in colour where a Nosemic one is milky white. If you see amoeba rather than rice grains, it's the much rarer *Amoeba Disease* which has the same symptoms.

Treatment

There are no medicines licensed to treat nosema in the UK. Fumagillin is now considered too toxic for things in the human food chain.

It's generally a symptom of weak colonies, poor nutrition, or bee strains which have had resistance inadvertently bred out. Can you improve these factors?

Nosema is killed by sterilising comb at 49°C for 24 hours (it may warp) or with 80% acetic acid's fumes for a week[376] - this assumes you want to re-use infected comb. Tools are sterilised with a blowtorch, sodium carbonate or harsher chemicals.

Nosema resistant lines of bees have been found in many populations. In Britain, even commercial beekeepers generally use natural selection for nosema - it's so rare here (because our bees are now resistant) that if a colony succumbs, it's considered to have improved the overall gene stock.

Further reading on N. ceranae: for a deep dive into this organism, the following paper covers aspects most sources don't mention such as behavioural changes of infected bees, amplified effects in the presence of insecticides, possible developing resistance to fumagillin etc:

The Role of Nosema ceranae (Microsporidia: Nosematidae) in Honey Bee Colony Losses and Current Insights on Treatment by Marin-Garcia, Peyre, Ahuir-Baraja, Garijo & Llobat (*Veterinary Sciences*, 2022). doi.org/10.3390/vetsci9030130, www.mdpi.com/2306-7381/9/3/130

Perhaps the take-home insight from this paper is:

"However, by the time the beekeeper detects visible symptoms, the colony is practically dead."

---

[375] A recent nomenclature change to the genus Vairimorpha after over a century of being known as Nosema has proven controversial, and confusing. A third type, *Nosema neumanni* has been recently (2017) identified in Uganda.

[376] nationalbeeunit.com leaflets *Fumigating Comb* and *Hive Cleaning and Sterilisation*.

Paralysis viruses

Several viruses with similar symptoms, affecting only adults. In practise, beekeepers can consider the Acute and Chronic ones as a single disease, and ignore the Slow one. One or two lethargic bees in a colony isn't worth panicking about. But at high levels, this will probably overwhelm the colony and spread to others – ask a bee inspector for advice.

Bee Paralysis Viruses (BPVs)

	(Local variants:)		
Acute (ABPV)	Acute (ABPV) Israeli (IAPV) Kashmir (KBV)	Rapid onset. Weakens hives. Rare	Vectored by varroa Trembling, shiny black body, bloating, K-wings Lethargy, crawlers
Chronic (CBPV)		Slower buildup. Can be serious. Becoming more common.	NOT vectored by varroa Trembling, shiny black body, bloating, K-wings, nibbled wings Odd reactions to smoke Shivering cluster on top bars Lethargy, crawlers
Slow (SBPV)		Slow buildup. Minor problem. Widespread.	Vectored by varroa. Paralyses front 2 pairs of legs. NO shinyness / trembling / bloating / K-wings

Symptoms

See chart above. You are looking for a combination of symptoms: dark bodies, K-wings, crawling **and** trembling. At that point, open the hive and look for trembling bees clustering in unusual places.

Nibbled wings arise because healthy bees harass infected ones trying to re-enter the hive.

Ambiguous symptoms:

Paralysis resembles poisoning, but poisoned bees aren't greasy, and typically spin on their backs (most pesticides are neural toxins). Poisoning creates a thin carpet of spread out dead bees, whereas paralysis victims are more lethargic and disoriented, and form a pile of dead bees directly outside the hive entrance[377]. There's a saying that if you remove pile of dead bees under an entrance and come back the next day to find no more dead bees, it was poisoning.

Crawling may be starvation or cold or lost young bees, but crawlers with paralysis will also tremble and have black, greasy bodies. Tracheal mites also cause trembling crawling.

Shiny / greasy bodies is due to hair loss. BUT this can also be due to robbing; or gathering honeydew - the hairs get stiffened by dry sticky stuff and snap off: or rubbing against rough

Shiny bees with ABPV.
Image © 2018 Andrew Bax

---

377 Budge, Simcock quoting Andy Wattam, in BBKA News #228 May 2021.

forage like borage: such bees are active, not paralysed crawlers. It can also be due to overheating during hive transport (bees vomit up honey over each other) or spraying with sugar water - the bees appear hairless then because the hairs are stuck flat.

Discussion

The Roots commented[378] "This is a disease that is **much more prevalent and virulent in warm than in cold climates.**" I've not seen this correlation mentioned elsewhere.

ABPV is an ambiguous term, referring both to a specific virus and the family including its sisters IAPV and KBV.

CBPV is unusual in several ways. It is *not* spread by varroa. But it has been found in other insects which are in occasional contact with honeybees, like a commercial bumblebee and *Formica rufa* ants. It's a rather shapeless, amorphous virus unlike any others we know of, which tend to have a regular structure; and is as yet unclassified (nothing similar known to group it with). It comprises 2 strands of RNA joined at one end, there are 4 variants (ABCD) but as yet no one's seen one made of 2 halves from different variants.

CBPV re-emerged in the UK in 2007. It seems to cluster round commercial apiaries with imported queens. Doesn't survive in UK over winter, but continually re-imported.

CBPV prevalence seems closely correlated with the number of colonies in a country, which decreased in the UK after WWI. It has increased exponentially in the last decade, mainly in high density apiaries[379].

Bailey grouped CBPV symptoms into *Type I* (K-wings, et alia) and *Type II* (crawlers, et alia) but researchers no longer find this distinction useful. Americans talk about *Syndrome (stage) I,* trembling crawlers, and *Syndrome II,* greasy black (hairless) "robbers" (trying to get back in the hive they've been thrown out of), but again these are simply different stages of the same thing.

It is not uncommon for colonies with CBPV to be queenless. Queens can catch the virus and die too.

Treatment

No real treatment options. The bees either recover or they don't. CBPV is the most lethal.

But you can take some actions to reduce spread to other colonies:

- Typically spread by adult-adult contact, or faeces. CBPV is specifically found **not** to spread by transferring brood combs between hives.
- **Don't** shook swarm them. This approach is discredited. The disoriented infected bees make their way back to **any** nearby hive, spreading it further. The success rate of this technique is only 50% but if you just leave colonies alone, 60% recover!
- ABPV and SBPV can be vectored by varroa, so varroa resistant bees or miticides may help.
- Probably made worse by crowded conditions as it can spread by contact, so it's suggested that you add boxes if the hive is congested. For example if bad weather keeps everyone in for a few weeks. But in the only case of Acute Bee Virus I've heard of personally, there were piles of dead bees and the survivors were then overrun by wax moths, so what they really needed was an amount of comb appropriate to the area they could patrol.

---

378   1910 edition of *The ABC and XYZ of Bee Culture*, p.134, an American work (still in publication - now on its 42nd edition).

379   Online lecture in 2020 by Norman Carrick, Sussex University

- Requeening is the standard recourse and, if it does not clear up, your only sure cure before it affects healthy hives (i.e. some strains of bee are resistant, others seem to have lost their natural resistance).

Another spreading mechanism is walking through infected faeces on the hive floor, so a dubious idea being tried by bee farmers (2021) is to *remove the floor*, just leave an open space until the bees improve, and have the hive at least 12" above ground level. Another argument for this practise is that dying adults drop out of the hive and are unable to re-enter, thus reducing viral load. You can only do this with a strong colony which can repel robbers. As I don't see faeces on my hive floors I suspect this may be an artefact of intensive bee farming with non hygienic bees.

Poisoning

Symptoms

A characteristic spinning-on-their-back "dying fly" symptom is pretty specific to poisoning. But, it may not be due to *pesticide* poisoning. It can be from fighting. Bee / wasp venom acts as a neurotoxin on bees.[380] Poisoning can be beekeeper-induced, e.g. miticides plus heat.

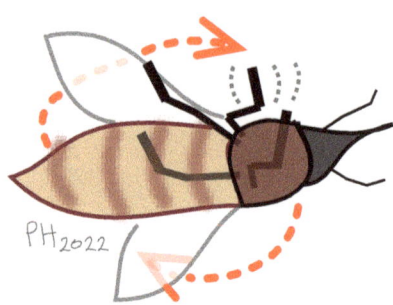

Watch for bees spinning on their back

Another stong indicator is a **thin** carpet of spread out dead bees, whereas paralysis virus victims are more lethargic and disoriented, and form a **pile** of dead bees directly outside the hive entrance.

However, there's no real definitive symptom, because there are a huge variety of poisons (insecticides, fungicides, herbicides, paint fumes etc) which attack different receptors, and symptoms vary with dosage.

"Tongues extended" is often quoted as a sign of poisoning, but many bees die with tongues out. It is usually a sign of starvation or gut biota imbalance preventing nutrient absorption (the bee is reflexively trying to feed). Norman Guiver[381] tells me tongues have to be *fully* out to indicate poisoning, and adds, "Piles of dying bees outside your hives is a direct pesticide hit from a local crops using neonics or other insecticides like dimethrin in spring or late summer. Dead bees in the bottom of the hive = slower death from herbicides in tank mixes with [slower acting but still deadly] fungicides."

Dead larvae upright in the cells, and turned brown but not collapsed like they do with foul brood, is possibly a sign of fungicide poisoning[382].

Discussion

Slow acting pesticides are spread rapidly through a hive by trophallaxis, so nearby spraying can kill an entire hive, not just the foragers.

---

380 *The Honeybee, its Natural History, Physiology and Management*, 1870, Bevan & Munn - while describing **vertigo** mentions Mr Golding's observation of spinning after being stung.
381 Norman had a lot of problems with pesticide poisoning and helped set up www.bee.watch which warns beekeepers if participating farmers in their area are spraying.
382 *Honey Days* by bee farmer Oliver Field, p.34 - writing in 1990, noted this symptom correlated with bees collecting water (spray) from crops sprayed with fungicide. Adult bees unaffected.

If you want to send a sample for analysis, get it quickly. Most pesticides and fungicides - except neonics - deteriorate rapidly in sunlight: about 50% is gone after half a day. The B-GOOD-project.eu is developing cheap lateral flow devices to detect some poisons.

Colonies might suffer high losses from poison, then recover. After recovery, you may notice your bees no longer fly in the direction where the poison is: perhaps a learned response, perhaps because scouts checking sprayed areas now don't return.

Robbing

Symptoms

Bees fight in a characteristic posture, curling their sting towards the target

- Frenzied activity at entrance
- Fighting between bees at the entrance
- Wasps also rob hives in Autumn / Winter
- Bees hovering in front of entrance with legs dangling: fly towards entrance and back off, change angle and dart in again to avoid guards, rather than landing and *walking* in. Not the smooth figure-8 flight of orienting bees, or random-looking collision-avoidance movement of a traffic jam 'bee cloud'.
- Bees running extraordinarily rapidly on the combs (to avoid challengers; visible through windows)
- Robbers don't carry food *in*, they have no pollen and their abdomens are not distended. They may look shiny - guard bees have chewed their hair off.
- Sprinkle icing sugar on the bees at the entrance, then watch your other hives. If white bees show up there, it confirms they are robbing. Returning foragers will go into the hive and rest a while, and be licked clean by other bees, whereas robbers will be quick in and out and will emerge still white shortly after being dusted.
- **Silent robbing:** Sometimes robbing is **covert** - it looks like normal activity; no frenzy and no defence. (These robbers probably smell like the hive they are robbing.) Silent robbers fly straight in. There is a continuum of behaviour between silent and frenzied robbing.
- Robbers continue flying very early / late in the day.
- You can confirm it is robbing by catching bees with distended abdomens **leaving** the hive as they climb the wall above the entrance, and gently pressing their abdomens. Bees leaving the hive should be empty. Robbers will regurgitate honey. (Thanks to Tomás Nevado for that tip.)

Discussion

- Tends to happen to weak hives, like nucs, failing colonies and swarms which didn't manage to build up to a strong colony by Autumn.
- The most important factor for repelling robbers is **morale**. A small queen-right hive will vigorously repulse invaders whilst a large one with a failing queen is apathetic.

- Robbing is often triggered by beekeepers taking too much honey in Autumn, making these colonies feel robbed and under pressure to take risks restocking for winter.
- Robbing is often triggered by feeding syrup.
- It is much harder to **stop** robbing than **prevent** it happening in the first place by providing an easily defensible entrance, like a small hole leading directly to brood comb, which is seething with nurse bees and any intruder must force their way into a mob in a restricted, bee space (between combs). This deters most wasps. But hives with bottom entrances are practically open to intruders, especially on cold mornings because wasps can fly and enter at low temperatures while bees are still clustering for warmth in the nest.

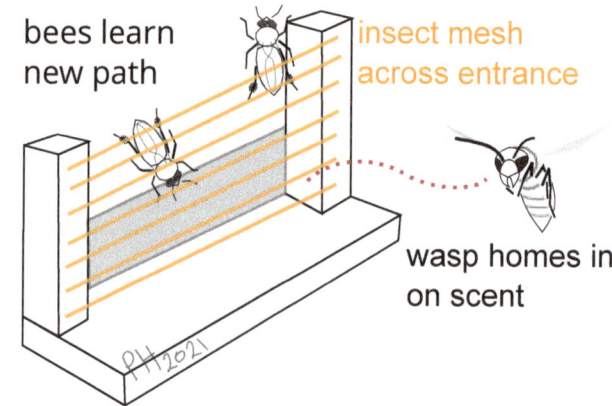

Robber screens work on misdirection. Easy to make, but little effect.

- Filipe Salbany hasn't seen robbing among the Blenheim bees (70+ wild colonies in ancient oaks), implying bees aren't naturally that stressed.

Treatment

Here's the standard modern advice. None of these techniques work well. Once robbing starts it is very difficult to stop.

- Move the hive being robbed.
- Block the entrance of the *robbing* hive for a couple of days (ensure they have ventilation!)
- **Reduce the entrance** to something easily defensible, like 2cm hole.
- Create a tunnel entrance (good deterrent vs wasps).
- Hives with comb oriented **cold way,** or brood comb immediately behind the entrance, are much less susceptible.
- "Robber screens" are widely recommended but only delay the robbers for 24 hours, I don't understand why people even mention them.
- (Unverified:) Robbers follow smells, so can be misdirected to a **large** entrance sealed with mesh, while the inhabitants use a second, small entrance.
- Feeding the suffering colony immediately is not wise – it can just increase the frenzy of the robbers.
- Wasps rob once nectar sources run out. The only way to stop them is to find and kill their source nest. They can fly in colder weather than bees and can rob while bees have to cluster.
- Since wasps prefer unguarded entrances, and are really put off by a mass of bees immediately behind the entrance: if there is a void with no comb behind your hive entrance, consider rearranging the combs or removing a box so the first thing the wasps encounter is comb heaving with bees. Only do this late in the year, when the colony is not going to need any more room to expand into.
- **Don't open hives for any reason in wasp robbing season,** which is August for me.

Propolis entrance reducer built by bees

Entrance modification examples

The best entrance system available is undoubtedly the one offered at beeportals.co.uk , based on multiple small round shielded entrances, instead of one large slot. Discussed further in chapter 17 under *Alternative Entrances.*

An emergency periscope entrance tunnel can be made out of cardboard. Jan Benson reports this works extremely well with a giant sized Toblerone tube, but a side effect is you feel quite sick from eating all the chocolate.

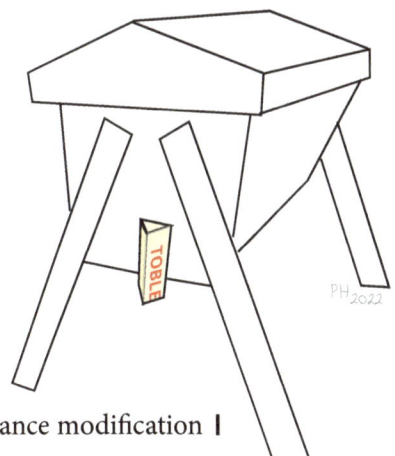

Periscope entrance modification

LEFT: wasps will be reluctant to rob this strong colony. MIDDLE: simple entrance reduction for colony a few months old. RIGHT: temporary tunnel entrance about one bee high is a very good deterrent.

Elegant historical solutions

- *The Bee Keeper's Manual*, by Henry Taylor 1860 suggests closing the hive entrance completely and confusing the robbers with smoke; if this fails, remove the hive under attack and replace with a box containing wormwood (absinthe) leaves which are repugnant to bees; or even *swap the two hives (robbers and robbed) around,* which is so devious I am overcome by admiration.

- Moses Quimby in *Mysteries of Beekeeping Explained*, 1853 points out it is usually a weak hive being robbed and if you wait until loads of robbers are inside... then block the entrance for 24 hours... the robbers unite with the weak colony and end up strengthening it!

- Pettigrew, in his *Handy Book of Bees*, 1875 relates finding a (strong) hive was being robbed by a neighbour's hive. He moved the victims 2 miles way and placed another hive in their place that night - which vigorously resisted the robbers. The robbers seemed to communicate to their entire colony "the game is up" and the robbing ceased almost instantly.
- If you feel brave: White, in *A Complete Guide to the Mystery and Management of Bees*, 1852 p.48 describes how on seeing a hive being robbed because the guards were letting the robbers in and out as if they were "their own family…. I laid my finger upon them, and presed them softly against the hive, which made them enraged, and then they soon began to fight the plunderers very frely, and at length, beat them quite off."

Sacbrood

Usually a minor problem only affecting a few brood

Symptoms

Larvae die **after capping.** Workers then uncap them, revealing larvae in a characteristic posture, sometimes described as like a "Chinese slipper" or gondola, on their backs with **tails** sticking up and out of cell. (Normally, if you uncap a cell, you would expect to see the head.)

Uncapped victim - yellowish and turning brown, rather than a healthy pearl-white

During decomposition they shrivel a bit, turn from normal white to yellow then brown.

Skin thickens so the remains can easily be removed from the cell intact with tweezers; if punctured contents are (infectious) brownish watery liquid with granular bits.

Bees uncap dead ones and you sometimes see sunken and discoloured caps, and caps with small **ragged** holes where this process has begun. Unlike Foul Brood, there is no melted goo in the cell or foul smell because the larva liquidises inside a **leathery sac.**

The sac adheres to the end of a toothpick and can be easily extracted from the cell.

Stages of decomposition, based on photos from Agriculture Australia.

Discussion, treatment

Also called SBV (Sac Brood Virus). Not too serious, no medicine available; generally managed by the bees but kills brood. Virus remains viable in honey for 4 weeks.

A Korean variant, kSBV is devastating Asian honeybees but does not seem to cross-infect European honeybees, sparking research into transferring this resistance to *A cerana*.

Sacbrood is more common in warm weather but wind reduces its incidence[383].

---

383   *Identifying the climatic drivers of honey bee disease in England and Wales,* Rowland, Rushton, Shirley, Brown & Budge, Scientific Reports (2021)

SHB - Small Hive Beetle

Notifiable pest in UK and Europe - you MUST contact Bee Inspector

Symptoms

- Comb covered in slime

- Hives develop a distinctive "rotten oranges" smell when the larvaes' faeces causes honey to ferment.

- Pretty obvious when hive is opened.

- Adult beetle has distinctive club shaped antennae. Its size varies with nutrition.

- Larva has two rows of spines down back.

- Can cause colonies to abscond.

Slimed honey covered in SHB larvae. Image courtesy Prof James D Ellis, university of Florida,Bugwood.org source: forestryimages.org/browse/detail.cfm?imgnum=5025044

Discussion *and* treatment

An African pest now in America and Australia, reached Italy 2014 (another reason not to import bees) but is unlikely to be a big issue in Britain: it may not survive our winters, because its larvae have to pupate in soil outside the hive, but it breeds fast and *could* become established; it is found in Canada. SHB is considered a minor pest in Africa.

Its larvae prefer **honey and pollen combs,** rather than brood areas. They don't leave webs like wax moth and hide from light, so a small infestation is difficult to spot. Using **pollen patties**[384] increases SHB numbers enormously; contrariwise, if hives are surrounded for several metres by impermeable membrane it reduces numbers significantly (the larvae pupate in soil).

Australians and Americans find it is a big issue in their hottest / tropical areas, but only an occasional nuisance in e.g. Canberra. The key, as with wax moth, is to have large colonies and **remove large areas of unpatrolled comb where the creatures can hide.** This is an extremely important point.

Intensive management actually promotes this pest. Opening a hive triggers the SHB to start breeding! Also it has been shown that it can smell the plume of an open hive downwind and will fly 10km to reach them. Pollen patties fed to bees are like SHB booster shots, the beetles lay eggs in them and they quickly seethe with beetle maggots.

If you do have SHB, use traps which can be slid in and out of the base of a hive without opening it. They can be filled with oil (drowns them) or diatomaceous earth (dessicates them, much less messy than oil). There's no known chemical treatment, but lately Americans have found they can trick SHB into eating diatomaceous earth with *murder sauce*, a mix of ~75%

Adult, with worker bee for scale.

---

384 A mix of fats and pollen, or low grade pollen substitute like soy flour, fed to boost brood laying out of season or where a large apiary has stripped an area of pollen like an industrial vacuum cleaner.

shortening (solid fat for baking), ~15% diatomaceous earth, a few drops of eucalyptus oil (attractant, enhances the kill rate) and a few drops of food colouring, traditionally red (so you can see it's well mixed). A few teaspoons of this are smeared in the corners / edges of the bottom board. It seems to be ignored by bees and very effective at killing adult beetles.

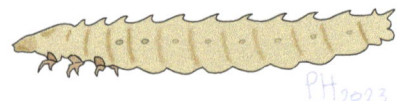

The larvae have 2 rows of spines down their backs.

Bees harass SHB and have been seen grabbing them and flying them out of hives, but SHB are very slippery.

Global warming will gradually make this more of a problem. Australians use chickens or guinea fowl in apiaries to eat the larvae - though chickens also eat the bees.

Tim Malfroy, an Australian Warré bee farmer, advises "If bearding is significant, nadir a new box with top-bars/frames and bearding will cease. Significant bearding leaves the colony vulnerable to small hive beetle attack."

It has long been known in Africa, that the smaller local beetles need a landing board to enter a hive[385]. SHB can't hover like a bee.

Although this is exploited by entrance guards like the one at guardianbee.com (which intriguingly further misdirects the beetles with red light), these products tend to be designed to be fitted to the intrinsically vulnerable slot-like entrances at the base of a conventional hive, which tend to have a shelf in front as part of the hive stand. More elegant surely to simply have a small entrance in a vertical wall…

Another African observation is that bees propolise their entrances down to one bee space in SHB areas.

Parasitic nematodes spread on soil around hives *(not in the hives)* have been trialled by the National Bee Unit - these infect the beetle larvae when they leave the hive to pupate in soil. The NBU found the commercially available *S. kraussei* and *S. carpoforcapsae* were pretty effective in Britain.

In NE Texas, hives in woods are overrun by SHB but hives in full sun survive. This is disputed by other American beekeepers. The heat presumably stresses the bees too though.

A possibly significant observation by an American is that a bee tree nest seemed to survive despite a heavy SHB infestation[386].

---

385 *Beehive designs for the Tropics*, Prof. G F Townsend, University of Guelph (1984), page 3. And as I mention elsewhere, without a landing board, ill bees exiting a hive fall out and do not return.

386 'Mill-J' in thread beesource.com/threads/hive-in-a-tree-not-a-swarm.373305/ . They remark that their bee-vac sucked bees from this nest along with SHB(!) which then infested the hives he put them in. *"If my hives had a quarter of the beetles that that bee tree had, I'd think they were goners. Yet that particular bee tree didn't seem effected by them. I saw zero beetle damage. Go figure."*

Shrew

See: mouse, above

A single shrew will kill a hive over winter.

Shrews are smaller than mice and can get through mouse guards. They're not rodents, they are hyperactive and eat voraciously, mainly insects and earthworms. Some have a venomous bite, and some use a crude echolocation (like a bat). So, basically, perfectly adapted to live in hives.

All British shrews are protected[387], but it seems legal to kill them if simply found in your hive.

Their droppings are longer than mouse ones, and spiral in shape. Mice chew heads off but shrews tend to eat the thorax.

Skunks and opossums

In America, skunks and opossums will scratch at a hive entrance to lure out bees and then eat large numbers, leaving characteristic scratch marks. The solution is said to be an entrance higher than they can stand on their hind legs.

Slugs and snails

Very occasionally, large slugs slime their way into a hive. Bees seem unable to sting through their thick skins. I have seen a 15cm long yellow one in a hive, the owner was lucky it had not slimed up a comb. It had to be removed by hand.

However, wrapping copper tape around a hive's leg stands is pretty effective if you stop vegetation forming a "bridge" past it. Which is why all my hives are on **legs**.

Pettigrew claimed snails were left alone by bees and could persist in a (skep) for months, but I've never seen one in my (wooden) hives.

Cut grass left below hives rots and attracts slugs.

| Slug or snail slime in a deadout

Snot brood

A nonspecific descriptive term sometimes used by American beekeepers to describe melted, discoloured brood from an undetermined cause (EFB, sacbrood, etc).

Also known as IBDS, Ideopathic Brood Disease Syndrome (Ideopathic means "disease of unknown cause"). Once thought to be caused by parasitic mite syndrome, but can occur with no mites.

It's not clear if it is a new disease or simply misidentifaction of others.

---

387   Schedule 6 of the Wildlife and Countryside Act (1981) forbids trapping, poisoning, shooting, exploding, electrocuting or deliberately running over them. Unless in your hive.

Spiders

Attitudes vary. Losses to spiders seem low these days - because there aren't many now - but Victorians considered them a significant issue. Remove webs near bee flight paths. American beekeepers occasionally report black widows inside hives.

Starvation

Symptoms

- Bees are listless, don't forage even when other hives are foraging (this may also be due to queenlessness)
- Activity shuts down before other hives' in dearths as bees conserve what they've got
- Heft the hive: is it abnormally light?
- No stores. Peel back the top cloth and look at the top edge of combs - this is where the last honey is stored. If even the brood combs lack top stores, they're starving.
- Drones excluded / beard of drones / piles of mature drone bodies below entrance (also happens after mating season)
- Bodies of bees have tongues fully extended (an ambiguous symptom, other causes possible)
- Dead bees with heads **deep** in cells (bees do occasionally happen to die with heads in cells from other causes, but then they are usually hanging part way out).
- Robbing other hives (typically after beekeeper takes too much honey)

Empty cells and dead adult bees **completely** inside cells indicates starvation.

Discussion

Starvation amplifies other stressors.

Dead bees outside a hive may also be heat / thirst stress.

In Britain we worry about nectar deaths. But pollen starvation is also possible:

- in large apiaries
- where honeybees did not co-evolve with the vegetation (America, Australia etc).

Treatment

You can feed syrup (see chapter 13). Suitable pollen is so plentiful in Britain and Europe you are unlikely to need pollen patties / substitutes.

I've collected swarms which just needed continual feeding. Syrup just seemed to be converted into more hungry bees. I adopted a ruthlessly Darwinian approach some years ago and now only feed when I feel the issue is not their genetics.

Stearin poisoning (larvae dying)

Very rare

Symptoms

Abnormally large numbers of young larvae die in comb, giving a spotty brood pattern.

Only happens when comb is built on [contaminated] foundation.

Discussion

If you can't figure out what disease is increasing mortality, this is a remote possibility. Wax adulteration is becoming more common.

There was a large scale die off of larvae in Belgium, Germany and neighbouring countries in 2016. The common factor was foundation from a supplier who had bought 'beeswax' in good faith from China. Some was bulked up with paraffin (which led to foundation melting when hot) and some with up to 21% stearin, which is used in soap and candles, and toxic to bees. Stearin also made the wax so hard, brood could not emerge.

Some ways of confirming wax adulteration are described at www.beeculture.com/facts-about-beeswax

Stone Brood

Very rare, not serious - for bees

Potentially harmful to humans, especially immunocomprised people (eg on chemotherapy).

Symptoms

These generally appear **after** capping.

Larval skin covered by powdery yellow, brown, green or black depending on species of fungus (*Aspergillus fumigatus, A. flavus* or *A, niger*). Final stage is a hard **greenish** mummified larva, difficult to crush, which the workers cannot remove from the comb.

Discussion, treatment

Fungus. Rare, implying a stress disease. Notifiable in some countries, but not UK.

The antifungal Fumidil B, used to treat nosema, was originally made from *A. fumigatus*. It lost its license for use in the UK in 2012 - simply because the market was too small to justify the manufacturers' expenses of keeping it licensed.

It's so rare that I've been unable to find much mention of it. It's seen as an opportunistic infection of hives weakened by other factors. I assume that these fungi are inhibited by warmth.

Distinguishing it from Chalk Brood

Stone Brood	Chalk brood
Larvae usually end up green (but sometimes black)	Larvae mummify first as white, then darken to black - never green
Not ejected	Floor / entrance littered with chalk brood mummies
Fungus erupts from behind the head.	Fungus erupts from the rear

Termites

These can be a huge problem. They eventually find their way to hives at ground level. They eat the wood of the hive, not the honey. Creosoting legs helps, but it dissipates over years and it is not usually uniform in penetration, so termites eat through the inside of wooden legs. Hives on metal stands have become more common, with oil or grease used to deter termites, but these are tenacious insects and they build bridges with their bodies to access the target.

Bases can be dipped in creosote, grease ringed and also further protected by inverting a tin, as ants and termites do not seem to be able to figure out going "up and over". Termites have problems getting through hardened propolis, so once a hive is old and propolised, they are less attracted to the wood.

Often the termites are simply starved of other forage - wood. Africans hang log hives from trees partly to avoid termites, though this doesn't stop ants.

Insecticides which kill termites also harm bees.

Toads

Historical sources mention toads lurking under hives and picking off the occasional returning forager, which are tired and easily caught. They don't seem an issue now.

Tracheal ("acarine") mites

See also: Isle of Wight Disease

Present in UK and US but no longer a major issue

<u>Symptoms</u>

- A lot of little clusters of bees in the hive instead of one large cluster
- Bees crawling round landing board with wings sticking up at funny angles ("K-wings" - note this term refers to ALL FOUR wings sticking out, not just 2, which is seen occasionally)
- May be accompanied by adult bees with distended abdomens
- Bees crawling away from hive or up grass stems, unable to fly
- Affects mainly young bees (under 9 days old) - still furry

View down thorax after pulling head off bee. You should be able to see this with a strong magnifying glass. Everything should be near-white. Left trachea are infected. Infected trachea may take on a beige tint.

<u>Confirmed by</u>

- Dissection under a low magnification stereo microscope. Two needles pin a bee by the chest to a cork, so the head and collar can be removed, enabling the spiracles to be examined.
- Heavy infestations can be seen with good eyesight or a magnifying glass, by pulling off a bee's head and looking at the first two thoracic tracheae. They should be pearly white. Mites cause discolouration.
- People often assume they are seeing a paralysis virus and don't think to check this.

Discussion

A tiny mite, *Acarapis woodi* which blocks some spiracles (breathing tubes) of bees. A large infestation will kill bees, perhaps by suffocation, but it only occupies specific spiracles. Some literature calls them HBTM (honey Bee Tracheal Mites).

Blamed, almost certainly falsely, for a mass British die-off in 1910-20 (see *Isle of Wight Disease*) - it was first observed at this time and appeared to spread rapidly, and these mites can vector the APV and CBPV viruses; but the mites are highly evolved for life in bees so presumably were always around at some level.

Bees seemed to rapidly adapt and rebound, possibly there was heavy selection pressure for...

- smaller spiracles?
- Stiffer guard hairs blocking entry?
- A heritable grooming behaviour with the middle legs[388]?
- It was never the real problem?

For whatever reason, perhaps simply because miticides commonly used against the larger varroa mite also kill it, it is no longer a problem in the UK/USA. Roger Patterson notes that in Britain, it only seemed to occur in Italian colonies[389]. Resistance to tracheal mites is now so widespread here, that when I went on a BBKA microscopy course, they couldn't show us samples and the instructor said he'd checked thousands of bees and never seen it[390].

Reached America around 1980, was a problem, but has subsided there too[391]. Currently the only place it is causing havoc is among Japanese *Apis cerana japonica*.

Note that there were no acaricides in 1920 except perhaps tobacco - this is yet another proof that bees can **rapidly** adapt to parasites.

A similar mite, *Acarapis externus* infests just the neck region.

Treatment

Female tracheal mites have to migrate between trachea and can only survive a few hours outside a bee. Grease patties (a mix of vegetable shortening and powdered sugar) placed above brood frames are effective - the shortening disrupts the mite's ability to find a new host.

A 2 week exposure to menthol will allegedly deal with these mites, though as a pungent chemical it will disrupt the colony's pheromone balance.

The mites can't survive without hosts - hives left empty for one week don't need sterilising.

---

388 www.glenn-apiaries.com/principles.html

389 He has 70 years' experience and mentions this in *Beekeeping: Challenge what you are told* p. 211.

390 In an article in the April 1988 issue of Beekeepers Quarterly A.C. Waring discusses the mite and how some colonies with crawling etc symptoms **do** have the mite but others do **not** display symptoms even though the mite can be found in dissected bees. He points out that the symptoms of Isle of Wight disease are identical to paralysis virus.

391 American Bee Journal Jan 2023 - *Where did the Tracheal Mites go?* by Alison McAfee - says they arrived there in 1984 but after an initial surge are now rare. It also mentions there is no proof acaricides help despite some vendors claiming so.

Tropilaelaps

Notifiable pest in UK and Europe - you MUST contact Bee Inspector

A tiny mite, which harms *Apis mellifera* (European honeybee) hives in Asia and is a potential pest the National Bee Unit is concerned about, because it does similar harm to, but reproduces much faster than varroa. Where varroa might kill a hive in 1-2 years, this mite will kill one in about 3 - 5 months. Native Asian bees such as *Apis cerana* can cope with it.

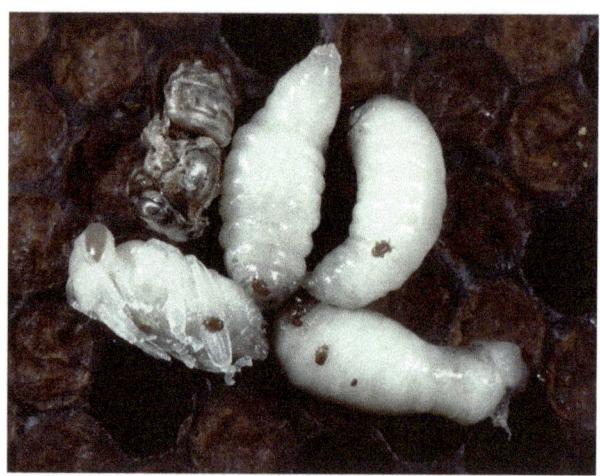

Tropilaelaps mites. About the same size as a pollen mite so confusion is possible, but Tropilaelaps scuttle incredibly fast between cells. Image by Denis Anderson, CSIRO, CC BY 3.0, https://commons.wikimedia.org/w/index.php?curid=35486198

Symptoms

- Your colonies develop a strange bald brood pattern (patches of bald brood).

- Deformed bees: kinked / missing legs and antennae, misshapen abdomens, irregular wings. The damage is caused by the mite's bites. **Fun fact:** *Tropilaelaps* have a double mouth like *Alien* with a feeding tube which pokes out once they've chewed a hole in the bee.

- Gently blowing over uncapped brood causes the mites to come out and run over the comb.

- Alcohol washes don't work - they don't seem to dislodge these mites, and unlike other mites and insects, this one *sinks* rather than floats in alcohol.

- Icing sugar rolls seem to dislodge them, though, and they can then be identified under magnification[392].

- The "comb bump" test - **very unreliable and kills many brood in the frame** - shake the adult bees from a capped brood comb. Firmly bump one side of the frame over a white surface, ideally a pan or greasy paper as tropilaelaps scuttle away rapidly. Turn the frame and repeat for a total of 4 bumps. Dislodged mites can allegedly be counted on the white surface. In practice, people report this rarely works.

- Mites can allegedly be seen if you uncap brood. The young ones are white and nearly motionless. But in practice, you generally damage the brood and their fluids hide the mites.

---

392 Maggie Gill, Victoria Tomkies and Dan Etheridge fact-checked a lot of the practical knowledge on a trip to Thailand funded by beediseasesinsurance.co.uk and published their findings in *BBKA News* July 2023 p. 222-224, discrediting the bump test, alcohol wash and uncapping test promoted in literature up until then, but discovering the trick of blowing over brood comb. Further work of theirs published in the Nov 2023 *BBKA News* showed that without live bees to feed on, *most* die within 24 hours; but if there are live bees present, a few mites can survive 120 hours - even on dead larvae! (And they discovered some samples later which survived even longer - unpublished data, personal communication) The implication is that a few find feeding spots on adults and can survive indefinitely without brood.

Discussion

There are four types of Tropilaelaps mite and two of them can parasitise *A. mellifera*. They live mainly in brood cells.

Our good luck so far:

It's not in Britain or Europe as of 2023

Whether they are any threat to hygienic *Apis mellifera* hives which can deal with varroa mites like *Apis cerana* does is unknown.

But don't get complacent:

- It's following the same migration pattern as varroa, so is expected in the US, Europe and Australia imminently. It's already in southern Russia.
- Early estimates that it will starve in 1.5 - 9 days without brood to feed on, limiting its spread, seem optimistic (see footnote, previous page)
- I would rate it as much more dangerous than the feared Small Hive Beetle, whose larvae need to pupate outside the hive where they are vulnerable to cold.
- A small global rise in temperature could allow it to establish even in Britain.
- It has evolved not to kill its natural host, the giant honeybee *A. dorsata* - but it can extract much more body mass from those without killing them. It does much more serious damage to *A. mellifera*.
- Miticides developed for varroa only work for a few months vs Tropilaelaps and it's thought this is because those miticides are only for the phoretic phase (when outside the sealed brood cell), so most Tropilaelaps are only exposed to a sub lethal level. And their genome turns out to have a ridiculous number of genes devoted to neutralising poisons, so they adapt really quickly.
- Unlike the near-clonal varroa, both males and females leave the cell so they mate with unrelated mites which accelerates gene mixing.
- *A. dorsata* have a "groom me" dance which *A. mellifera* does not have.
- Like varroa, it spreads and amplifies viruses. Certainly DWV and Black Queen Cell Virus, probably others not yet tested for.
- It needs constant brood to feed on so was initially expected to die in *current* British winters... but it turns out it can survive harsh, broodless Korean winters so that hope is lost. The inside of hives is quite warm.

Treatment

- Brood breaks - e.g. cage queen for 9 days and remove brood, or walk-away splits.
- Miticides may work - but expect rapid spread of resistance.
- Research[393] into using the larger varroa mite as a biological control (!!!) was inconclusive, but didn't find many cohabited cells.

---

393 *Competitive effect of Varroa destructor and Tropilaelaps mercedesae in Apis mellifera brood cells* Dongwon Kim, Chuleui Jung Andong National University, Republic of Korea

Varroa mites

Low numbers are normal.

See also: DWV

Symptoms *("varoosis")*

- Deformed wings (varroa spread and amplify DWV, Deformed Wing Virus)
- Dead mites on floor
- Mites on shoulders of adult bees

A severe (i.e. American) infestation is termed **parasitic mite syndrome:**

- All the above become common
- White varroa excrement (guanine) inside brood cells, white crystals like coarse salt; and white rims to brood cells
- Colony may become aggressive
- Off centre, ragged holes in capped brood.

Discussion

Runaway varroa numbers are a *symptom* of underlying stresses, not the direct *cause* of colony collapse. You need to address possible stressors, typically:

- Wrong bee stock (no hygienic traits)
- No brood breaks (overfeeding / overprolific queen)
- Cold hive diverting effort from hygienic practises

I actually forgot to mention varroa in the first draft of this book - they're a non-issue. To get this floor debris picture, I had to comb through four hives to find varroa. I'm not doing anything clever, I'm just not creating conditions they can readily thrive in.

But the arguments are being overtaken by events...

The Royal Society has published a metastudy[394] confirming that bees are developing resistance to varroa worldwide.

- African, Asian and Africanised (South American) honeybees are resistant to varroa.

Floor debris, with worker & drone for scale. A: wax capping, brood. B: wax capping, honey. C: wax scales dropped by bee before use. **D: varroa mites** (colour darkens with age - babies are translucent white). E: pollen. F: propolis. G: wax moth excrement.

96% of varroa usually hang on **under** the bee, only 1% on top as shown in most photos*. So looking at the backs of bees on comb is not a good indicator of mite levels. Looking at the underside of bees through windows is better.
*Ramsey et al, doi.org/10.1073/pnas.1818371116

---

394 *Parallel evolution of Varroa resistance in honeybees: a common mechanism across continents?* Grindrod & Martin (2021) Proceedings of the Royal Society B. royalsocietypublishing.org/doi/10.1098/rspb.2021.1375

- Imported European bees are resistant in: Cuba; the Primorsky region of Russia; Fernando de Noronha.
- These populations are all in areas where the locals had no access to miticides, and had to rely on natural selection. It wasn't rocket science, they just bred from survivors.
- Low-intervention beekeepers using local swarms in Britain, France, Ireland, and increasing numbers elesewhere, don't see varroa problems. The Swedish island of Gotland[395] is well studied.
- It's almost as if ectoparasite resistance is not just a *latent* ability in all mellifera, it's the *default* setting; but this resistance is absent in high intensity beekeeping, perhaps due to inadvertently deselecting it while breeding for other characteristics, or due to some feature of management style.

As BIBBA members tend to be very interested in commercial beekeeping and their management differs only in using local bees, yet rarely need to treat[396], this indicates that the biggest factor in suppressing varroa is using local bees.

Numbers are typically assessed by the *mite drop per day,* monitoring how many fall through a mesh floor onto a sticky floor[397]. 2-3 / day is OK, 10/day a bit alarming; but what really matters is not the absolute numbers but: *is the mite drop increasing day by day.*

A small level of continuous control turns out to be far more effective than even large levels of one-off control (see graph).

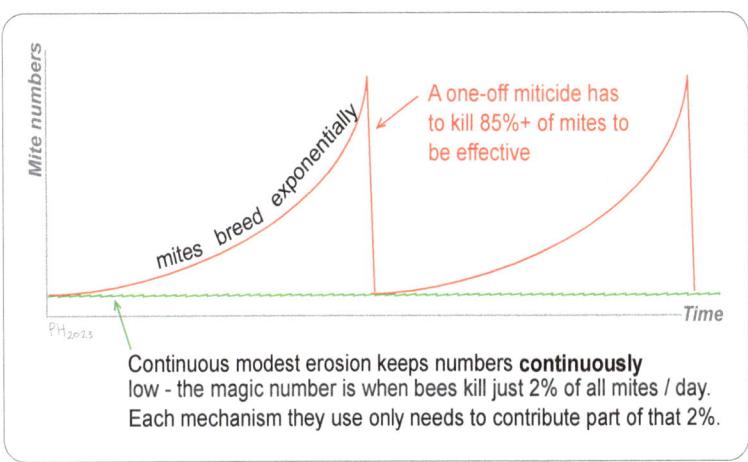

Continuous modest erosion keeps numbers **continuously** low - the magic number is when bees kill just 2% of all mites / day. Each mechanism they use only needs to contribute part of that 2%.

But even really hygienic colonies will struggle with huge mite loads, e.g. another colony dies with lots of mites nearby and your bees rob this "mite bomb". So keep an eye out for symptoms like deformed wings.

- You can tidy your house twice a year, or a little bit every day.
- Total mite numbers over the year - cumulative damage to the colony - is the area under each line.
- Based on modelling by Gareth John.

Location (weather) affects varroa. Scotland has a 2 month brood break, whilst mild Cornwall has none. Varroosis is more prevalent in regions of higher temperature, and reduces with increasing wind and rain[398].

---

395   Natural varroa mite-surviving Apis mellifera honeybee populations, Barbara Locke, 2015
396   I understand Jo Widdicombe of BIBBA is a bee farmer and treats most of his bees, but some are non treatment. He finds losses are higher with non treatment.
397   You buy these, or can use a piece of paper smeared with vaseline. Leave in place for e.g. 3 days, count the mites (tedious) and calculate average drop per day. Use the same size paper in the same position (under the brood nest). Mite drop is low during brood breaks. A newer, faster method is an alcohol wash, drowning about 100 bees - the mites float.
398   *Identifying the climatic drivers of honey bee disease in England and Wales*, Rowland, Rushton, Shirley, Brown & Budge, Scientific Reports (2021) analysing 317,000 apiary visits(!) by Inspectors over 10 years

An upside of varroa?

Arriving in Britain in 1992, it wiped out most of the hives in the country in just a few months.

This brought home the dangers of unrestricted importation of bees; how migratory beekeeping spreads pests; and probably gave the residual feral population a breathing space to bounce back, with reduced flooding of areas by foreign drones. It acted as a *keystone parasite,* reducing the numbers of badly adapted bees which had got out of balance.

Treatment approaches

- Natural selection (varying from the hardline "Bond" leave-and-let-die strategy I use, to more managed flavours like "Darwinian")[399]
- Targeted conventional breeding[400]
- Brood breaks -natural (swarms; forage breaks without feeding)
- Brood breaks - inadvertent (honey binding preventing laying, eg in small volume hives / natural cavities)
- Brood breaks - forced (queen confinement)
- Shook swarm - drastic method for runaway mite numbers (see chapter 13)
- Drone culling. Sometimes referred to as *biotechnical control* to distract from the fact this is deliberately raising a mass of animals to get infested, then killing them. (Left alone, bees practise a much more highly targeted form.) Another term for drone culling is *colony castration.*
- *Propolis (see below)*
- *Eco-floors (see below)*
- *Small cell theory (see below)*
- *Miticides (see below)*
- *A combination of techniques*

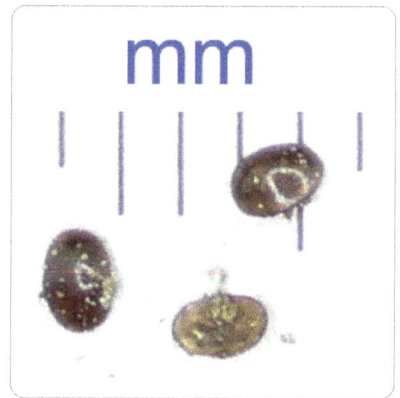

Dark brown mites are mature females. White ones are immature, or moulted carapaces. Males are white, smaller and rarer.

---

399 Kefuss' **Bond** strategy has high initial losses. **Soft Bond** means stop treating the most resistant colonies in a conventional apiary and if they survive, raise queens from them. **Darwinian** as proposed by Seeley aims for negative selection, i.e. eliminate the worst queens, and use of apicentric principles like well insulated hives - a form of **directed** selection rather than completely hands-off natural selection; Neuman & Blacquière discuss a more guided conventional selection, (*Limitations to Darwinian Black Box honey bee selection,* 2020, https://doi.org/10.1007/s10530-019-02001-0) and propose splits to propagate resistant colonies, which tends to reduce genetic diversity. Kefuss has experimented with **accelerated Bond strategy** where frames of infested brood are deliberately introduced to a colony!

400 Trailblazed by Ron Hoskins who began using instrumental insemination to propagate his survivor colonies in the mid 1990s. Already an experienced breeder, he avoided inbreeding after fixing traits by crossing queens with closely related drones, he crossed with *local swarms* he caught. Pure race breeders are greatly limited by inbreeding.

Propolis[401]

Heating the hive (pyroxia / fever / hyperthermia treatment) - Bees can take ~43C for a few minutes, varroa die ~38C. *Downsides:* Potential unreinforced comb collapse issues. Queens' sperm stores may be degraded. Complex - why not just use well insulated hives? Bees run brood areas at a slightly elevated temperature in the presence of varroa. Even if it does work - does not help select for resistant bees, they are still dependent on human help. *Question:* If pyroxia works, why do hives in Southern Europe have varroa? Their bees have to work hard to *cool* their hives. *Further information:* Dr Wolfgang Wimmer in Austria has developed a hyperthermia chamber into which you transfer just frames of capped brood - www.varroa-controller.com

Eco-floors (covered with leaves, compost etc) populated with commensal predatory *Stratiolelaps scimitus* mites[402] or book scorpions[403]. *Questions:* if it's desirable, why do tree bees keep floors scupulously clean *once comb gets within 3-5cm of the floor?* Perfect hiding place for Small Hive Beetle and wax moths. My hives seem pretty clear of varroa without these predators. I am unaware of any control experiments to demonstrate the varroa abscence was due to these populated eco-floors; success may rather be due to genetics, management or hive type, because the proponents tend to be very committed to local bees, low intervention, and apicentric hives.

Small cell theory. As discussed in chapter 11.3, *Reading Comb*, I'm sceptical of this but the proponents have noticed interesting correlations. For example Stephan at english.resistantbees.es/?p=279 found if he allowed natural comb, their bees made 15% drone cells *on each comb*, which acted to attract the mites away from worker brood, and seemed able to somehow suppress mite egg laying in the drone cells. An interesting argument against using foundation, if true for all bees.

Entrance position: Lazutin remarks[404] that floor level entrances increase mite loads, because bees pick up mites dislodged from above.

A combination of techniques[405].

The following approaches are **ineffective**: icing sugar (i.e. dusting an entire hive - particularly problematic in high humidity - bees end up sticky - note hive atmospheres are humid! Bees

---

401   Significantly reduces deformed wings on bees with varroa. *Honeybees use propolis as a natural pesticide against their major ectoparasite*, Pusceddu et al, doi.org/10.1098/rspb.2021.2101 - so old brood comb good.

402   Primary proponent: Geert Steelant (Netherlands – website: www.biobestrijding.nl/varroa-mite-treatment-with-predatory-mites). This mite is readily available by mail order as a biological control for aphids. Canadian researchers found this mite can be scattered into hives and reduces varroa levels without side effects to bees, honey etc: Rondeau S, Giovenazzo P, Fournier V (2018) *Risk assessment and predation potential of Stratiolaelaps scimitus (Acari: Laelapidae) to control Varroa destructor (Acari: Varroidae) in honey bees*. PLoS ONE 13(12): e0208812

403   Championed by Torben Schiffer (Germany – website: beenature-project.com) A really old British beekeeper (started in the 1960s) told John Haverson that book scorpions were in every hive until we began using miticides in the 90s, i.e. the chemicals destroyed the predators which had been our first line of defence. Perhaps they still inhabit wild tree colonies. **Further information:** Thomas Gfeller and Petra Studers' project in Berne; and www.dheaf.plus.com/warrebeekeeping/beier_1951_pseudoscorpions_english.pdf

404   *Keeping Bees with a Smile*, Lazutin & Sharaskin, part III, 'The Hive Entrance'

405   Westerham Beekeepers detail a mix that works for them at westerham.kbka.org.uk/natural-beekeeping

subjected to intensive *'sugar rolls'* to measure mite levels die in a week); bee gym[406]; FGMO (Food Grade Mineral Oil, aka paraffin - petroleum products are generally toxic to bees! And its vapour can explode near smokers!); sea salt (no control experiments to back up claims).

Appropriateness of miticides

Sometimes you have to treat, for example if you are starting with non-resistant stock and things get out of hand, or you are located next to a commercial *bee-yard* (an American apiary). I advocate *reducing* the use of miticides but recognise they are sometimes necessary. They should be a tool, not a crutch.

Sophisticated miticide users use IPM - Integrated Pest Management - only using miticides when needed, not routinely. They monitor mite levels.

Oxalic acid datasheets bear these warning symbols and advise wearing gloves and masks for good reasons.

Miticides - notes

These are eye-wateringly smelly.

Consequences:

- Hives get downright aggressive towards e.g. humans while they can't smell their queen[407].
- Bees can't identify their own hive, leading to drifting and fighting.
- The nurse bees normally detect damaged and diseased brood by smell, and eject them - 'hygienic behaviour'. This has been observed since at least 1964[408], in colonies with AFB, well before varroa were on the scene. It seems likely that hygiene is very common, *but beekeepers are suppressing it by using miticides.*

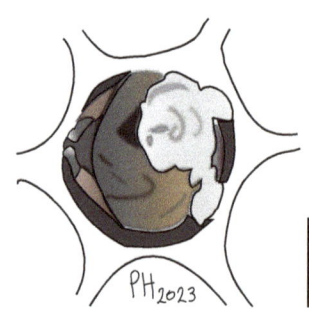

Newly hatched bees with severe varoosis may die **head down** in cells with white mite frass visible on their rears.

---

406 A controlled test on 20 colonies and found the Bee Gym had no measurable effect on numbers of mites dropped or damaged. *The effect of the 'Bee Gym™' grooming device on Varroa destructor mite fall from honey bee (Apis mellifera) colonies,* Pattrick, Block & Glover (2017). Incidentally, people have tried using queen excluders as a "scratching post" but found they gradually damaged wings.

407 Speculation: if workers can't smell queen and brood, there would be an increased incidence of drone laying workers - but I can't find any data on historical trends in DLWs.

408 Rothenbuhler, *Behaviour genetics of nest cleaning in honeybees* (1964). Another series of famous papers by Spivak and various co-authors studied brood ejection in response to EFB, varroa etc from 1998 on. The point is - hygienic behaviour seems widespread and longstanding.

Also:

- Miticides are toxic to bees[409] - it's a matter of dosage. People often observe several hundred dead bees after using miticides, this is "normal".[410]

- The volatile acids are dangerous to *us*. Particularly our eyes.

- Mites evolve resistance to miticides - rapidly. E.g: organophosphate *coumaphos*, the pyrethroid *fluvalinate* and recently *Amitraz*.

- They don't solve anything, just buy time

- They affect drone fertility and lead to early queen failure[411]

- They accumulate in wax - long term brood / queen exposure

- They remove selection pressure for the mite to become less virulent, so as not to kill their host, instead selecting for mites which breed *fast* before being killed

Mite frass (white dots) in a deadout. This frame uses black plastic foundation so it is easy to see. Image © Dr Rachael Bonoan, USA

Essential oils are pungent and have the same disrupting effect as miticides. All thymol does is make bees grumpy and less tolerant, it doesn't trigger magic specific behaviour.

Personal experience. In my first couple of years' beekeeping, with Buckfast bees in Oxfordshire, I saw huge mite counts and used miticides. I switched to local bees, and stopped feeding. After a couple of years mite drops had plummetted to 1-4 a day, I no longer saw deformed wings, and I stopped bothering with mite counts.

---

409 See e.g. doi: 10.3390/insects9020055 - tests on 5 acaricides - fluvalinate and coumaphos are particularly lethal to bees. A wider 2018 review, *Effects of synthetic and organic acaricides on honey bee health: A review* by Erik Tihelka (Bristol) DOI:10.26873/SVR-422-2017 details effects varying from adult death, brood death, loss of fertility, learning difficulties, larval weight loss, colony mortality(!), egg mortality queen losses; just like humans being treated with **chemotherapy,** the immune system is weakened so the patient is more vulnerable to disease; and discusses e.g. thymol is essentially harmless below 27C, but above this causes high bee mortality. The point is that miticides inevitably cause some problems, especially if applied incorrectly, but if you're next to a big apiary of commercial bees they may be your only option.

410 There are occasional queries on forums where people ask why an otherwise thriving hive died. Sometimes this is immediately after using **oxalic acid** vapour. The idea these may be linked is dismissed as their other hives survived. This is an example of *knowledge fade*. Ron Hoskins told me that back in 1993/4, *beekeepers began using* **formic acid** *to treat mites, this was fraught with problems as it wasn't just hazardous to the human applying it, it was easy to kill a colony by overdosing it. Acids are coming back into fashion partly because it's one instantaneous dose, though dosage and temperature is critical.* Beekeepers seem to be relearning this, e.g. thewalrusandthehoneybee.com/thoughts-on-formic-pro but view it as acceptable. Incidentally, mite strips placed above winter clusters can kill the bees by simply driving them to the cold walls.

411 In 1993 Ron Hoskins observed a sudden collapse in his queens' fertility and he realised there was **only one thing** that had changed, which was that he was now using miticides to kill varroa. So he went treatment-free, and their fertility returned. In 2007 academia caught up with the publication of *The Effects of Miticides on the Reproductive Physiology of Honey Bee* (Lisa Marie Burley) showing the drones had been affected – their lifespans and sperm viability are reduced by every miticide tested. In effect they were "shooting blanks", and as a drone only becomes fully fertile in the last few days of its life, a few days' reduction in drone-lifespan has a huge impact. Ron also found colonies treated with Checkmite (coumaphos) failed to produce viable queens.

Varroa - summary:

I *don't know* why my own bees have no varroa problems, I haven't done control experiments to find out why. I simply interfere with them as little as possible.

Stop Press - new discovery: It's just been found that some *brood* have a genetically inherited way of reducing varroa numbers, in addition to anything the nurse bees do[412]. The mechanism is unknown, but could involve disrupting the pheromone mating cues the mites use.

Afterthought: Varroa reproduction is hindered by high nest humidity[413]. But Small Hive Beetle are said to *prefer* high humidity for reproduction. If true, hive design may end up trying to trade off these competing requirements.

Orderly uncapping/recapping (round, central holes) and chewing (by nurse bees) of "faulty" brood - almost certainly due to varroa damage. Photo © Craig Turner

Wasps

See also: Robbing, above.

Common, sometimes serious

- Attack hives opportunistically
- Entrance modifications sometimes help
- May have to find & kill wasp nest
- Wasps are useful, don't kill needlessly
- Arguably, if a colony cannot defend itself with a small entrance, it is best out of the gene pool

Wasp signs

From images supplied by Susanna and John Blackmore

Typical dismembered debris below hive entrances

Symptoms

Some years wasps run out of other food sources and risk attacking bee colonies. Other years they don't, and they are useful at keeping pests down, so don't jump to kill them. Danger signs are:

- *Spring / summer:* they kill bees near the hive, remove head and wings and legs, and fly off with the high-protein middle bit for their young;
- *Autumn:* they grab bees on the landing board, rip their heads off to suck out the honey / nectar in their abdomen(!!!) and leave the body there.

Mice also leave dismembered bodies, but *inside* the hive.

---

412 *Host brood traits, independent of adult behaviours, reduce Varroa destructor mite reproduction*, Scaramella et al (2023) doi.org/10.1016/j.ijpara.2023.04.001

413 *High Humidity in the Honey Bee Brood Nest Limits Reproduction of the Parasitic Mite Varroa jacobsoni*, Kraus & Velthuis (1996) - very clear effect above 80% RH, but no effect seen from high nest temperatures

Discussion

Lifecycle: why wasps can change from garden ally to hive enemy in Autumn: Wasp queens hibernate alone over winter. In Spring they build a small nest and raise a first generation of workers, who then do the work while the queen concentrates on laying, until the colony peaks around 8,000 wasps (for the common and German wasps we usually see in the UK).

These nests gather large amounts of protein to raise their larvae in Spring and early Summer: the adults just need carbohydrates[414]. So adult wasps are extremely good for your garden because they kill huge numbers of aphids, caterpillars and other pests. They are also pollinators as they visit flowers for nectar and spread pollen.

So during Spring and Summer wasps are generally a good thing.

This changes in late Autumn as the queen dies (the nests only last one year) and the nectar sources dry up. The workers are now desperate for fuel and zero in on rotting fruit and alcoholic beverages (thus have a reputation for aggression). One potential source of food is honey, so they probe hives – which up to now they have largely left alone as too risky to rob. If they manage to get in and eat stores, they will smell like the hive's own bees and be able to re-enter more easily. Also they seem to guide other wasps to the food source, and the number of wasps spirals to indefensible levels: it takes 2-3 bees to overcome one wasp. It is best to prevent them getting in in the first place.

Wasp problems vary enormously from year to year. Their numbers are very susceptible to poor weather because unlike bees, they don't have stores to tide them over drought and heavy rain.

Around here, we tend to see wasp problems peak at the end of August - then temporarily stop as the ivy flow begins (it's the last common nectar source here). It's easier to get nectar from flowers than well guarded hives! But once the ivy stops, the wasps are back, though fortunately many have died off by then. Gareth John has a *useful rule of thumb for our area:* don't open hives for any reason between the last week in July to the first week in September, when the ivy flow starts, and it's safe to open hives again.

Of course, wasps don't read the books. In 2021, we saw wasps raiding hives and carrying off brood (more accurately, bits of brood!) in October. I asked Karol, a wasp expert on beekeepingforum.co.uk why they were doing this so late in the year. He replied that this was unusual but not uncommon. 2021's cold Spring had delayed wasp development. And now, some wasp nests were entering a second round of brood laying. Wasps sometimes raise brood as late as November in Britain. With little other insect prey around, they zero in on hives.

| Wasps robbing a deadout (through glass window)

---

414   The adults get much of their carbohydrate fuel from their larvae, which regurgitate a sticky goo the adults eat. The babies feed the adults. Weird.

Treatment: *anti-wasp strategies*

- See: **Robbing** section for some techniques.
- Some people kill queen wasps they see in Spring on the grounds that fewer nests means less problem later in the year. This is certainly a good idea for Asian Hornets but as wasps are beneficial early in the year, I advise not to.
- In August I begin reducing entrances to smaller hives. This deters wasps because the bees can guard a small entrance more easily – in particular, tunnels make them nervous. I've tried robber screens, which are meshes over the obvious entrance but if wasps have started targeting a hive they only seem to delay problems for a day. Be careful not to reduce entrances too much for large colonies: they need to breathe, and a permanent cloud of flyers outside may indicate traffic jamming.
- Remove piles of dead bees.
- Wasps are more flexible than bees and some are smaller. Watch to see if they are getting in through an unguarded crack, for example under the roof or through the floor. (*Poorly made WBCs whose parts don't fit together properly are essentially indefensible, but snug fitting ones are fine.*)
- I leave wasps alone until September. At that point I *only kill their nests if I notice them creating a serious problem for my hives*. They're going to die anyway.
- Ethical point: wasps have no options. Humans do. We don't just have programmed instincts. No need to kill something if it is not causing a problem.
- Once you kill the problem wasp nest, there's no need to kill others. Wasp colonies, like bees, learn and remember food sources; and not every colony risks raiding beehives.
- If the above doesn't work I escalate to placing wasp traps, both near and far from the hives.

Fallen fruit near hives attracts wasps and hornets

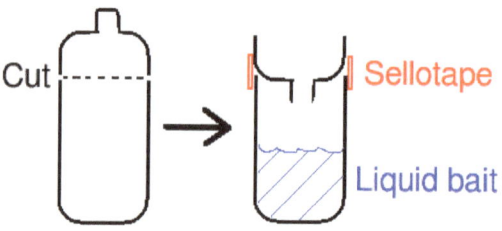

A simple wasp trap made from a plastic drink bottle.

Wasp traps - not necessarily a good idea

- **Traps *attract* wasps so only use them if necessary.** Theoretically, a good trap won't have many wasps in it - it catches all the scout wasps before they can go tell their friends. Don't put traps anywhere near hives unless the wasps are already attacking. You might choose to put one elsewhere to lower the background population of wasps, but in my experience in "bad wasp years" you won't make a dent in the population raiding your hives.
- **Non specificity:** Wasp traps catch *many* other insects too, and no one is sure they really help – once a wasp nest targets a hive it doesn't stop until you find and kill the wasp nest.

- **Bait liquid - early in the year**: water, with a mix of fruit (strawberry jam / marmalade have strong scents, or you can use ribena) and meat (they are particularly attracted to fish, for example fish skin). *Important*: add a drop of washing up liquid to minimise suffering (it lowers surface tension so the wasps drown quickly).
- **Bait liquid - late in the year**: meat is no longer craved. Switch to a sugary / alcoholic bait like cheap cider, plus a drop of washing up liquid. I've heard vinegar works too. Do **not** use honey or sugar as these can attract bees[415]. The cheap plastic bottle trap illustrated works fine: the principle is the wasps always fly up to escape, but the exit is not at the highest point.
- Wasp traps should be upwind of a beehive. If downwind, the wasps may ignore it and go for the attractive hive scent (its *plume*).
- In the USA you can buy wasp pheromones to attract wasps to traps, but I feel that is a bit unsporting.
- In Italy, wasps and hornets damage citrus fruit orchards, and the operators bait their traps with ammonia and a slice of anchovy. Bees are not interested in this smell.

Finding wasp nests

Tricky. Look for *movement* - wasps flying into / from:

- Compost heaps
- Air bricks in house walls
- Roofs – typically at the edges, or under loose tiles

Some wasps also dig in underground and these nests are really tricky to spot, being covered by leaves. They may be under a hedge, or a log. I've also found their distinctive egg-shaped nests hidden behind ivy very near my hives, or in trees.

The nest will be within 300 meters - they're short range flyers. They're territorial – I've seen 2 wasps fighting on top of a hive, presumably: "this is OUR food source!" They raid each others' nests.

A few final words on wasp-bee interactions.

- In my experience, the most important deterrent seems to be the morale of the bees. I've seen small young colonies with just 2 or 3 zealous guards unbothered by wasps who instead harried a larger but less vigorous hive – this is perhaps related to how much queen pheromone is motivating the sentries.
- Varroa treatments like thymol and formic acid prevent bees from detecting robbers by scent, so inhibit their ability to repel intruders.
- Some people say wasps are more of a problem in Spring/Summer because they communicate the location of rich food sources back to their nest, whereas in Autumn, it's every wasp for herself. However my experience is that late Autumn is the problem season.

Further reading: waspbane.com - apart from selling an excellent wasp trap, there are loads of practical tips on wasps for beekeepers.

---

415   Curiously, traps smelling of fruit and alcohol don't seem to catch bees, even though I've come across multiple references to bees eating the juice of overripe figs (Br Adam visiting Calabria) / windfall apples & raspberries (Bevan) / melons (internet). Basically, any syrup above 15% sugar. I suspect they don't appear to go in wasp traps because scouts who find them never report back.

Wax moth

See also: Bald Brood

More friend than foe

The Lesser and Greater wax moth (*Achroia grisella* and *Galleria mellonella*) are common in hives. The Greater is the one that damages woodwork, chewing into wood to pupate, and it eats the Lesser, but the Lesser can live in colder regions. The bees usually handle both.

But things **can** get out of hand in weak hives, towards the end of a colony's life, or if you have comb stored over winter.

Lesser wax moth larva.

Symptoms:

On hive floor / inspection board / in roof:

- white grubs, secured by a thread at one end. You will often see a couple of these; the problems begin if numbers spiral out of control.

On combs:

- Silk webs.
- Small black cylindrical faeces.
- Lines of uncapped brood with traces of silk ("bald brood")
- young bees webbed together.

Greater WM larva and adults. Females are much larger than males.

On woodwork:

- [Greater Wax Moth only] shallow grooves about 4mm wide carved into frames, top bars and hive walls where the wax moth pupate. In really bad infestations this can riddle and ruin the woodwork. And poly hives.

Discussion / treatment:

Wax moths and bees have a *commensal* relationship, both benefit. Moth larvae dispose of old, rotten brood comb and are not a problem for a healthy hive: whilst bees have difficulty chewing through silk-reinforced old comb [though bees do manage to uncap cocoons when hatching], moth larvae slice through silk easily *and digest it*. The remnant

Badly mothed comb (with yellow pollen spread around) showing webs and black tubular faeces. I would burn or bin this.

crumbs are sterile[416]. If the comb is well populated, the bees will stop them, by biting the larvae.

The larvae need the nutrients in **dark comb** and aren't very interested in honeycomb or new comb. They need propolis, the proteins in silk, pollen. So they eat **dark brood** comb and almost completely ignore new white honeycomb - though they will eventually damage that if they've eaten everything else. The larvae dislike being daubed in honey (whereas lesser WM larvae eat it).

Beekeepers who spin frames and then store the combs over winter for re-use in Spring, kill WM eggs by

- fumigation with SO2 (burning sufur) or ethanoic (acetic) acid
- freezing for 48 hours
- big operations gamma-irradiate
- Bt (a bacterium, some strains are GM) - kills WM larvae

Lesser WM

Lesser WM, and web trail in brood comb characteristic of both WMs

Greater WM - much broader and about 50% longer. (Also some varroa near the number "20")

Mothballs are no longer used (carcinogenic).

Bald - uncapped - brood will usually develop normally, but sticky webbing across brood cells can prevent hatching bees emerging. So as the moths head for dark, older comb it is important there are enough bees to patrol these.

Greater WM males are significantly smaller than females.

Warm weather boost: Moth activity is slow in winter. But in summer these opportunists can overwhelm a weak hive with a population surge in just days. If a hive has more comb than it can patrol (ie after swarming, numbers reduced) wax moths can get out of control.

---

416   NBU leaflet Fact_30_wax_moth.pdf: when they digest old comb they destroy foul brood: cases of AFB fell in New Zealand after greater wax moths were inadvertently imported.

Fun facts:

- You may well have fed waxworms to pets or wild birds. These are wax moth larvae - someone deliberately raises them!
- The Greater WM can digest polythene (!) and has been known to damage poly hives.
- The crumbs left over after they eat comb are a strong swarm lure.
- The larvae always anchor themselves with a silk safety line so you can't rely on them falling out if you tip a comb. Even if you do knock them on the ground, they can wriggle 50m to a hive.
- Will Hanrott spotted a swarm ejecting WM larvae from their new home.

Woodpeckers

Symptoms:

Fist-sized hole in wall in particularly harsh winters. Nest area destroyed.

Discussion:

In Britain the Green Woodpecker, an insectivore, sometimes raids hives in winter, if frozen ground prevents them eating their normal prey, ants. It is probably a learned behaviour as it isn't universally seen. After it creates a hole, rats sometimes follow.

I've not heard of it happening to Warré or horizontal Top Bar Hives, probably because these don't have convenient ledges to perch on whilst pecking. WBCs are said to be ignored too. This may be because they typically only drill if they hear a response to an initial tap, and thick or double walled hives muffle this so are less at risk.

Poly hives are particularly vulnerable.

Treatment:

The hive is wrapped in chicken wire, so they can't get near enough to the walls, or plastic sheeting to create a continuous smooth surface covering protruding perching ledges.

John Carnegie found an easier and reliable option is to hang CDs from branches near the hives. The flickering as they rotate puts off the woodpeckers.

Review 12

Your hive has dozens of larvae and adult bodies outside. How would you distinguish between disease and starvation?

Activity at your hive entrance has dwindled away and hard crumbly white pellets are outside the entrance. Is this simply queen failure?

What would trigger you to contact a bee inspector?

Chapter 15
Hive post mortems

Refer to your notes for clues

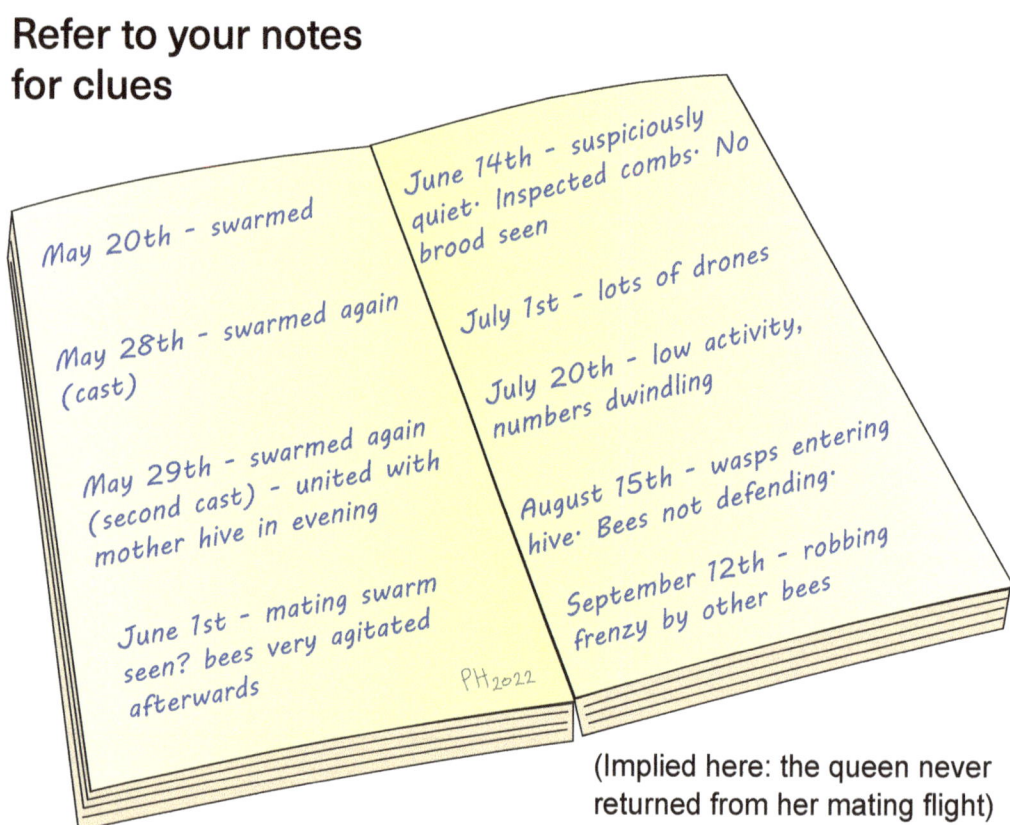

(Implied here: the queen never returned from her mating flight)

A dead colony (a 'deadout') should be examined to determine cause of death.

It is also a fantastic learning opportunity.

A side aim is to harvest any honey before it is robbed!

If a disease, etc is implied refer to chapter 14.2, *Disease Specifics*

Collapses can be **multi causal with no clear single cause.** Once a colony drops below critical mass it can collapse in a week, faster than undertakers remove bodies, **leaving a carpet of bodies on the hive floor**[417].

---

417   Beware glib assumptions like *"piles of bees on the floor implies either starvation, or a fast acting disease which killed bees faster than they could crawl out / be ejected."* Such clues are all somewhat ambiguous **indicators**, *not* **rules.**

External clues

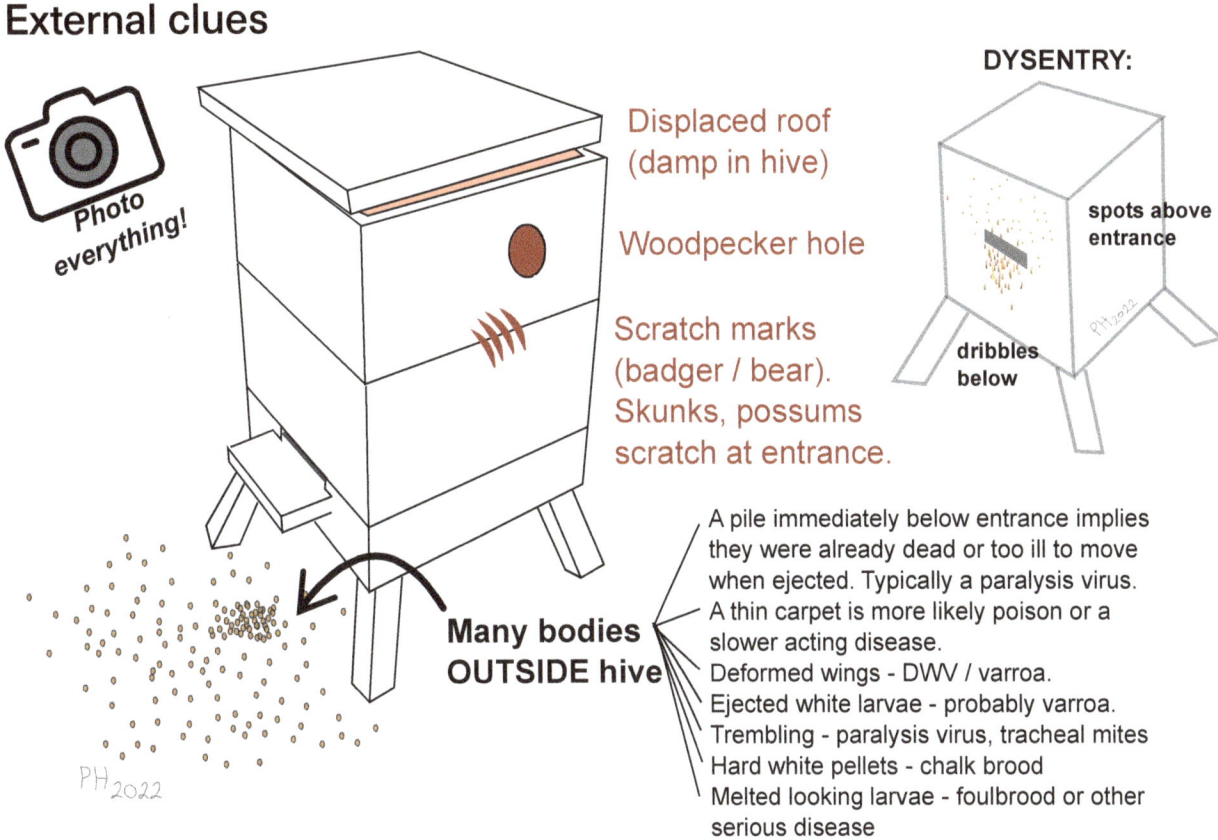

- Displaced roof (damp in hive)
- Woodpecker hole
- Scratch marks (badger / bear). Skunks, possums scratch at entrance.

DYSENTRY: spots above entrance, dribbles below

Many bodies OUTSIDE hive
- A pile immediately below entrance implies they were already dead or too ill to move when ejected. Typically a paralysis virus.
- A thin carpet is more likely poison or a slower acting disease.
- Deformed wings - DWV / varroa.
- Ejected white larvae - probably varroa.
- Trembling - paralysis virus, tracheal mites
- Hard white pellets - chalk brood
- Melted looking larvae - foulbrood or other serious disease

Photo everything!

Fresh dismembered bodies implies a predator - see *Mouse, Shrew, Wasp* in the Disease / Pest chapter.

Poisoning: tongues sticking out is a **non specific symptom.** *Most* bees die with their tongues out. Poisoning is extremely rare in Britain, more common in America. Most poisoned bees die on the wing away from the hive. Most "poisoning" turns out to be a paralysis virus. But if poisoning is suspected, *do not touch these bees with bare hands.* Put 200+ bodies in ziploc bag in freezer to show bee inspector, who may ask you to send them for analysis. If possible, select bodies carrying pollen, which can be examined microscopically to see what they were foraging on.

Note time, weather conditions. Refer to *Poison* in Disease / Pest chapter.

Heft the hive. If very light, could be starvation, whereas lots of honey in a winter deadout implies the colony died at the start of winter.

Internal clues

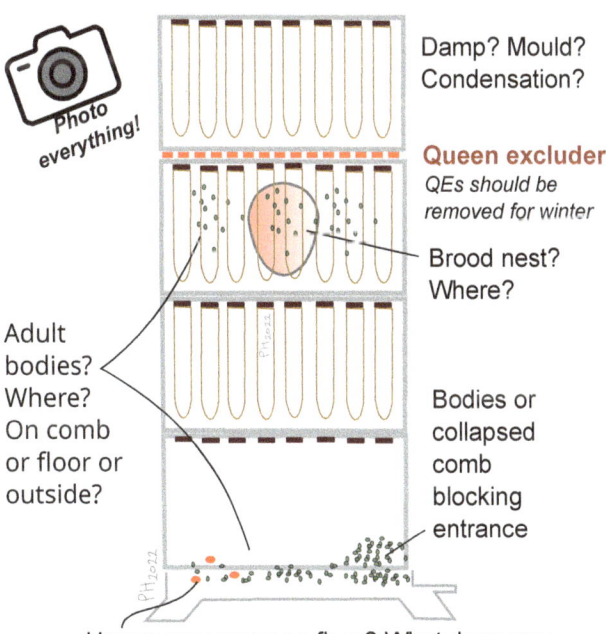

Photo everything!

- Damp? Mould? Condensation?
- **Queen excluder** *QEs should be removed for winter*
- Brood nest? Where?
- Adult bodies? Where? On comb or floor or outside?
- Bodies or collapsed comb blocking entrance

How many varroa on floor? What does wax debris look like? Freeze debris and save for a few days - you will think of questions later!

Bees die of old age in hives which go queenless and dwindle over months. Many young, furry bees implies a faster cause. Bees on a floor may be covered in wax crumbs; this is simply wax moth detritus, post mortem.

Varroa migrate to the last few living bees so finding a couple of bees with mites means little. I'd be puzzled if I did *not* see a couple of varroa. Look at the floor (dead mites) to gauge levels of infestation. Many dead bees with shrivelled, deformed abdomens and / or wings is a sign of varroosis, as is white crystals in brood cells.

Mould implies cold: eg corners, floor: can you improve insulation?

A soggy quilt, or damp top board (depending on hive type) covered in black stains implies a lack of ventilation trapping moisture - this leads to water dripping on the cluster, chilling them and promoting mould. Quilts can be too dense. This scenario is most likely for large colonies.

Consider recent weather. Lots of dead bees on the floor may simply mean poor weather stopped them flying out over winter.

Investigate mysterious cell contents

Pull cells apart with tweezers, toothpicks and a magnifying glass.

- What's under mould?
- What's under cell cappings?
- What state are brood in?

Followup actions

- Write up observations
- Do you need second opinion
- Do you need to contact bee inspector
- Harvest honey
- Remove bulk of detritus to limit wax moth damage to woodwork. Scorching is only necessary if the hive was diseased; propolis and wax are strong swarm lures. No one ever cleaned a tree cavity: you want to *retain* the microbiome. Keep a couple of brood combs as swarm lures for next year, but wrap them in plastic and freeze for 48 hours to kill parasites.

Analysing ambiguous deadouts: example

Winter starvation. Tight cluster of dead bees surrounded by empty cells. The central bees are black - being damper they rotted post mortem. The floor of the hive would have many more bodies. The more mould, the longer they have been dead. If you don't see a tight cluster, dissect for evidence of tracheal mites (see entry in Disease section - but they're rare now).

3 hives were found dead in February. They had been flying in January, each contained 5kg stores, masses of dead bees on the floor (not outside) and had been confined by poor weather. A neighbour's hives had identical symptoms. A number of people discussed these (total experience, about 140 years).

Comb clues

Queen cells
Many clues from shape etc. Refer to "Reading Comb" chapter

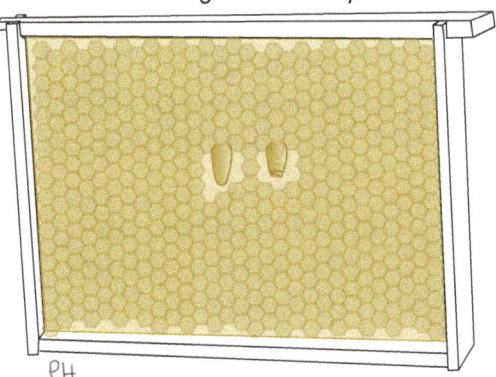

Drone layers

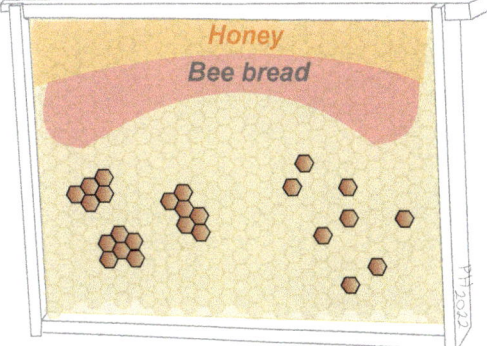

Isolated patches of drone brood: failing queen (out of sperm)

Scattered, individual drone brood: laying workers

Dysentry
Dribbles of brown faeces. (Nosema tends to be just spots, healthy poo is yellow)

Starvation
Heads so deep in cells, only tips of stings are visible. In winter, clustered for warmth.

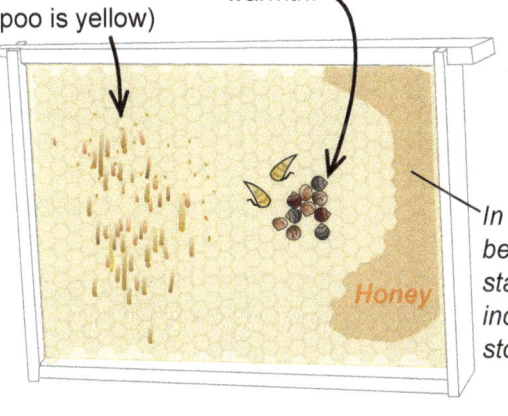

In winter, bees may starve inches from stores

Dead brood

If no brood in summer - probably queen missing / queen failure

If brood bodies seem odd: check colour, state of bodies, then refer to "Brood disorder ID chart" at start of "Diseases & Pests" chapter

Mould, webs:
In brood comb:
  Chalk brood
  Wax moth larvae
  (Rarely): spiders; stonebrood
  (After a cold, damp period): often simply
      damp pollen stores etc (not a bee disease)
On honeycomb:
  Small Hive Beetle larvae - accompanied by slime

| Note: you can simultaneously have a drone laying queen and drone laying workers.

- Not queen failure - large patch of worker brood.
- Not isolation starvation - remaining cluster is on honey.
- The remaining cluster was probably too small to generate enough heat to keep warm and likely died of cold.
- Adults mouldy - but only brood have fungal diseases, so this is irrelevant, just a random mould growing through already-dead bees.

- Floors absolutely covered in dead bees.
- No deformed wings or varroa.
- Lots of small maggots, probably from flies in one hive (one is red in this picture) - we decided this was unrelated, opportunistic laying of eggs by flies in already-dead bodies.

| Both photos © 2021 Andrew Bax

Conclusion: probably paralysis virus; the apiary experienced ABPV 6 years before. Weather-confined bees were unable to eject bodies (bodies outside hive in winter are a *good* sign, colony actively ejecting its dead). Possible complications from gathering excessive ivy honey in September, reducing their options - when clustering for warmth, bees are more insulated if they can lurk in empty cells. Hive owner reported quilts above bees were damp but not mouldy, so excessive moisture was not an issue, but they are reviewing the hives' insulation. Hives need sterilising and comb burning after honey extracted.

Note that there were a lot of different clues pointing in different directions, because new problems kick in as a hive dies, and even experts can't be 100% sure of the root cause. I myself initially assumed the many black bodies on the hive floors were simply rotted, concentrated on the more complete bodies and overlooked paralysis viruses. Obviously they can strike in winter too.

Points of note

A common winter death mechanism is simply *too few bees*. You need a critical mass to keep warm, raise the next generation of brood etc.

But *too many bees* can also be a problem, they generate a lot of humidity and in damp winters this condenses in the hive leading to excessive damp. Large colonies need more ventilation in winters; if they have a quilt, check if it is soggy.

Marie Celeste Syndrome

A term for a hive where all the bees have mysteriously vanished. This could be due to CCD, where stores remain; or absconding, where the stores are sometimes relocated to the new home, or robbed out by other bees / wasps.

Ambiguous deadout: further example

A hive died. At first glance it resembled starvation: empty combs and a pile of dead bees inside. But this made no sense - it was the end of Autumn and the hives next to it were bulging with stores.

The combs were all empty - even most of the pollen was gone - but pristine, no sign of robbing (ripped open caps), everything was left in good order.

On examining the dead bees there was no sign of disease / varroa / wing damage, and (big clue!) they were all young and furry. (This key evidence would have been obscured if I used eco-floors.) There was no sign of capped drone brood (laying workers / failing queen) or any drones or mini drones among the bodies. There were about 12 capped worker cells, which I concluded were simply the last few worker brood chilled in their cells because no one was covering them.

Conclusion: the foragers suddenly defected to a different hive, taking the stores with them, having suddenly lost their queen and finding they had no options (eggs to raise new queen). The last generation of young bees stayed in the hive because by the time they were old enough to venture outside, everyone who *could* fly had left. The children didn't know where to go and huddled in place until they'd eaten anything that remained, then starved.

Chapter 16
Breeding, Genetics and Rapid Adaptation

Races, lineages and mixes; colour; insect DNA; drone genetics; supersisters and other curious family relationships; mutants and intersex bees; hyperpolyandry and obligate outbreeding to maximise gene mixing; diploid drones; chromosomal crossover; imports dirupt sequences; negative vs positive selection; aggression of crosses; rapid changes from queen selection and correspondingly rapid reversion to stable local ecotype; my largely-leave-alone breeding strategy; natural outbreeding vs commercial inbreeding; how human and bee reared queens differ; local landrace lessons; conclusions; considerations before adopting a leave-alone open mating strategy

> *"Bees' winning formula is their inconsistency"* - Dr C.C. Miller

Here we look at races; how mating and heredity works in bees; how humans breed bees - and some flaws in their assumptions and methods - and come to some conclusions about the **how, why** and **what** of the facts behind bee breeding.

This will give you a level of understanding similar to that of a *queen rearer*, but with an emphasis on theory and implications rather than practical queen rearing techniques, which you can learn from many books or conventional queen rearing courses. However, once you appreciate the subtle but powerful nuances of artificial selection, and its unintended fallout, you may well favour leave-alone natural selection.

There's some technical stuff here, about how insect DNA and mating operates differently to mammals'. Don't be fazed - it's so weird it's fun.

Races (breeds)

European *Apis mellifera* populations were isolated for 2.5M years - millions of generations! - by physical barriers, diverging into distinct races adapted for specific environments. When the glaciers retreated 23,000 years ago, *Amm* spread across cool northern Europe. In the last few hundred years, these relatively mild tempered, easily farmable races have been exported around the world.

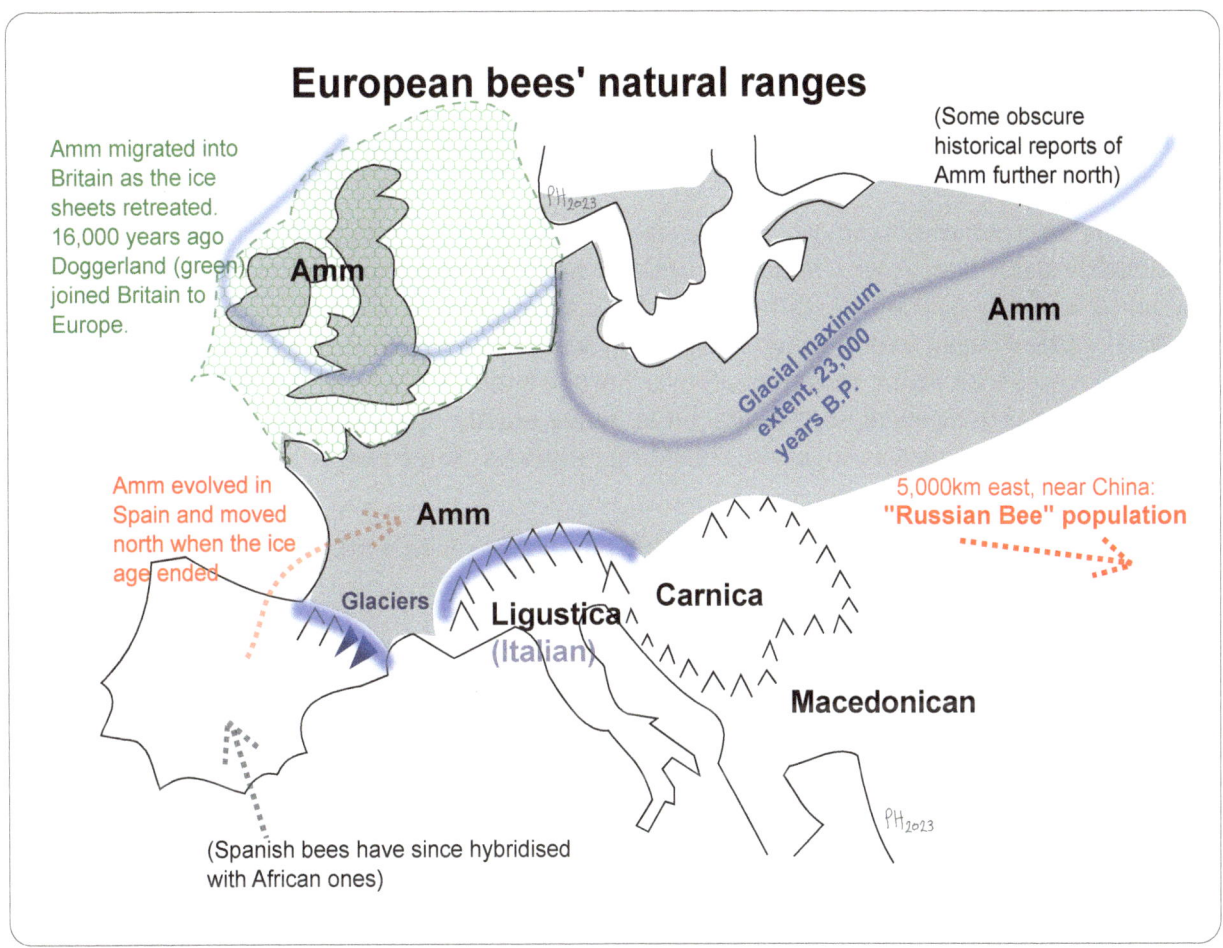

Due to global warming, wild bees can survive further north each year.
Showing every race would clutter the map. There are more.

Significant races

Amm (*Apis mellifera mellifera*, the Black Bee) has many ecotypes. Some are brown. It's been disparaged by commercial breeders pushing their own bees for over a century, and has been deliberately replaced in large areas (notably most of Germany), but it is persistent and hardy and making a comeback. It has several adaptations for long damp winters.

Around Europe, the Middle East and Asia there are genetically distinct **Spanish**, **Sicilian**, **Cypriot**, **Caucasian** etc populations of Western honeybees. Most of the southern European ones were trialled in Britain by the Victorians who concluded they gave aggressive crosses and died a lot in our climate, though **Italians** (*A m ligustica*) can just about hang on if you feed them a lot. **Carniolans** (*A m carnica*) from the Alps are popular in Germany. Some breeders in Cyprus actually raise genetically British bees, allowing them to supply exceptionally early packages / nucs to bee farmers in Britain in Spring - much like Californian breeders raise early bees for America - but sometimes the bees they send are British queens with a package of Cypriot workers to get them started, which raises the possibility of drone laying workers spreading Cypriot genes here.

Mixed races

'**Russian Bees**' are a genetically stable population founded by a mix of bees imported to the Primorsky region of eastern Russia by boat from 1870, and the trans-Siberian railway after 1904. Their DNA is a mix of **Carnica, Carpathian, Italian, Amm** and **Caucasian (Ukrainian)**, but sources disagree on which is dominant. In Primorsky they encountered *varroa* and developed resistance to it. For this reason they're becoming popular in America, though may not be the best for honey farming.

Buckfasts are not native to anywhere. They're an artificial cross, but unlike Russian bees not a particularly stable one, as we will discuss under "aggression of crosses" below. They are primarily Italian (specifically *ligurian*, north Italian) and Black Bee (*Amm*), with traces of many other races. The breed began with Brother Adam of Buckfast Abbey around 1920, who spent 70 years trialling crosses with exotic bees from all over Europe and north Africa. Buckfast make-up was refined over this period, so it is difficult to define exactly[418]. These days the Federation of European Buckfast Breeders supervises breeding programs. Some people apply the term to any home-bred multi-hybrid.

Local landraces (ecotypes) are **genetically stable** populations with a broad gene pool, adapted to a very specific region. They may be a local variant of a pure race, or a genetic blend of several, in places where races' geographical ranges overlap.

Africa has many tough pests and European bees don't thrive there, so locals use one of the many distinct subspecies there, which are much tougher - and more defensive - than European bees. They rapidly became resistant to *varroa* when it was imported. You hear mainly about *A m scutellata* - the one which crossed with European bees to give the aggressive **Africanised Honey Bee** in the Americas. In some regions African bees migrate hundreds of km seasonally in huge combined swarms with up to 60 queens. Of particular note is the **Cape Bee,** *A m capensis* which has a unique ability for workers to lay fertile eggs, clones of the worker mother (*thelytoky*). Its workers readily invade and parasitise other species' nests: the number of active foragers dwindles as their clones lay more clones, the original queen dies and the colony collapses. South Africans refer to the importation of this bee to the wrong region as the *capensis calamity* (widespread colony collapses).

Asia has many species of bee, such as *Apis cerana* and *A dorsata*, mentioned elsewhere in passing but their management is beyond the scope of this book. They're not Western honeybees (*A mellifera*) and have very different behaviours, for example they abscond and migrate a lot.

---

418   There are few definitive analyses of Buckfast DNA, they're usually described as "very variable". This short Japanese paper doi:10.1080/23802359.2018.1450660 analysed DNA from a Northern Irish Buckfast bee, finding the Italian strain dominated. The current beekeepers at Buckfast Abbey, Clare Densley and Martin Hann have analysed Brother Adam's records and find the Buckfast bee changed over time - Br Adam added French *Amm* to his original "B1" line in 1940, and a Greek cross in 1960. The Abbey no longer gets royalties and many people now label their own crosses "Buckfast". They now use local bees. (2022 Zoom lecture, *Beekeeping at Buckfast Abbey*, BIBBA.com)

Lineages

Where should one draw the distinction between different types of bee? Beekeepers tend to use the informal, imprecise term *race*. Strictly speaking, the taxonomic order could be written *species, subspecies, race, strain, local ecotype (landrace)* - but only *species* and *subspecies* are official terms.

In parallel to this, scientists sometimes refer to *lineages*. This isn't really relevant to practical beekeeping - it is part of a debate about how bees evolved and they occasionally reclassify races into different lineages as new DNA evidence is uncovered[419]! So I suggest you ignore this unless you are a bee researcher, but for the record there are 5 lineages -

A - African - 11 subspecies

C - Carniolan, Italian and some more obscure subspecies

M - Mellifera (western Europe), 3 subspecies

O - Oriental (7 subspecies but not Asian bees like *A. cerana* - those are different *species*)

Y - Ethiopian (3 subspecies)

Hybrids, mutts and mongrels

English-speaking beekeepers tend to use the word *hybrid* to mean a cross between two subspecies. Strictly speaking, a hybrid is a cross between two *species*, and usually infertile (e.g. mules). This can cause confusion when precise language is required, talking to scientists and foreigners.

Mongrel and *mutt* are terms some queen rearers use to disparage local landraces which don't have a recognised name like *ligustica*. Since their makeup is irrelevant if they are a stable genetic population, and better adapted to their area, and healthier than imports, the terms tell us more about the rearer - and his product - than the local bee.

Hybrid vigour (heterosis) occurs when two races are combined and create an artificial diploid, with excellent, but different virtues coming from both sides. Example: mules and hinnies are tough donkey/horse hybrids. This is different to simply using outbreeding to undo the *inbreeding depression* of single-race breeders, it's an actual boost beyond either race's normal abilities. If fertile, these crosses don't tend to breed true (like F1 vegetable seeds). Buckfasts in my experience lack hybrid vigour, they're merely prolific, and susceptible to varroa.

Hybrid vigour

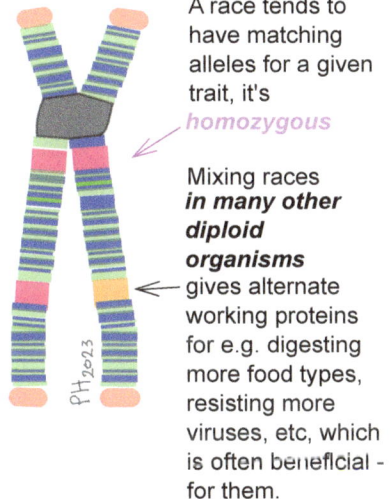

A race tends to have matching alleles for a given trait, it's *homozygous*

Mixing races **in many other diploid organisms** gives alternate working proteins for e.g. digesting more food types, resisting more viruses, etc, which is often beneficial - for them.

An F1 cross of two breeds won't breed true: its offspring only get one of these two great traits.

---

419 In the course of writing this book (4 years) the "accepted history" has evolved twice. The latest DNA meta-analysis is *Multiple mitogenomes indicate Things Fall Apart with Out of Africa or Asia hypotheses for the phylogeographic evolution of Honey Bees*, by Stephen Carr (2023), doi.org/10.1038/s41598-023-35937-4 . He found some discrepancies between what mtDNA and nuclear DNA implied about origins, notes that many early papers were based on one sample bee (DNA testing was very expensive) and suggests modern bees actually originated in northern Europe about 780,000 years ago and migrated down right into Africa - not the other way round!

Colour

Bees *thrive* on variation in their colony. Seeing non-uniform colouring in a hive indicates a healthy spread of traits. This photo fom a hive in Oxford is typical, but some colonies have much more varied colouration. As a rule of thumb, bees from a breeder will be much more uniform in appearance; breeders use this to spot outbreeding.

Colour is actually a poor indicator of race because, like our own eyes, many genes affect colour. But it conveys *some* information. I've noticed that if I catch a very uniformly patterned, light coloured swarm, they are less likely to survive their first year - because they're commercial bees which are too dependent on feeding and miticides.

My local bees are more like those in the photo, often somewhat darker. My queens start off bright yellow-orange but within a few months are at least as dark as the one in this photo.

This queen's children have a healthy amount of pattern variation

Cordovan bees are not a race, but a colour. The black chitin - exoskeleton - in bees with this mutation is brown, so the bees look lighter than their siblings because their stripes and shoulders are brown / tan instead of black; eyes may be pink. They're usually Italian stock, and only bred in the USA.

Bee DNA

Male honeybees have 16 **single** chromosomes. Females have 16 **pairs.** This is completely different to mammalian DNA, which operates on pairs for both sexes. Furthermore, bees can't easily be penned in so most queens mate with whoever they like. This has many consequences for anyone trying to breed bees.

Terminology

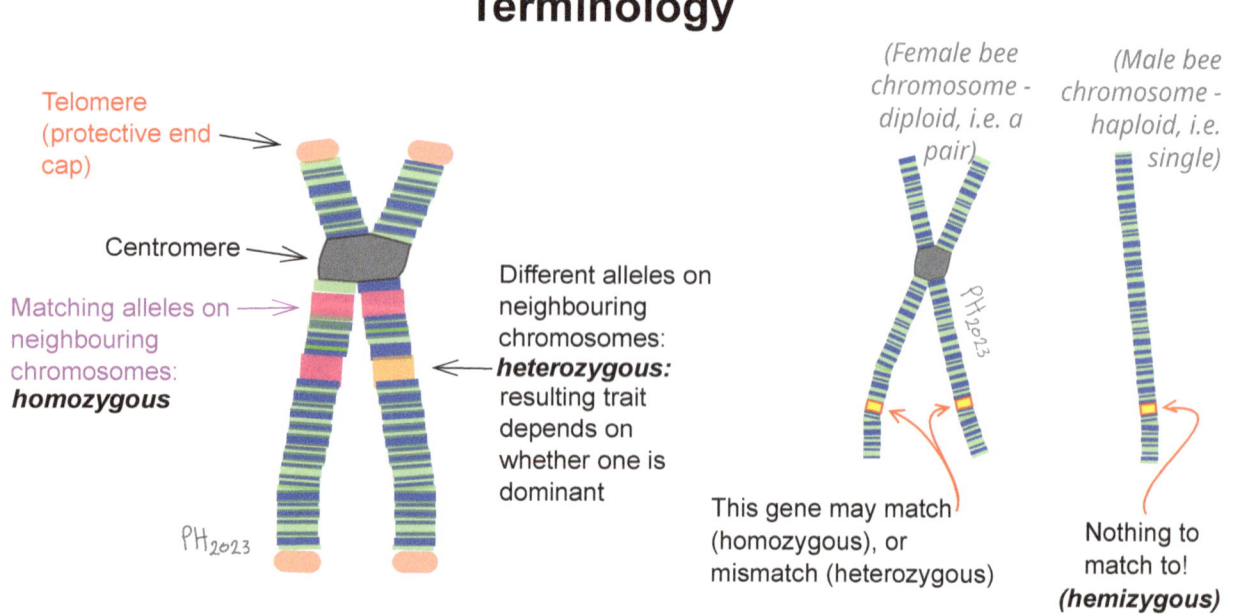

Chromosome / DNA structure

Other DNA, outside nucleus

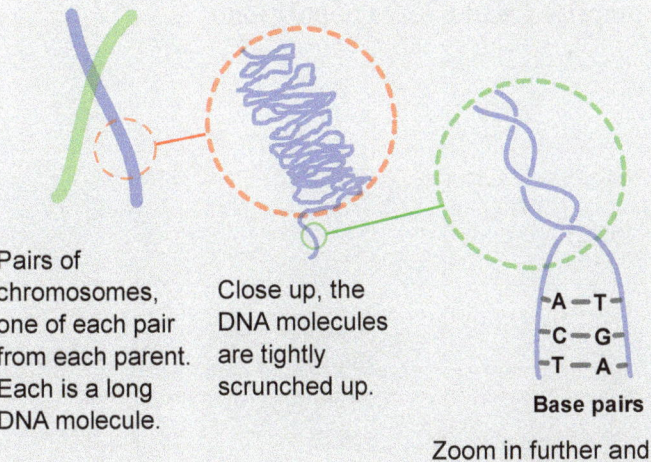

Pairs of chromosomes, one of each pair from each parent. Each is a long DNA molecule.

Close up, the DNA molecules are tightly scrunched up.

Base pairs

Zoom in further and you find the DNA molecules are double helixes many base pairs long.

Mitochondrial DNA is in rings (bacterial origin?) and is only inherited from the mother

Self replicating loops or strands of DNA or RNA called plasmids also float around cells; they are common in bacteria. Some contain useful genes.

Drone genetics

Drones have no father. Their 16 single chromosomes (haploid) come solely from their mother. When she laid their egg, she didn't fertilise it by addition of a sperm.

Yes, you read that right. No chromosomes from a male parent → male bee.

Before a queen lays an egg in a cell, she measures it with her antennae; and for small worker cells adds 5-10 sperm, from the store in her spermatheca organ. For large drone cells she just lays the unfertilised egg.

Drones' lack of fathers can be useful - it's why workers *who never mated* can lay drones, and it has some mind bendingly subtle implications for human breeders like: drones don't have dominant/recessive genes because they don't have chromosome pairs.

Early research indicated that the children of a queen tended to be born in batches from the same drone father, but more recent, more accurate research[420] with DNA technology has shown that the drones' sperm is in fact mixed together quite randomly in the queen's spermatheca over a timescale of weeks. The ratio of

Female honeybee

Male honeybee

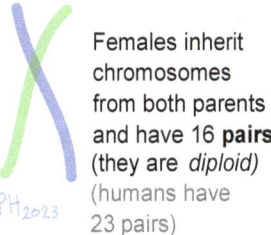

Females inherit chromosomes from both parents and have 16 **pairs** (they are *diploid*) (humans have 23 pairs)

Males develop from unfertilised eggs and with no father, have only 16 **single** chromosomes *(haploid)*

Drone brothers aren't identical

Queen: diploid, 16 pairs of chromosomes

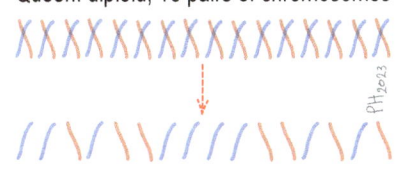

Each haploid drone son inherits a **random** set of 16 chromosomes

Similarly, the children of a single drone - who are of course all diploid, female - are not clones as some think. They inherit a random selection of their mother's chromosomes.

---

420 *Does Patriline Composition Change over a Honey Bee Queen's Lifetime?* Brodschneider et al (2012), doi: 10.3390/insects3030857

subfamilies do change slightly over timescales of months. The mix of children present in an open-mated hive is never more than ~30% from one father, and usually no more than 16% from one drone – though some drones' children only comprise 1% of a hive's population.

Brothers, sisters, and more

This is where it gets really mind bending!

Because a queen has offspring from several drone matings, her worker children may be **supersisters** (same drone father, sharing 75% of their genes); **sisters** (brother drone fathers, so the offspring's genes have 50% in common) or **half sisters** (different drone fathers, or *patrilines*, 25% same genes, like human sisters[421]).

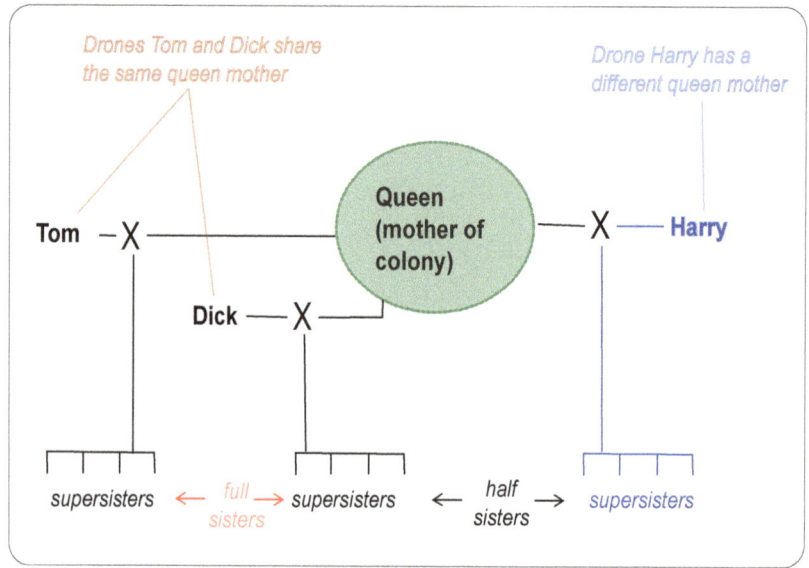

Drone relationships are even weirder!

- A Queen is both mother and sister of her own drones.

- And **her** mother queen is the 'father' of her drones.

- Drones have no [male] fathers.

- In effect, drone genes skip a generation.

Lacking chromosome **pairs,** drones inherit pure characteristics - which helpfully prunes lethal genes from the population.

It can also be useful for human breeders. For example, most breeding programs are based on identifying a handful of "breeder queens" with excellent, inheritable qualities. From these one rears and sells "production queens" - and crucially, any sons (drones) of these daughters will have a subset of her genes - *because drones have no fathers* - and spread the traits the beekeeper thinks desirable. (Of course, neighbouring beekeepers may not be happy if they see aggressive crosses.)

An interesting detail here - the *production* queens are put to work and stimulated to lay at maximum rate, and run out of eggs in 1-2 years. But the original *breeder* queen (grandmother of the production hives) is kept in a small, low stress hive, and outlives her daughters by several years.

Breeder queens cost hundreds of euros. The queens advertised for tens of euros are production queens. Production queens from *breeders* are consistent and often sold out before the season starts.

---

421   Not 50%! Human siblings get 50% of their mother's genes and 50% of their father's, but each human sibling gets a *random selection* of the chromosome-pairs from each parent so on average each human sister shares **25%** of her genes with her sister. But for bees, the father *only has single chromosomes to share,* altering the percentages.

Mutants and mixes

All species have occasional chromosomal mixups, things can go wrong. Careful observers may see these at hive entrances:

- **Rare: Small drone** - worker sized. Sign of a drone laying hive. The workers, or a queen who has run out of sperm, are laying their unfertilised eggs in worker cells, which develop into drones.

- **Very rare: White eyed drone** - very rare mutants. Because drones only have 1 set of chromosomes, if they have a mutation it is not masked by a working copy. (If a human mother has a faulty, recessive gene it can be masked by a working copy from the father.) So sometimes blind drones with white, red and green splotched eyes, or other defects are spotted. Probably a sign of minor inbreeding. No significant impact on hive survival, but the drones are blind so once they emerge from the hive to mate they just crawl around on the ground.

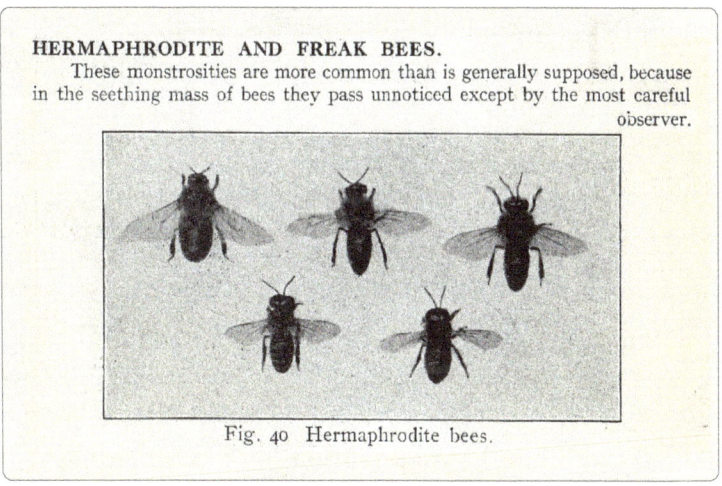

White Eyed drone. Photo © 2019 Jon Darvill, Australia.

- **Only visible by their abscence: Diploid drones.** About one in 20 cells in healthy brood comb is empty. This is where the nurse bees spotted and ate an intersex baby[422]. These holes are in worker brood comb: these diploid "drones" are really abnormal *workers*, not drones - they are diploid (16 *pairs* of chromosomes).

Researchers have rescued and hand-reared diploid drones to find out what they look like: if they're not constrained by worker sized cells, slightly larger than normal drones[423]. More on DD's below.

Variable frequency: This image of **intersex bees** using very outdated language is from the 1930 volume *Bee-Keeping New & Old* by William Herrod-Hempsall. He describes the samples thus:

1. Worker head, drone body
2. Worker head & abdomen, drone thorax, legs & wings
3. Worker abdomen (rear), drone elsewhere
4, 5 Drone heads, worker elsewhere

Numbers vary between hives. Rothenbuhler's 1954 book *Hereditary aspects of Gynandromorph occurrence in honey bees* mentions some colonies produce hundreds or even thousands of them.

HERMAPHRODITE AND FREAK BEES.
These monstrosities are more common than is generally supposed, because in the seething mass of bees they pass unnoticed except by the most careful observer.

Fig. 40 Hermaphrodite bees.

---

422 This conveniently 'just happens' to match the requirements for empty cells for heater bees, i.e. adults who use them as anchor points when the brood temperature is a bit low. These heater bees vibrate their wing muscles to generate heat. Some beekeepers think brood frames should be solid brood - this actually implies lack of space to lay.

423 honeybee.drawwing.org/book/diploid-drones?page=3. Researchers reared diploid drones (which is tricky as they may be cannibalised).

Obligate outbreeders: hyperpolyandry and the the diploid drone disaster

Diploid drones arise because there is a sex "master switch" gene called *csd* (complementary sex determiner) which comes in a number of variants (*alleles*). In a healthy population, there is only an occasional match[424].

If the *csd* alleles on adjacent chromosomes in a pair match (see diagram), you get diploid drones.

If inbreeding reduces the pool of alleles, the number of matches, thus DD's shoots up. You begin seeing lots (e.g. 50%) of empty brood cells and the colony can't raise enough workers to keep the colony going. The colony collapses very suddenly.

This is a huge problem for professional bee breeders and they keep careful records of who mated with whom, keeping several lines of bees going in parallel or buying breeder queens from other breeders.

The tipping point is when the number of allele variants drops to ~6[425]. A normal wild population has ~20.

This is a huge incentive for bees to **maximise** the mixing of genes in the apiary. Hives with a variety of drone-fathers survive better[426]. This is why, given the opportunity, queens open mate with 13+ drones, termed *polyandry*. Particularly promiscuous queens are described as *hyperpolyandrous*.

(Female bee chromosome: diploid, i.e. a pair)

Bee sex is determined by **number** of chromosomes (diploid vs haploid), but... females have a sex related gene with (at least) 23 variants (alleles) and these *must not match* on paired chromosomes or the result is a sterile *diploid drone*

---

424 It used to be thought there were ~20 sex alleles in a local population, but Lechner et al (2014) argue there may be as many as 53 locally and 116-145 worldwide (doi:10.1093/molbev/mst207).

425 *Sex determination and the evolution of polyandry in honey bees* (2002), Tarpy & Page, DOI:10.1007/s00265-002-0498-7 . Their experiments and analysis showed that once the number of viable brood dropped below 72% (i.e. diploid drones exceeded 28%) the chances of surviving winter dropped suddenly from above 80% to sub-40%. In bibba.com/june-2024-bibba-monthly/ Dorian Pritchard (a geneticist) explains this corresponds to 6 or less csd alleles in the population. Each queen carries 2 versions of the csd allele (she's diploid, and if they were identical she'd be a diploid drone) so a viable population for an area, or breeding operation, needs *at least* 3 queen lines, and in practise you need more for improved resilience.

426 *Queen promiscuity lowers disease within honeybee colonies* (2006), Seeley & Tarpy, doi:10.1098/rspb.2006.3702 - colonies with multiple drone fathers more resistant to AFB. The work was expanded in 2009 DOI:10.1007/s00040-009-0016-2 by Invernizzi,Peñagaricano, & Tomasco showing different patrilines had differing resistances to Chalkbrood and AFB. In 2015, Delaplane, Pietravalle, Brown & Budge published *Honey bee colonies headed by hyperpolyandrous queens have improved brood rearing efficiency and lower infestation rates of varoa mites*, PLoS ONE. Delaplane points out that bees have evolved polyandry, with occasional hyperpolyandry, as a first order mixing of genes; this is done at drone congregation areas where there may be drones from 200 hives. During oogenesis, meiosis then shuffles genes more finely than in any other land animal. That's three stages of extreme gene shuffling - it's obviously a significant evolutionary advantage to them. He argues that breeders, by selecting for useful traits, are unintentionally selecting out others which are also useful, but which we are not aware of. Of course, that's what "leave alone" beekeeping is all about - trusting natural selection.

   Inbreeding also results in the loss of disease resistance alleles - *Why Honey Bees are Dying: A Forgotten Cause* (Jaques van Alphen, 2023 doi: 10.20944/preprints202310.1544.v1)

Paradigm shift - negative selection

Consider natural selection as *eliminating the weakest*, rather than *survival of the fittest*.

This is *negative selection*, letting undesirables die and breeding from survivors. It retains a large pool of genes.

Queen breeders are inbreeders

Commercial breeders use *positive selection*, choosing a few "best" queens and discarding many fit colonies.

Beekeepers then prop up these less adaptable creatures with sugar feeds, miticides etc - survival of the feeblest.

They don't introduce **new** traits, they "reinforce" "desirable" characteristics by *suppressing* their opposite. For example, to make queens more prolific they select queens which *don't* turn egg laying *off* in dearths[427].

However, as many bee traits aren't located on single genes, and some genes are physically near others and tend to move with their neighbours, emphasising a character by deletion of an opposite tendency can mean other traits are affected.

A *sophisticated* **breeder** (a highly skilled specialist, distinct from a straightforward *queen rearer*) needs at least 100 hives so they can select for multiple factors like docility, propolis collection, swarminess[428], honey yield (which can be tricky - how do you know the high yield hive isn't simply quietly robbing from its neighbour? That's surprisingly common)[429]. Reinforcing these traits is difficult as bees tend to outbreed, and because some desirable traits are dominant[430]. Inbreeding is used to reinforce a trait - but you have to know what you're doing. Preparation for queen rearing begins at least two years ahead so you can evaluate the *daughter* queens' traits and decide which "grandmother" queen gives consistently good descendant colonies.

A skilled bee breeder's lines breed true, whereas amateur queen rearers often jump to conclusions about one-off successes - which are often just a random combination which doesn't breed true.

Trait heritability
h2

0.38	Calmness on comb
0.37	Gentleness
0.27	Honey yield
0.18	Hygienic behaviour
0.06	Swarminess

A simple Mendelian single-allele trait is easy to fix in the queen line, but some involve 30 pairs of genes! Breeders need to be careful not to reinforce bad traits while trying to fix tricky ones.

---

427   Usually how prolific a queen is, depends mainly on stimulative feeding and how large the hive is. Even back in 1770 Wildman commented (*Management of Bees*) if you hive bees in a large cavity it excites them to build more comb / lay more eggs. And it's well known that a swarm placed in a hive seems to program its laying thereafter to how big the original cavity was.

428   Swarminess is associated with prolific, fast breeding; and particularly the Carniolan strain of bees. Victorian users of the original British Black bee (Amm) commented on how swarmy their bees became after they introduced Italian and Carniolan bees to their apiaries. I've noticed my bees, isolated from imports and completely unselected, have become a lot less swarmy in the last couple of years, and other low-intervention beekeepers have remarked on this...

429   Real breeders consider subtleties like, is a colony really producing lots of excess honey, after accounting for all the syrup it was fed?

430   Dominant genes can be *trickier* to reinforce and fix in a population. You can't immediately tell if a breeder queen has one or two copies (one in each chromosome set) because recessive alleles will be masked if the matching one on the other chromosome is dominant. You want queens that are homozygous (two identical alleles at this chromosome location) so they pass at least one copy to all offspring.

Some lines of bees are advertised as resistant to chalkbrood, foulbrood etc. Since wild bees tend to be resistant to these (or they die), it is truer to say such lines are ones where such traits have not been inadvertently bred out[431]. Some breeders are even using artificial insemination with sperm from just one drone (SDI - Single Drone Insemination) which is the fastest way to ensure rare disease resisting alleles die out!

If a colony you buy fails, it is easy for the supplier to blame external factors like other beekeepers. People tend to believe the supplier because they paid good money for a 'premium' product, and can't admit they made a bad choice - what economists call the *sunk cost fallacy* or psychologists, *cognitive dissonance*.

Breeders may advertise bees with "wondertraits" but, once sold to normal beekeepers, such traits are rapidly diluted by open mating. Even if all local bees were wiped out, as some breeders seem to desire, the wondertraits would rapidly decay because of open mating with other breeders' bees (Italians x Buckfasts, etc). Breeding races outside their natural range is a dead end[432]: it's genetic pollution. Landraces don't have one supertrait like superlative grooming - they have a bit of everything, maintained by extensive outbreeding, and this **is** typically stable.

Even the breeders stress their strains will die without regular beekeeper interventions: and if they need help to survive, they're clearly not fit. It's a short term fix, not a solution for "better bees".

Line versus hybrid breeding

Line breeding is when you breed from a pool of one pure race. This inevitably leads to inbreeding and the need to swap genetic material with a similar breeder (a different *queen line*).

A more sophisticated, and much more complicated system is crossing several races in a controlled manner, for hybrid vigour. This tends to require artificial insemination techniques and very careful record keeping for consistent results. Many hybrid breeders get their pure stock from line breeders, rather than raise multiple distinct queen lines themselves.

Chromosomal crossover

Not only do queens typically mate with 13+ drone-fathers, there is an extra degree of gene shuffling in bees - at *chromosomal crossover*, where genes swap between chromosomes during meiosis. I.e, bee father/mother DNA strands swap sections between each other during oogenesis. Just to be clear, this happens in all diploid animals, but to an unusual degree in bees.

Chromosomal crossover may help repair DNA damage; an alternative theory is that it promotes diversity.

Chromosomal crossover

Diploid chromosome **pair**, e.g. human, female honeybee etc.

A close look reveals that the father / mother chromosomes are finely shuffled. This is seen in all diploid organisms, but in honeybees it occurs on a massive scale. **This extra level of gene shuffling aids super rapid adaptation.**

---

431 Jaques van Alphen puts this more scientifically in his 2023 paper, *Why Honey Bees Are Dying: A Forgotten Cause* DOI:10.20944/preprints202310.1544.v1

432 The breakthrough research proving the advantage of local bees was J. Louveaux et al's 1966 work *Les modalities de l'adaptation des abeilles (Apis mellifica L.) au milieu naturel.* showed how **local** French bees were adapted to population peaks in time for distinct local nectar flows. Taking colonies from Paris to Landes or vice versa, their population surges were poorly timed and they did worse than back home. Colonies from hybrids had intermediate population surges which weren't suitable for either locale. This side effect, where local colonies are made worse by importing foreigners, is called **outbreeding depression.**

Massive reshuffling is risky as all genes need to be retained in the species' standard order to work, but is appropriate for fast breeding organisms in very changeable environments such as insects, whose small size and limited foraging range makes them vulnerable to localised weather, forage and pest variations from year to year.

Imports disrupt sequences?

Disclaimer: the following conclusions are my own views, not accepted science (yet), but offer a plausible mechanism for F2 aggression.

Early, *Mendelian* genetics looked at simple binary traits, with just 2 values. Mendel derived the basic laws of inheritance, dominant / recessive genes etc from observations on e.g. the size of pea plants (tall vs. dwarf).

But most genes are not binary switches. They're more like sliders; and many traits arise from the **interaction** of **many** genes, like the resonances of a piano's notes forming a chord.

Some of these *non Mendelian* gene combinations seem strongly conserved across all bees. Others vary a bit between races - it doesn't really matter metabolically if races have a slightly different wing vein pattern.

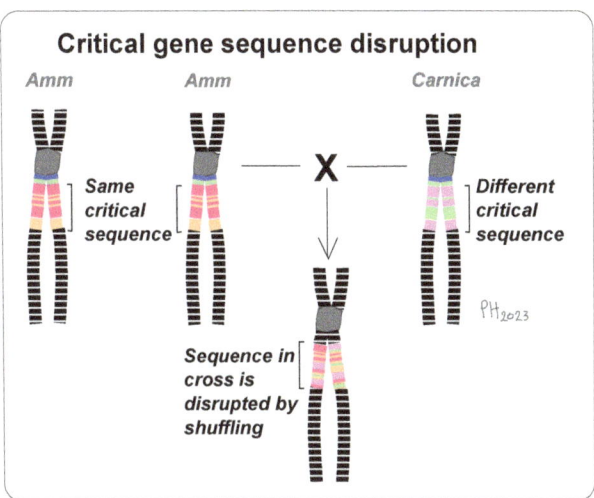

[Speculation!] If all males and females in an area are one race, mating doesn't corrupt key sequences when chromosomes swap sections during chromosomal crossover. Mixing races breaks up key sequences.

Finely balanced gene sequences need to remain in a certain sequence. I contend that such sequences are strongly conserved - **within races**. If two *Amm* mate, the offspring is "within normal parameters". But if Amm crosses with, say, Carnica, then because of chromosomal crossover (see above) the critical sequences can get jumbled.

One important example is aggression. Each race is pretty calm when pure. The docility breeds true. But over their 2.5M years of separation, they seem to have evolved different gene sequences to moderate this. Mixed race crosses can be pretty grumpy.

Furthermore - and again this is controversial, but my view is: it's looking increasingly like importing foreigners **doesn't** introduce useful new traits because most traits are formed from **groups** of genes. It's only if you have a dominant background population who all have the same alleles along some key sequence that it isn't corrupted by chromosomal crossover. Mixing races breaks useful sequences up, which could explain why bees are grumpy until the damage is repaired.

This also explains why breeders find it so difficult to fix stable traits. It is very easy to de-tune a trait that depends on multiple alleles being just so. As soon as those bees outbreed with your local bee (whether that is a stable local landrace or descendants of bees from a different breeder), the trait gets disrupted. It would be particularly difficult to fix a new trait in Buckfasts, which are already a tightrope balance of mixed genes and have more genetic diversity than pure races.

It may also be relevant to why bees relocated from their home area lose varroa resistance traits.

Good theories have testable predictions. Sequence disruption effects would imply that people who cross, say, Italian and Caniolan queens to get "the best traits of both" would actually get severely substandard performers. This may have passed unnoticed due to the aggression of many such crosses resulting in abandonment of such trials, and observer bias. It is interesting to note that it took a decade for Brother Adam to create his first acceptable B1 generation of Buckfasts, and he kept tinkering for decades, where BIBBA breeders get gentle, productive, healthy bees in 2 years from local stock, often simply by weeding out the worst.

Aggression of crosses

The main honey bee stocks sold commercially in Britain are **"local"**, **Buckfast** (this term is now used for any home made multi-cross), **Italian**, and occasionally **Carniolan**. If Buckfast, Italian or Carniolan cross mate with each other, or with local landraces or **Amm**, it is common for the "F2 generation" to be very defensive - said to be a 1 in 4 chance, though of course most people have more than one hive and it only takes one 'hot' hive for the apiary to be unpleasant. There's no consensus why, but it usually happens when the queens have 50/50 mixed lineage (F1, thus her offspring are F2).

This is how the infamous "killer bees" of South America arose: a cross between European and African bees --> "Africanised Honeybee" or AHB[433].

Why are AHB bad? — *They're an aggressive hybrid.*

But don't breeders keep importing foreign bees - creating an aggression timebomb? — *That's different, those are highly productive.*

But AHB make unbelievable amounts of honey. And they're resistant to varroa and famously healthy. — *And free. So just No.*

For *European* honeybees, the bad temper follows the queen - you can swap queens between hot & calm hives and see the temper follow the queen, though it may take a couple of days for a large colony[434]. So it's thought it's to do with the queen's pheromones.

There is a fascinating example of doublethink here - American beekeepers, and some States' laws, consider AHB a Problem, whilst they actively spread non local bees to "improve yields" and.... increase aggression. AHB colonies produce *massive* honey harvests and are varroa and disease resistant, so should be considered as potentially useful tools!

You can get F2 aggression when any races are crossed; e.g. *Amm x carnica* is well known for this; it happens 1 generation earlier with Buckfasts, because their 'pure' queens are already an F1 mix, giving them a particularly bad reputation.

You can mix *Amm* from all over the British Isles without this effect, and British landraces (typically 40%+ Amm genetically) can be mixed with *Amm* peacefully, because all *Amm* are in the same *temperament group*, as Beowulf Cooper termed it.

---

433 To be precise, AHB has an exaggerated response to sting pheromone, which is just one aspect of most bees' defensive behaviour. Those people I've met who have encountered AHB all say their reputation for aggression is overstated. Also AHB seem to calm down once they have dominated an area, e.g. Puerto Rico. This seems to imply they are mainly aggressive on the edges of their colonisation range where they mix with other races (disrupted key sequences..?). There are some articles claiming the The Puerto Rican population suddenly became gentler after hurricanes Irma and Maria, but the reasoning for this is unclear and locals were already reporting calmer bees before the hurricanes.

434 Roger Patterson says he's seen this temper swap about 15 times, and I've read the same elsewhere for *European* bees. However for AHB, the mechanism is different - the workers have an exceptionally low trigger threshold for responding to alarm pheromone. So if one bee stings you or is squashed in an inspection the whole hive attacks, resulting in hundreds or even thousands of bees dying after stinging you. However, areas with established AHB seem to be experiencing a calming as the AHB gene pool stabilises at a less "wasteful" trigger threshold.

Should you find you have an aggressive or weak hive, the 'best' solution is usually to kill the queen and allow the colony to raise and open-mate a new one. This is negative selection - not breeding from the human-arbitrary 'best' but removing a very few queens of the worst colonies each year. This aids the removal of the least fit but allows natural selection to select the most fit, which maintains a broad genetic base. You will rapidly end up with the local landrace, and calm bees. Often this happens in one generation.

For many years, queen sellers maintained there were no local bees and blamed "local mongrels" for aggressive crosses. (Blaming a local survivor population they claimed didn't exist.) Lacking better information, buyers felt locked in to a specific queen supplier in order to retain traits and docility.

Is temperament transmitted by drones?

Lab analysis of DNA appears to show that the largest factor influencing aggression is: how many others in the colony carry heightened-aggression alleles, as if they all set each other on edge[435].

More practically, breeders do have some useful rules of thumb. There are credible accounts[436] that confirm transmission of aggressive behaviour through the male line. This makes sense evolutionarily: if it was a trait inherited from the queen then all bees in the hive would have the same defensive threshold and that could be Very Bad[437].

Because aggression is carried by drones, breeders can churn out pure Italian, Caniolan or mixed race timebombs which are calm when bought. Purchasers *and neighbouring beekeepers* then find the next generation is aggressive: the imported colony's new queen open mates with local bees, which are often from a different temperament group, and the purchasers blame the local bees for the problem, because people rarely blame themselves. Of course it's these purchasers themselves who are importing the problem, *and* making neighbouring colonies aggressive, *and* reducing the general fitness of the area. The ethical responsibility lies with the importer.

---

435 *Genomic regions influencing aggressive behavior in honey bees are defined by colony allele frequencies,* Avalos et al (2020).

436 E.g. Simmins in *A Modern Bee Farm* (1887) p.34 & following. Based in Brighton, England he was experimenting with *Amm*, Italian, Carniolan, and Cyprian. He talks of apiaries with 200-300 hives so was exteremely experienced. ROB Manley (*Honey Farming*, 'Breeding Bees' chapter) confirms it through observation. Br Adam found Carniolan drones produced gentle crosses with other races but Carniolan queens crossed with non Carniolan drones produced aggressive offspring (*In Search of the Best Strains of Bee*, p.174).

437 Insight by Gareth John, conversation 19/1/2020

Rapid reversion / selection symmetry

Breeding can't create new genes[438] but it can accelerate changes. Beekeepers are often surprised how quickly they can "improve" their bees' characteristics just by raising queens from their "best" existing hives. Typically improvement can be achieved in two generations, but rapid selection works both ways! The corollary is that it's generally agreed that imported bees revert to the local landrace in two years[439].

The fact that crosses are notorious for aggression illustrates how bees can shuffle their genes massively in a generation.

WHY is bee gene shift so amazingly rapid?

No one knows what the real cause is, but my intuition is, F2 aggression and its rapid dilution is telling us something very fundamental about rapid adaptation.

The following is a **speculative** hypothesis, but it helps visualise and explain why bees seem to change radically in 2-3 generations: perhaps rapid "reversion to the local mean" is a feature of a colony *repairing disruption to key sequences*. These will be most indeterminate (disordered, detuned) when mixed **50/50** from different races.

Whatever the reason, whilst some beekeepers say natural selection to create varroa resistant bees[440] will take years, others simply allowing local swarms to open mate. It takes about 2 years for the nature of your bees to shift once you go hard core no treatment / only emergency feeding / minimum intervention.

2 years is remarkably fast. Isn't that at most 2 generations? So surely, 25% of the foreign genes are still there? Well, not necessarily. Even without **access to DNA lab tech, I can see the results of gene shuffling in swarms, and they do regress in 1-2 years**. I conclude:

- Really stressed colonies are known to supersede more than once a year, thus **less than** 25% unfit genes remain after 2 years.

- Genes can't dilute faster than 50% per generation so lots of learning / adaptation is going on. Unlike himan breeders, colonies try *many things in parallel*. A wide genetic pool helps.

- DNA tests indicate wild bees are often genetically identical to local managed stocks - but it's what they do with it (epigenetics and behaviours) that matters. The honeybee genome was only mapped in 2006 and researchers can see methylation and other indicators of epigenetic

---

438 Viable, useful mutations are vanishingly rare, partly because insects that rear young kill any that smell weird. Most evolution is from new combinations of existing genes.

439 Local bee landraces are defined more by behaviour than genetics. A century ago Oxfordshire bees were black, now they are stripy orange things, but they survive unmanaged despite varroa and intermittent forage. Bees appear to have recessive behaviours which can emerge as needed. There may also be area-specific gut microbiota. Latent grooming and other behaviours "spontaneously" arose around the world when varroa arrived [Metastudy: *Parallel evolution of Varroa resistance in honey bees; a common mechanism across continents?* Grindrod & Martin, 2021 doi.org/10.1098/rspb.2021.1375], implying that rather than "varroa resistance" traits being some new mutation, they are present in **all** bee populations but are more heavily suppressed (inadvertently selected out) in inbred ones sold by breeders. "Hygienic behaviour" - ejecting infected larvae - is not novel behviour, it was researched in foulbrood / chalkbrood colonies in 1938 (Park et al), 1942 (Woodrow), Rothenbuhler (1960s), etc. Rothenbuhler found "hygienic behaviour" is a recessive trait, only passed on to about 1/3 of daughter colonies, so bees don't waste resources having it "always switched on".

440 Different styles are called *Hard / Soft Bond Method* or *Darwinian Beekeeping*, see chapter 14.2 under *varroa*

activity, but don't yet understand how it works in insects. It would be significant if they found bee epigenetics varies between areas, but we're a few years away from knowing that.

- Requeening with an import simply doesn't wipe local microadaptations for long. If the colony survives, hyperpolyandry ensures local traits are rapidly re-expressed.
- Maybe pest resistance is not just latent but the **default** and is just inadvertently selected out by breeders, or disrupted when other bees are imported, or suppressed by humans diverting bees to making honey. This would explain why so many wild survivor colonies are around.

Rapid adaptation is well known in biology - for example pests developing resistance to insecticides[441]. In unstable climates (extreme pressure), evolution will push variation to a limit where useful new trait combinations are selected ASAP. I suggest - *and note this is unproven and controversial* - that in bee terms **colonies under pressure will pump out more child-swarms**[442]. To put this another way, *excessive swarming and queen cells may be a sign of stress*. This would explain why leave-alone beekeepers see reduced swarming *and number of queen cells* after a few years - unfit genes have been shaken out.

Temper vs. calmness

It's worth noting that when talking about breeding, some people use *calmness* to refer to something other than grumpiness.

- **Temper** refers to how likely the bees are to sting you - are they docile, defensive or downright aggressive.
- **Calmness** [on the comb] can refer to **runniness** - do they run around chaotically when exposed to light. It makes it hard to inspect a comb and more significantly, if the workers are runny, so is the queen - and she may run out of the hive, or into weird areas like feeders. This is a big issue if you open hives regularly.

Commercial breeders try to minimise both these traits.

My own breeding strategy

Rather than start from an inbred commercial population, with potentially useful survival traits bred out, I aim to create a broad, non inbred survivor population.

Fig. 581 Cutting out natural queen cells.

Driving a skep by drumming

Images from *Beekeeping New and Old* (1930)

Old photos often show handling with no protection

---

441 This paper discusses some genetic mechanisms if you're interested: *Rapid adaptation in a fast-changing world: Emerging insights from insect genomics*, McCulloch & Waters (2022), doi: 10.1111/gcb.16512

442 Spendthrift is a rational strategy amid chaos. Don't bother with huge honey stores if a beekeeper is likely to steal them - take what you can and run. Google "the marshmallow experiment" - people don't always delay gratification for a bigger reward later if their life experience indicates you can't rely on planning or promises. Unstable lives --> the rational response is immediate gratification. So perhaps bees raise more babies (swarms) to get ahead of the thefts.

This is a **long term** strategy to create a wide genetic pool of locally adapted bees; from which bees suitable for honey farming can be selected later, if desired. Once the background survivor population has resistant traits, there is a general herd immunity even if these traits are weakened by selecting for others.

This is a **qualitative** strategy whereas a breeder would assess and select specific traits **quantitatively.**

I use open mating. I am in a rural area with several wild colonies, but it's a bit of an isolated oasis so I import one swarm a year from survivor colonies 5-10 miles away to avoid inbreeding. My bees seem to be steadily getting more vigorous each year, and swarminess has dropped significantly. Their colour has stabilised towards the dark end of Oxfordshire's stripy orange-and-brown bees, but varies a lot, a few are "black" bees but clearly there's a huge mix of genes. I view this as having a huge experimental breeding program on my doorstep, doing sophisticated selection for me.

Even in just 2 hive types, my eight colonies vary significantly in age and size, foraging and swarming behaviour, honey accumulation, comb building etc. It's not just genetic - a late cast has very different foraging options and develops very differently to an early prime swarm. This is ideal for my amateur research apiary.

Most people measure the worth of bees by honey yield, but this varies primarily with both forage and thus weather. So it is very easy to ascribe a good year to one's queen selection, when actually it could simply be that the local strain is reasserting its genes in your stock.

Fit colonies benefit from stress because bad weather, say, disrupts their pests like wasps *more* while they coast through on stores and a brood break. Wasps lack winter honey stores and new wasp queens start nests from scratch each year, so stop-start spring weather knocks back their nest development, so not every year is a "wasp year."

Natural outbreeding vs commercial inbreeding

Minimisation of genetic bottlenecks and retention of maximum traits

Professor Tom Seeley looked at survivor colonies in the Arnot Forest before and after varroa arrived. He found[443] that populations which survived were descended from just a few queens (by examining mitochondrial DNA, passed down female lines), i.e. varroa had indeed killed most of the original wild colonies. But the *nuclear* DNA showed almost as much variation as in the pre-varroa populations. Genetic diversity had been preserved despite a near-extinction event.

However, the surviving Arnot colonies are more similar in their paternal makeup to a random sample of US bees than the original wild colonies', so one cannot simply assert that wild survivor populations are less vulnerable to gene loss[444]. Too small a population will die out if it is isolated.

---

443 *A survivor population of wild colonies of European honeybees in the northeastern United States: investigating its genetic structure* (Seeley, Tarpy, Griffin, Carcione, Delaney (2014)

444 I initially over-interpreted Seeley's results and concluded that wild survivor populations were less susceptible to inbreeding. I put this to geneticists Dorian Pritchard (Northumberland) and Keith Browne (U Galway) who both explained why there was insufficient evidence to back up that assertion. So I was wrong. It's important to consider facts that contradict one's biases. Browne's own analysis of unmanaged Irish colonies' genetic variation, *Investigation of Free Living Honeybees in Ireland* (2020), doi.org/10.1080/00218839.202 0.1837530 shows no sign of a genetic bottleneck (unlike the Arnot population) and, again, matches the genetic signature of surrounding managed *Amm* colonies, reinforcing the point that wild colonies don't have significantly different makeup. (Though interestingly, these Irish survivors were all remarkably pure *Amm*, despite the presence of other races in a minority of Irish hives.)

Key point: Pests and diseases can't evolve resistance to the natural control methods - analysis of the Arnot genome showed that natural resistance uses an <u>entire suite of small traits</u>[445] that keep changing in emphasis as successive queens mate with new drones.

This isn't unique to bees - it's a widespread phenomenon. Tasmanian devils are rapidly adapting to a novel transmissible cancer, and genome analysis indicates this is through small changes to hundreds of genes[446]. Conservation projects were set up captive breeding populations in the 2000s - but scientists now worry reintroducing those will *dilute* the wild population's survival traits.

Local bees come with free hybrid vigour, because they open mate. It's precisely BECAUSE crosses are NOT uniform - a spread of abilities within the family - that such colonies thrive. For example, some bees, dubbed 'hit squads', have been seen to specialise in witchfinding infested larvae (better sense of smell?) or obsessively allogrooming[447]; but you don't want your entire workforce doing one job.

Unstable commercial supertraits

An example is **uncapping/recapping**[448]. Some varroa resistant colonies do a lot of uncapping & ejection of infested larvae (hygienic trait) and breeders have selected for this, by deliberately killing zones of brood comb and watching the response. Some beekeepers buying these report that with the next generation of queens, the colonies sometimes throw out *too many* pupae, infested or not, and the colony collapses. The trait has run out of control.

In contrast, after John Harbo at the USDA found an allele for a similar trait, **varroa sensitive hygiene (VSH)** he set up harbobeeco.com which helps breeders fix this in their queens, and whilst it can be diluted if the bees open mate, it doesn't seem to cause problems if the entire population has it.

In general, a single superpower is an evolutionary dead end. The pest will evolve a countertrait.

---

445   A followup paper to the previous reference, and a much easier read, really brings this out. Key phrase: *634 sites showing significant differences* [between pre and post varroa populations of bees]. In other words naural resistance uses many small shifts in parallel. *Museum samples reveal rapid evolution by wild honey bees exposed to a novel parasite.* Mikheyev, Tin, Arora & Seeley (2015) DOI:10.1038/ncomms8991

446   *Contemporay and historical selection in Tasmanian devils (Sarcophilus harrisii) support novel, polygenic response to transmissible cancer.* Stahlke et alia (2021) doi.org/10.1098/rspb.2021.0577

447   Moore et al (1995), Kolmes 1989 *Grooming specialists among worker honey bees,* Dorian Pritchard 2016 doi.org/10.1080/00218839.2016.1196016 - Pritchard mentions "Group cleaning" was first observed in *A. cerana* colonies by Peng *et al* (1987), then in *Am ligustica* by Fries, Wei, Shi, Chen (1996) *Grooming behaviour and damaged mites* Apidologie 37: 3-11

448   A new and very informative website covers this, and related traits - www.varroaresistant.uk . An image of such behaviour is in chapter 14.2 under Varroa.

Human and Bee reared queens are different

There is a reason to be sceptical of queens chosen by humans. Most of the time the bee breeder selects a random larva, and this is almost bound to be one of the common patrilines in the hive. But when the bees themselves select a worker larva for an emergency queen, they seem to go to great lengths to avoid random (common) lines.[449] Instead, the workers prefer to promote certain rare patrilines which have been dubbed "royal patrilines".

This is a very interesting result, not just because it seems the bees can spot such differences, and choose according to circumstance; but because beekeeper-created queens use the "wrong queens".

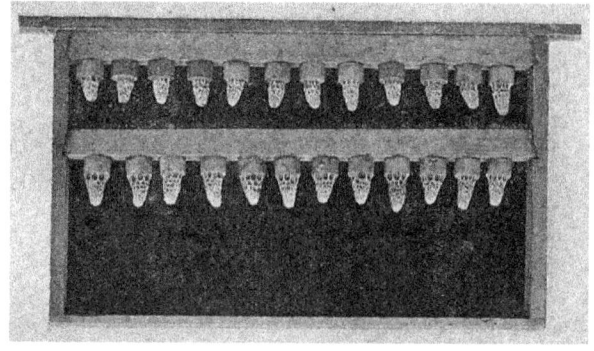

Mass queen rearing systems still resemble this 1930 photo from Bee Keeping New & Old. Of interest: note how the cells vary in size. Queens are encouraged to lay in these cells, or young larvae are transferred ('grafted') in by a human, and raised in closely monitored colonies or incubators. If you just want a handful of queens, excess sealed queen cells can be taken from a colony about to swarm, or you can split colonies.

Furthermore, if you rear a batch of queens without allowing the colony to select a few favourites (normally the workers control which ones emerge first, some queens fight each other, others may escape with casts) you are bypassing a fitness test.

These weaknesses accumulate over generations.

A remarkable coincidence

In retrospect, my aims and my bees' aims regarding varroa were the same. When I stopped trying to manage and monitor mites, and got out of their way, their gene shuffling rapidly pulled the required tools out of deep storage, almost as if they had millions of years' experience of handling similar issues.

Or I could have spent years trying to control their breeding for the same result.

Local landrace lessons

The fact that these ecotypes remain good tempered unless crossed with imports, shows that these are stable, pure races. This is emphasised by the way bees revert to an area's Type in 2 years - there is a selection pressure. The only "mongrels" around are Buckfasts and the crosses caused by importing other races.

---

449 Withrow & Tarpy, *Cryptic "Royal" subfamilies in honey bee (Apis mellifera) colonies,* PLoS One 2018, DOI: 10.1371/journal.pone.0199124. The DNA analysis results were strong (patrilines present **only** in **emergency** queens but so rare in the normal workers they weren't detected). It's presumably an evolved behaviour which ensures maximum genetic diversity - and interestingly, the opposite of what humans do: we'd promote one of our own subfamily! But creation of an emergency queen is rare, and the only occasion the bees (workers) can choose the patriline. When colonies raise normal, planned queens (for swarming, rather than an emergency queen) the father is random, because the workers can't choose which patriline the original queen laid - they're not promoting a worker larva. **Note:** The *Miller method* of queen rearing gives the workers the choice., using shaped comb with many edges to promote the production of masses of emergency queen cells.

Another paradigm: colonies experiment a lot; survivors share strategies (drones). In effect there is an involuntary population-level altruism, or evolutionary randomisation-and-filtering going on: the sacrifice of entire colonies for radical experiments maintains optimum colony density whilst creating a mechanism to conserve once-useful traits in recessive (dilute) form over millenia. This doesn't mean a given colony will act suicidally - just that there's a reason for extreme behaviour, like a novel foraging pattern or weather response, and if it's not actively countersurvival it will be retained in the area's "gene bank". Due to the extreme level of genetic variation in bees, and capacity to absorb losses, such extremes occur more often than in most species.

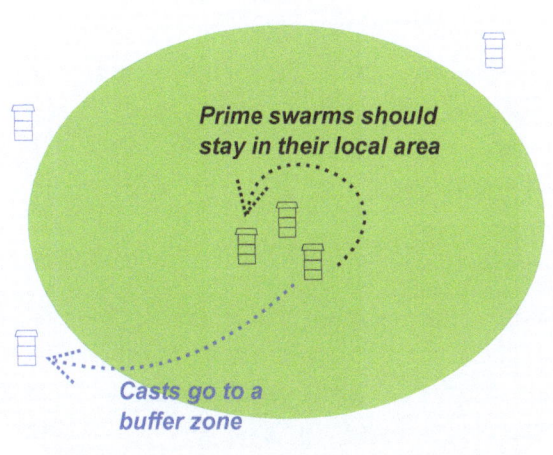

Considerations before leaping into a hard "leave and let die" strategy

Brutal Darwinian selection is unfeasible in areas with few wild survivor bees. Germany systematically replaced their original population with non local bees since the 1940s and now find this '*Hard Bond*' method doesn't work, because their local survivor traits are too weak. There, a more gradual breeding program is necessary, tailing off treatments (negative selection, weeding out the worst to retain a wide genetic pool).

Breeders use this scheme. Prime swarms' queens lay pure queens and drones, so are retained in the core zone. In casts, even if the queen breeds with outsiders, the drones are still pure because drones have no fathers, and form a genetic shield around the core area. Always try to keep prime swarms near their origin point to retain local adaptations.

If a population crashes too rapidly, the remaining gene pool may be too small to be viable, or it may lose resilience. Commercial beekeepers can't afford to go Hard Bond with all their hives simultaneously.

Most, if not all of the known survivor populations evolved when bees were just left alone: Russian (Primorsky) bees; Brazil (AHB); South Africa (scutellata); Cuba; French, British and Irish wild colonies, etc. Beekeepers then selected for desirable traits from survivor colonies - *not the other way round*.

The Swedish island of Gotland is sometimes held up as an example of problems when a survivor gene pool is too small. It was an early attempt (1999) to create an isolated varroa-resistant population by dumping a load of colonies from around Sweden (for maximum genetic variety), deliberately infested with mites, in a bee-free area and just seeing who survived. This was the original Hard Bond "leave and let die" experiment. Initially, after 4 years, they appeared to be struggling, the original 150 colonies dwindling to 13. At this point people assumed that was the final result and lost interest; the figure of 13 colonies *after four years* was still being quoted in 2020. The problem with this argument is that all the other survivor populations took over 4 years to recover from varroa, so the question is: how are the Gotland bees doing now (2024)? The local beekeeper and inspector tell me this data is being collated by Barbara Locke at Uppsala university, who implies she will be publishing something relevant soon - which should be interesting.

The bottom line: If a swarm from a local wild colony survives treatment-free in your apiary, it has useful genetic traits and is worth conserving.

Apiaries are a tougher test on such bees than living in a small isolated cavity of their choice - wild nests tend to have lower varroa loads because the bees can propolise without limit, optimise comb, swarm when they want (brood / parasite break), are remote from infested colonies, are generally more defensible than a ground level hive etc.

Breeding: conclusions

You can enhance desirable characteristics by selective breeding remarkably rapidly, and if you use local stock you can avoid inbreeding because the local gene pool has a wide base. Exclusively using non-local stock risks genetic bottlenecking, and dependence on your supplier. Justifying selective breeding by saying "everyone else is doing it" is avoiding responsibility for diminishing the gene pool and spreading aggression.

I recognise that bee farmers often have to specialise to stay competitive, so some buy queens. If you need consistent traits for mass bee farming, bear in mind most queen *rearers* are just raising queens as a sideline, and don't have the same grasp of subtleties as skilled bee *breeders*. Either way, such commercially inbred bees are specialist insects, not generalists - what they excel in is inevitably paid for in survival skills and adaptability in the face of unplanned challenges. So they'll need more looking after. If you do have to buy in queens, try to find a source of your local (national) race bred for commercial applications, not a hybrid. It will breed more true and give more consistent harvests in your climate, and after interbreeding with locals.

I don't bother with queen rearing as I prioritise survival, and learn more by using variable open mated local swarms. Which, despite propaganda to the contrary, are invariably calm. But I *am* involved in breeding - I'm just letting natural selection do the complicated bits for me.

Review 13: genetics

How would you explain the relationship between a drone and a worker from the same queen to a non beekeeper?

List three downsides of inbreeding bees

There's adaptation in: bees; their gut microbes; varroa; and the viruses varroa carry. Can you say which is the dominant effect?*

*Varroa mate with siblings.

Chapter 17
Revisiting assumptions

Do large colonies really make more honey? Up to a point...

This core belief is worth reviewing. It's known that the majority of foraging is done by the oldest bees. So smaller colonies of local bees (which ramp up laying before anticipated forage surges) have been promoted by some *Amm* users, pointing out that *they seem to get the same honey yield as with imported prolific bees.*

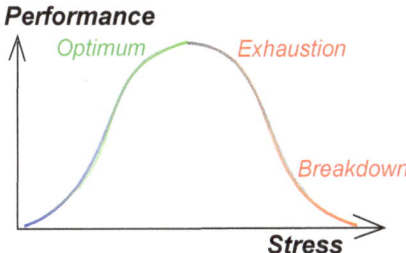

Diminishing returns: the Stress Curve applies to many disciplines

Unless you are surrounded by constant, massive forage (i.e. a bee farmer moving hives between crops) the gains from huge, prolific colonies are marginal. Raising brood is resource-intensive, and it's thought the reason winter bees live longer is largely due to raising few brood. If bees live longer or can divert just a few days' lifespan from brood rearing to foraging in Autumn, it could make a big difference to the colony energy balance. Remember, *most of the foraging is done by the oldest foragers.* Moreover, with a longer lived worker population, it is easy to see that a less prolific queen can give rise to a larger overall hive population than a more prolific queen with shorter lived bees!

John George concluded in 1928[450], colonies can be Too Large. He actually got more honey from a medium sized WBC than a larger hive, and remarked the larger one was more trouble to manage - more defenders, more heavy parts to lift - the numerous bees and larvae needed more food, leaving less for him. Nor did it prevent swarming. Note this would be in conditions we consider unlimited forage.

This topic has been hotly debated on the American-centric beesource.com forum, where people concluded a 6 frame Langstroth colony made about 2/3 the honey of a 12-frame one. The point was made that if you have 2 medium sized colonies, you have more genetic variation and more resilience in your apiary than one huge one, less back strain, and fewer stings[451]; but you also have a bit more management. If you really want to maximise honey, you need to ensure all your colonies are at maximum size just before the big nectar flows, and that's what the craft of commercial bee management is all about.

---

450   *Bee Keeping New & Old* by Herrod-Hempsall (1930) p.442
451   Larger colonies are more assertive; reassembling a huge hive more bees can emerge to defend it.

Swarming is said to reduce honey harvests by reducing the number of bees. However, one reason people perform splits (artificial swarms), is to... get more honey. It depends how you manage your splits / swarms.

Interestingly, if a "large" colony is actually mainly nurse bees, they don't do much foraging; and such colonies don't store honey in times of big flow, food going in is used immediately. This is a point very few new beeks understand. 8 frames of bees does not always mean your supers are instantly full of honey.

Do swarms really reduce honey yield?

Late swarms can be very large: without a brood break, the hive population grows unnaturally large, especially if they have been given stimulative feeding from early Spring. When the swarm leaves, almost all flyers can be enticed / excited to leave with the swarm, including the ones that would normally have hatched *after* a swarm left to keep the hive running. The colony will be slow to rebuild and may even struggle. The conventional beekeeper concludes that swarming is a problem.

But *early* swarms are welcomed by some. The break in brood rearing allows the bees to rapidly store honey[452]. Large prime swarms gather simply astonishing amounts of honey in the two weeks after swarming. The combined honey gathered by the parent hive and the swarm can considerably exceed that which would likely have been gathered by the parent hive if it had not swarmed.

Also, whilst a *migratory honey farmer* needs continuous high bee numbers, a static hive which has lost some bees needs *less* food over the summer dearth (this assumes you use unstimulated bees tuned to the local forage rhythm!). Fewer mouths to feed. The mother hive then rebuilds numbers for Autumn forage.

Snelgrove developed various swarm management techniques which take advantage of this opportunity. In effect, allowing the creation of a swarm in the same hive as the parent, separated by the Snelgrove board. This can give very high honey yields without actual swarming.

Thus, rote intervention is highly likely to create the worst of all worlds: over large swarms, unbalanced parent hives and reduced honey surplus.

Pseudo-dwindling and outstations

Gareth John observes his apiary *very* carefully. He has seen new colonies whose queen fails sometimes relocate back to their mother colony, moving stores and *apparently* dwindling.

He's also seen casts sometimes acting as *outstations* for mother colonies: sometimes the cast has *no queen at all* (because the queen does not mate) and just builds comb and gathers honey, then *the bees return, with the honey, to the mother colony*. This behaviour is well documented in ants, where outstations are termed satellite nests.

This leaves honeycomb in a prepared cavity, and as it is not brood comb it is not eaten by wax moth over winter. Next Spring... this is where the prime swarm goes.

---

452   Some beekeepers artificially induce brood breaks during flows to maximise honey by caging the queen. Roger Patterson, *Beekeeping: Challenge what you are told* p. 177 discusses this old and little known technique. It recognises that huge continuously-laying colonies don't give the biggest honey crops most British years: excessive brood are a *burden* in marginal environments. The skeppist Pettigrew, in his *Handy Book of Bees* (1875) devotes chapter XIX to comparing the swarming versus non swarming systems, and concludes allowing swarms usually produces more honey.

On a related note, it's widely reported that gangs of scouts sometimes clean and prepare cavities well before colonies swarm into them[453].

Ivy and OSR: conflicting stories

There are contradictory tales about OSR. The key thing to understand is that there are hundreds of varieties, many very recently bred, so old advice needs re-evaluation. Some flower in very early Spring, others in Summer. Some are attractive to bees - **others are not.** It is largely wind pollinated but yields can be boosted a few percent by insect pollination.

OSR was hailed as the saviour of British beekeeping when farmers began growing it in huge amounts. In Canada *millions* of acres of the *canola* cultivar are pollinated in a short window by *millions* of hives. Extensive monocultures breed huge pest populations and farmers used neonicotinoids against flea beetles, leading to tension with beekeepers. Beekeepers won in Britain and the EU, and OSR production fell here.

Ivy is about the last major nectar source here, blooming in September, but although other beekeepers say it is a major forage source I have noticed that my own bees do not consistently forage on it, even when ivy within metres of the hives is covered in other short range insects. I think the bees use it as a nectar source of last resort, and if they find something nicer within a mile or so they'll use that.

However my bees do love its pollen and gather that prodigiously, and if everything else has stopped flowering they will pack in the the nectar. Which can be a problem if the temperature drops suddenly - this triggers the bees to stop processing and cluster: the hive can get too damp in winter and the unripe nectar ferments, potentially causing dysentery.

Canadians in canola dominated areas find their bees overwinter better on sugar syrup than OSR honey, which crystallises rock hard.

It seems for both canola and ivy, you can have too much of a good thing.

Lime (linden) trees

You hear various stories that lime nectar is toxic to bees, it makes them drunk, you find dead bees under lime trees. But they were the dominant tree across Europe as Amm evolved there and are recognised as good forage. Obviously people are only seeing part of the story.

Koch and Stevenson reviewed the literature[454] and concluded the nectar is not toxic, but there is some evidence it contains caffeine which the bees get addicted to. They return to the trees even once the nectar is exhausted, and eventually starve – but this may be due, not to caffeine addiction but lack of other forage (desperation). Thus you sometimes find masses of honeybees and especially bumblebee bodies under lime trees. This indicates a lack of alternative forage in the area. (i.e. lime trees are sometimes planted in otherwise forage-poor cities.) The phenomenon is only associated with certain types of lime tree (see the paper for details).

In addition when lime nectar *is* flowing, the bees gorge themselves so have the characteristic drunk / tired flight patterns when they get back to the hive.

---

453 Gareth John, Oxfordshire; Julian Wormauld, Wales; other reports of swarms staying outside a cavity for a few days before moving in after rotten comb etc from a previous nest has been cleaned out.

454 *Do linden trees kill bees? Reviewing the causes of bee deaths on silver linden (Tilia tomentosa)*, 2017, Royal Society doi.org/10.1098/rsbl.2017.0484

They also note the late flowering of the tree coincides with the end of the lifecycle of many bumblebees, leading to deaths from old age as they forage.

The paper does not consider honeydew from lime trees; nor the factor that commercial bees overbreed for areas with local forage gaps. The locally adapted bees at Blenheim Forest monitored by Filipe Salbany have access to many types of lime tree, and don't have problems even under the "problem" varieties.

Process of Acute Enforced Adaptation

Filipe Salbany coined this term for the phenomenon where colonies which are moved seem to panic and forage solely for honey, not pollen. This can lead to problems.

Alternative entrances

There are hives that use *vertical* slots, tunnels and periscopes, even downward-pointing floor entrances. A more common variation is multiple **small** entrances, which are worth discussing. They are flexible, offering better defense against robbers, better ventilation and reduced traffic jams.

Dual entrances are actually the normal arrangement worldwide in cool climates; they were common in British skeps (see images in chapter 12), but in modern British National hives, the upper one has evolved into a ventilation hole in the crown board, which bees can't easily use as an entrance. In American Langstroths, the upper one doubles as a safeguard against suffocation when lower entrances get covered in snow. In Eastern Europe hives may have up to 4 at different

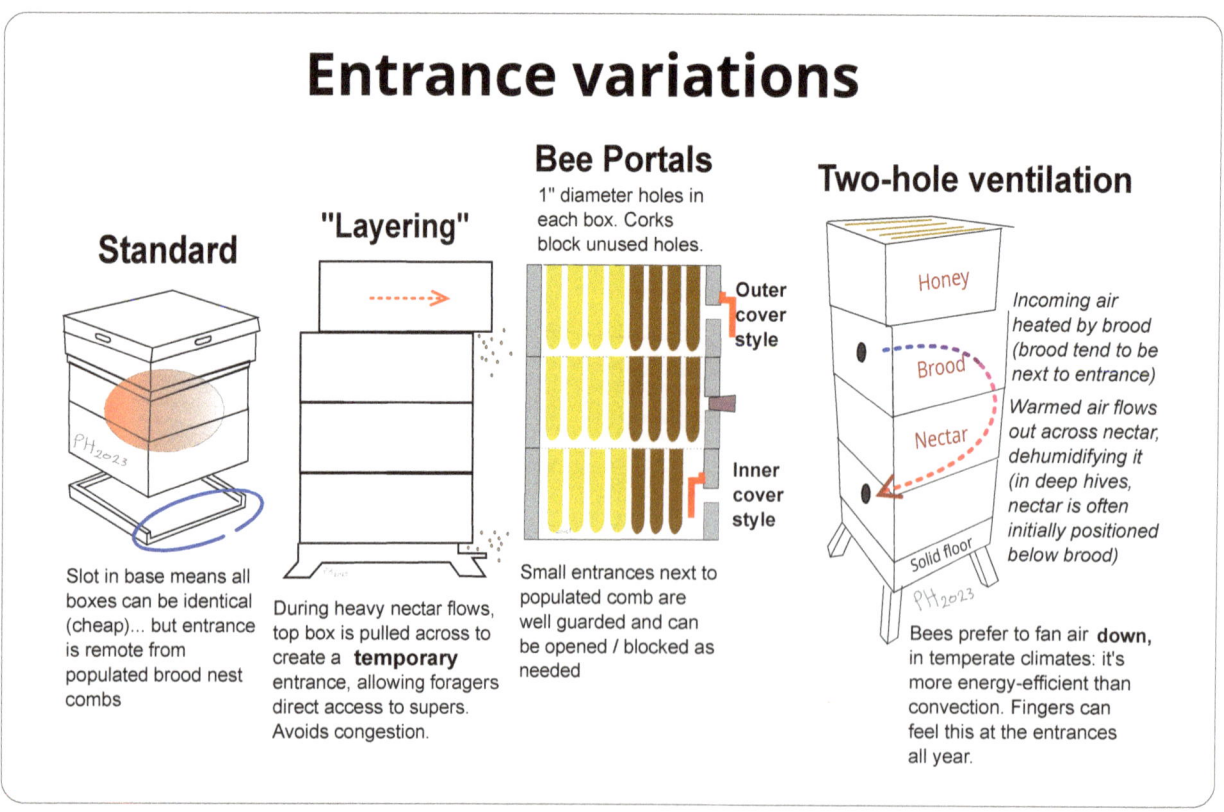

> This form of two-hole ventilation is regulated by the bees, because their populated comb is next to the entrance. An added hole right at the top of the hive creates an uncontrolled, chilling chimney effect. Unsurprsingly, bees often block the holes in crown boards with comb.

levels. Foragers tend to take pollen to the entrance nearest brood.

Bees sometimes chew their own second entrances in weak spots of wood. Extra entrances work best next to comb (and thus bees - guards, fanners).

Lazutin in Russia found dual entrance hives suffered much less from damp[455] due to much better ventilation. For large colonies a small upper / large lower entrance is best, but small colonies can't guard the lower one and produce less moisture, implying the lower entrance should be smaller and covered in protective mesh.

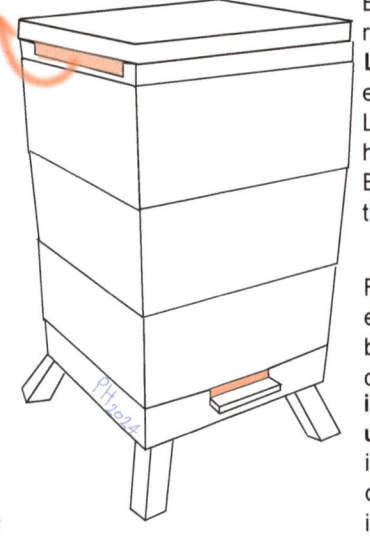

Problematic dual entrances

Bees need to regulate airflow. **Large** upper entrances, like this Langstroth design, haemorrhage heat. Easily seen with a thermal camera.

Floor level entrances are below populated comb and often **ignored and unguarded** if a more convenient route is available.

Apart from guarding issues, another potential issue with multiple entrances is seen in Langstroths. These have a ridiculously large upper entrance and they lose far too much heat through them in winter. A 1" diameter hole is fine, and the bees will then optimise it with propolis.

You can adapt standard hives (framed stacks-of-boxes) to use holes in each box. The holes are typically one inch diameter holes, scorched to make them smooth, to avoid wing damage, and blockable by a cork. The concept has been extended by beeportals.co.uk to make the entrances more defensible, whilst avoiding the traffic jams and ventilation problems associated with long tunnels and periscope style entrances[456].

If you have a floor level entrance, beeiqsolutions.com sell the Hive Gate, a plastic tube inside the hive which guides incoming robbers into or near the main cluster of bees. It works with solid and mesh floors.

The Condensing Hive: insulate, don't ventilate

The concept of a sealed, well insulated cavity with a small entrance is generally known as a *condensing hive*. Like your mouth, hives don't need big entrances to breathe, as bees force air through if needed by fanning; and whilst they need some oxygen, it's not as much as a warm blooded mammal which needs to keep around 37C all the time.

Small entrances help trap heat in the hive by ensuring condensation occurs *inside* - a big entrance allows nest atmosphere to exit before releasing its latent heat. Small entrances also help retain the antimicrobial nest atmosphere and, of course, they help in hive defence.

The key principles are: **managing humidity burns much more energy than regulating hive temperature.** And regulating hive air temperature is easier with a solid floor. But there needs to

---

455 *Keeping Bees with a Smile*, Lazutin & Sharaskin, part I, Winter Ventilation of the Beehive. Interestingly, he states that with his horizontal hives, air is expelled from the **upper** entrance, proven by hoarfrost forming there, which is opposite to the diagram above ('2 hole ventilation') based on Gareth John's observations with deeper hives in England. Different hives, climates, and time of year - there's no rule for all hives.

456 Bees seem to like tunnels 6-10cm long - no need to make ventilation difficult. Even in 1864 *The Times Bee Master* remarked: "Simple hives with simple entrances are best: long tortuous entrance tunnels just makes for irritable bees."

be a way for condensed water to escape, like running out of a bottom entrance, or holes in the floor.

Evaporating nectar to create honey uses a tremendous amount of fuel due to the *heat of vaporisation* (i.e. water takes ages to boil). In a natural nest or condensing hive, the humidity is allowed to re-condense, releasing heat energy which is retained inside the hive, and the water runs out the hive entrance or wicks out along the tree fibres.

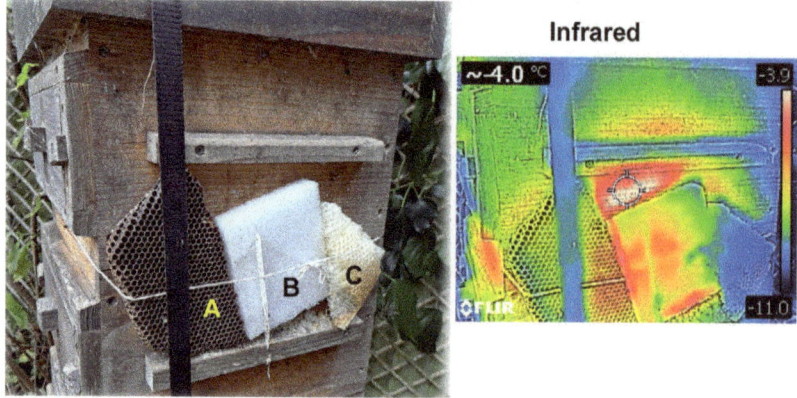

Any insulator helps. In infrared, we can see this Warre's winter cluster glowing clearly (white / red) through 25mm cedar walls. Swapping the materials around showed **B** - waste plastic packing material dramatically reduced heat loss; empty combs **A** and **C** helped slightly, though A (old brood comb) was significantly better than C (honeycomb). Another reason to keep old comb!

In a hive with an open mesh floor, or too many entrances, the humidity is vented before it can recondense, so the bees never get the heat back and have to gather much more fuel (nectar). Basically open mesh floors and poor insulation throw away honey.

A typical strong colony might occupy 60 litres, or 0.06 cubic metres, with the honey ripening done at 40°C (they warm honeycombs during this process). The graph shows there is only about 0.06 x 50, i.e. **3 grams of water vapour** in the hive at any one time. Such a hive might typically lose 500 grams' weight in one night as they evaporate excess water. If they can only warm the combs to 35°C, the air will carry away *30% less moisture (!!!)* which means they need to work significantly harder (fan more air) to remove water from nectar to make honey. This is another reason **a well insulated hive makes more honey** [457].

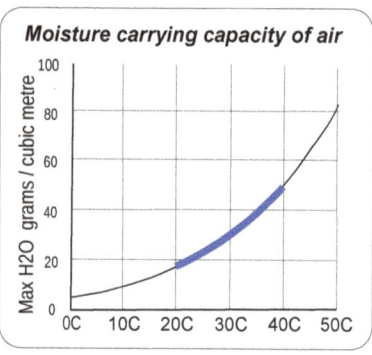

The humidity will always condense somewhere, but ideally you don't want it to drip on the bees and chill them, so a really well insulated roof is key so it condenses on walls, which ideally are roughened to promote propolis deposition. As long as the cavity is >10C, propolis vapour will sterilise it and suppress mould. Water condensing on a propolised surface is safe for bees to drink. A small entrance means the bees also retain the nest atmosphere of pheromones and antimicrobial propolis, rather than it being flushed out of a mesh floor.

It's possible for hives to overheat in full sun so some people argue mesh floors are useful ventilation, but it depends on conditions, like a very large crowded colony in a hot climate with no shade. But Britain has a lot of damp, chilly days when bees would like to maintain a *low humidity* hive - *without an open mesh* floor - because damp air chills you faster than dry air because cold water absorbs much more energy than cold air[458].

---

457   This is based on lectures by Torben Schiffer.
458   Because unlike the simple diatomic springs of O2 and N2, the three atoms in H2O can store energy in more ways - by twisting and bouncing against different partners and hydrogen-bonding to nearby H2O molecules.

On the subject of humidity, Guy Thompson points out, it's possible bees may actually switch between two nectar dessication techniques:

If a hive is **overheating,** fanning saturated humid air out cools the hive rapidly and cools the hive faster. In effect you divert excess heat by boiling water and venting the steam out of the hive door.

But if the hive **needs heating,** it is thermally more efficient to dispose of water in liquid form than as a humid airstream, i.e. let it condense against a cool wall and dribble out the entrance, since the latent heat of evaporation is thereby not lost but kept within the hive. The specific heat capacity of **water** is around 4 joules/gram/°C but the latent heat of **vaporisation** is a whopping 230 J/g .
A closed internal vapour condensation cycle is much more energy efficient than a venting cycle.

Scanning with binoculars for bee nests in their natural habitat. This oak in Blenheim Woods is ~800 years old and supports many species on its surface and in cavities. Nests are almost always high up, in living wood, shaded, smaller than commercial hives, **extremely** well insulated and surrounded by tons of thermal mass, and long lived (10% winter fail rate) even in areas infested with varroa.

Incidentally, the National is designed as a condensing hive, it's easy in our mild winters. Americans went for ventilation.

Finding wild nests

There are loads, but *people never look up.* 90% are well above head height - though a few are lower or, rarely, in ground level cavities. Walking around quietly, i.e. **alone**, keep an ear out for massed buzzing above you. It's usually foragers on a tree in bloom, but sometimes it's a nest. If you find one, look for more nearby. Nests can occur in clusters: don't assume you've found "the nest".

On finding one in a building, ask the property owner to contact you if a swarm emerges. And if you collect a swarm look for the mother nest, which is usually actively flying within sight *if you look up,* and again, give your details to the property owner. They sometimes ring you a week later (a cast) or the next year. Swarms from a nest tend to land in the same position.

Nests in trees are trickier to spot: bees' colouration is camouflaged against bark, and there is a lot of visual clutter and obscuration from branches and leaves. They're easier to find in late Spring before trees fully leaf up. Traffic levels and noise are often low, because they're not continuously trucking in excess nectar for a beekeeper. But there are some tricks, collectively known as *beelining*[459]. Often this involves a special 2-compartment box to trap and release foragers and thus triangulate on their nest location, though this doesn't work well in forests, where bees dodge round trees. Some people attach a streamer to the bees to make them more visible in flight - indigenous Australians used cobwebs on their native bees. Position the sun behind trees of interest so the flickering of wings is clearer. Nests are rarely in dead trees, and may only be buzzy a few hours a day, so try walking at different times of day. If you're not sure which tree they're in but you know it's near, press your ear to trees - you can sometimes hear the buzzing of the nest conducted through wood.

Most wild nests near me are in old, crumbly houses, with entrances just under the eaves or in chimneys. **If you see a pile of dead bees on a pavement, look up**. Bees like slate roofs and old Listed buildings. **Look along the roof** so it appears as a line, you will then see the bees in silhouette against the bright sky. This picture is a Victorian house near me with 5 colonies in its roof / chimney!

There are choke points where they can be found - bee highways around obstacles; sunny glades; watering spots. Water foragers are particularly useful as they choose the nearest source to their nest, drink to bursting and fly straight home.

The very best beeliners unconciously pick up on patterns and subtle clues others miss. They initially calm themselves, often with a ritual, to attune to an area's spirit, becoming more aware and receptive.

If you find a free living nest, please consider logging it at freelivingbees.com (community / story based) and honeybeewatch.com (citizen science survey). These websites deliberately obscure the exact location because *there are beekeepers who go around killing them - illegally.*

It's now generally accepted that despite varroa, there are at least tens of thousands of unmanaged colonies around the British Isles; and you would expect similar numbers in suitable parts of the rest of the European honeybee's home range, like France[460].

---

459   Tom Seeley's book *Following the Wild Bees* is about these techniques. Wasp researchers use similar ones - wasplining! One way of distinguishing wasp nests is, wasps fly silently. Tropical *honeyguide birds* actively lead humans to bee nests for a share of the loot - they eat the grubs and wax.

460   Some countries have problems. Spain is suffering from climate change and Asian Hornets. Many German nest sites (old houses) were trashed by WW2 and their forests are now mainly young conifers for lumber. But bees are adaptable. A recent metastudy of 200 papers concluded that worldwide, there are more wild colonies than managed ones, though most are in South America and Africa. *Density of wild honey bee, Apis mellifera, colonies worldwide,* Visick & Ratnieks (2023), doi.org/10.1002/ece3.10609

British wild bees were declared extinct twice, firstly when Isle of Wight disease decimated hives in the 1920's (actually the main die-off was *imported* bees), and again after varroa's arrival in 1992 with people confidently (mis)quoting a flawed 2012 study[461]. A common statement was "any wild nest is an escaped swarm from an apiary, they never survive over winter." This may be because such escapees did indeed perish over winter: Roger Patterson, Carl Kolyer and others have done many cutouts in Britain and noted that any nests which came through a winter invariably had dark queens.

But by ~2018, a tipping point was reached because so many people were stating on forums, "I don't care what the self declared experts say, I see several wild nests near me". There followed a few years of shouting as sceptics' assertions slid from "if there are such colonies, they only last 1 year" to 2 years... 3 years... 5 years[462]; and now no one seriously questions that we have a self sustaining population of tens of thousands of wild survivor colonies in the UK alone.

Lessons to learn:

- The alarmists initially won the argument: simple ideas propagate faster than complex truths.
- Few people thought to systematically go look for colonies and check the claim (a handful did, and were ignored).
- Few people actually read scientific papers: most just repeat a soundbite that supports their agenda.

I see 7-10 unmanaged, long lived colonies in my village. One lived 19 years.

America isn't Europe

This map explains why advice varies across Europe and North America.

Europe is further north than most of the USA - Britain is north of the entire USA! Climates are radically different, tending to be stabler, but more extreme in the US.

So bear in mind I am writing in England. Britain is relatively mild (USDA plant hardiness zones 6 to 9), with a long growing season and plenty of rain all year. The Gulf Stream warms the western coast[463] - we're a tepid swamp.

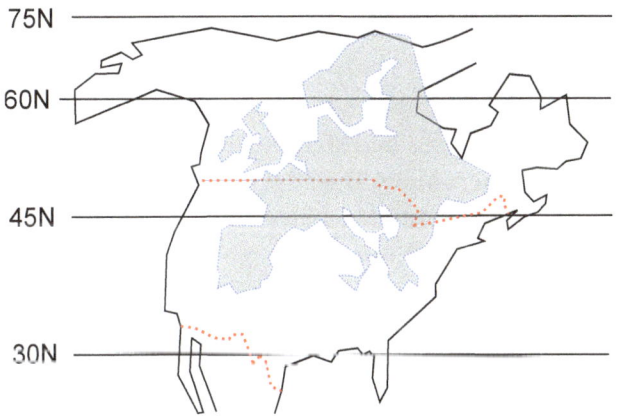

---

461 Thompson, C. E. (2012) *The health and status of the feral honeybee (Apis mellifera sp) and Apis mellifera mellifera population of the UK*. Thompson was writing her PhD thesis and was presumably a little inexperienced. In retrospect it is obvious the regions where she chose to look for wild bees, because they lacked beekeepers, were going to be poor bee country - they were conifer forest or clear felled land recovering from conifer afforestation. More detail: www.naturalbekeepingtrust.org/wild-bees. A more significant finding in this paper was that wild nests' genetics were essentially the same as those in nearby apiaries, and Amm was consistently the largest (49%) part of all UK sites' genome though she studied just a small number of such colonies. The "no wild bees" proponents took this as proof that the wild ones were escapees from apiaries, but what it really indicated was that bees will mate with whoever is nearby, the gene flow works both ways. More recent work in Serbia (*Free living vs managed bees - a population genetic approach,* Davidović, Patenković et al, EurBee 9 European Congress of Apidology, 2022) indicates wild colonies are more related to each other than managed colonies, implying a stable, preferentially conserved background of local genetics.

462 A disingenuous and misleading comparison: if you requeen a managed hive every 1-2 years *it's not the same colony*.

463 For now. If the Atlantic Conveyor, the Gulf Stream's engine, fails as some predict, we'll be more like similar-latitude Newfoundland.

The USA faces problems other areas don't, such as massive queen failure[464].

British / European geography doesn't favour huge megafarms. That's good for bees in many ways. We *may* have more gardens (not my field of knowledge) and we certainly use fewer pesticides.

The relatively stable continental weather systems lead to much more predictable, larger honey crops in America and Europe than the fickle British maritime climate.

US almond pollinators can claim money for replacement packages from the federal ELAP scheme if they lose more than 17.5% of their bees. This is on top of the $100 per colony for pollinating almonds. This incentivises working bees hard until they die. This is an example of the *cobra effect,* making a situation worse by compensating failure.[465]

American bees have a much smaller genetic pool to call upon[466] *if you buy queens.* However, bees have been there 400 years and this is easily long enough for local ecotypes to arise in isolated areas.

Comb quirks

Once you begin really looking at comb, you begin noticing all kinds of subtleties in its construction, like the round (not hexagonal) cells and shallow reinforcing cells at right angles to the others along the edge of a free hanging comb. Natural comb is a dynamic tapestry which the bees keep reconfiguring: each cell and bit of brace comb is a customised tool fulfilling a specific purpose. This tends to be overlooked by users of uniform sheets of flat foundation.

Large drone brood cells don't fit right next to small worker ones, so there is a band about 2 cells wide where the cells between are stretched into intermediate sizes.

Stretched and non hexagonal cells are very common at the ends of top bars. Regular hexagon arrays are mostly seen on foundation.

Comb edge.

---

464 Unique American factors include the postal system! Dataloggers showed 20% of shipped packages were exposed to high temperatures, sometimes over 40C. Lab experiments reproducing this showed sperm viability in a mated queen dropped from 90% to 20%, which could explain "why beekeepers have to requeen every 6 months not 2 years now". *Colony Failure Linked to Low Sperm Viability in Honey Bee (Apis mellifera) Queens and an Exploration of Potential Causative Factors,* Pettis et al, 2016 - https://doi.org/10.1371/journal.pone.0147220   Pettis made the point (online lecture 2021) that this is another reason for using local bees, and he is struck by British queens regularly lasting 2 years, he 'even' saw a 4 year old one in Australia.

465 See Wikipedia's incidentally hilarious entry on *Perverse Incentive*

466 In studies conducted in 1993-1994 and in 2004-2005, U.S. commercial queen producers self-reported the production of close to 1 million queens for sale from around 600 and 500 queen "mothers", respectively (Schiff and Sheppard 1995, 1996; Delaney et al 2009). Further information at www.researchgate.net/publication/303873161_Status_of_breeding_practices_and_genetic_diversity_in_domestic_US_honey_bees

When two combs merge, sometimes the cells are at different angles[467]! One comb may have a flat edge towards the top, the other a corner. In this case the bees "roll" the cells in a transition band until the angles match. Occasionally you will see a 5 or 7 sided cell in these transitions, and cells with 4 to 9 sides have been reported[468].

Different bee races pack honey with an air gap under the cap (capped honey appears white - "dry cappings"), or with the wax cap touching the wet honey behind (appears dark). Russian researchers found[469] that the air gap significantly improves insulation - it's a cold weather adaptation in *Amm* and *carnica* bees.

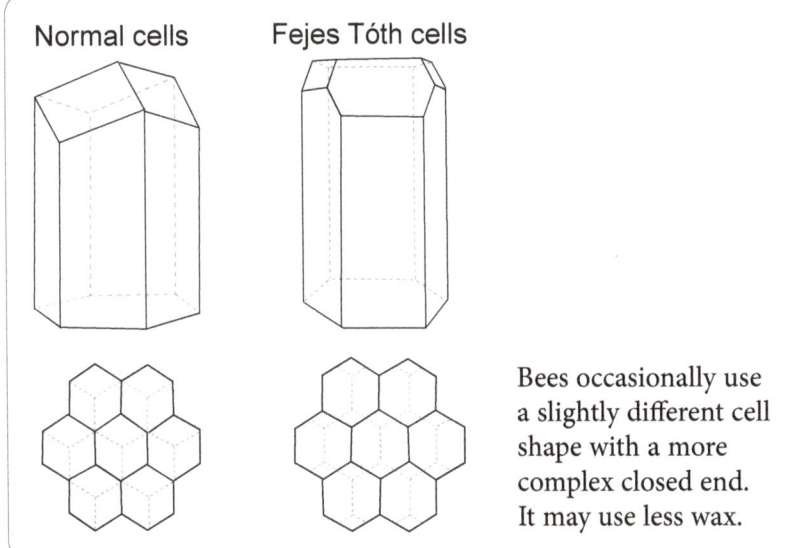

Bees occasionally use a slightly different cell shape with a more complex closed end. It may use less wax.

You've had some right cowboys in here, mate

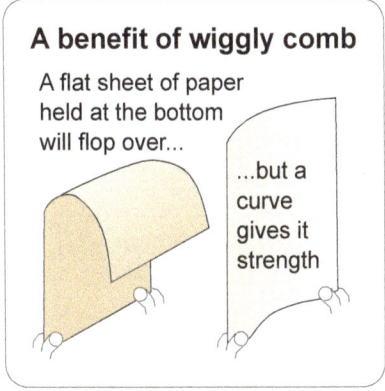

A benefit of wiggly comb

A flat sheet of paper held at the bottom will flop over...

...but a curve gives it strength

Some thoughts to ponder

Bees build up a static charge when flying. Does fanning precipitate dust, helping sterilise the hive atmosphere? This is the principle behind electrostatic precipitators.

Warm clusters raise Spring brood earlier - this phenomenon has been known and exploited for decades to ensure lots of early bees for pollination etc.[470]

I've preached the benefits of well insulated hives - so do they promote brood raising before forage is ready? No. At least mine don't. They respond appropriately to external cues like day length, because they are local; and available forage (nectar *and* pollen), because I don't feed them.

Bees avoid other colonies' brood comb. - This is a very dramatic, strong effect. You see it when merging two nests: eventually one nest area depopulates as its bees die... and the remaining bees

---

467 *Imperfect comb construction reveals the architectural abilities of honeybees* (2021) Smith, Napp & Petersen, PNAS doi.org/10.1073/pnas.2103605118 . Not open access. Another paper discussing non standard cells is *Stigmergy versus behavioral flexibility and planning in honeybee comb construction* Gallo and Chittka (2021) https://doi.org/10.1073/pnas.2111310118

468 This has been known for a long time, Tickner Edwardes mentioned it in 1916 in *Lore of the Honeybee* p. 136

469 СВЕТЛАЯ ПЕЧАТКА ОБЛЕГЧАЕТ ПЕРЕДВИЖЕНИЕ ПЧЕЛ ПО СОТАМ, LEBEDEV V.I.1, KASYANOV A.I. (2018) or in English, "Dry capping faciliates moving of the lone bees on the cold combs." elibrary.ru/item.asp?id=37103435 and let your browser translate.

470 Quimby noted large (warm) clusters raise brood earlier (*Mysteries of Beekeeping Explained*, chapter 3 p.38-39); Storch mentioned the restricted space boom. It's frequently mentioned by poly hive users.

avoid it. I've also seen it when using bait comb in a hive, the swarm will build their nest as far from it as possible. You can also see bees avoid comb treated heavily with miticides.

But conventional beekeepers routinely swap frames between hives. Imagine how that distorts bee decisions...

Do bees really orient comb N-S? Perhaps they simply prefer cold way - and most beekeepers orient their hives with the entrance pointing vaguely South.

Slow and steady wins the race? Bees are endurance athletes - a huge flying range compared to wasps, which are sprinting ambush predators. It's well known in sports science that you can do more work, overall, if you stay below your Critical Power fatigue threshold, above which you tire rapidly. It seems conceivable therefore that, all else being equal, running hives as hard as possible may give *less* honey over a year. And of course overtired workers can't attend to cleaning, hygiene, propolis gathering and other healthy habits.

Hexagons are not uniquely a feature of wax use. This wasp nest is made from pulped wood.

Honey vs Swarm oriented management - there is an interesting conclusion in Julian Lubieniecki's book *A complete practical guide for beekeepers*[471]:

◆ Short flows favour honey-oriented management

◆ Long continuous flows favour swarm-oriented management (lots of little hives of bees)

True? Perhaps. Worth thinking on? Yes.

Business analogies

Abrupt changes cause havoc.

Marginal Gains Rule - it's easy to make lots of small marginal gains - and they don't just add up linearly, they *multiply together* giving you exponential improvement. Like, allow propolis, add insulation, fewer interventions, better diet. This works in reverse too - cumulative stressors multiply.

Metrics and targets (mite counts, selecting queens for fecundity etc) can constrain more fundamental, important factors like health, inbreeding and flexibility.

---

471   Lubieniecki was Russian and wrote this in 1859. This gem was noticed by GregB on Beesource.com

Chapter 18
Summing up, moving on

General insights - a bonfire of assumptions

Bee health is **largely dependent on how well they can express natural behaviour.** A big factor is the use of **local** bees, and many disorders are symptoms of **over**management[472].

Every area of human knowledge is a chaotic mass of Experts with Opinions and very narrow Knowledge. In reality, our "knowledge" is a current working best guess. Most of what we "know" of bees is under unnatural conditions. Alas, conventional beekeeping is driven by fear, and people crave easy answers: "Miticide? Great. Symptom solved for now." However, simple can also be sophisticated. Leave the bees enough honey, and reduce meddling, and your bees can sort out most health issues. Delegate!

Naturally evolved varroa immunity is stable. Survivor populations have shifted *many* traits a *bit*, which is a more robust "fix" than one massively reinforced characteristic[473]. Leave-alone management isn't a danger to other colonies because if ours were mite ridden, they'd be dead. But bear in mind, observational beekeeping is a holistic approach. Simply going treatment-free will not control varroa if the bees are still pushed to their limits producing honey.

Intriguing findings

There are **lots** of quiet low-intervention beekeepers, prioritising health not wealth. People have converged on this solution from multiple paths, from bee farming to the default do-nothing approach, and find it less prone to problems. Whereas conventional beekeepers' associations find ~40% of new joiners drop out within 2 years, the rate is lower for leave-alone beekeepers so the proportion grows.

Fact-checking this book, it really came home to me how most true innovation is from the fringes, where people have rejected rules and play with ideas[474]. The conventional beekeeping community has been essentially reactive, waiting for someone to create a magic bee (hint: it's in front of

---

472 David Graeber's book *Bullshit Jobs: A Theory* argues that much human activity is nonproductive. In contrast, bees do everything for a reason; natural selection is ruthless.

473 See Tom Seeley' *Progress report on three years of treatment-free beekeeping,* Natural Bee Husbandry No.16, Summer 2020 and American Bee Journal August 2020, where he compares how different "mite resistant" strains fared.

474 Countries such as Britain, France, and Ireland have a relaxed attitude to beekeeping regulation, permitting amateur experiments to proliferate.

you). The various programs to breed mite resistant bees are years behind amateurs. Some of these experiments fail; others lead to new insights. Hobbyists with time, actually *see* what they're looking at, and allow it to develop, where commercial beekeepers view irregularities as a problem.

Extensive Beekeeping, common in Africa (and historically, European skeppists), is an alternative business model to most bee farmers' *Intensive Beekeeping*. Rather than concentrating your assets in complex equipment under continuous management producing a high return on investment, you have lots of cheap hives left to themselves, producing less per hive but with lower overheads.

The decrease in swarminess noted by many low-intervention beekeepers after a couple of years, suggests that *excessive swarming is partly a response to a stressful environment*[475]. It makes sense that evolutionarily, colonies under stress will throw off many swarms in an effort to find a balance that thrives (because each queen will mate with a lot of drones, so each colony is likely to be quite different). Once they find a stable equilibrium, the gene shuffling is turned down[476], if only because the failures die out rapidly. It's the entrepreneurs' "failing smart" strategy - *fail fast, fail often*.

This can't be the whole story as swarminess can definitely be a response to crowding, and this reduction in swarminess may simply be due to queens of commercial origin outbreeding with less prolific locals. But it's quite a noticeable effect.

Next time you swamp a hive with pungent miticide, reflect on this: AI researchers find the intelligence (adaptability) of a robot swarm is based fundamentally on communication…

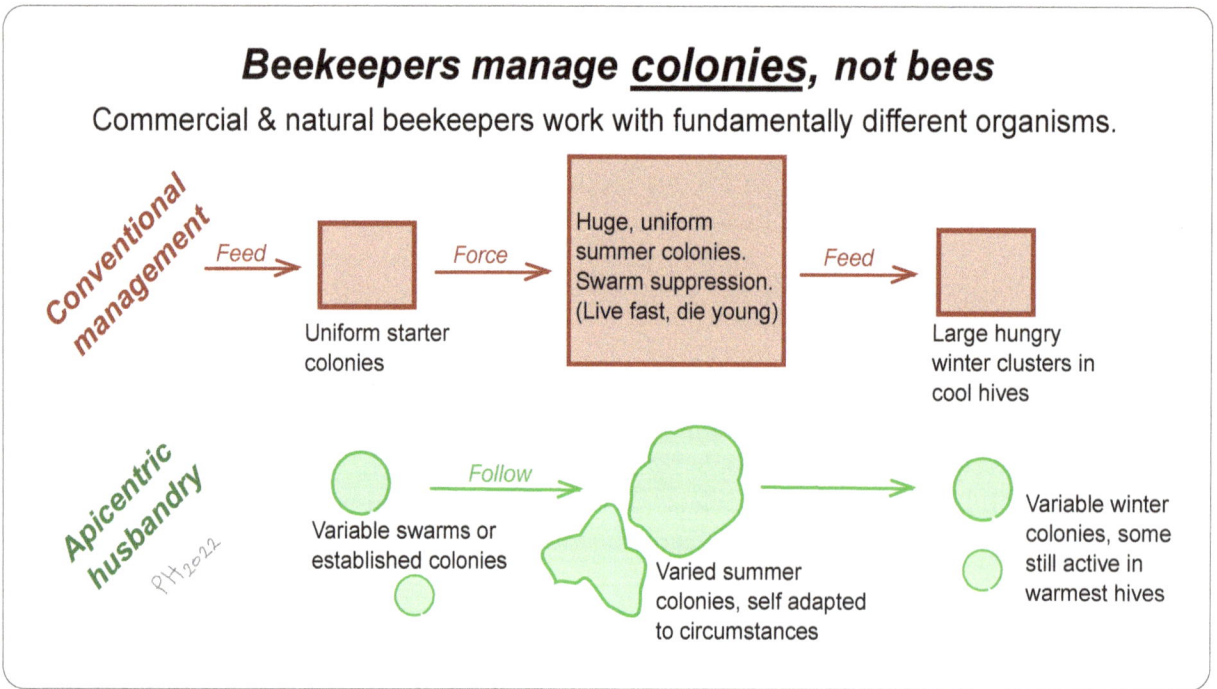

Commercial management requires a consistent, predictable path every year - early & big, suppress swarms, near clone queens etc. It's a live fast, die young, get things done, fix it later paradigm.

Low interventionists instead delegate management to the semi-wild colony, which reacts differently every year as it proactively adapts and fixes issues early for you.

---

475 As far back as 1978, Lazutin noted stress promotes excessive swarming (*Keeping Bees with a Smile*, p.70-71).

476 Bacteria, yeast, cancer cells and plants massively increase their (random) mutation rate under stress - a process called *"hypermutation"* or *"stress induced mutation"*. A good layman's overview of the subject is www.wired.com/2014/01/evolution-evolves-under-pressure/

Part 3   Chapter 18   Summing up, moving on

Commercial beekeeping: shades of grey

If this were a novel, we'd need a villain for dramatic tension and everything would work out neatly at the end. But this is real life. What we do potentially affects people dependent on an income from bees, and pollinating the food crops we eat. They need a certain kind of prolific bee, we desire a leave-alone self managing type. So who's the villain?

The conservators of beekeeping lore were generally commercially motivated - beekeepers with other priorities are a modern aberration[477]. Before varroa we were all treatment-free. I hope, though, that this book gives some readers ideas on alternative management options. After all, if you have a remote apiary, wouldn't it be convenient for it to be low management? You can delegate to the bees. Overcontrol is fragile[478]. Every hour saved is one you can use elsewhere, and now you *know* TF beekeeping is possible. Honey yield is primarily dependant on forage.

Some fringe voices talk up a culture war between opposing styles. They crave attention, exaggerate the threat of The Other and push a narrative of false choices[479].

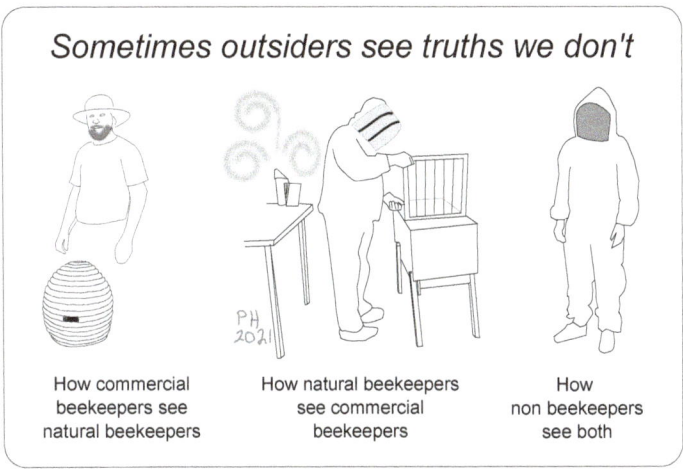

Sometimes outsiders see truths we don't

How commercial beekeepers see natural beekeepers | How natural beekeepers see commercial beekeepers | How non beekeepers see both

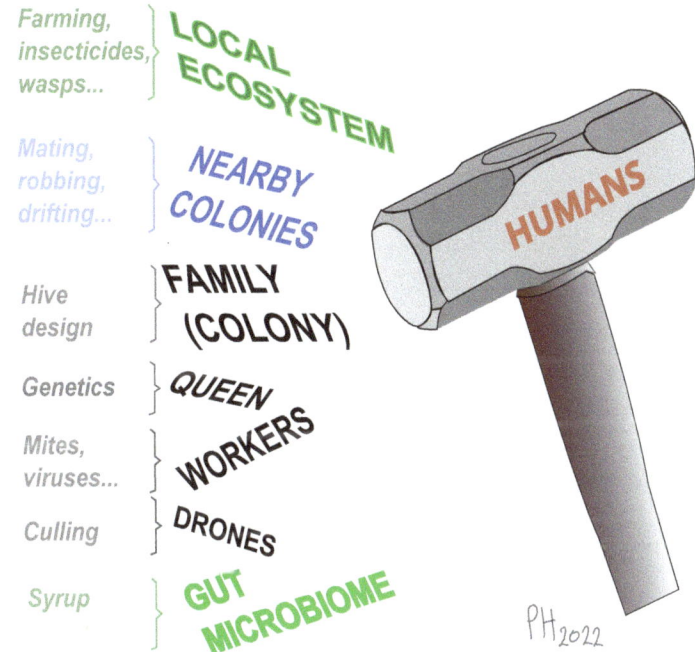

In a finite world, one must be aware of the impact of one's actions. Bees don't just "create money from air" without consequence.

---

477   Difficult to be precise, but I think the movement took off with the publication of key works like Phil Chandler's *The Barefoot Beekeeper* around 2007, somewhat earlier in the USA and the 1970s in Russia (Lazutin). Don't project an idealised romantic illusion onto Olde Time Beekeeping. It was about income and food.

478   "*A loose network absorbs shock; a tightly coupled one transmits it.*" - Thomas Homer-Dixon, complex systems expert. Example: Just In Time supply chain failures during the pandemic. Build in options and delegate to bees where possible. They have flourished for millions of years by biding their time on stores and capitalising on opportunities.

479   Anger is inversely proportional to facts. People often blame a malignant presence for their misfortunes, like the African Azande people's fascinating self-consistent logic explaining things in terms of witchcraft and sorcery. In Sociology, *Group Threat Theory* and *Integrated Threat Theory* assert that perceived threat and reactions increase as the size of an out-group increases. The solution is similar to merging casts peacefully: mingle, don't stay in isolated groups.

The truth is, we overlap in most things we do: most of us are pretty relaxed about what other beekeepers do around them. I know that if I get a swarm from a commercial hive, it will rapidly open-mate with my bees and be pretty much native within 2 years[480].

This reflects what I've learned from bees. A variety of approaches is *useful*, no need to squabble, it's not a competitive hobby. We learn more by trying many strategies in parallel. For example, bee farmers are the only beekeepers with enough hives to spot real trends, like the massive fall in numbers of swarms in South Africa. They are adept at diagnosing disease and don't over-inspect - only needing to examine a couple of brood frames per hive to spot abnormalities (in just 60-120 seconds). They have encyclopaedic knowledge about forage. They can teach you a lot!

Bee farmers are in a low margin economic trap and cannot afford radical experiments[481]: fundamentally, trends in low cost agribusiness (like cheap imports[482] and pollination needs) drive intensification, not bee farmers per se.

The real risks for bees are from **very** large scale commercial operations - what some term the *Bee-Industrial Complex*. A general economic principle is, as time goes on, the most efficient - lowest margin - operations dominate and there is a race to the bottom in employee / bee benefits. Marginal cost savings *and their impact on bees across a country* are amplified enormously when a handful of operations run hundreds of thousands of hives. I'm not just talking about spreading disease or unsuitable bees, but legislative change following lobbying by commercial interests.

Looking ahead

Writing in 2024, it's pretty easy to see some **known** issues coming down the line:

Doom and gloom: The next decade will likely feature a lot of disruption from Asian Hornets and Small Hive Beetle in Europe and Britain. New Zealand's beekeeping industry will contract as manuka becomes the Last Big Thing and other sources come on stream and undercut them. As I was writing this book, Australian beekeeping went into self-destruct mode when varroa was found in 2022 - the authorities activated a scorched-earth firebreak policy, killing 30,000 hives before abandoning it in 2023, realising the bee farmers they were trusting to self-report infection were the vectors.

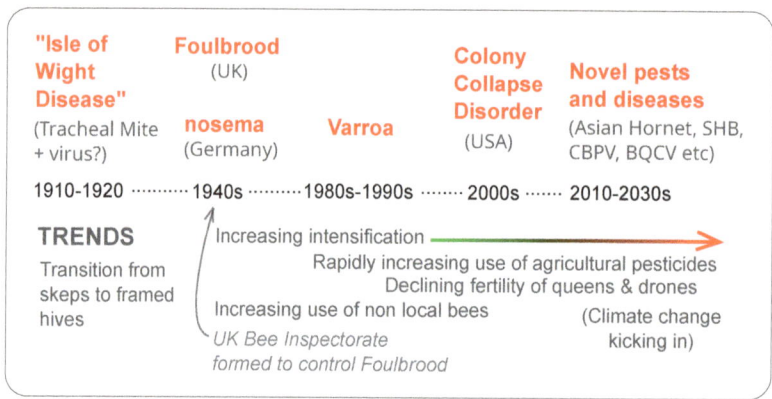

Institutional memory follows a pattern of repeating errors every 15-20 years across many industries. Beekeeping keeps pushing things to collapse.

---

480   Modern landscapes support fewer colonies. During population crashes from varroa etc, isolated wild populations may have *benefited* from some new blood being introduced to avoid diploid drone (inbreeding) collapse.

481   Shipping hives to remote locations and managing them there is *expensive*. Human accomodation, labour costs etc... you're talking 5-10% profit on pollination contracts, while imports undercut your honey sales. Full time bee farmers need at least 500 hives to make a living: a huge capital investment and more than one person can run. One more perturbation, like a new pest, could drive many out of business.

482   There is a massive amount of honey fraud. One grey area is vacuum drying nectar - see tinyurl.com/m56xxyrm. More background at analytica.co.nz , www.honeyap.org The opposite approach is taken by melibio.com, whose refreshingly honest Unique Selling Point is "this is a honey substitute **not** made by bees".

Evolution of beekeeping economics

Honey → Wax (candles) → Honey → Pollination →
- Luxury goods?
- Health / wellness services and products?
- Pharma products?
- Conservation?
- Environmental monitoring?
- Rewilding / ecosystem + climate stabilisation?
- Targeted delivery of vaccines to specific crops?

PH₂₀₂₄

| The honey market is currently under pressure from low cost imports and fake honey

But more positively, varroa resistance will be widespread in areas where a local population persists. In the last five years, this has become generally accepted. Factional disputes will dissolve as new data is assimilated. Locally adapted bees will be a Big Thing. Hive design will improve. The existence of a huge population of genuine, genetically distinct and stable wild bees in some countries[483] will be accepted and BIBBA will grow much larger than its current 2,000 members. Commercial breeders will attempt to claim credit for solving the varroa issue, but the main driver will be naturally arising esistance in local populations.

This will become clear in Australia, because the authorities hadn't thought far enough ahead - there were no miticides approved for use. And it's clear now, varroa was noticed in 2022 but really arrived around 2020, hitch-hiking in on smuggled bees i.e. the local bees have already had ~4 years to adapt. This has created a continent-sized natural experiment with ideal conditions for bees to rapidly express resistance.

Commercial beekeepers are experimenting too: they stand to lose most if one more perturbation collapses their business model. Examples include trialling mass queen banking of local queens over winter; using warm poly hives for earlier brood raising, so they don't need to import bees to satisfy their need for very early large colonies; mimicking swarms' brood breaks by temporarily caging queens; and truly local bees (not Buckfasts or imports).

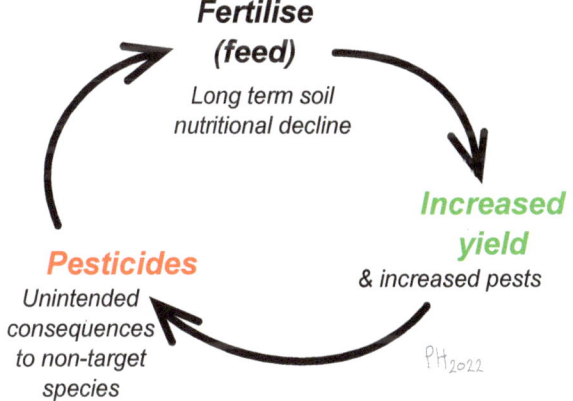

| Intensive agriculture has parallel dependencies to commercial apiculture's feed / treat cycle. Converting to organic farming takes years to undo damage. Fortunately beekeepers are less trapped: less sunk cost, not tied to specific fields, and can change part of their operation in parallel.

---

483 British / European operations have several pest resistance factors lacking in America & Australia, so will adapt first. They're smaller, operate over shorter ranges, more suitable pollen and have *much* wider genetic diversity. Certain European countries have a good balance of regulation: we've more or less snuffed out foulbrood and control pesticides much more stringently, while leaving beekeepers freedom to experiment. Some countries like Germany though have tight regulation and a shortage of suitable old houses / trees.

As the health benefits of undoing 150 years of industrial inbreeding become apparent, there will be a selection pressure on beekeepers, especially if the resilience extends to the next wave of pests[484]. The majority of new beekeepers will be aware of, and incorporate at least some lower-intervention beekeeping principles - if only because the proportion taking it up primarily for money is falling as honey fraud, climate change etc reduces profits. Even now, it's clear there are lots of hidden low-intervention beekeepers just keeping their heads down.

Institutional gridlock

The craft is often taught through a series of exams with syllabuses carefully refined over a century. This has upsides, like recognised qualifications; but introducing radically alternative practises to an interlocking system is challenging.

So don't worry about the apparent inflexibility of official institutions. It just means the bureaucratic framework can't rapidly stretch to accept practices which contradict current teaching. Many members, examiners and bee inspectors are very interested in learning from alternative approaches. Others are more cautious about the risks of non-canon ideas or are more interested in the commercial aspects. That's OK, we all learn from each other.

It would be simplest for BKAs to simply accept the contradictions to orthodox training and have a *parallel* training module, explaining it's a different balance point, like a wild colony. That avoids having to rewrite the existing modules.

The tightrope trap

Commercial, high yield beekeeping is a balancing act. Running your hives hot gives little margin for error.

If bees are not allowed to swarm and the queen is unnaturally prolific, then like a car with no brakes they have few options if they run into trouble. In a sudden dearth, they can starve quickly. Without brood breaks they're vulnerable to runaway disease and pest problems. This is why such hives need constant inspections.

Consider this system stability diagram. Imagine the colony as a ball resting in the dip at the top of a curve. The dip for the red curve is shallow: its ball can roll out of its stable position with just a little nudge. Once perturbed out of their rest state, the red colony numbers can crash rapidly (thus the steep edges): overmanagement has removed their

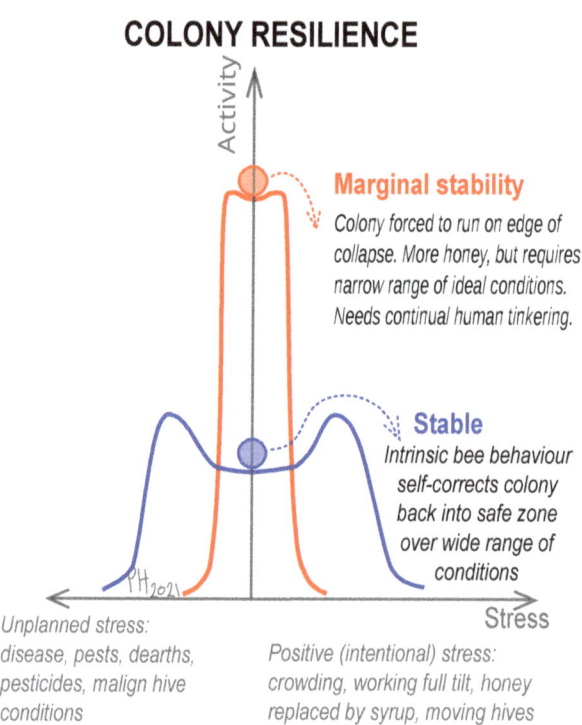

---

484   Jacques van Alphen suggests everyone should use their local bee, i.e. *Amm* in northern Europe, with migratory beekeepers using selectively bred strains and hobbyists with static hives using open mating, to preserve rare alleles.

fallback options. Don't be fooled by the size of the colonies - such bees are surviving, not thriving.

Low-impact beekeeping is like the blue curve, more safety margin, a wealth of health. Although there are fewer bees per hive present *continuously,* so less suitable for commercial migatory beekeeping, there are fewer mouths to feed in the *local* forage gaps. The wider shape of the curve reflects that there is a greater range of stable conditions - and more strategy options because things decline slower. The higher "lips" round the stable area indicate it is more difficult to perturb into decline.

To sum up, low intervention beekeeping isn't difficult, it's not in competiton with commercial beeks and nobody's harmed. Beekeeping isn't a competitive sport: we're all in it together, and a spread of approaches is appropriate.

Review 14

Write down five take-aways you have learned from this book

1.

2.

3.

4.

5.

What do you intend researching next?

What have you learned about yourself?

Further reading

In addition to subject-specific references scattered throughout this book, I suggest the following are key works as eye opening pathfinders.

There are many blogs, forums etc on the internet but these tend to echo back the search terms you input, and recycle half truths. Books take you to areas you hadn't previously considered.

General easy reading

The Buzz About Bees, by Professor Jürgen Tautz, Springer, ISBN 978-3-540-78727-3. Lots of high quality photos, very readable, explains *why* bees do things.

Honeybee Democracy by Professor Tom Seeley, Princeton University Press, ISBN 978-0-691-14721-5 - many insights into colony life and how they decide things.

The Lives of Bees by Professor Tom Seeley, Princeton University Press, ISBN 978-0-691-16676-6 - how bees live in the wild.

Keeping Bees with a Smile, Fedor Lazutin & Leo Sharashkin, New Society Publishers, ISBN 9780865719279 - some overlap with this book but different insights; focused on cold weather hives in Russia and North America.

Natural Bee Husbandry journal: naturalbeekeepingtrust.org/natural-bee-husbandry explores paths less trodden - some high quality research and thought provoking articles.

A deeper dive

NBKT science page: naturalbeekeepingtrust.org/the-science-p2.

The Biology of the Honey Bee, Mark Winston, Harvard University Press, ISBN 0-674-07408-4.

Treatment-Free Beekeeping, by David Heaf. IBRA / Northern Bee Books, ISBN 978-1-913811-00-6.

Beekeeping: Challenge what you are told, by Roger Patterson, self published on behalf of BIBBA.com - a beekeeper with 70 years' experience (framed hives) does a deep dive into common beekeeper errors, bee behaviour and Things That Just Aren't True. Vast range of topics. Straight talking, no punches pulled.

Uncovering lost lore

Old and foreign books and magazines contain novel insights. If you find lost pearls of wisdom, please spread them.

Afterwords

Books like this may give the impression, everything worked from day one for me. That is not the case, I muddled through conflicting management advice; some swarms were duds, and no colony lasts forever. Mistakes were made. So don't despair if you have a few deadouts, because we all have some.

Don't take my writings as a recipe. *"The Tao that can be described is not the real Tao[485]."* Better to concentrate on *why* and let you figure out the best *how* for your situation.

I hope the broad sweep of the book introduces you to many new concepts, helps you sort wisdom from the tsunami of misinformation, and hopefully, spot new links, questions, and mysterious holes no-one ever considered (why *don't* they do this...?).

If anyone has more tips, particularly on non invasive inspections, I'd love to hear them. Who knows, there may be a second edition of this some day. I can be contacted at **NaturalBeek@gmail.com**. I welcome corrections, and your own observations, but will not reply to abusive mails.

To sum up -

- Stop worrying about the Rules.
- There's no One Correct Answer.
- Guide yourself by moral principles which Feel Right.
- Don't panic: the bees 'know' what they're doing.
- We can learn from bees' response to uncertainty: they choose flexibility over one rigid response.

You're doing all right, actually.

Evaluating advice

Here's a succinct table to quickly assess advice. The lower the score, the more dubious the information.

You should, of course, apply the same standards to this book.

Remember that even if you disagree with someone's beliefs, their *experiences* are still valid.

---

485   *Tao Te Ching*, verse 1

Information source	Rating
Source stands to profit from their advice (e.g. sells queens)	-20
Appeals to Authority, but vague on exactly which Authority	-5
Quotes information from social media without checking it[486]	-5
States things in terms of certainty (absolutist information is simplified)[487]	-10
Research scientist in this field	+20
Ignores challenging questions, answers a different question	-10
Responds with aggressive "wall of words" distraction tactics - "accuse and confuse" - answers questions with questions, jams your brain with unverifiable statements, avoids answering the question. Interested in dominating a group with their narrative, not dialectic [calm debate to estabish truth][488]	-10
Knowingly lies[489]	-10
Traduces others	-10
Blames others for own problems	-5
Demands alternative views are suppressed	-10
Reputation among other local beekeepers	+10 to -10
Accuses you of lying / denies your direct observations	-20
Quotes supporting sources…	+5
…but on checking them they do not say what he says!	-10
(On a forum) Many thousands of posts, mainly insulting others[490] (status oriented.)	-5
(On a forum) uses their own name rather than posting anonymously	+10
Uses emotive language	-5
Claim matches / contradicts something you know to be true	+5 to -5
Trust your gut. Something doesn't seem right.	-15
Statement doesn't make sense ("word salad")	-5
Source is an "AI" ("anything repeated a lot on the internet is true")	-10

---

486   Social media is awash with clickbait from scammers who make money on adverts. The more extreme the story, the more views it gets. *Social media is a very unreliable source.* **Warning signs:** fake sources tend to have only been around a short time and have very few posts. Images are often re-used and re-labelled, you can check if they have been used before with reverse image search engines like Tineye.

487   Jumping to instant conclusions, believing blindly in Authority and never backing down / changing your mind, is not responsible behaviour when you're caring for living creatures.

488   Hostile responses are sometimes an attempt by dogmatists to drown out a good point before their own brain processes it. To convince themselves, not you.

489   Police interviewers resolve conflicting statements by looking for inconsistent statements, checking evidence / records ("street work"), and assume *"one lie implies all lies"*

490   A useful rule of thumb is, the more toxic they are, the less relevant

Beekeeping specific factors	Trust Index
>5 years' experience	+10
> 5 hives	+5
Bee Inspector	+10
Misguided beliefs about swarms & unmanaged bees ("always feed, requeen"; "local bees are worthless")	-5
Uses truly local bees / queens (not imports / hybrids bred locally)	+5
Treats for varroa	-10
Certainty about causes of CCD or Isle of Wight Disease (no one really knows)	-10
They consider several stings per inspection normal	-5
Their primary motivation is honey ("backwards beekeeping")	-5
Based in different country to you (I'm in England)	-5
Lives near you	+5
They feed supplements, but cannot give a clear reason why - they trust adverts	-5
Using euphemisms for killing bees[491] (indicates self deception)	-2
Quotes soundbites from references funded by someone who stands to profit (gullible)	-5
What bee diseases they have seen in their hives, how often, and what they ascribe these to. These should be rare events!	-10 if regularly seen

---

491   An *alcohol wash* is where you shake 200-300 bees with alcohol to see how many mites drop off. The bees die. This is a popular way to check mite levels in America. *Sugar rolls* are less immediately lethal but simply take longer. Other euphemisms relate to "*replacing*" queens and using drone combs as "*traps*", attempting to convey a sense of banality and routine. If bees do this - fine, that's their programming. But humans have *choice*.

Acknowledgements

"Individually, we are one drop. Together, we are an ocean" – Ryunosuke Satoro

Reference works always emerge from a deep and wide pool of experience and a group effort.

My deep thanks go to all the people who have so generously shared their experience and knowledge with me over the years. Below I mention some by name, but I have gained something from every beekeeper I have ever met. It has been a privilege learning from them all.

In particular, Gareth John, Filipe Salbany and Ron Hoskins patiently answered endless questions on obscure points. Being able to draw on their collective ~170 years of beekeeping experience was invaluable.

| Filipe and Gareth inspect a skep

But a whole community of extraordinary folk have influenced this book. Polymaths John Haverson, David Heaf, and Roger Patterson generously shared their experience and insights. Martyn Townsend and Steve Donohoe gave valuable input on poly hives. Several scientists such as Derek Mitchell, Giles Budge, Alexandra Valentine, Dorian Pritchard, and a number of others replied clearly and comprehensively to enquiries from a complete stranger. They provided more insight and detail than could be covered here but from which I benefitted and have attempted to distil for this work. Any errors or omissions remaining in this work are my own.

Many papers, forums and previous works have been mined for data, knowledge and observations. And my local network of treatment-free beekeepers, OxNatBees, has proven to be a great resource for the non-judgemental group pooling of observations and refining ideas.

Numerous photographers, credited below their images, have generously allowed their stunning pictures to be used.

As the manuscript came together, various sections were checked by my wife Lynne, Gareth John, Helen Nunn, Ingo Scholler, Arnold Desandere, John Woods, Chris Park, and Guy Thompson who gave tremendous feedback and critiques on many aspects.

I would also like to thank Jeremy Burbidge, my publisher at Northern Bee Books, for his encouragement and support throughout; and my designer Simon Paterson who has transformed my rough manuscript into a beautiful book with his skills in typography and layout.

And now, to mention the most important contributors of all, I will wander down the garden to spend some time thanking my bees for all the fascinating years of learning, wonder and joy they have gifted me, and to tell them that their book has finally been published.

Appendix: Citizen Science

There's a huge lack of hard data in bee science, but it's straightforward to gather. For example, our local Treatment-Free (TF) group, OxNatBees, records winter losses and compares these with the British averages published by the BBKA:

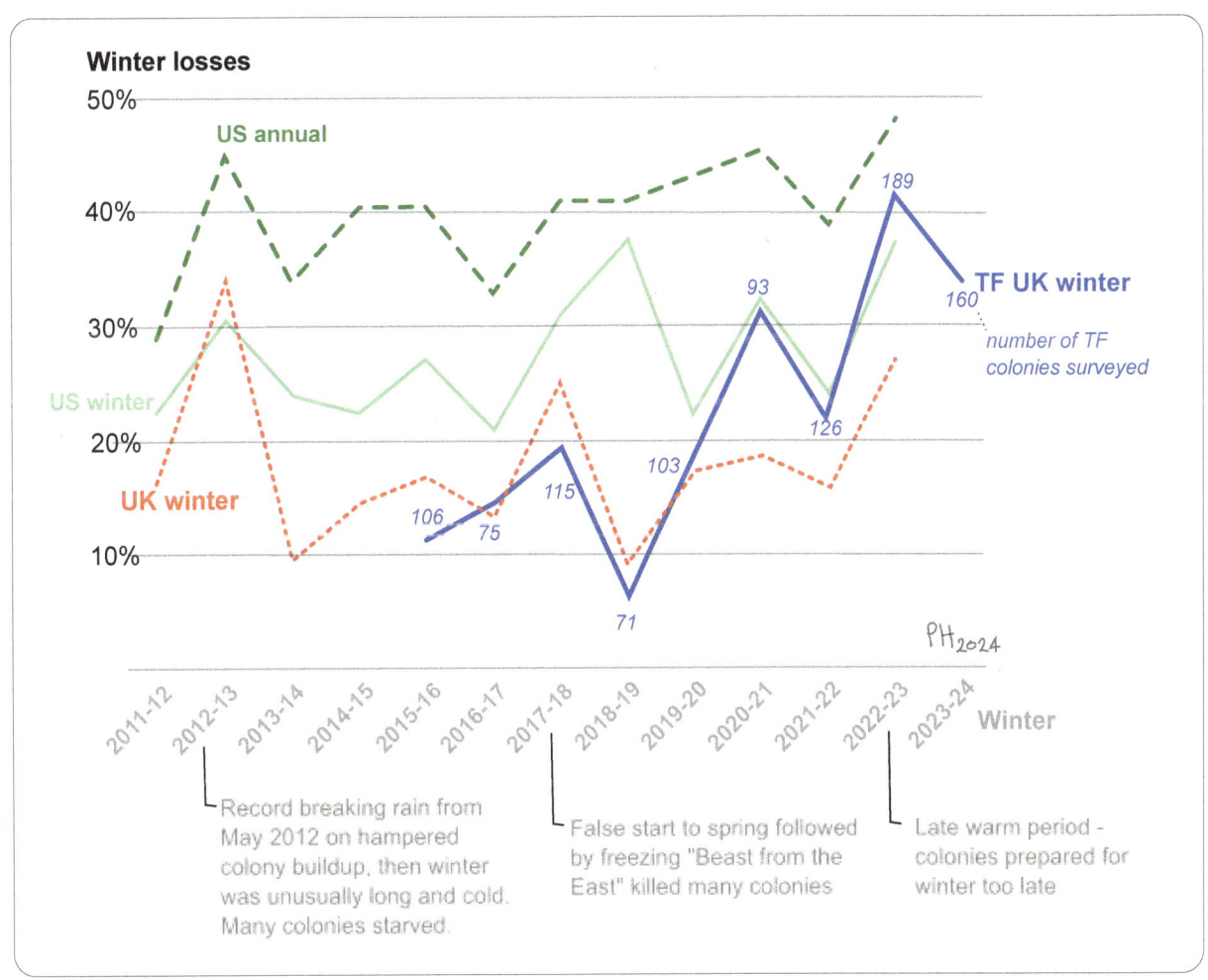

Commentary

Losses vary a lot, in step with "conventional" losses - and weather. With typical losses around 33%, in effect we run sustainable "herds" of mixed ages where young swarms replace losses.

The UK and US surveys are by the BBKA, COLOSS and BeeInformed, covering thousands of hives. American losses vay wildly between states, and are always higher than Europe's. The TF results are centred around Oxfordshire, and are from Warré, TBH, National, WBC hives and a few exotics like Einraumbeutes and skeps, totalling 1038 datapoints over 9 years, which is significant.

Beware false comparisons here. TF hives are almost never fed, and *never requeened*. A requeened hive is *not the same colony* - conventional loss reports mask this. Another data bias here is that our TF members are incentivised to report losses, as we ask our members "who needs swarms?" around the same time as this survey. In high loss years we hear of and see extraordinary loss rates in nearby commercial apiaries, which never seem reflected in BBKA figures.

Microclimate effects are important. For example, our high losses in 2022-23 were almost all in the (damp) area around Oxford.

Our English results contrast sharply with those of a major American study[492], which recorded higher losses in chemical-free hives.

If you wish to join a scientific study, remember any hive can have a super with inspectable frames added above it - even a log hive.

---

492 *A longitudinal study demonstrates that honey bee colonies managed organically are as healthy and productive as those managed conventionally*, Underwood et alia (2023), doi.org/10.1038/s41598-023-32824-w . Here "organic" refers to "only" using formic acid, oxalic acid or thymol. The interesting part is their third chemical-free group of 96 hives, not mentioned in the paper's title, spread across 8 farms in Pennsylvania or W Virginia. The experiment ran for 3 years. The chemical-free hives were Langstroths using uniform small-cell (4.9mm) foundation, and initially given queens grafted from a hive which had been treatment-free for 7 years - i.e. these queens were artifically produced, no swarms were involved. Colonies were inspected every 2 weeks in summer, monthly in winter. Honey was routinely harvested. About the only thing in common with our English low-intervention style was that the chemical-free hives were not routinely fed. The chemical-free hives had about a 70% annual loss rate; the other two management systems averaged 23% *winter* losses. The loss rates don't seem to be stated in the same form, but after 3 years, only one chemical-free colony remained, which is curious as the founder colony was 7 years old and we'd expect nearer 30 survivors in England; there were 29 and 38 of the other cohorts' 96+96 hives left.

Index

ABPV *see Paralysis Viruses*
Absconding 60, 99, 106, **126,** 259, 293-294
Acarapis woodi, Acarapis externus
   see Tracheal mite
Acarine *see Tracheal mite*
Addled brood..286
AFB *see Foulbrood*
Aggression
   Chapter 6 - stressors and safety 16, 51, 55, 62, 72, 384
   see also Genetics
AmFV, *Apis mellifera* Filamentous Virus...286
Amoeba disease............................**286,** 283, 314
Antibiotics... 48, 293, 311
Ants 19, 32, 221, 270, **286,** 317, 328
Artificial swarm.....................................249, 374
Asian hornet, *Vespa velutina* 287-290, 311, 388
Autopsy (hive) *see Post Mortem*
Badgers............................... 60, 71, 174, 290, 347
Bait hive 121, 127-128
Bald brood....................159, 195, **290,** 330, 342
Beard **150-152,** 153, 169, 326
Bears.......................71, 74, 174, 230, **291,** 347
Beepocalypse *see CCD*
Bee
   Bee bob 128-129
   Bee bread............26, 163, 168, 184, 189, 194, **195-199,** 207
   Bee escape 132, 133, 216
   Bee space............................ **183-185,** 231, 290
Bee inspectors..........**66-67,** 138, 267-268, 274, 302, 307, 310
Beelining 379-380
Beepocalypse *see CCD*

Bee races and types
   AHB (Africanised Honey Bee)..............59, 67, 71, 79, 105, 333, 354, 364, 371
   Amm (A.m. mellifera, Black Bee) 41, 64, 65, 81, 87, 144, 183, 185, 194, 195, 205, 219, 225, 274, 295, 300, 311, **353,** 354, 361, 381, 383
   Buckfast 65, 90, 312, 313, 338, **354,** 364, 370
   A. cerana 170, 171, 204, 228, 229, 269, 289, 311, 329, 330, 354, 369
   Cape *(A.m. capensis)* 204, **354**
   Carniolan (A.m. carnica) 64, 65, 67, 105, 194, 205, 225, **353-354,** 355, 361, 364, 365, 383
   Cordovan 356
   A. dorsata 331, 354
   hybrids 64, 354, **355,** 362, 364, 369
   Italian (A.m. ligustica)........... 29, 65, 90, 145, 205, 299, **353,** 361, 364
   Russian................ 81, 277, **353-354,** 371
   A. m. scutellata............................. 60, 354, 371
Bee suit *see Protective clothing*
BIAS..180, 250
Blenheim forest bees............153, 320, 376, 379
Bond strategy................................**334,** 366, 371
Botulism.. 81, 86, 267
BQCV - Black Queen Cell Virus 291
Braula flies..284, 292
Breeding *see Genetics*
Brood
   Bald brood, Chalk brood, Stonebrood - *see specific index entries*
   ejected 159, 161, 347
Canola *see OSR*

Castes   *see Drones, Queens, Workers*
CBPV   *see Paralysis Viruses*
CCD, Colony Collapse Disorder.......... 13, 280, **293-294**
Chaining   *see Festooning*
Chalk brood ... 61, 86, 136, 271, 273, 274, 275, 285, **294-295**, 309, 360, 366
Chilled brood...... 159, 160-161, 215, 273, 285, **296**, 303
Cold way.... 18, **49**, 50, 118, 211, 221, 320, 384
Comb
  reading comb 183-199
  brood comb... 18, 27, 135, 144, 177, **183-193**, 204-209, 277, 309, 314, 335, 343, 359, 378, 383
  honeycomb...... 27, 49, 50, 135, 150, 183-186, **194-195**, 196, 204, 215, 242, 292, 323
  old / black comb 129, **184-187**, 205, 271
  comb collapse/brace comb 50, 51, 54, 56, 160, 183, 184, 186, 203, **208**, 243, 244, 256, 347
  unusual comb 382-384
Condensing hive.................................... 377-379
Crystallisation   *see Honey*
Cut-outs.... 117, 186 (comb images), **258-259**, 275, 381
CWV, Cloudy Wing Virus 296
Dancing floor................................ 48, 171, 183
Darwinian beekeeping................. 334, 366, 371
Deadouts........ 51, 125, 187, 196-198, **260-261**, **346-351**
Death's-head Hawk Moth............ 153, 170, **297**
Debris..... 56, **158-165**, 175, 203, **332**, 338, 348
Degeneracy / Degenerate hives 144
Demaree.. 250
Disease - *general points: Chapter 14.1; alphabetic listing: chapter 14.2*
Dividing   *see Splitting colonies*
DNA   *see Genetics*
Drones...
  20, 27, **28-34**, 47, 54, 70, 130, 155, 187, 190, 258, 337, 357
  diploid............................ 26, **359-360**, 388
  congregation areas (DCAs).................32, 360
  exclusion................................ 34, 150, 152
  genetics 357-360, 365, 371
  from laying queens and workers..... 29-30, 33, 47, 145, 149, 162, 181, 191, 193

Drumming 224, 226, **245-246**, 249
Dwindling.............................125, 261, 291, 315
  Pseudo-dwindling................................... 374
DWV, Deformed Wing Virus...... 86, 156, 161, 279, 281, **297-298**
Dysentery   *see also Nosema* 29, 159, 274, **298-300**, 310-311, 349
Earwigs 300
Eco-floors 53, 56, 163, 202, 213, 335, 351
Ecotype 64, 353, 354, 370, 382
EFB   *see Foulbrood*
Einraumbeute   *see Hives*
Egg 25, 29, 163, **193**, 357, 358
Fat bodies 29
Feeding.................... 55, 56, 97-98, 124, 208, **233-239**, 307
Ferals   *see Wild bees*
Festooning... 166
Flies 157, 292, 301
  Assassin fly (Robber fly).................... 301
  Dragonfly.. 301
Fondant.................... 208, 233, 235, 238,
Forage........... 16, 17, 19, 21, 23, 45, **57-63, 68**, 90, 96, 135, 229, 373, 375
Forager 16, 17, 19, 21, 28, 29, 40, 61, 87, 108, 136, 143, 145, 156, 257, 380
Forced abscond................................... 259
Foul brood (AFB, EFB) 267, 271, 273-276, 279, **285, 302-310**, 366
Foundation ... 31, 33, **47-48**, 49, 150, 155, 186, 188, 190, 193, 244, 327
Frames....... 45, **47-50**, 54, 88-89, 134, 276, 277
Frisch, Karl von 16
Fungi and moulds 97, 158, 163, 196-198, 201, 236, 275, 277, **278-279**, 378
  fungicides 86, 318
  spore collection 163, 270
  see also Stamets, Paul; Chalk brood; Stone brood
Genetics............ 32, 41, 64, 101, 239, 261, 334, **352-372**, 381, 382
  Bee DNA.. **356-362**
  Breeding........... 12, 64, 86, 101, 268, 270, 272, 291, 353-354, **360-372**
  Diploids and Haploids....... 355, 356, 357, 360
  Inbreeding 251, 334, 355, **361-362**, 368-370
  Mutations..................................... 359

Temperament and Aggression.....72, 253-254, **363-365,** 367
'Ghost' bees 148
Golden Hive 52, 56
  see also Einraumbeute
Gotland survivor bees.........................333, 371
Grafting...370, 398
Growdowns 260
Half sisters 358
Heaf, David 51, 52, 124, 188, 209, 217, 225, 392, 396
Heddon split... 250
Hemizygous.. 356
Heterozygous 356
Hives..............................44-56, 88-90, 376-379
  Alveary...227
  Cathedral 53-54
  Drayton 53
  Einraumbeute 45, 46, **52-54**
  Hefting 172
  Hexagonal see Cathedral
  Langstroth...........45, 49, 50, 54, 55, 204, 234, 373, 376, 377
  Layens..52, 53
  Lazutin 45, **53**
  Log 228-231, 242
  Lune Valley....................................45, **53**
  National 52, 54, 55, 89, 137, **231-232,** 376, 379
  Poly **55-56,** 97, 383
  Skep................. 194, **218-227,** 245-246, 276, 386, 396
  TBH.......51, 90, 122, 136, **200-208,** 234, 241, 243-244, 250
  Warré.................45, 50, 51, 96, 97, **209-217,** 239, 241
  WBC..54, 55, 340, 373
  Zeidler (Tree)........................45, 52, 228-231
Hoffman spacing 47, 184, 231, 277
Homozygous 355, 356, 361
Honey............................... **84-86,** 90, 373-376
  crystallisation........ 37, 85, 136-137, 195, 196, 234, 299
  fermentation 85, 137, 168, 195, 235, 236, 299, 375
  harvesting.............. **132-137,** Warre: 215-217
  unripe 133, 137, 168
Honey binding..................... **126,** 189, 235, 334

Honeydew.............148, 167, 234, 299, **310-311**
Honeydew Flow Disease 310-311
Horizontal transmission...............251, 272-273
Hoskins, Ron..........31, 101, 164, 191, 298, 334, 337, 396
Huber, François 146, 153, 191, 193
Humidity 87, 144, 206, 210, 235, 315, 338, 351, 377-379
Hybrids and hybrid vigour 64, 105, 353-354, **355,** 362-365
Hybrid breeding 362
Hygienic behaviour 31, 40, 91, 144, 161, **164,** 185, 189, 194, 198, 238, 270, 275, 277, 281, 295, 298, 309, 318, **333,** 336, 366, 369
Hyperpolyandry 360
IAPV *see Paralysis Viruses*
IBDS *see Snot brood*
Insulation.........20, 52, 54, 56, **88-90,** 136, 163, 173, 197, 214, 223, **231, 377-379,** 383
Isle of Wight disease 13, 225, **312-313,** 300, 329, 381
Ivy 136-137, 160, 195, 207, **375**
John, Gareth 30, 31, 77, 91, 114, 119, 130, 151, 153, 155, 170, 211, 212, 232, 247, 261, 333, 340, 365, 374, 375, 377, 396
KBV *see Paralysis Viruses*
Kefuss, John.................................... 275, **334**
Killing bees.......47, 67, 225-227, **262,** 334, 380, 388
K Wing..................................**155-156,** 159, 268
Landrace 64, 354, 355, 362-366, 370
Larva *see Brood*
Layens hive *see Hives*
Lazutin, Fedor................ 53, 277, 335, 377, 386, 387, 392
Lifespan
  of bees 27, 40, **41,** 135, **236-237,** 296, 311, 314, 337
  of colonies 261
Lime / Linden trees.......................145, 375-376
Lineage 355
Line breeding 362
Log hive *see Hives*
Martin, Professor Stephen.....13, 290, 333, 366
Merging *see Uniting*
Mice..313-314
  mouse guards...97
Mite bomb 165, 324

Mites, exotic 301-302
see also Varroa; Tracheal; Tropilaelaps
Miticides 48, 64, 72, 85, 116, 126, 273, 293, 318, **333**, 335, **336-338**, 384
Moving hives.. 255-258
Mutants   *see Genetics*
Mutilla europaea   *see Wasp*
Nadiring................................... 209, 212, 213, 218
National hive   *see Hives*
Neglected Drone Brood................................. 314
Nematodes... 302
Neonicotinoids   *see Pesticides*
Nosema 31, 86, 236, 237, 272, 273, 279, 280, 291, 298, **314-316**
Nucleus 23, 102
Nurse bees 26-28, 42, 191, 207, 215, 247, 336, 359
Open mating 360, 362, 365, 366, 368, 369, 371, 372, 390
Opossum   *see Skunk*
OSR, Oil Seed Rape / canola ... **37**, 73, 86, 137, 167, 206, 299, **375**
OTC (aka OxyTetraCycline, Terramycin) 279, 303, **307-309**
Outstations   *see Satellite Nests*
Package 33, **102-103**, 226, 248, 382
Pagden split 250
Paralysis viruses 156, 277, 293, 310, 312, **316-318**, 329, 347
Parasite insights 21-23, 56, 74, 86, 88, 103, 165, 201, 251, 269, 272-274, 281, 291, 329, 333-334, 369
Park, Chris................................ 219-223, 227, 396
Pesticides 36, 37, 48, 61, 62, **85**, 130, 242, 268, 293, **316**, **318-319**
Pheromones................... 15, 29, 33, 38, 39, 106, 107, 128, 131, 154, 197, 205, 236, 249, 286, 341, 364, 378
  alarm 70, **71**, 76, 168, 364
  Nasonov..................... 106, 114, 142, 149, 151
Poison 22, 78, 88, 120, 155, 240, 262, **318-319**, 327, 347
  see also Pesticides, Venom
Pollen   *see also Bee bread* 25, 28, 39, 57, 61, 63, 97, 134, **145-146**, 148, 160, 163, 164, 183, 195-197, 235, 239, 273
Poly hive   *see Hives*

Pollen binding... 126
Poo   *see Dysentery*
Post mortem (hive) *346-351*
Princesses ... 34-36, 99, 100-101, 107-108, 125, 156, 247
Prokopovych, Petro.. 54
Propolis.................... 50, 51, 77, **86-87**, 97, 146, 147, 153, 154, 155, 184, 198, 231, 269, 275, 309, 321, 332
Protective clothing 76-77
Pseudoscorpions...................................... 284, 335
Queen
  bank... 219, 389
  cage / clip 38, 179, **244**
  cells 26, 108, 183, 187, **191-193, 251-253**, 291, 367, 370
  clipping................... 31, **37-38, 100,** 108, 130
  emergency 25, **36**, 183, **192**, 250, 370
  excluder................. 27, 30, 31, 39, 54, 223, 255
  finding 253-255
  intercaste 38
  piping 107, **169-170**, 297
  requeening 17, **37**, 39, 72, 103, 251-253, 262, 272, 382
Queen line   *see line breeding*
Quilt 96, 209, **210,** 348
Rap test (a.k.a. Knock test) 144, **171**
Rescue bar 208
Robbing........... 72, 143, 144, 194, 247, **319-322**
Rolling the bees 75
Royal jelly 26, 28, 193, 314
Russian scion.. 128
Sacbrood 285, 322-323
Salbany, Filipe 38, 60, 77, 92, 107, 123, 153, 157, 223, 241, 293, 320, 376, 396
Satellite nests (outstations).......................... 374
SBPV   *see Paralysis Viruses*
Schiffer, Torben......... 41, 87, 92, 155, 167, 230, 275, 335, 378
Seeley, Professor Tom 16, 33, 130, 151, 165, 191, 334, 360, 368, 369, 380, 385, 392
Sharashkin, Leo 53, 335, 377, 392
SHB   *see Small Hive Beetle*
Shook swarm....................... 244-245, 251, 317
  'shake and starve' technique.............. 308, 309
Shot brood.. 189

Shrew 325
Skep 118, 194, **218-227**, 245, 246, 255, 276, 300, 386, 396
Skunk & Opossum 161, 325, 347
Slugs and Snails 211, 221, 235, 325
Small bees 149, 187-188, 190
 see also Queens, intercaste
Small cell theory 187-188
Small Hive Beetle 160, **323-324**, 335
Smells 167-168
Smokers 74, 239-241
Snelgrove 251, 374
Snot brood 325
Snow 29, 121, 173, 214, 230, 376
Soldier bees 75
Somerville, Matt 230
Sounds 168-172, 297
Spermatheca 35, 357
Spiders 58, 128, 221, 326
Splitting (dividing) colonies 103, **249-251**, 272
Stamets, Paul 279
Starvation ... 39, 65, 96, 127, 143-144, 206, 275, 290, **326**, 348
 isolation starvation 48, 206
Stearin poisoning 48, 327
Stings 78-81
 anaphylaxis 79-80
 remedies 80-81
 venom 78-80, 318
Stone brood 285, 327
Storch 175, 181
Sump *see Eco floor*
Super, supering 50, 54, 88, 130, 194, 214, 217, **218**, 232, 398
Supersedure 36, 183, 192, 251, 280
Superorganism 15-18, 269, 273
Supersister 358
Swarms 16, 17, 20, 34, **99-131**, 225, 272, 367, 371, 374, 384
 artificial 249
 catching 113-120
 hiving 120-123
 primes 100, 101, 104, 105, 118
 casts 107-108, 118
 lure 127-130
 end of life 129, 280

 false 129
 mating 129
 starvation 119, 127, 129
Syrup 97, 119, 137, 160, **233-239**, 299
Tanging 131
Taranov split 251
Termites 328
Terramycin *see OTC*
Thelytoky 30, 354
Three foot / three mile rule 257
Tools 110, 178-179, 224
Top bars 47, 203, 204, 208, 212
Top Bar Hive *see Hives*
Tracheal mite 272, **328-330**
Trap-out 260
Trophallaxis 22
Tropilaelaps mite 284, 330-332
Uncapping / Recapping 290, 338, 369
Uniting colonies 83, 131, **247-249**
Usurpation 100, 129
Varroa mite 13, 22, 83, 164, 245, 267, 270, 277, 281, 284, 285, **332-338**, 354, 366, 369-372, 385
Varroa Sensitive Hygiene (VSH) *see Hygienic Behaviour*
Veil *see Protective clothing*
Venom *see stings*
Ventilation and fanning 50, 88, 142, 149, 287
Vertical transmission 272-273
Viruses (general) 277
Waggle dance 16, 48
Warm way *see Cold Way*
Warré hive *see Hives*
Washboarding 87, 142, 155
Wasps 338-341
 Exotic varieties 302
 Mutila europaea 302
Water 57, 58, 62, 85, 88, 97, 146, 377-379
Wax 28, 48, 130, 160, 162, 164, 184, 225, **242-243**, 327
Wax moth 135, 162, 196, **342-344**
Wild bees 274-275, 379-381
Wild nests 86, 89, 275, 379-381
Winter preparation 97-98
Workers 28-29

www.ingramcontent.com/pod-product-compliance
Ingram Content Group UK Ltd.
Pitfield, Milton Keynes, MK11 3LW, UK
UKHW051854110225
454949UK00007B/66